ENGINEERING MECHANICS

STATICS

SECOND EDITION

ANTHONY BEDFORD ■ WALLACE FOWLER

UNIVERSITY OF TEXAS AT AUSTIN

ADDISON-WESLEY

An imprint of Addison Wesley Longman, Inc.

Menlo Park, California • Reading, Massachusetts • Harlow, England
Berkeley, California • Don Mills, Ontario • Sydney • Bonn • Amsterdam • Tokyo • Mexico City

Senior Acquisitions Editor: Michael Slaughter
Associate Editor: Susan Slater
Editor in Chief: Chuck Iossi
Editorial Assistants: Colleen Kelly, Chandrika Madhaven
Production Manager: Pattie Myers
Art and Design Supervisor: Kevin Berry
Production Editors: Diane Williams, Caroline Jumper
Production Assistant: Kamila Storr
Copy Editor: Bruce Emmer
Proofreader: Brian Jones
Cover Design and Cover Illustrator: Yvo Riezebos
Text Design: Wilson Graphics & Design (Kenneth J. Wilson)
Frontmatter Design: Vargas Williams Design (Juan Vargas)
Layout: Vargas Williams Design (Edie Williams)
Illustrator: James A. Bryant
Technical Artists: Precision Graphics
Compositor: American Composition & Graphics, Inc.
Printer and Binder: R.R. Donnelley and Sons, Willard
Cover printer and separator: Phoenix Color Corporation

Photo Credits

Chapter 1: Opener, Mark Segal/Tony Stone Images/Chicago, Inc.; 1.4, Dennis Mitchell/Allsport Photography, Inc.; P1.22, NASA.
Chapter 2: P2.125 (background photo), NASA.
Chapter 3: Page 84 (upper), ITAR-TASS; (lower) Tom Pantages; 3.24, NASA.
Chapter 6: 6.3, Brownie Harris/The Stock Market; 6.15 and 6.19, Marshall Henrichs; 6.17, Pierre Berger/Photo Researchers, Inc.
Chapter 9: 9.4, Courtesy of Uzi Landman; 9.20(a), Courtesy of SKF Industries.
Chapter 10: 10.15, Warner Dieterich/The Image Bank; 10.19, G + J Images/The Image Bank; 10.19(a), Steve Niedorf/The Image Bank.

Library of Congress Cataloging-in-Publication Data
Bedford, A.
    Statics, engineering mechanics / Anthony Bedford and Wallace
Fowler. —2nd ed.
      p. cm.
    Includes index.
    ISBN 0-201-18070-7
    1. Statics.   I. Fowler, Wallace   II. Title.
TA351.B43   1998
620.1'03—dc21

98-18289
CIP

2 3 4 5 6 7 8 9 10—DOW—02 01 00 99 98

Addison Wesley Longman, Inc.
2725 Sand Hill Road
Menlo Park, CA 94025

Anthony Bedford is Professor of Aerospace Engineering and Engineering Mechanics at the University of Texas at Austin. He received his B.S. degree at the University of Texas at Austin, his M.S. degree at the California Institute of Technology, and his Ph.D. degree at Rice University in 1967. He has industrial experience at the Douglas Aircraft Company and TRW, and has been on the faculty of the University of Texas at Austin since 1968.

Dr. Bedford's main professional activity has been education and research in engineering mechanics. He is the author or coauthor of many technical papers on the mechanics of composite materials and mixtures and of two books, *Hamilton's Principle in Continuum Mechanics* and *Introduction to Elastic Wave Propagation.* From 1973 until 1983 he was a consultant to Sandia National Laboratories, Albuquerque, New Mexico.

He is a licensed professional engineer and a member of the Acoustical Society of America, the American Society for Engineering Education, the American Academy of Mechanics, and the Society for Natural Philosophy.

Wallace Fowler is Paul D. and Betty Robertson Meek Centennial Professor in Engineering at the University of Texas at Austin. He also holds the title of University Distinguished Teaching Professor in Aerospace Engineering and Engineering Mechanics. Dr. Fowler received the B.A. in mathematics and the M.S. and Ph.D. in engineering mechanics at the University of Texas at Austin and has been on the faculty there since 1965. In 1981–1982, he was Distinguished Visiting Professor in Astronautics and Computer Science at the United States Air Force Academy. Since 1991, he has been Associate Director of the Texas Space Grant Consortium.

Dr. Fowler's areas of teaching and research are dynamics, orbital mechanics, and spacecraft/mission analysis and design. He is author or coauthor of many technical papers on spacecraft dynamics and on engineering education. He has received numerous teaching awards at the local, regional, and national levels, including the AIAA-ASEE Distinguished Aerospace Educator (Leland Atwood) Award, the ASEE Fred Merryfield Award for Excellence in Teaching Engineering Design, and the University of Texas Chancellor's Council Outstanding Teaching Award. He is a registered professional engineer. He is a fellow of the American Institute of Aeronautics and Astronautics (AIAA) and the American Society for Engineering Education (ASEE). He served two terms (four years) as Vice President for Member Affairs of ASEE.

Our original objective in writing this book was to present the foundations and applications of statics as we do in the classroom. We used many sequences of figures, emulating the gradual development of a figure by a teacher explaining a concept. We stressed the importance of visual analysis in gaining understanding, especially through the use of free-body diagrams. Because inspiration is so conducive to learning, we based many of our examples and problems on a variety of modern engineering applications. With encouragement and help from many students and fellow teachers who have used the book, we continue and expand upon these themes in this edition.

Goals and Themes

Enhanced Visualization

We help students visualize the developments of solutions, especially the drawing of free-body diagrams:

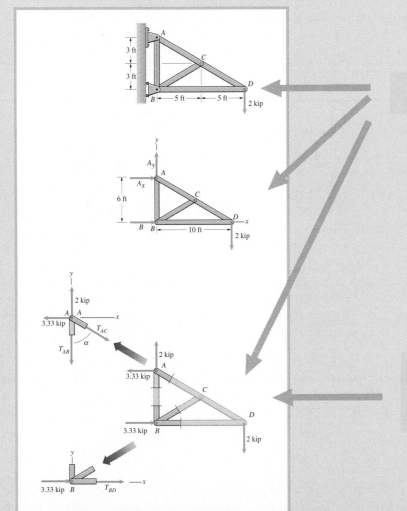

Using sequences of figures clarifies the steps of a solution.

Ghosting is used to emphasize the parts of an object that are isolated in obtaining free-body diagrams.

Examples that Teach

The design of our examples helps students learn how to approach problems, solve them, and critically judge the results:

Many examples contain "Strategy" sections showing the preliminary planning needed to begin a solution. What principles and equations apply? What must be determined, and in what order?

The solution is then described in detail, using sequences of figures when needed to clarify the steps.

Example 2.15

Suppose that you pull on the cable OA in Fig. 2.37, exerting a 50-N force $\mathbf{F}$ at O. What are the components of $\mathbf{F}$ parallel and normal to the cable OB?

STRATEGY

Resolving $\mathbf{F}$ into components parallel and normal to OB (Fig. a), we can determine the components by using Eqs. (2.26) and (2.27). But to apply them, we must first express $\mathbf{F}$ in terms of scalar components and determine the components of a unit vector parallel to OB. We can obtain the components of $\mathbf{F}$ by determining the components of the unit vector pointing from O toward A and multiplying them by $|\mathbf{F}|$.

SOLUTION

The position vectors from O to A and from O to B are (Fig. b)

$$\mathbf{r}_{OA} = 6\mathbf{i} + 6\mathbf{j} - 3\mathbf{k} \text{ (m)},$$

$$\mathbf{r}_{OB} = 10\mathbf{i} - 2\mathbf{j} + 3\mathbf{k} \text{ (m)}.$$

Their magnitudes are $|\mathbf{r}_{OA}| = 9$ m and $|\mathbf{r}_{OB}| = 10.6$ m. Dividing these vectors by their magnitudes, we obtain unit vectors that point from the origin toward A and B (Fig. c):

$$\mathbf{e}_{OA} = \frac{\mathbf{r}_{OA}}{|\mathbf{r}_{OA}|} = \frac{6\mathbf{i} + 6\mathbf{j} - 3\mathbf{k}}{9} = 0.667\mathbf{i} + 0.667\mathbf{j} - 0.333\mathbf{k},$$

$$\mathbf{e}_{OB} = \frac{\mathbf{r}_{OB}}{|\mathbf{r}_{OB}|} = \frac{10\mathbf{i} - 2\mathbf{j} + 3\mathbf{k}}{10.6} = 0.941\mathbf{i} - 0.188\mathbf{j} + 0.282\mathbf{k}.$$

The force $\mathbf{F}$ in terms of scalar components is

$$\mathbf{F} = |\mathbf{F}|\mathbf{e}_{OA} = (50)(0.667\mathbf{i} + 0.667\mathbf{j} - 0.333\mathbf{k})$$

$$= 33.3\mathbf{i} + 33.3\mathbf{j} - 16.7\mathbf{k} \text{ (N)}.$$

Taking the dot product of $\mathbf{e}_{OB}$ and $\mathbf{F}$, we obtain

$$\mathbf{e}_{OB} \cdot \mathbf{F} = (0.941)(33.3) + (-0.188)(33.3) + (0.282)(-16.7)$$

$$= 20.4 \text{ N}.$$

The parallel component of $\mathbf{F}$ is

$$\mathbf{F}_\text{p} = (\mathbf{e}_{OB} \cdot \mathbf{F})\,\mathbf{e}_{OB} = (20.4)(0.941\mathbf{i} - 0.188\mathbf{j} + 0.282\mathbf{k})$$

$$= 19.2\mathbf{i} - 3.8\mathbf{j} + 5.8\mathbf{k} \text{ (N)},$$

and the normal component is

$$\mathbf{F}_\text{n} = \mathbf{F} - \mathbf{F}_\text{p} = 14.2\mathbf{i} + 37.2\mathbf{j} - 22.4\mathbf{k} \text{ (N)}.$$

DISCUSSION

You can confirm that two vectors are perpendicular by making sure their dot product is zero. In this example,

$$\mathbf{F}_\text{p} \cdot \mathbf{F}_\text{n} = (19.2)(14.2) + (-3.8)(37.2) + (5.8)(-22.4) = 0.$$

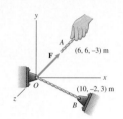

Figure 2.37

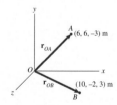

(a) The components of $\mathbf{F}$ parallel and normal to OB.

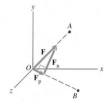

(b) The position vectors $\mathbf{r}_{OA}$ and $\mathbf{r}_{OB}$.

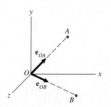

(c) The unit vectors $\mathbf{e}_{OA}$ and $\mathbf{e}_{OB}$.

Many examples conclude with "Discussion" sections pointing out properties of the solution, or commenting on alternative solution methods, or pointing out ways to check answers.

Engineering Design

We include simple design considerations in many examples and problems without compromising emphasis on fundamental mechanics. Optional examples titled "Application to Engineering" provide more detailed discussions of the uses of statics in engineering design:

Example 4.9

Application to Engineering

Rotating Machines

The crewman in Fig. 4.25 exerts the forces shown on the handles of the coffee grinder winch, where $\mathbf{F} = 4\mathbf{j} + 32\mathbf{k}$ N. Determine the total moment he exerts (a) about point O, (b) about the axis of the winch, which coincides with the x axis.

A specific engineering application is first described and analyzed.

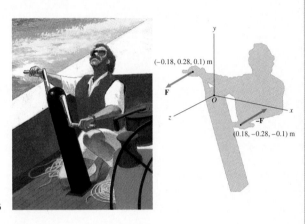

(−0.18, 0.28, 0.1) m

$\mathbf{F}$

(0.18, −0.28, −0.1) m

Figure 4.25

STRATEGY

(a) To obtain the total moment about point O, we must sum the moments of the two forces about O. Let the sum be denoted by $\Sigma\,\mathbf{M}_O$. (b) Because point O is on the x axis, the total moment about the x axis is the component of $\Sigma\,\mathbf{M}_O$ parallel to the x axis, which is the x component of $\Sigma\,\mathbf{M}_O$.

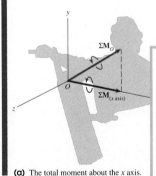

(a) The total moment about the x axis.

Notice that this is the result given by Eq. (4.4): Since $\mathbf{i}$ is a unit vector parallel to the x axis,

$$\Sigma\,\mathbf{M}_{(x\,\text{axis})} = (\mathbf{i} \cdot \Sigma\,\mathbf{M}_O)\,\mathbf{i} = 17.3\mathbf{i} \text{ (N-m)}.$$

DESIGN ISSUES

The winch in this example is a simple representative of a class of rotating machines that includes hydrodynamic and aerodynamic power turbines, propellers, jet engines, and electric motors and generators. The ancestors of hydrodynamic and aerodynamic power turbines—water wheels and windmills—were among the earliest machines. These devices illustrate the importance of the concept of the moment of a force about a line. Their common feature is a part designed to rotate and perform some function when it is subjected to a moment about its axis of rotation. In the case of the winch, the forces exerted on the handles by the crewman exert a moment about the axis of rotation, causing the winch to rotate and wind a rope onto a drum, trimming the boat's sails. A hydrodynamic power turbine (Fig. 4.26) has turbine blades that are subjected to forces by flowing water, exerting a moment about the axis of rotation. This moment rotates the shaft to which the blades are attached, turning an electric generator that is connected to the same shaft.

A "Design Issues" section then discusses design implications of the application and places it in a broader engineering context.

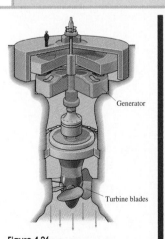

Generator

Turbine blades

Figure 4.26
A hydroelectric turbine. Water flowing through the turbine blades exerts a moment about the axis of the shaft, turning the generator.

Relevant Applications

We place statics within the context of engineering practice
by including applications from many fields of engineering:

3.53 The device shown is towed beneath a ship to measure
water temperature and salinity. The mass of the device is 130
kg. The angle $\alpha = 20°$. The motion of the water relative to the
device causes a horizontal drag force D. The hydrostatic pressure distribution in the water exerts a vertical "buoyancy"
force B. The magnitude of the buoyancy force is equal to the
product of the volume of the device, $V = 0.075$ m³, and the
weight density of the water, $\gamma = 9500$ N/m³. Determine the
drag force D and the tension in the cable.

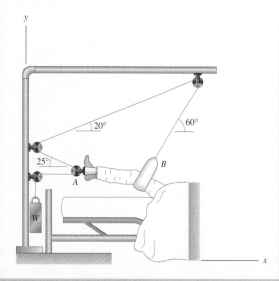

4.163 The engine above the airplane's fuselage exerts a thrust
$T_o = 16$ kip, and each of the engines under the wings exerts a
thrust $T_U = 12$ kip. The dimensions are $h = 8$ ft, $c = 12$ ft, and
$b = 16$ ft. If you represent the three thrust forces by a force **F**
acting at the origin O and a couple **M**, what are **F** and **M**?

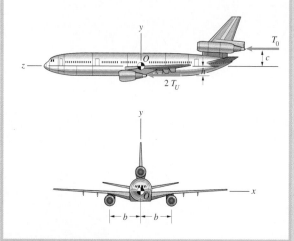

3.96 The system shown is called Russell's traction. If the
sum of the downward forces exerted at A and B by the
patient's leg is 32.2 lb, what is the weight W?

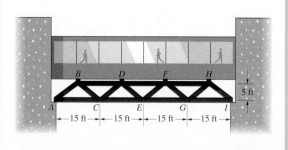

6.22 The Warren truss supports a walkway between two
buildings. The walkway exerts vertical 12-kip loads at B, D,
F, and H. Model the supports at A and I as roller supports.
Determine the axial forces in members BC, CD, and CE.

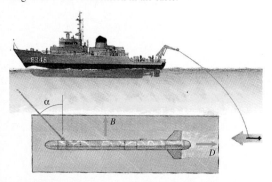

Consistent Use of Color

To help students recognize and interpret elements of figures, we use consistent identifying colors:

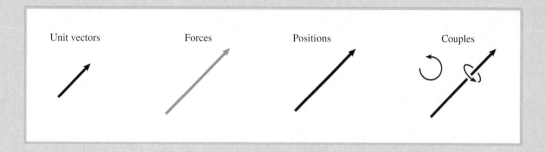

Computational Mechanics

Some instructors prefer to teach statics without requiring the use of a computer. Others use statics as an opportunity to introduce students to the use of computers in engineering, having them either write their own programs in a lower level language or use higher level problem-solving software. Our book is suitable for each of these approaches. We provide optional, self-contained "Computational Mechanics" sections with examples and problems designed for solution by a programmable calculator or computer. See Examples 3.8 on p. 117 and 5.13 on p. 262.

New to the Second Edition

Positive responses from users and reviewers have led us to retain the basic organization, content, and features of the first edition. During our preparation of this edition, we examined how we presented each concept, example, figure, summary statement, and problem. Where necessary, we made changes, additions, or deletions to simplify and clarify the presentation. In response to requests, we made the following notable changes:

- We have added an "Application to Engineering" example on Rotating Machines in Chapter 4.
- We have revised our discussion of loads in Chapter 6.
- Chapter 7 has been reorganized. We cover distributed forces on beams earlier than in the first edition, as an application of centroids.
- Friction is now covered earlier, in Chapter 9.
- Chapter 10, now called "Internal Forces and Moments," covers shear forces and bending moments in beams, cables subjected to distributed and discrete loads, and forces exerted by pressure.
- We have added new examples where users indicated more were needed. Many of the new examples continue our emphasis on realistic and motivational applications and engineering design.
- We have revised many of the existing problems and have added more than 200 new ones. As with the examples, many of the new problems focus on placing statics within the context of engineering practice.

Commitment to Students and Instructors

In revising the textbook and solutions manual, we have taken precautions to ensure accuracy to the best of our ability. Reviewers examined each stage of the manuscript for errors. We have each solved the new problems in an effort to be sure that their answers are correct and that they are of an appropriate level of difficulty. David Hartman of Northern Arizona University also checked the new text, examples, and problems. Any errors that remain are the responsibility of the authors. We welcome communication from students and instructors concerning errors or areas for improvement. Our mailing address is Department of Aerospace Engineering and Engineering Mechanics, University of Texas at Austin, Austin, Texas 78712. Our electronic mail address is abedford@mail.utexas.edu.

Supplements

PRINTED SUPPLEMENTS

Instructor's Solutions Manual   The manual for the instructor, completely revised by Wallace Fowler, contains complete step-by-step solutions to all problems. Each solution includes the problem statement and the associated art.

Design Problems Supplement   This is a paperback supplement of approximately 65 pages and over 100 design problems that are based on problems from the textbook. This product makes it possible to introduce students to the concepts of engineering design in the context of statics.

Additional Problems Set   As a service to aid instructors and students in having the greatest variety and quantity of problems, we have developed an additional problem set. This supplement consists of approximately 500 problems (roughly 200 from *Statics* and 300 from *Dynamics*) using the art from the texts. The problems were written and checked by the authors and are intended to provide a fresh look at the concepts presented.

SOFTWARE SUPPLEMENT

Working Model® Simulations   Approximately 100 problems and examples from the text have been recreated on disk as Working Model simulations. These simulations have been constructed to allow the student to change the values of variables and observe the results. The student can explore physical situations in a "what if" manner and thereby develop deeper conceptual insights than possible through quantitative problem solving alone. A site license for these simulations is available free to adopters, or they can be purchased by students bundled with the text for a nominal additional charge.

Acknowledgments

Many students and teachers have given us insightful comments on the first edition. The following colleagues reviewed the first edition or the manuscript of the second edition and made many valuable suggestions.

Edward E. Adams
Michigan Technological University

Raid S. Al-Akkad
University of Dayton

Jerry L. Anderson
Memphis State University

James G. Andrews
University of Iowa

Robert J. Asaro
University of California, San Diego

Leonard B. Baldwin
University of Wyoming

Gautam Batra
University of Nebraska

Mary Bergs
Marquette University

Spencer Brinkerhoff
Northern Arizona University

William (Randy) Burkett
Texas Tech University

Donald Carlson
University of Illinois

Major Robert M. Carpenter
U.S. Military Academy

Douglas Carroll
University of Missouri, Rolla

Namas Chandra
Florida State University

James Cheney
University of California, Davis

Ravinder Chona
Texas A&M University

Anthony DeLuzio
Merrimack College

Mitsunori Denda
Rutgers University

James F. Devine
University of South Florida

Craig Douglas
University of Massachusetts, Lowell

Marijan Dravinski
University of Southern California

S. Olani Durrant
Brigham Young University

Estelle Eke
California State University, Sacramento

William Ferrante
University of Rhode Island

Robert W. Fitzgerald
Worcester Polytechnic Institute

George T. Flowers
Auburn University

Mark Frisina
Wentworth Institute

Robert W. Fuessle
Bradley University

William Gurley
University of Tennessee, Chattanooga

John Hansberry
University of Massachusetts, Dartmouth

W. C. Hauser
California Polytechnic University, Pomona

R. Craig Henderson
Tennessee Technological University

James Hill
University of Alabama

Allen Hoffman
Worcester Polytechnic Institute

Edward E. Hornsey
University of Missouri, Rolla

Robert A. Howland
University of Notre Dame

Joe Ianelli
University of Tennessee, Knoxville

Ali Iranmanesh
Gadsden State Community College

David B. Johnson
Southern Methodist University

E. O. Jones, Jr.
Auburn University

Serope Kalpakjian
Illinois Institute of Technology

Kathleen A. Keil
California Polytechnic University, San Luis Obispo

Seyyed M. H. Khandani
Diablo Valley College

Charles M. Krousgrill
Purdue University

B. Kent Lall
Portland State University

Kenneth W. Lau
University of Massachusetts, Lowell

Norman Laws
University of Pittsburgh

William M. Lee
U.S. Naval Academy

Donald G. Lemke
University of Illinois, Chicago

Richard J. Leuba
North Carolina State University

Richard Lewis
Louisiana Technological University

Bertram Long
Northeastern University

V. J. Lopardo
U.S. Naval Academy

Frank K. Lu
University of Texas, Arlington

K. Madhaven
Christian Brothers College

Gary H. McDonald
University of Tennessee

James McDonald
Texas Technical University

Jim Meagher
California Polytechnic State University, San Luis Obispo

Lee Minardi
Tufts University

Norman Munroe
Florida International University

Shanti Nair
University of Massachusetts, Amherst

Saeed Niku
California Polytechnic State University, San Luis Obispo

Harinder Singh Oberoi
Western Washington University

James O'Connor
University of Texas, Austin

Samuel P. Owusu-Ofori
North Carolina A&T State University

Venkata Panchakarla
Florida State University

David J. Purdy
Rose-Hulman Institute of Technology

Colin P. Ratcliffe
U.S. Naval Academy

Daniel Riahi
University of Illinois

Charles Ritz
California Polytechnic State University, Pomona

George Rosborough
University of Colorado, Boulder

Robert Schmidt
University of Detroit

Robert J. Schultz
Oregon State University

Patricia M. Shamamy
Lawrence Technological University

Sorin Siegler
Drexel University

L. N. Tao
Illinois Institute of Technology

John Tomko
Cleveland State University

John Valasek
Texas A&M University

Thomas J. Vasko
University of Hartford

Mark R. Virkler
University of Missouri, Columbia

William H. Walston, Jr.
University of Maryland

Reynolds Watkins
Utah State University

Charles White
Northeastern University

Norman Wittels
Worcester Polytechnic Institute

Julius P. Wong
University of Louisville

Constance Ziemian
Bucknell University

The people of Addison Wesley Longman and their colleagues helped us at every stage with professionalism, patience, humor, and friendship, especially Kevin Berry, Chuck Iossi, Dan Joraanstad, Caroline Jumper, Colleen Kelly, Chandrika Madhavan, Rob Merino, Pattie Myers, Susan Slater, Michael Slaughter, Kamila Storr, Janet Weaver, Diane Williams, and Edie Williams.

Finally we thank our wives, Nancy and Marcia, for cheerfully accepting the sacrifices inherent in such a project.

Anthony Bedford and Wallace Fowler
January 1998
Austin, Texas

Contents

3 Forces   83

4 Systems of Forces and Moments

5 Objects in Equilibrium 207

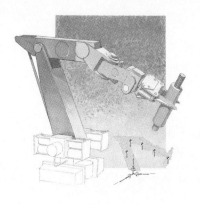

6 Structures in Equilibrium 271

7 Centroids and Centers of Mass 327

8   Moments of Inertia   387

9   Friction   439

10   Internal Forces and Moments   497

11 Virtual Work and Potential Energy 551

Appendixes 580

ENGINEERING MECHANICS

STATICS

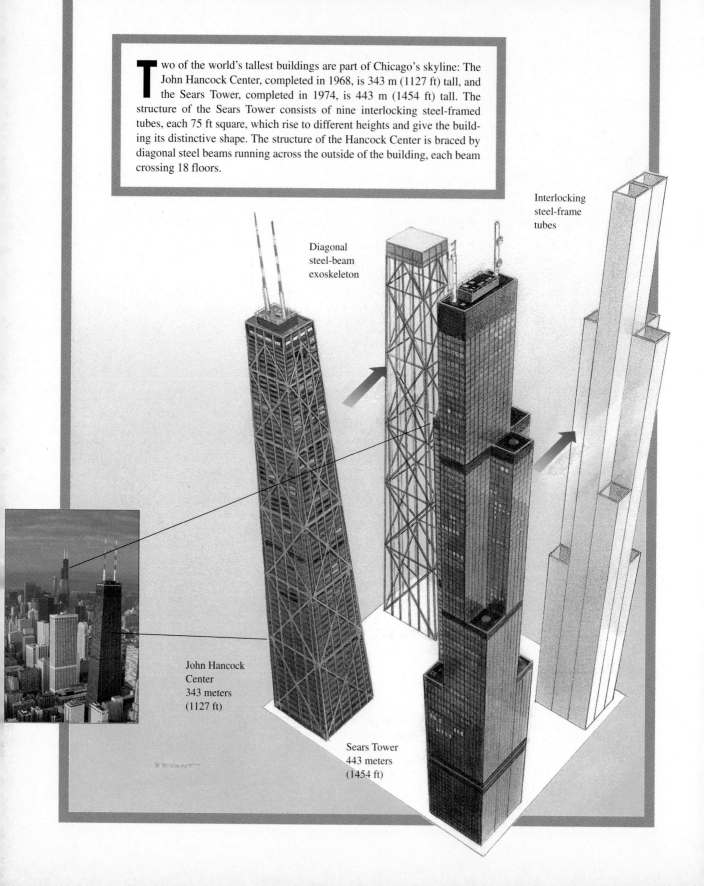

Two of the world's tallest buildings are part of Chicago's skyline: The John Hancock Center, completed in 1968, is 343 m (1127 ft) tall, and the Sears Tower, completed in 1974, is 443 m (1454 ft) tall. The structure of the Sears Tower consists of nine interlocking steel-framed tubes, each 75 ft square, which rise to different heights and give the building its distinctive shape. The structure of the Hancock Center is braced by diagonal steel beams running across the outside of the building, each beam crossing 18 floors.

Interlocking
steel-frame
tubes

Diagonal
steel-beam
exoskeleton

John Hancock
Center
343 meters
(1127 ft)

Sears Tower
443 meters
(1454 ft)

BRYANT

Chapter 1

Introduction

THE innovative structures of the John Hancock Center and the Sears Tower—both conceived by the same person, engineer Fazlur Khan (1929–1982)—are two solutions to the problem of supporting the wind and gravity loads on a large building. During each step of the design of these structures, the engineers relied on, and were guided by, the principles of statics. Statics is one of the sciences underlying the art of structural design.

1.1 *Engineering and Mechanics*

How do engineers design complex systems and predict their characteristics before they are constructed? Engineers have always relied on their knowledge of previous designs, experiments, ingenuity, and creativity to develop new designs. Modern engineers add a powerful technique: They develop mathematical equations based on the physical characteristics of the devices they design. With these mathematical models, engineers predict the behavior of their designs, modify them, and test them prior to their actual construction. Aerospace engineers used mathematical models to predict the paths the space shuttle would follow in flight. Civil engineers used mathematical models to analyze the response to loads of the steel frame of the 1454-ft Sears Tower in Chicago.

Engineers are responsible for the design, construction, and testing of the devices we use, from simple things such as chairs and pencil sharpeners to complicated ones such as dams, cars, airplanes, and spacecraft. They must have a deep understanding of the physics underlying these devices and must be familiar with the use of mathematical models to predict system behavior. Students of engineering begin to learn how to analyze and predict the behavior of physical systems by studying mechanics.

At its most basic level, mechanics is the study of forces and their effects. Elementary mechanics is divided into **statics**, the study of objects in equilibrium, and **dynamics**, the study of objects in motion. The results obtained in elementary mechanics apply directly to many fields of engineering. Mechanical and civil engineers who design structures use the equilibrium equations derived in statics. Civil engineers who analyze the responses of buildings to earthquakes and aerospace engineers who determine the trajectories of satellites use the equations of motion derived in dynamics.

Mechanics was the first analytical science; consequently fundamental concepts, analytical methods, and analogies from mechanics are found in virtually every field of engineering. For example, students of chemical and electrical engineering gain a deeper appreciation for basic concepts in their fields such as equilibrium, energy, and stability by learning them in their original mechanical contexts. In fact, by studying mechanics they retrace the historical development of these ideas.

1.2 *Learning Mechanics*

Mechanics consists of broad principles that govern the behavior of objects. In this book we describe these principles and provide examples that demonstrate some of their applications. Although it is essential that you practice working problems similar to these examples, and we include many problems of this kind, our objective is to help you understand the principles well enough to apply them to situations that are new to you. Each generation of engineers confronts new problems.

Problem Solving

In the study of mechanics you learn problem-solving procedures you will use in succeeding courses and throughout your career. Although different types of problems require different approaches, the following steps apply to many of them:

- Identify the information that is given and the information, or answer, you must determine. It's often helpful to restate the problem in your own words. When appropriate, make sure you understand the physical system or model involved.

- Develop a *strategy* for the problem. This means identifying the principles and equations that apply and deciding how you will use them to solve the problem. Whenever possible, draw diagrams to help visualize and solve the problem.

- Whenever you can, try to predict the answer. This will develop your intuition and will often help you recognize an incorrect answer.

- Solve the equations and, whenever possible, interpret your results and compare them with your prediction. This last step is a *reality check*. Is your answer reasonable?

Calculators and Computers

Most of the problems in this book are designed to lead to an algebraic expression with which to calculate the answer in terms of given quantities. A calculator with trigonometric and logarithmic functions is sufficient to determine the numerical value of such answers. The use of a programmable calculator or a computer with problem-solving software such as *Mathcad* or *TK! Solver* is convenient, but be careful not to become too reliant on tools you will not have during tests.

Sections headed "Computational Mechanics" contain examples and problems that are suitable for solution with a programmable calculator or a computer.

Engineering Applications

Although the problems are designed primarily to help you learn mechanics, many of them illustrate uses of mechanics in engineering. Sections headed "Application to Engineering" describe how mechanics is applied in various fields of engineering.

We also include problems that emphasize two essential aspects of engineering:

- *Design.* Some problems ask you to choose values of parameters to satisfy stated design criteria.

- *Safety.* Some problems ask you to evaluate the safety of devices and choose values of parameters to satisfy stated safety requirements.

Subsequent Use of This Text

This book contains tables and information you will find useful in subsequent engineering courses and throughout your engineering career. In addition, you will often want to review fundamental engineering subjects, both during the remainder of your formal education and when you are a practicing engineer.

The most efficient way to do so is by using the textbooks with which you are familiar. Your engineering textbooks will form the core of your professional library.

1.3 *Fundamental Concepts*

Some topics in mechanics will be familiar to you from everyday experience or from previous exposure to them in physics courses. In this section we briefly review the foundations of elementary mechanics.

Space and Time

Space simply refers to the three-dimensional universe in which we live. Our daily experiences give us an intuitive notion of space and the locations, or positions, of points in space. The distance between two points in space is the length of the straight line joining them.

Measuring the distance between points in space requires a unit of length. We use both the International System of units, or SI units, and U.S. Customary units. In SI units, the unit of length is the meter (m). In U.S. Customary units, the unit of length is the foot (ft).

Time is, of course, familiar—our lives are measured by it. The daily cycles of light and darkness and the hours, minutes, and seconds measured by our clocks and watches give us an intuitive notion of time. Time is measured by the intervals between repeatable events, such as the swings of a clock pendulum or the vibrations of a quartz crystal in a watch. In both SI units and U.S. Customary units, the unit of time is the second (s). The minute (min), hour (hr), and day are also frequently used.

If the position of a point in space relative to some reference point changes with time, the rate of change of its position is called its *velocity*, and the rate of change of its velocity is called its *acceleration*. In SI units, the velocity is expressed in meters per second (m/s) and the acceleration is expressed in meters per second per second, or meters per second squared (m/s^2). In U.S. Customary units, the velocity is expressed in feet per second (ft/s) and the acceleration is expressed in feet per second squared (ft/s^2).

Newton's Laws

Elementary mechanics was established on a firm basis with the publication in 1687 of *Philosophiae naturalis principia mathematica*, by Isaac Newton. Although highly original, it built on fundamental concepts developed by many others during a long and difficult struggle toward understanding (Fig. 1.1). Newton stated three "laws" of motion, which we express in modern terms:

1. *When the sum of the forces acting on a particle is zero, its velocity is constant. In particular, if the particle is initially stationary, it will remain stationary.*

2. *When the sum of the forces acting on a particle is not zero, the sum of the forces is equal to the rate of change of the linear momentum of the*

Figure 1.1

Chronology of developments in mechanics up to the publication of Newton's *Principia* in relation to other events in history.

The figure shows a timeline with historical events on the left and developments in mechanics on the right:

Historical Events	Date	Developments in Mechanics
Peloponnesian War	400 B.C.	Aristotle: Statics of levers, speculations on dynamics
		Archimedes: Statics of levers, centers of mass, buoyancy
Roman invasion of Britain	0	
		Hero of Alexandria: Statics of levers and pulleys
		Pappus: Precise definition of center of mass
	A.D. 400	
		John Philoponus: Concept of inertia
Coronation of Charlemagne	800	
Norman conquest of Britain		
Signing of Magna Carta	1200	
		Jordanus of Nemore: Stability of equilibrium
		Albert of Saxony: Angular velocity
Bubonic plague in Europe		Nicole d'Oresme: Graphical kinematics, coordinates
		William Heytesbury: Concept of acceleration
	1400	
Printing of Gutenberg Bible		
Voyage of Columbus		Nicolaus Copernicus: Concept of the solar system
		Dominic de Soto: Kinematics of falling objects
		Tycho Brahe: Observations of planetary motions
Founding of Jamestown Colony	1600	Simon Stevin: Principle of virtual work
		Johannes Kepler: Geometry and kinematics of planetary motions
Thirty Years' War		Galileo Galilei: Experiments and analyses in statics and dynamics, motion of a projectile
Pilgrims' arrival in Massachusetts		
Founding of Harvard University		René Descartes: Cartesian coordinates
		Evangelista Torricelli: Experiments on hydrodynamics
	1650	Blaise Pascal: Analyses in hydrostatics
Settlement of Carolina		
		John Wallis, Christopher Wren, Christiaan Huyghens: Impacts between objects
Pennsylvania grant to William Penn		
		Isaac Newton: Concept of mass, laws of motion, postulate of universal gravitation,
Salem witchcraft trials		analyses of planetary motions
	1700	

particle. If the mass is constant, the sum of the forces is equal to the product of the mass of the particle and its acceleration.

3. *The forces exerted by two particles on each other are equal in magnitude and opposite in direction.*

Notice that we did not define force and mass before stating Newton's laws. The modern view is that these terms are defined by the second law. To demonstrate, suppose that we choose an arbitrary object and define it to have unit mass. Then we define a unit of force to be the force that gives our unit mass an acceleration of unit magnitude. In principle, we can then determine the mass of any object: We apply a unit force to it, measure the resulting acceleration, and use the second law to determine the mass. We can also determine the magnitude of any force: We apply it to our unit mass, measure the resulting acceleration, and use the second law to determine the force.

Thus Newton's second law gives precise meanings to the terms **mass** and **force**. In SI units, the unit of mass is the kilogram (kg). The unit of force is the newton (N), which is the force required to give a mass of one kilogram an acceleration of one meter per second squared. In U.S. Customary units, the unit of force is the pound (lb). The unit of mass is the slug, which is the amount of mass accelerated at one foot per second squared by a force of one pound.

Although the results we discuss in this book are applicable to many of the problems met in engineering practice, there are limits to the validity of Newton's laws. For example, they don't give accurate results if a problem involves velocities that are not small compared to the velocity of light (3×10^8 m/s). Einstein's special theory of relativity applies to such problems. Elementary mechanics also fails in problems involving dimensions that are not large compared to atomic dimensions. Quantum mechanics must be used to describe phenomena on the atomic scale.

Newtonian Gravitation

Figure 1.2

The gravitational forces between two particles are equal in magnitude and directed along the line between them.

Another of Newton's fundamental contributions to mechanics is his postulate for the gravitational force between two particles in terms of their masses m_1 and m_2 and the distance r between them (Fig. 1.2). His expression for the magnitude of the force is

$$F = \frac{Gm_1m_2}{r^2},$$ (1.1)

where G is called the universal gravitational constant.

Newton calculated the gravitational force between a particle of mass m_1 and a homogeneous sphere of mass m_2 and found that it is also given by Eq. (1.1), with r denoting the distance from the particle to the center of the sphere. Although the earth is not a homogeneous sphere, we can use this result to approximate the weight of an object of mass m due to the gravitational attraction of the earth,

$$W = \frac{Gmm_E}{r^2},$$ (1.2)

where m_E is the mass of the earth and r is the distance from the center of the earth to the object. Notice that the weight of an object depends on its location relative to the center of the earth, whereas the mass of the object is a measure of the amount of matter it contains and doesn't depend on its position.

When an object's weight is the only force acting on it, the resulting acceleration is called the acceleration due to gravity. In this case, Newton's second law states that $W = ma$, and from Eq. (1.2) we see that the acceleration due to gravity is

$$a = \frac{Gm_E}{r^2}.$$ (1.3)

The **acceleration due to gravity at sea level** is denoted by g. Denoting the radius of the earth by R_E, we see from Eq. (1.3) that $Gm_E = gR_E^2$. Substituting this result into Eq. (1.3), we obtain an expression for the acceleration due to gravity at a distance r from the center of the earth in terms of the acceleration due to gravity at sea level:

$$a = g\frac{R_E^2}{r^2}. \tag{1.4}$$

Since the weight of the object $W = ma$, the weight of an object at a distance r from the center of the earth is

$$W = mg\frac{R_E^2}{r^2}. \tag{1.5}$$

At sea level, the weight of an object is given in terms of its mass by the simple relation

$$W = mg. \tag{1.6}$$

The value of g varies from location to location on the surface of the earth. The values we use in examples and problems are $g = 9.81$ m/s^2 in SI units and $g = 32.2$ ft/s^2 in U.S. Customary units.

Numbers

Engineering measurements, calculations, and results are expressed in numbers. You need to know how we express numbers in the examples and problems and how to express the results of your own calculations.

Significant Digits   This term refers to the number of meaningful (that is, accurate) digits in a number, counting to the right starting with the first nonzero digit. The two numbers 7.630 and 0.007630 are each stated to four significant digits. If only the first four digits in the number 7,630,000 are known to be accurate, this can be indicated by writing the number in scientific notation as 7.630×10^6.

If a number is the result of a measurement, the significant digits it contains are limited by the accuracy of the measurement. If the result of a measurement is stated to be 2.43, this means that the actual value is believed to be closer to 2.43 than to 2.42 or 2.44.

Numbers may be rounded off to a certain number of significant digits. For example, we can express the value of π to three significant digits, 3.14, or we can express it to six significant digits, 3.14159. When you use a calculator or computer, the number of significant digits is limited by the number of digits the machine is designed to carry.

Use of Numbers in This Book   You should treat numbers given in problems as exact values and not be concerned about how many significant digits they contain. If a problem states that a quantity equals 32.2, you can assume its value is 32.200. . . . We express intermediate results and answers in the examples and the answers to the problems to at least three significant digits. If you use a calculator, your results should be that accurate. Be sure to avoid round-off errors that occur if you round off intermediate results when making a series of calculations. Instead, carry through your calculations with as much accuracy as you can by retaining values in your calculator.

1.4 *Units*

The SI system of units has become nearly standard throughout the world. In the United States, U.S. Customary units are also used. In this section we summarize these two systems of units and explain how to convert units from one system to another.

International System of Units

In SI units, length is measured in meters (m) and mass in kilograms (kg). Time is measured in seconds (s), although other familiar measures such as minutes (min), hours (hr), and days are also used when convenient. Meters, kilograms, and seconds are called the **base units** of the SI system. Force is measured in newtons (N). Recall that these units are related by Newton's second law: One newton is the force required to give an object of one kilogram mass an acceleration of one meter per second squared:

$$1 \text{ N} = (1 \text{ kg})(1 \text{m/s}^2) = 1 \text{ kg-m/s}^2.$$

Because the newton can be expressed in terms of the base units, it is called a **derived unit**.

To express quantities by numbers of convenient size, multiples of units are indicated by prefixes. The most common prefixes, their abbreviations, and the multiples they represent are shown in Table 1.1. For example, 1 km is 1 kilometer, which is 1000 m, and 1 Mg is 1 megagram, which is 10^6 g, or 1000 kg. We frequently use kilonewtons (kN).

Table 1.1 The common prefixes used in SI units and the multiples they represent.

Prefix	Abbreviation	Multiple
nano-	n	10^{-9}
micro-	μ	10^{-6}
milli-	m	10^{-3}
kilo-	k	10^{3}
mega-	M	10^{6}
giga-	G	10^{9}

U.S. Customary Units

In U.S. Customary units, length is measured in feet (ft) and force is measured in pounds (lb). Time is measured in seconds (s). These are the base units of the U.S. Customary system. In this system of units, mass is a derived unit. The unit of mass is the slug, which is the mass of material accelerated at one foot per second squared by a force of one pound. Newton's second law states that

$$1 \text{ lb} = (1 \text{ slug})(1 \text{ ft/s}^2).$$

From this expression we obtain

$$1 \text{ slug} = 1 \text{ lb-s}^2/\text{ft}.$$

We use other U.S. Customary units such as the mile (1 mi = 5280 ft) and the inch (1 ft = 12 in.). We also use the kilopound (kip), which is 1000 lb.

In some engineering applications, an alternative unit of mass called the pound mass (lbm) is used, which is the mass of material having a weight of one pound at sea level. The weight at sea level of an object that has a mass of one slug is

$$W = mg = (1 \text{ slug})(32.2 \text{ ft/s}^2) = 32.2 \text{ lb},$$

so 1 lbm = (1/32.2) slug. When the pound mass is used, a pound of force is usually denoted by the abbreviation lbf.

Angular Units

In both SI and U.S. Customary units, angles are normally expressed in radians (rad). We show the value of an angle θ in radians in Fig. 1.3. It is defined to be the ratio of the part of the circumference subtended by θ to the radius of the circle. Angles are also expressed in degrees. Since there are 360 degrees (360°) in a complete circle, and the complete circumference of the circle is $2\pi R$, 360° equals 2π rad.

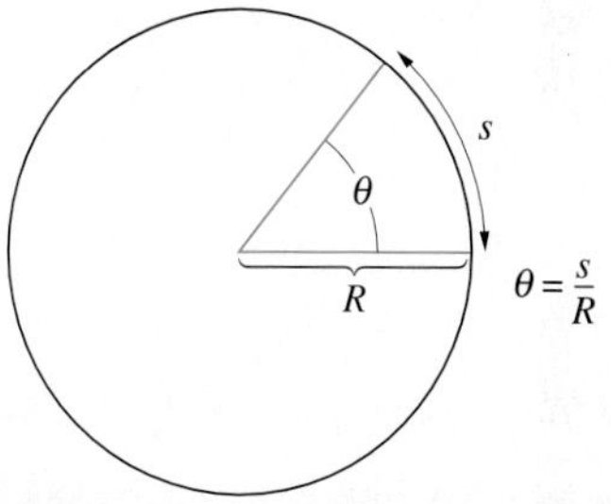

Equations containing angles are nearly always derived under the assumption that angles are expressed in radians. Therefore when you want to substitute the value of an angle expressed in degrees into an equation, you should first convert it into radians. A notable exception to this rule is that many calculators are designed to accept angles expressed in either degrees or radians when you use them to evaluate functions such as sin θ.

Figure 1.3
Definition of an angle in radians.

Conversion of Units

Many situations arise in engineering practice that require you to convert values expressed in units of one kind into values in other units. If some data in a problem are given in terms of SI units and some are given in terms of U.S. Customary units, you must express all of the data in terms of one system of units. In problems expressed in terms of SI units, you will occasionally be given data in terms of units other than the base units of seconds, meters, kilograms, and newtons. You should convert these data into the base units before working the problem. Similarly, in problems involving U.S. Customary units, you should convert terms into the base units of seconds, feet, slugs, and pounds. After you gain some experience, you will recognize situations in which these rules can be relaxed, but for now the procedure we propose is the safest.

Converting units is straightforward, although you must do it with care. Suppose that we want to express 1 mi/hr in terms of ft/s. Since one mile equals 5280 ft and one hour equals 3600 seconds, we can treat the expressions

$$\left(\frac{5280 \text{ ft}}{1 \text{ mi}}\right) \quad \text{and} \quad \left(\frac{1 \text{ hr}}{3600 \text{ s}}\right)$$

as ratios whose values are 1. In this way we obtain

$$1 \text{ mi/hr} = 1 \text{ mi/hr} \times \left(\frac{5280 \text{ ft}}{1 \text{ mi}}\right) \times \left(\frac{1 \text{ hr}}{3600 \text{ s}}\right) = 1.47 \text{ ft/s}.$$

We give some useful unit conversions in Table 1.2.

Table 1.2   Unit conversions.

Time	1 minute	=	60 seconds
	1 hour	=	60 minutes
	1 day	=	24 hours
Length	1 foot	=	12 inches
	1 mile	=	5280 feet
	1 inch	=	25.4 millimeters
	1 foot	=	0.3048 meter
Angle	2π radians	=	360 degrees
Mass	1 slug	=	14.59 kilograms
Force	1 pound	=	4.448 newtons

Example 1.1

Figure 1.4

If an Olympic sprinter (Fig. 1.4) runs 100 meters in 10 seconds, his average velocity is 10 m/s. What is his average velocity in mi/hr?

SOLUTION

$$10 \text{ m/s} = 10 \text{ m/s} \times \left(\frac{1 \text{ ft}}{0.3048 \text{ m}}\right) \times \left(\frac{1 \text{ mi}}{5280 \text{ ft}}\right) \times \left(\frac{3600 \text{ s}}{1 \text{ hr}}\right)$$

$$= 22.4 \text{ mi/hr}.$$

Example 1.2

Suppose that in Einstein's equation

$$E = mc^2$$

the mass m is in kilograms and the velocity of light c is in meters per second.
(a) What are the SI units of E?
(b) If the value of E in SI units is 20, what is its value in U.S. Customary base units?

STRATEGY

(a) Since we know the units of the terms m and c, we can deduce the units of E from the given equation.
(b) We can use the unit conversions for mass and length from Table 1.2 to convert E from SI units to U.S. Customary units.

SOLUTION

(a) From the equation for E,

$$E = (m \text{ kg})(c \text{ m/s})^2,$$

the SI units of E are kg-m^2/s^2.

(b) From Table 1.2, 1 slug = 14.59 kg and 1 ft = 0.3048 m. Therefore

$$1 \text{ kg-m}^2/\text{s}^2 = 1 \text{ kg-m}^2/\text{s}^2 \times \left(\frac{1 \text{ slug}}{14.59 \text{ kg}}\right) \times \left(\frac{1 \text{ ft}}{0.3048 \text{ m}}\right)^2$$

$$= 0.738 \text{ slug-ft}^2/\text{s}^2.$$

The value of E in U.S. Customary units is

$$E = (20)(0.738) = 14.8 \text{ slug-ft}^2/\text{s}^2.$$

Example 1.3

George Stephenson's *Rocket* (Fig. 1.5), an early steam locomotive, weighed about 7 tons with its tender. (A ton is 2000 lb.) What was its approximate mass in kilograms?

Figure 1.5

STRATEGY

We can use Eq. (1.6) to obtain the mass in slugs and then use the conversion given in Table 1.2 to determine the mass in kilograms.

SOLUTION

The mass in slugs is

$$m = \frac{W}{g} = \frac{14{,}000 \text{ lb}}{32.2 \text{ ft/s}^2} = 434.8 \text{ slugs.}$$

From Table 1.2, 1 slug equals 14.59 kg, so the mass in kilograms is (to three significant digits)

$$m = (434.8)(14.59) = 6340 \text{ kg.}$$

Problems

1.1 The value of π is 3.141592654.... What is its value to seven significant digits?

1.2 What is the value of e (the base of natural logarithms) to five significant digits?

1.3 Determine the value of the expression $1/(2 - \pi)$ to three significant digits.

1.4 The opening in a soccer goal is 24 ft wide and 8 ft high. Use these values to determine its dimensions in meters to three significant digits.

1.5 The dimensions of the Boeing 777-200 aircraft are length = 209 ft 1 in., wingspan = 199 ft 11 in., and height (bottom of wheels to top of vertical stabilizer) = 60 ft 6 in. Express these dimensions in meters to three significant digits.

1.6 Suppose that you have just purchased a Ferrari Dino 246GT coupe and you want to know whether you can use your set of SAE (U.S. Customary unit) wrenches to work on it. You have wrenches with widths w = 1/4 in., 1/2 in., 3/4 in., and 1 in., and the car has nuts with dimensions n = 5 mm, 10 mm, 15 mm, 20 mm, and 25 mm. Defining a wrench to fit if w is no more than 2% larger than n, which of your wrenches can you use?

P1.6

1.7 The 1829 *Rocket*, shown in Figure 1.5, could draw a carriage with 30 passengers at 25 mi/hr. Determine its velocity to three significant digits: (a) in ft/s; (b) in km/hr.

1.8 High-speed "bullet trains" began running between Tokyo and Osaka, Japan, in 1964. If a bullet train travels at 240 km/hr, what is its velocity in mi/hr to three significant digits?

1.9 In December 1986, Dick Rutan and Jeana Yeager flew the *Voyager* aircraft around the world nonstop. They flew a distance of 40,212 km in 9 days, 3 minutes, and 44 seconds.
(a) Determine the distance they flew in miles to three significant digits.
(b) Determine their average speed (the distance flown divided by the time required) in kilometers per hour, miles per hour, and knots (nautical miles per hour) to three significant digits.

1.10 Engineers who study shock waves sometimes express velocity in millimeters per microsecond (mm/μs). Suppose the velocity of a wavefront is measured and determined to be 5 mm/μs. Determine its velocity: (a) in m/s; (b) in mi/s.

1.11 Geophysicists measure the motion of a glacier and discover it is moving at 80 mm/year. What is its velocity in m/s?

1.12 The acceleration due to gravity at sea level in SI units is g = 9.81 m/s^2. By converting units, use this value to determine the acceleration due to gravity at sea level in U.S. Customary units.

1.13 A *furlong per fortnight* is a facetious unit of velocity, perhaps made up by a student as a satirical comment on the bewildering variety of units engineers must deal with. A furlong is 660 ft (1/8 mile). A fortnight is 2 weeks (14 nights). If you walk to class at 2 m/s, what is your speed in furlongs per fortnight to three significant digits?

1.14 The cross-sectional area of a beam is 480 in^2. What is its cross-sectional area in m^2?

1.15 At sea level, the weight density (weight per unit volume) of water is approximately 62.4 lb/ft^3. Use this value to determine the mass density of water in kg/m^3.

1.16 A pressure transducer measures a value of 300 lb/in^2. Determine the value of the pressure in pascals. A pascal (Pa) is one newton per meter squared.

1.17 A horsepower is 550 ft-lb/s. A watt is 1 N-m/s. Determine the number of watts generated by (a) the Wright brothers' 1903 airplane, which had a 12-horsepower engine; (b) a modern passenger jet with a power of 100,000 horsepower at cruising speed.

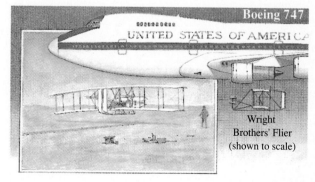

Boeing 747

Wright Brothers' Flier (shown to scale)

P1.17

1.18 In SI units, the universal gravitational constant $G = 6.67 \times 10^{-11}$ N-m^2/kg^2. Determine the value of G in U.S. Customary base units.

1.19 If the earth is modeled as a homogeneous sphere, the velocity of a satellite in a circular orbit is

$$v = \sqrt{\frac{gR_E^2}{r}},$$

where R_E is the radius of the earth and r is the radius of the orbit.
(a) If g is in m/s^2 and R_E and r are in meters, what are the units of v?
(b) If $R_E = 6370$ km and $r = 6670$ km, what is the value of v to three significant digits?
(c) For the orbit described in (b), what is the value of v in mi/s to three significant digits?

1.20 In the equation

$$T = \frac{1}{2}I\omega^2,$$

the term I is in kg-m^2 and ω is in s^{-1}.
(a) What are the SI units of T?
(b) If the value of T is 100 when I is in kg-m^2 and ω is in s^{-1}, what is the value of T when it is expressed in terms of U.S. Customary base units?

1.21 The aerodynamic drag force D exerted on a moving object by a gas is given by the expression

$$D = C_D S \frac{1}{2}\rho v^2,$$

where the drag coefficient C_D is dimensionless, S is a reference area, ρ is the mass per unit volume of the gas, and v is the velocity of the object relative to the gas.
(a) Suppose that the value of D is 800 when S, ρ, and v are expressed in SI base units. By converting units, determine the value of D when S, ρ, and v are expressed in U.S. Customary base units.
(b) The drag force D is in newtons when the expression is evaluated using SI base units and is in pounds when the expression is evaluated using U.S. Customary base units. Using your result from (a), determine the conversion factor from newtons to pounds.

1.22 The Lockheed-Martin X-33 reusable launch test vehicle, when fully fueled, will weigh 273,300 lb at sea level. Its weight at sea level with its fuel expended will be 62,700 lb.
(a) Determine its mass in slugs when it is fully fueled and with its fuel expended.
(b) Determine its weight in meganewtons at sea level when it is fully fueled and with its fuel expended.

P1.22

1.23 The acceleration due to gravity is 13.2 ft/s^2 on the surface of Mars and 32.2 ft/s^2 on the surface of the earth. A woman weighs 125 lb on earth. To survive and work on the surface of Mars, she must wear life-support equipment and carry tools. What is the maximum allowable weight on earth of the woman's clothing, equipment, and tools if the engineers don't want the total weight on Mars of the woman and her clothing, equipment, and tools to exceed 125 lb?

1.24 A person has a mass of 50 kg.
(a) The acceleration due to gravity at sea level is $g = 9.81$ m/s^2. What is the person's weight at sea level?
(b) The acceleration due to gravity on the surface of the moon is 1.62 m/s^2. What would the person weigh on the moon?

1.25 The acceleration due to gravity at sea level is $g = 9.81$ m/s^2. The radius of the earth is 6370 km. The universal gravitational constant $G = 6.67 \times 10^{-11}$ N-m^2/kg^2. Use this information to determine the mass of the earth.

1.26 A person weighs 180 lb at sea level. The radius of the earth is 3960 mi. What force is exerted on the person by the gravitational attraction of the earth if he is in a space station in orbit 200 mi above the surface of the earth?

1.27 The acceleration due to gravity on the surface of the moon is 1.62 m/s^2. The radius of the moon is $R_M = 1738$ km. Determine the acceleration due to gravity of the moon at a point 1738 km above its surface.

Strategy: Write an equation equivalent to Eq. (1.4) for the acceleration due to gravity of the moon.

1.28 If an object is near the surface of the earth, the variation of its weight with distance from the center of the earth can often be neglected. The acceleration due to gravity at sea level is $g = 9.81$ m/s^2. The radius of the earth is 6370 km. The weight of an object at sea level is mg, where m is its mass. At what height above the surface of the earth does the weight of the object decrease to $0.99mg$?

1.29 The centers of two oranges are 1 m apart. The mass of each orange is 0.2 kg. What gravitational force do they exert on each other? (The universal gravitational constant $G = 6.67 \times 10^{-11}$ N-m^2/kg^2.)

1.30 At a point between the earth and the moon, the magnitude of the earth's gravitational acceleration equals the magnitude of the moon's gravitational acceleration. What is the distance from the center of the earth to that point to three significant digits? The distance from the center of the earth to the center of the moon is 383,000 km, and the radius of the earth is 6370 km. The radius of the moon is 1738 km, and the acceleration due to gravity at its surface is 1.62 m/s^2.

Geodesic domes, lattices of beams assembled in repeating geometric units, are much lighter than conventional structures of equal volume. The Spaceship Earth building near Orlando, Florida, contains 1450 beams in repeating triangular units that form a sphere 50 m (164 ft) in diameter. To analyze forces in the members of such structures, you must resolve the force vectors into components in three dimensions—a technique discussed in this chapter.

Chapter 2

Vectors

TO describe a force acting on a structural member, you must specify both the magnitude of the force and its direction. To describe the position of an airplane relative to an airport, you must specify both the distance and direction from the airport to the airplane. In engineering we deal with many quantities that have both magnitude and direction and can be expressed as vectors. In this chapter we review vector operations, resolve vectors into components, and give examples of engineering applications of vectors.

Vector Operations and Definitions

Engineers designing a structure must analyze the positions of its members and the forces acting on them. When designing a machine, they must analyze the velocities and accelerations of its moving parts. Positions, forces, velocities, and accelerations are **vectors**. These and many other physical quantities important in engineering, such as electromagnetic fields, can be represented by vectors and analyzed by vector operations. Here we review fundamental vector operations and definitions.

2.1 Scalars and Vectors

A physical quantity that is completely described by a real number is called a **scalar**. Time is a scalar quantity. Mass is also a scalar quantity. For example, you completely describe the mass of a car by saying that its value is 1200 kg.

In contrast, you have to specify both a nonnegative real number, or **magnitude**, and a direction to describe a vector quantity. Two vector quantities are equal only if both their magnitudes and their directions are equal.

The position of a point in space relative to another point is a vector quantity. To describe the location of a city relative to your home, it is not enough to say that it is 100 miles away. You must say that it is 100 miles west of your home. Force is also a vector quantity. When you push a piece of furniture across the floor, you apply a force of magnitude sufficient to move the furniture and you apply it in the direction you want the furniture to move.

We will represent vectors by boldfaced letters, $\mathbf{U}$, $\mathbf{V}$, $\mathbf{W}$, . . . , and will denote the magnitude of a vector $\mathbf{U}$ by $|\mathbf{U}|$. In handwritten work, a vector $\mathbf{U}$ can be represented by the symbols $\overline{U}$, $\vec{U}$, or $\underset{\sim}{U}$. A vector is represented graphically by an arrow. The direction of the arrow indicates the direction of the vector, and the length of the arrow is defined to be proportional to the magnitude.

For example, consider the points A and B of the mechanism in Fig. 2.1(a). We can specify the position of point B relative to point A by the vector $\mathbf{r}_{AB}$ in Fig. 2.1(b). The direction of $\mathbf{r}_{AB}$ indicates the direction from point A to point B. If the distance between the two points is 200 mm, the magnitude $|\mathbf{r}_{AB}| =$ 200 mm.

The cable AB in Fig. 2.2 helps support the television transmission tower. We can represent the force the cable exerts on the tower by a vector $\mathbf{F}$ as shown. If the cable exerts an 800-N force on the tower, $|\mathbf{F}| = 800$ N.

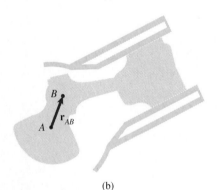

(a)

(b)

Figure 2.1

(a) Two points A and B of a mechanism.
(b) The vector $\mathbf{r}_{AB}$ from A to B.

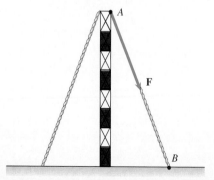

Figure 2.2

Representing the force cable AB exerts on the tower by a vector $\mathbf{F}$.

2.2 *Rules for Manipulating Vectors*

Vectors are a convenient means for representing physical quantities that have magnitude and direction, but that is only the beginning of their usefulness. Just as you manipulate real numbers with the familiar rules for addition, subtraction, multiplication, and so forth, there are rules for manipulating vectors. These rules provide you with powerful tools for engineering analysis.

Vector Addition

When an object moves from one location in space to another, we say it undergoes a **displacement**. If we move a book (or, speaking more precisely, some point of a book) from one location on a table to another, as shown in Fig. 2.3(a), we can represent the displacement by the vector **U**. The direction of **U** indicates the direction of the displacement, and $|\mathbf{U}|$ is the distance the book moves.

Suppose that we give the book a second displacement **V**, as shown in Fig. 2.3(b). The two displacements **U** and **V** are equivalent to a single displacement of the book from its initial position to its final position, which we represent by the vector **W** in Fig. 2.3(c). Notice that the final position of the book is the same whether we first give it the displacement **U** and then the displacement **V** or we first give it the displacement **V** and then the displacement **U** (Fig. 2.3d). The displacement **W** is defined to be the sum of the displacements **U** and **V**:

$$\mathbf{U} + \mathbf{V} = \mathbf{W}.$$

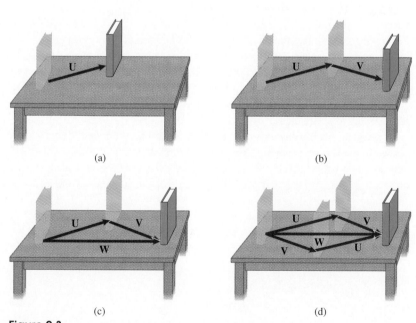

(a)

(b)

(c)

(d)

Figure 2.3

(a) A displacement represented by the vector **U**.
(b) The displacement **U** followed by the displacement **V**.
(c) The displacements **U** and **V** are equivalent to the displacement **W**.
(d) The final position of the book doesn't depend on the order of the displacements.

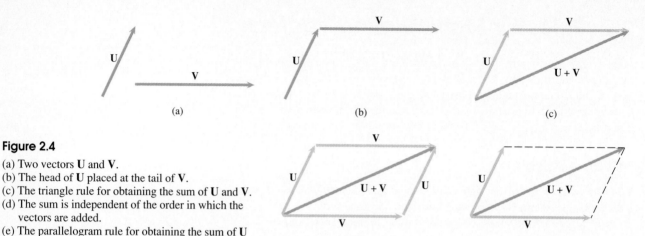

Figure 2.4

(a) Two vectors **U** and **V**.
(b) The head of **U** placed at the tail of **V**.
(c) The triangle rule for obtaining the sum of **U** and **V**.
(d) The sum is independent of the order in which the vectors are added.
(e) The parallelogram rule for obtaining the sum of **U** and **V**.

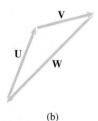

(a)

(b)

Figure 2.5

(a) The sum of three vectors.
(b) Three vectors whose sum is zero.

The definition of vector addition is motivated by the addition of displacements. Consider the two vectors **U** and **V** shown in Fig. 2.4(a). If we place them head to tail (Fig. 2.4b), their sum is defined to be the vector from the tail of **U** to the head of **V** (Fig. 2.4c). This is called the **triangle rule** for vector addition. Figure 2.4(d) demonstrates that the sum is independent of the order in which the vectors are placed head to tail. From this figure we obtain the **parallelogram rule** for vector addition (Fig. 2.4e).

The definition of vector addition implies that

$$\mathbf{U} + \mathbf{V} = \mathbf{V} + \mathbf{U} \qquad \text{Vector addition is commutative.} \qquad (2.1)$$

and

$$(\mathbf{U} + \mathbf{V}) + \mathbf{W} = \mathbf{U} + (\mathbf{V} + \mathbf{W}) \qquad \text{Vector addition is associative.} \qquad (2.2)$$

for any vectors **U**, **V**, and **W**. These results mean that when you add two or more vectors, you don't need to worry about the order in which you add them. The sum is obtained by placing the vectors head to tail in any order. The vector from the tail of the first vector to the head of the last one is the sum (Fig. 2.5a). If the sum is zero, the vectors form a closed polygon when they are placed head to tail (Fig. 2.5b).

A physical quantity is called a vector if it has magnitude and direction and obeys the definition of vector addition. We have seen that a displacement is a vector. The position of a point in space relative to another point is also a vector quantity. In Fig. 2.6, the vector $\mathbf{r}_{AC}$ from A to C is the sum of $\mathbf{r}_{AB}$ and $\mathbf{r}_{BC}$.

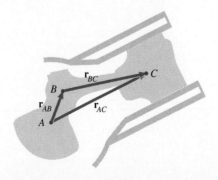

Figure 2.6

Arrows denoting the relative positions of points are vectors.

A force has direction and magnitude, but do forces obey the definition of vector addition? For now we will assume that they do. When we discuss dynamics we will show that Newton's second law implies that force is a vector.

Product of a Scalar and a Vector

The product of a scalar (real number) a and a vector $\mathbf{U}$ is a vector written as $a\mathbf{U}$. Its magnitude is $|a||\mathbf{U}|$, where $|a|$ is the absolute value of the scalar a. The direction of $a\mathbf{U}$ is the same as the direction of $\mathbf{U}$ when a is positive and is opposite to the direction of $\mathbf{U}$ when a is negative.

The product $(-1)\mathbf{U}$ is written as $-\mathbf{U}$ and is called "the negative of the vector $\mathbf{U}$." It has the same magnitude as $\mathbf{U}$ but the opposite direction.

The division of a vector $\mathbf{U}$ by a scalar a is defined to be the product

$$\frac{\mathbf{U}}{a} = \left(\frac{1}{a}\right)\mathbf{U}.$$

Figure 2.7 shows a vector $\mathbf{U}$ and the products of $\mathbf{U}$ with the scalars 2, -1, and 1/2.

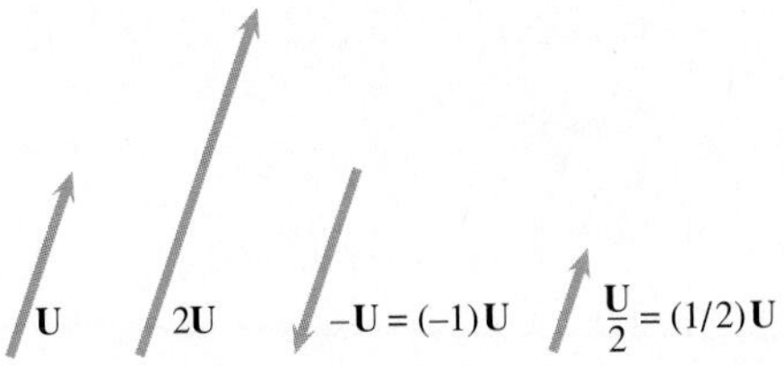

Figure 2.7

A vector $\mathbf{U}$ and some of its scalar multiples.

The definitions of vector addition and the product of a scalar and a vector imply that

$$a(b\mathbf{U}) = (ab)\mathbf{U}, \qquad \text{The product is associative with respect to scalar multiplication.} \qquad (2.3)$$

$$(a + b)\mathbf{U} = a\mathbf{U} + b\mathbf{U}, \qquad \text{The product is distributive with respect to scalar addition.} \qquad (2.4)$$

and

$$a(\mathbf{U} + \mathbf{V}) = a\mathbf{U} + a\mathbf{V} \qquad \text{The product is distributive with respect to vector addition.} \qquad (2.5)$$

for any scalars a and b and vectors $\mathbf{U}$ and $\mathbf{V}$. We will need these results when we discuss components of vectors.

Vector Subtraction

The difference of two vectors $\mathbf{U}$ and $\mathbf{V}$ is obtained by adding $\mathbf{U}$ to the vector $(-1)\mathbf{V}$:

$$\mathbf{U} - \mathbf{V} = \mathbf{U} + (-1)\mathbf{V}. \qquad (2.6)$$

Consider the two vectors $\mathbf{U}$ and $\mathbf{V}$ shown in Fig. 2.8(a). The vector $(-1)\mathbf{V}$ has the same magnitude as the vector $\mathbf{V}$ but is in the opposite direction (Fig. 2.8b). In Fig. 2.8(c), we add the vector $\mathbf{U}$ to the vector $(-1)\mathbf{V}$ to obtain $\mathbf{U} - \mathbf{V}$.

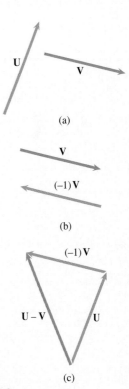

Figure 2.8

(a) Two vectors $\mathbf{U}$ and $\mathbf{V}$.
(b) The vectors $\mathbf{V}$ and $(-1)\mathbf{V}$.
(c) The sum of $\mathbf{U}$ and $(-1)\mathbf{V}$ is the vector difference $\mathbf{U} - \mathbf{V}$.

Unit Vectors

A **unit vector** is simply a vector whose magnitude is 1. A unit vector specifies a direction and also provides a convenient way to express a vector that has a particular direction. If a unit vector **e** and a vector **U** have the same direction, we can write **U** as the product of its magnitude $|\mathbf{U}|$ and the unit vector **e** (Fig. 2.9),

$$\mathbf{U} = |\mathbf{U}|\mathbf{e}.$$

Any vector **U** *can be regarded as the product of its magnitude and a unit vector that has the same direction as* **U**. Dividing both sides of this equation by $|\mathbf{U}|$:

$$\frac{\mathbf{U}}{|\mathbf{U}|} = \mathbf{e},$$

we see that dividing any vector by its magnitude yields a unit vector that has the same direction.

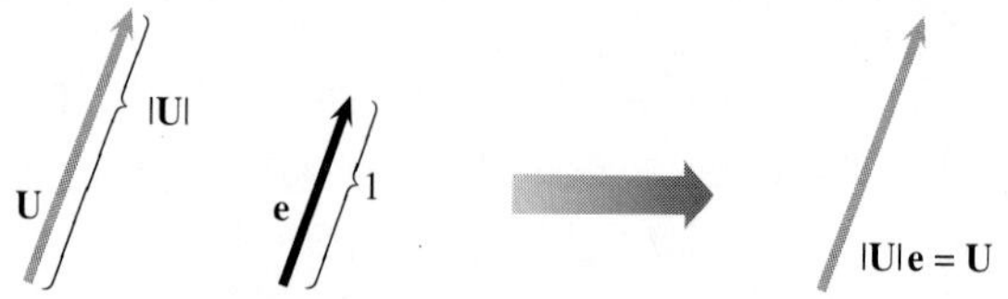

Figure 2.9

Since **U** and **e** have the same direction, the vector **U** equals the product of its magnitude with **e**.

Vector Components

When a vector **U** is expressed as the sum of a set of vectors, each vector of the set is called a **vector component** of **U**. Suppose that the vector **U** shown in Fig. 2.10(a) is parallel to the plane defined by the two intersecting lines. We can express **U** as the sum of vector components **V** and **W** that are parallel to the two lines, as shown in Fig. 2.10(b). We say that **U** is **resolved** into the vector components **V** and **W**.

Figure 2.10

(a) A vector **U** and two intersecting lines.
(b) The vectors **V** and **W** are vector components of **U**.

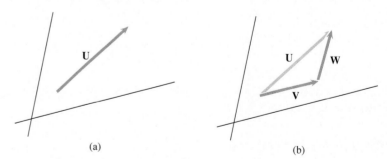

(a)   (b)

You can solve some vector problems by drawing vector diagrams to scale and measuring the results or by applying trigonometry to vector diagrams. We demonstrate these approaches in the following examples. We will then show in the next section that expressing vectors in terms of mutually perpendicular vector components results in a much simpler way to solve vector problems.

Example 2.1

The cables AB and AC in Fig. 2.11 help support the cantilevered roof of a sports stadium. The forces the cables exert on the pylon to which they are attached are represented by the vectors $\mathbf{F}_{AB}$ and $\mathbf{F}_{AC}$. The magnitudes of the forces are $|\mathbf{F}_{AB}| = 100$ kN and $|\mathbf{F}_{AC}| = 60$ kN. Determine the magnitude and direction of the sum of the forces exerted on the pylon by the cables (a) graphically and (b) by using trigonometry.

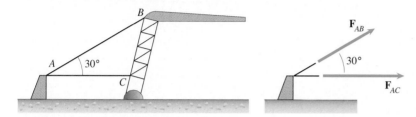

Figure 2.11

STRATEGY

(a) By drawing the parallelogram rule for adding the two forces with the vectors drawn to scale, we can measure the magnitude and direction of their sum.
(b) We will calculate the magnitude and direction of the sum of the forces by applying the laws of sines and cosines (Appendix A, Section A.2) to the triangles formed by the parallelogram rule.

SOLUTION

(a) We graphically construct the parallelogram rule for obtaining the sum of the two forces with the lengths of $\mathbf{F}_{AB}$ and $\mathbf{F}_{AC}$ proportional to their magnitudes (Fig. a). By measuring the figure, we estimate the magnitude of the vector $\mathbf{F}_{AB} + \mathbf{F}_{AC}$ to be 155 kN and its direction to be 19° above the horizontal.
(b) Consider the parallelogram rule for obtaining the sum of the two forces (Fig. b). Since $\alpha + 30° = 180°$, the angle $\alpha = 150°$. By applying the law of cosines to the shaded triangle,

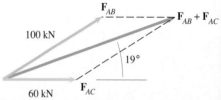

(a) Graphical solution.

$$|\mathbf{F}_{AB} + \mathbf{F}_{AC}|^2 = |\mathbf{F}_{AB}|^2 + |\mathbf{F}_{AC}|^2 - 2|\mathbf{F}_{AB}||\mathbf{F}_{AC}| \cos \alpha$$
$$= (100)^2 + (60)^2 - 2(100)(60) \cos 150°,$$

we determine that the magnitude $|\mathbf{F}_{AB} + \mathbf{F}_{AC}| = 154.9$ kN.

To determine the angle β between the vector $\mathbf{F}_{AB} + \mathbf{F}_{AC}$ and the horizontal, we apply the law of sines to the shaded triangle:

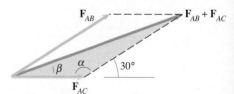

(b) Trigonometric solution.

$$\frac{\sin \beta}{|\mathbf{F}_{AB}|} = \frac{\sin \alpha}{|\mathbf{F}_{AB} + \mathbf{F}_{AC}|}.$$

The solution is

$$\beta = \arcsin\left(\frac{|\mathbf{F}_{AB}| \sin \alpha}{|\mathbf{F}_{AB} + \mathbf{F}_{AC}|}\right) = \arcsin\left(\frac{100 \sin 150°}{154.9}\right) = 18.8°.$$

DISCUSSION

Engineering applications of vectors usually require the precision of analytical solutions, but experience with graphical solutions can help you understand vector operations. Carrying out a graphical solution can also help you formulate an analytical solution.

Example 2.2

The force $\mathbf{F}$ in Fig. 2.12 lies in the plane defined by the intersecting lines L_A and L_B. Its magnitude is 400 lb. Suppose that you want to resolve $\mathbf{F}$ into vector components parallel to L_A and L_B. Determine the magnitudes of the vector components (a) graphically and (b) by using trigonometry.

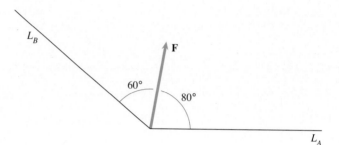

Figure 2.12

SOLUTION

(a) We draw dashed lines from the head of $\mathbf{F}$ parallel to L_A and L_B to construct the vector components, which we denote $\mathbf{F}_A$ and $\mathbf{F}_B$ (Fig. a). By measuring the figure, we estimate their magnitudes to be $|\mathbf{F}_A| = 540$ lb and $|\mathbf{F}_B| = 610$ lb.

(b) Consider the force $\mathbf{F}$ and the vector components $\mathbf{F}_A$ and $\mathbf{F}_B$ (Fig. b). Since $\alpha + 80° + 60° = 180°$, the angle $\alpha = 40°$. By applying the law of sines to triangle 1,

$$\frac{\sin 60°}{|\mathbf{F}_A|} = \frac{\sin \alpha}{|\mathbf{F}|},$$

we obtain the magnitude of $\mathbf{F}_A$:

$$|\mathbf{F}_A| = \frac{|\mathbf{F}| \sin 60°}{\sin \alpha} = \frac{400 \sin 60°}{\sin 40°} = 538.9 \text{ lb}.$$

Then by applying the law of sines to triangle 2,

$$\frac{\sin 80°}{|\mathbf{F}_B|} = \frac{\sin \alpha}{|\mathbf{F}|},$$

we obtain the magnitude of $\mathbf{F}_B$:

$$|\mathbf{F}_B| = \frac{|\mathbf{F}| \sin 80°}{\sin \alpha} = \frac{400 \sin 80°}{\sin 40°} = 612.8 \text{ lb}.$$

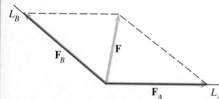

(a) Graphical solution.

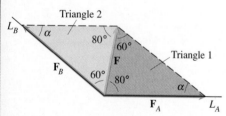

(b) Trigonometric solution.

Problems

Refer to the following diagram when solving Problems 2.1 through 2.7.

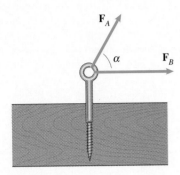

P2.1–2.7

2.1 The magnitudes $|\mathbf{F}_A| = 60$ N and $|\mathbf{F}_B| = 80$ N. The angle $\alpha = 45°$. Graphically determine the magnitude of the sum of the forces $\mathbf{F} = \mathbf{F}_A + \mathbf{F}_B$ and the angle between $\mathbf{F}_B$ and $\mathbf{F}$.
Strategy: Construct the parallelogram for determining the sum of the forces, drawing the lengths of $\mathbf{F}_A$ and $\mathbf{F}_B$ proportional to their magnitudes and accurately measuring the angle α, as we did in Example 2.1. Then you can measure the magnitude of their sum and the angle between their sum and $\mathbf{F}_B$.

2.2 The magnitudes $|\mathbf{F}_A| = 60$ N and $|\mathbf{F}_B| = 80$ N. The angle $\alpha = 45°$. Graphically determine the magnitude of the force $\mathbf{F} = 2\mathbf{F}_A - 3\mathbf{F}_B$ and the angle between $\mathbf{F}_B$ and $\mathbf{F}$.

2.3 The magnitudes $|\mathbf{F}_A| = 100$ lb and $|\mathbf{F}_B| = 140$ lb. The angle $\alpha = 40°$. Use trigonometry to determine the magnitude of the sum of the forces $\mathbf{F} = \mathbf{F}_A + \mathbf{F}_B$ and the angle between $\mathbf{F}_B$ and $\mathbf{F}$.
Strategy: Use the laws of sines and cosines to analyze the triangles formed by the parallelogram rule for the sum of the forces as we did in Example 2.1. The laws of sines and cosines are given in Section A.2 of Appendix A.

2.4 The magnitudes $|\mathbf{F}_A| = 60$ N and $|\mathbf{F}_B| = 80$ N. The angle $\alpha = 45°$. Use trigonometry to determine the magnitude of the force $\mathbf{F} = 2\mathbf{F}_A - 3\mathbf{F}_B$ and the angle between $\mathbf{F}_B$ and $\mathbf{F}$.

2.5 The magnitudes $|\mathbf{F}_A| = 100$ lb and $|\mathbf{F}_B| = 140$ lb. If α can have any value, what are the minimum and maximum possible values of the magnitude of the sum of the forces $\mathbf{F} = \mathbf{F}_A + \mathbf{F}_B$, and what are the corresponding values of α?

2.6 The magnitude $|\mathbf{F}_A| = 60$ N and the angle $\alpha = 45°$. If the magnitude of the sum of the forces $|\mathbf{F}_A + \mathbf{F}_B| = 180$ N, what is the magnitude of $\mathbf{F}_B$?

2.7 The magnitude $|\mathbf{F}_A| = 100$ lb and the magnitude $|\mathbf{F}_B| = 140$ lb. Suppose that the support acted on by the two forces can safely withstand a total force of 240 lb. What is the acceptable range of the angle α?

2.8 The force $\mathbf{F}$ of magnitude 8 kN lies in the plane defined by the intersecting lines L_A and L_B. Suppose that you want to resolve $\mathbf{F}$ into a vector component $\mathbf{F}_A$ parallel to L_A and a vector component $\mathbf{F}_B$ parallel to L_B. Determine the magnitudes of $\mathbf{F}_A$ and $\mathbf{F}_B$ (a) graphically and (b) by using trigonometry.

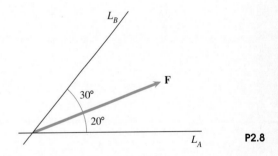

P2.8

2.9 The rocket engine exerts an upward force of 4 MN (meganewtons) magnitude on the test stand. If you resolve the force into vector components parallel to the bars AB and CD, what are the magnitudes of the components?

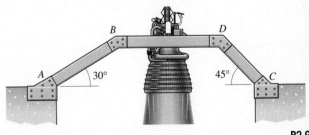

P2.9

2.10   The magnitude of the vertical force **F** is 6 kN. If you resolve **F** into vector components parallel to the bars AB and CD, what are the magnitudes of the components?

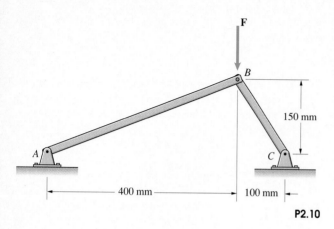

P2.10

2.11   A structural engineer determines that the truss in Problem 2.10 will safely support the force **F** if the magnitude of each of the vector components of **F** parallel to the bars does not exceed 20 kN. Based on this criterion, how large can the magnitude of **F** be?

2.12   The suspended weight exerts a downward 2000-lb force **F** at A. If you resolve **F** into vector components parallel to the wires AB, AC, and AD, the magnitude of the component parallel to AC is 600 lb. What are the magnitudes of the components parallel to AB and AD?

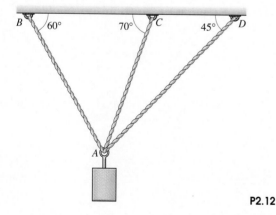

P2.12

2.13   The wires in Problem 2.12 will safely support the weight if the magnitude of the vector component of **F** parallel to each wire does not exceed 2000 lb. Based on this criterion, how large can the magnitude of **F** be? What are the corresponding magnitudes of the vector components of **F** parallel to the three wires?

2.14   Two vectors $\mathbf{r}_A$ and $\mathbf{r}_B$ have magnitudes $|\mathbf{r}_A| = 30$ m and $|\mathbf{r}_B| = 40$ m. Determine the magnitude of their sum $\mathbf{r}_A + \mathbf{r}_B$
(a) if $\mathbf{r}_A$ and $\mathbf{r}_B$ have the same direction.
(b) if $\mathbf{r}_A$ and $\mathbf{r}_B$ are perpendicular.

2.15   A spherical storage tank is supported by cables. The tank is subjected to three forces: the forces $\mathbf{F}_A$ and $\mathbf{F}_B$ exerted by the cables and the weight **W**. The weight of the tank $|\mathbf{W}| = 600$ lb. The vector sum of the forces acting on the tank equals zero. Determine the magnitudes of $\mathbf{F}_A$ and $\mathbf{F}_B$ (a) graphically and (b) by using trigonometry.

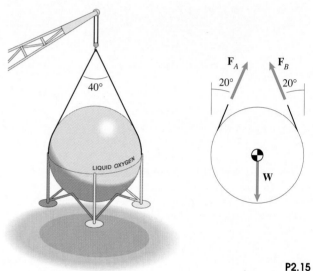

P2.15

2.16   The rope ABC exerts forces $\mathbf{F}_{BA}$ and $\mathbf{F}_{BC}$ on the block at B. Their magnitudes are $|\mathbf{F}_{BA}| = |\mathbf{F}_{BC}| = 800$ N. Determine $|\mathbf{F}_{BA} + \mathbf{F}_{BC}|$ (a) graphically and (b) by using trigonometry.

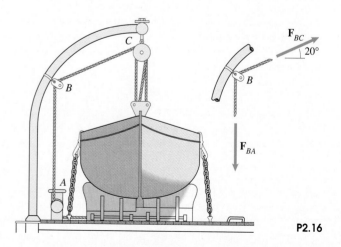

P2.16

2.17 Two snowcats tow a housing unit to a new location at McMurdo Base, Antarctica. (The top view is shown. The cables are horizontal.) The sum of the forces $\mathbf{F}_A$ and $\mathbf{F}_B$ exerted on the unit is parallel to the line L, and $|\mathbf{F}_A| = 1000$ lb. Determine $|\mathbf{F}_B|$ and $|\mathbf{F}_A + \mathbf{F}_B|$ (a) graphically and (b) by using trigonometry.

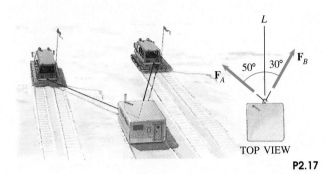

TOP VIEW

P2.17

2.18 A surveyor determines that the horizontal distance from A to B is 400 m and that the horizontal distance from A to C is 600 m. Determine the magnitude of the horizontal vector $\mathbf{r}_{BC}$ from B to C and the angle α (a) graphically and (b) by using trigonometry.

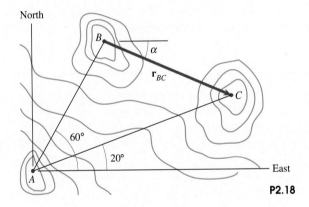

P2.18

2.19 The vector $\mathbf{r}$ extends from point A to the midpoint between points B and C. Prove that

$$\mathbf{r} = \frac{1}{2}\left(\mathbf{r}_{AB} + \mathbf{r}_{AC}\right).$$

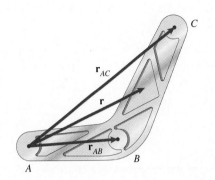

P2.19

2.20 By drawing sketches of the vectors, explain why

$$\mathbf{U} + (\mathbf{V} + \mathbf{W}) = (\mathbf{U} + \mathbf{V}) + \mathbf{W}.$$

Cartesian Components

2.3 Components in Two Dimensions

You will find that vectors are much easier to work with when you can express them in terms of mutually perpendicular vector components. Here we explain how to resolve vectors into cartesian components in two and three dimensions and give examples of vector manipulations using components.

Consider the vector **U** in Fig. 2.13(a). By placing a cartesian coordinate system so that **U** is parallel to the *x-y* plane, we can resolve it into vector components $\mathbf{U}_x$ and $\mathbf{U}_y$ parallel to the *x* and *y* axes (Fig. 2.13b),

$$\mathbf{U} = \mathbf{U}_x + \mathbf{U}_y.$$

Then by introducing a unit vector **i** defined to point in the direction of the positive *x* axis and a unit vector **j** defined to point in the direction of the positive *y* axis (Fig. 2.13c), we can express the vector **U** in the form

$$\boxed{\mathbf{U} = U_x\mathbf{i} + U_y\mathbf{j}.} \tag{2.7}$$

The scalars U_x and U_y are called **scalar components** of **U**. When we refer simply to the components of a vector, we will mean its scalar components. We will call U_x and U_y the *x* and *y* components of **U**.

The components of a vector specify both its direction relative to the cartesian coordinate system and its magnitude. From the right triangle formed by the vector **U** and its vector components (Fig. 2.13c), we see that the magnitude of **U** is given in terms of its components by the Pythagorean theorem,

$$\boxed{|\mathbf{U}| = \sqrt{U_x^2 + U_y^2}.} \tag{2.8}$$

With this equation you can determine the magnitude of a vector when you know its components.

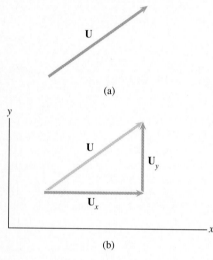

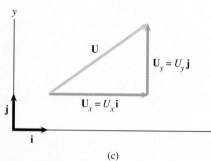

Figure 2.13

(a) A vector **U**.
(b) The vector components $\mathbf{U}_x$ and $\mathbf{U}_y$.
(c) The vector components can be expressed in terms of **i** and **j**.

Manipulating Vectors in Terms of Components

The sum of two vectors **U** and **V** in terms of their components is

$$\begin{aligned}\mathbf{U} + \mathbf{V} &= (U_x\mathbf{i} + U_y\mathbf{j}) + (V_x\mathbf{i} + V_y\mathbf{j}) \\ &= (U_x + V_x)\mathbf{i} + (U_y + V_y)\mathbf{j}.\end{aligned} \tag{2.9}$$

The components of **U** + **V** are the sums of the components of the vectors **U** and **V**. Notice that in obtaining this result we used Eqs. (2.2), (2.4), and (2.5).

It is instructive to derive Eq. (2.9) graphically. The summation of **U** and **V** is shown in Fig. 2.14(a). In Fig. 2.14(b) we introduce a coordinate system and resolve **U** and **V** into their components. In Fig. 2.14(c) we add the x and y components, obtaining Eq. (2.9).

The product of a number a and a vector **U** in terms of the components of **U** is

$$a\mathbf{U} = a(U_x\mathbf{i} + U_y\mathbf{j}) = aU_x\mathbf{i} + aU_y\mathbf{j}.$$

The component of $a\mathbf{U}$ in each coordinate direction equals the product of a and the component of **U** in that direction. We used Eqs. (2.3) and (2.5) to obtain this result.

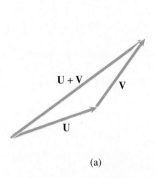

(a)

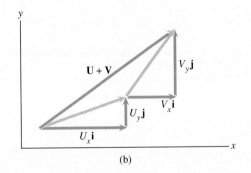

(b)

(c)

Figure 2.14
(a) The sum of **U** and **V**.
(b) The vector components of **U** and **V**.
(c) The sum of the components in each coordinate direction equals the component of **U** + **V** in that direction.

Position Vectors in Terms of Components

You can express the position vector of a point relative to another point in terms of the cartesian coordinates of the points. Consider point A with coordinates (x_A, y_A) and point B with coordinates (x_B, y_B). Let $\mathbf{r}_{AB}$ be the vector that specifies the position of B relative to A (Fig. 2.15a). That is, we denote the vector *from* a point A *to* a point B by $\mathbf{r}_{AB}$. You can see in Fig. 2.15(b) that $\mathbf{r}_{AB}$ is given in terms of the coordinates of points A and B by

$$\mathbf{r}_{AB} = (x_B - x_A)\,\mathbf{i} + (y_B - y_A)\,\mathbf{j}. \tag{2.10}$$

We can state this result as a simple rule: The x component of the position vector from a point A to a point B is obtained by subtracting the x coordinate of A from the x coordinate of B, and the y component is obtained by subtracting the y coordinate of A from the y coordinate of B.

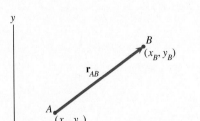

(a)

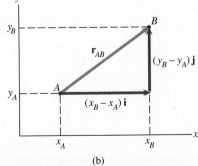

(b)

Figure 2.15
(a) Two points A and B and the position vector $\mathbf{r}_{AB}$ from A to B.
(b) The components of $\mathbf{r}_{AB}$ can be determined from the coordinates of points A and B.

In the following examples we show you how to manipulate vectors in terms of their components. In Example 2.3, vectors are given in terms of their components and the objective is to carry out operations, including determining the product of a scalar and a vector, adding vectors, and determining the magnitude of a vector. We then explore examples in which it is necessary to resolve vectors into components.

Example 2.3

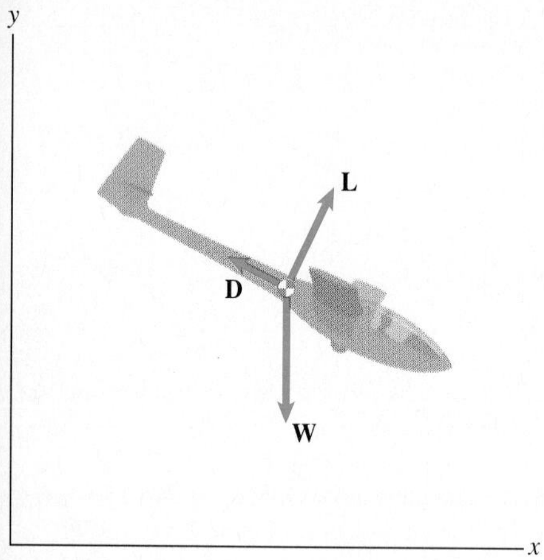

The forces acting on the sailplane in Fig. 2.16 are its weight $\mathbf{W} = -600\mathbf{j}$ (lb), the drag $\mathbf{D} = -200\mathbf{i} + 100\mathbf{j}$ (lb), and the lift $\mathbf{L}$.

(a) If the sum of the forces on the sailplane is zero, what are the scalar components of $\mathbf{L}$?

(b) If the lift $\mathbf{L}$ has the components determined in (a) and the drag $\mathbf{D}$ increases by a factor of 2, what is the magnitude of the sum of the forces on the sailplane?

STRATEGY

(a) By setting the sum of the forces equal to zero, we can determine the scalar components of $\mathbf{L}$. (b) Using the value of $\mathbf{L}$ from (a), we can determine the components of the sum of the forces and use Eq. (2.8) to determine its magnitude.

SOLUTION

(a) We set the sum of the forces equal to zero:

$$\mathbf{W} + \mathbf{D} + \mathbf{L} = \mathbf{0}.$$

$$(-600\mathbf{j}) + (-200\mathbf{i} + 100\mathbf{j}) + \mathbf{L} = \mathbf{0}.$$

Solving for the lift, we obtain

$$\mathbf{L} = 200\mathbf{i} + 500\mathbf{j} \text{ (lb)}.$$

(b) If the drag increases by a factor of 2, the sum of the forces on the sailplane is

$$\mathbf{W} + 2\mathbf{D} + \mathbf{L} = (-600\mathbf{j}) + 2(-200\mathbf{i} + 100\mathbf{j}) + (200\mathbf{i} + 500\mathbf{j})$$
$$= -200\mathbf{i} + 100\mathbf{j} \text{ (lb)}.$$

From Eq. (2.8), the magnitude of the sum is

$$|\mathbf{W} + 2\mathbf{D} + \mathbf{L}| = \sqrt{(-200)^2 + (100)^2} = 224 \text{ lb}.$$

Figure 2.16

Example 2.4

Hydraulic cylinders are used to exert forces in many mechanical devices. The force is exerted by pressurized liquid (hydraulic fluid) pushing against a piston within the cylinder. The hydraulic cylinder AB in Fig. 2.17 exerts a 4000-lb force $\mathbf{F}$ on the bed of the dump truck at B. Express $\mathbf{F}$ in terms of scalar components using the coordinate system shown.

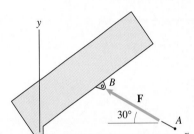

STRATEGY

When the direction of a vector is specified by an angle, as in this example, we can determine the values of the components from the right triangle formed by the vector and its components.

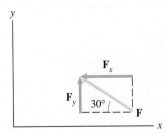

Figure 2.17

SOLUTION

We draw the vector $\mathbf{F}$ and its vector components in Fig. (a). From the resulting right triangle, we see that the magnitude of $\mathbf{F}_x$ is

$$|\mathbf{F}_x| = |\mathbf{F}| \cos 30° = (4000) \cos 30° = 3460 \text{ lb}.$$

$\mathbf{F}_x$ points in the negative x direction, so

$$\mathbf{F}_x = -3460\mathbf{i} \text{ (lb)}.$$

The magnitude of $\mathbf{F}_y$ is

$$|\mathbf{F}_y| = |\mathbf{F}| \sin 30° = (4000) \sin 30° = 2000 \text{ lb}.$$

The vector component $\mathbf{F}_y$ points in the positive y direction, so

$$\mathbf{F}_y = 2000\mathbf{j} \text{ (lb)}.$$

The vector $\mathbf{F}$ in terms of its components is

$$\mathbf{F} = \mathbf{F}_x + \mathbf{F}_y = -3460\mathbf{i} + 2000\mathbf{j} \text{ (lb)}.$$

The x component of $\mathbf{F}$ is -3460 lb, and the y component is 2000 lb.

(a) The force $\mathbf{F}$ and its components form a right triangle.

DISCUSSION

When you determine the components of a vector, you should check to make sure they give you the correct magnitude. In this example,

$$|\mathbf{F}| = \sqrt{(-3460)^2 + (2000)^2} = 4000 \text{ lb}.$$

Example 2.5

The cable from point A to point B exerts an 800-N force $\mathbf{F}$ on the top of the television transmission tower in Fig. 2.18. Resolve $\mathbf{F}$ into components using the coordinate system shown.

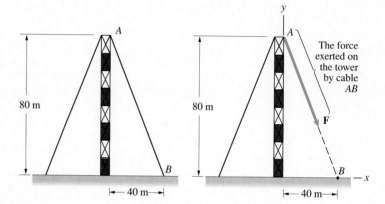

Figure 2.18

STRATEGY

We determine the components of $\mathbf{F}$ in three ways.

First Method   From the given dimensions we can determine the angle α between $\mathbf{F}$ and the y axis (Fig. a), then determine the components from the right triangles formed by the vector $\mathbf{F}$ and its components.

Second Method   The right triangles formed by $\mathbf{F}$ and its components are similar to the triangle OAB in Fig. (a). We can determine the components of $\mathbf{F}$ by using the ratios of the sides of these similar triangles.

Third Method   From the given dimensions we can determine the components of the position vector $\mathbf{r}_{AB}$ from point A to point B (Fig. b). By dividing this vector by its magnitude, we will obtain a unit vector $\mathbf{e}_{AB}$ with the same direction as $\mathbf{F}$ (Fig. c), then obtain $\mathbf{F}$ in terms of its components by expressing it as the product of its magnitude and $\mathbf{e}_{AB}$.

SOLUTION

First Method   Consider the force $\mathbf{F}$ and its vector components (Fig. a). The tangent of the angle α between $\mathbf{F}$ and the y axis is $\tan \alpha = 40/80 = 0.5$, so $\alpha = \arctan(0.5) = 26.6°$. From the right triangles formed by $\mathbf{F}$ and its components, the magnitude of $\mathbf{F}_x$ is

$$|\mathbf{F}_x| = |\mathbf{F}| \sin 26.6° = (800) \sin 26.6° = 358 \text{ N}$$

and the magnitude of $\mathbf{F}_y$ is

$$|\mathbf{F}_y| = |\mathbf{F}| \cos 26.6° = (800) \cos 26.6° = 716 \text{ N}.$$

Since $\mathbf{F}_x$ points in the positive x direction and $\mathbf{F}_y$ points in the negative y direction, the force $\mathbf{F}$ is

$$\mathbf{F} = 358\mathbf{i} - 716\mathbf{j} \text{ (N)}.$$

Second Method   The length of the cable AB is $\sqrt{(80)^2 + (40)^2} = 89.4$ m. Since the triangle OAB in Fig. (a) is similar to the triangle formed by $\mathbf{F}$ and its vector components,

$$\frac{|\mathbf{F}_x|}{|\mathbf{F}|} = \frac{OB}{AB} = \frac{40}{89.4}.$$

Thus the magnitude of $\mathbf{F}_x$ is

$$|\mathbf{F}_x| = \left(\frac{40}{89.4}\right)|\mathbf{F}| = \left(\frac{40}{89.4}\right)(800) = 358 \text{ N}.$$

We can also see from the similar triangles that

$$\frac{|\mathbf{F}_y|}{|\mathbf{F}|} = \frac{OA}{AB} = \frac{80}{89.4},$$

so the magnitude of $\mathbf{F}_y$ is

$$|\mathbf{F}_y| = \left(\frac{80}{89.4}\right)|\mathbf{F}| = \left(\frac{80}{89.4}\right)(800) = 716 \text{ N}.$$

Thus we again obtain the result

$$\mathbf{F} = 358\mathbf{i} - 716\mathbf{j} \text{ (N)}.$$

Third Method   The vector $\mathbf{r}_{AB}$ in Fig. (b) is

$$\mathbf{r}_{AB} = (x_B - x_A)\,\mathbf{i} + (y_B - y_A)\,\mathbf{j} = (40 - 0)\,\mathbf{i} + (0 - 80)\,\mathbf{j}$$
$$= 40\mathbf{i} - 80\mathbf{j} \text{ (m)}.$$

We divide this vector by its magnitude to obtain a unit vector $\mathbf{e}_{AB}$ that has the same direction as the force $\mathbf{F}$ (Fig. c):

$$\mathbf{e}_{AB} = \frac{\mathbf{r}_{AB}}{|\mathbf{r}_{AB}|} = \frac{40\mathbf{i} - 80\mathbf{j}}{\sqrt{(40)^2 + (-80)^2}} = 0.447\mathbf{i} - 0.894\mathbf{j}.$$

The force $\mathbf{F}$ is equal to the product of its magnitude $|\mathbf{F}|$ and $\mathbf{e}_{AB}$:

$$\mathbf{F} = |\mathbf{F}|\,\mathbf{e}_{AB} = (800)(0.447\mathbf{i} - 0.894\mathbf{j}) = 358\mathbf{i} - 716\mathbf{j} \text{ (N)}.$$

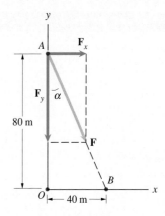

(a) Vector components of $\mathbf{F}$.

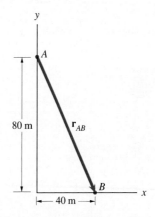

(b) The vector $\mathbf{r}_{AB}$ from A to B.

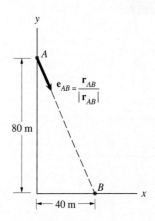

(c) The unit vector $\mathbf{e}_{AB}$ pointing from A toward B.

Example 2.6

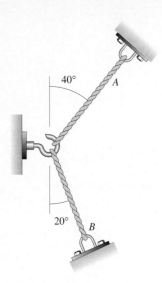

The cables A and B in Fig. 2.19 exert forces $\mathbf{F}_A$ and $\mathbf{F}_B$ on the hook. The magnitude of $\mathbf{F}_A$ is 100 lb. The tension in cable B has been adjusted so that the total force $\mathbf{F}_A + \mathbf{F}_B$ is perpendicular to the wall to which the hook is attached.
(a) What is the magnitude of $\mathbf{F}_B$?
(b) What is the magnitude of the total force exerted on the hook by the two cables?

STRATEGY

The vector sum of the two forces is perpendicular to the wall, so the sum of the components parallel to the wall equals zero. From this condition we can obtain an equation for the magnitude of $\mathbf{F}_B$.

SOLUTION

(a) In terms of the coordinate system shown in Fig. (a), the components of $\mathbf{F}_A$ and $\mathbf{F}_B$ are

$$\mathbf{F}_A = |\mathbf{F}_A| \sin 40° \, \mathbf{i} + |\mathbf{F}_A| \cos 40° \, \mathbf{j},$$
$$\mathbf{F}_B = |\mathbf{F}_B| \sin 20° \, \mathbf{i} - |\mathbf{F}_B| \cos 20° \, \mathbf{j}.$$

The total force is

$$\mathbf{F}_A + \mathbf{F}_B = (|\mathbf{F}_A| \sin 40° + |\mathbf{F}_B| \sin 20°) \, \mathbf{i}$$
$$+ (|\mathbf{F}_A| \cos 40° - |\mathbf{F}_B| \cos 20°) \, \mathbf{j}.$$

By setting the component of the total force parallel to the wall (the y component) equal to zero,

$$|\mathbf{F}_A| \cos 40° - |\mathbf{F}_B| \cos 20° = 0,$$

we obtain an equation for the magnitude of $\mathbf{F}_B$:

$$|\mathbf{F}_B| = \frac{|\mathbf{F}_A| \cos 40°}{\cos 20°} = \frac{(100) \cos 40°}{\cos 20°} = 81.5 \text{ lb}.$$

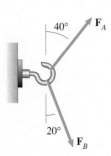

Figure 2.19

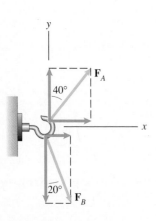

(a) Resolving $\mathbf{F}_A$ and $\mathbf{F}_B$ into components parallel and perpendicular to the wall.

(b) Since we now know the magnitude of $\mathbf{F}_B$, we can determine the total force acting on the hook:

$$\mathbf{F}_A + \mathbf{F}_B = (|\mathbf{F}_A| \sin 40° + |\mathbf{F}_B| \sin 20°)\,\mathbf{i}$$
$$= [(100) \sin 40° + (81.5) \sin 20°]\,\mathbf{i} = 92.2\mathbf{i}\ \text{(lb)}.$$

The magnitude of the total force is 92.2 lb.

DISCUSSION

You can obtain the solution to (a) in a less formal way. If the component of the total force parallel to the wall is zero, you can see in Fig. (a) that the magnitude of the vertical component of $\mathbf{F}_A$ must equal the magnitude of the vertical component of $\mathbf{F}_B$:

$$|\mathbf{F}_A| \cos 40° = |\mathbf{F}_B| \cos 20°.$$

Therefore the magnitude of $\mathbf{F}_B$ is

$$|\mathbf{F}_B| = \frac{|\mathbf{F}_A| \cos 40°}{\cos 20°} = \frac{(100) \cos 40°}{\cos 20°} = 81.5\ \text{lb}.$$

Problems

2.21 A force is given in terms of its scalar components by $\mathbf{F} = 4\mathbf{i} - 3\mathbf{j}$ (kN). What is its magnitude?

Strategy: The magnitude of a vector is given in terms of its components by Eq. (2.8).

2.22 A force is given in terms of its scalar components by $\mathbf{F} = F_x\mathbf{i} + F_y\mathbf{j}$. The component $F_x = 120$ lb. The component F_y is negative, and the magnitude of $\mathbf{F}$ is 150 lb. What is the component F_y?

2.23 A support is subjected to a force $\mathbf{F} = F_x\mathbf{i} + 80\mathbf{j}$ (N). If the support will safely support a force of magnitude 100 N, what is the allowable range of the component F_x?

2.24 If $\mathbf{F}_A = 600\mathbf{i} - 800\mathbf{j}$ (kip) and $\mathbf{F}_B = 200\mathbf{i} - 200\mathbf{j}$ (kip), what is the magnitude of the force $\mathbf{F} = \mathbf{F}_A - 2\mathbf{F}_B$?

2.25 If $\mathbf{F}_A = \mathbf{i} - 4.5\mathbf{j}$ (kN) and $\mathbf{F}_B = -2\mathbf{i} - 2\mathbf{j}$ (kN), what is the magnitude of the force $\mathbf{F} = 6\mathbf{F}_A + 4\mathbf{F}_B$?

2.26 Two perpendicular vectors $\mathbf{U}$ and $\mathbf{V}$ lie in the x-y plane. The vector $\mathbf{U} = 6\mathbf{i} - 8\mathbf{j}$ and $|\mathbf{V}| = 20$. What are the scalar components of $\mathbf{V}$?

2.27 A fish exerts a 200-N force $\mathbf{F}$ on the pole. Express $\mathbf{F}$ in terms of scalar components.

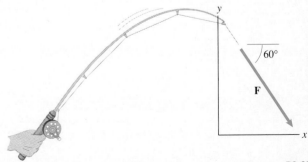

P2.27

2.28 A person exerts a 60-lb force **F** to push a crate onto a truck. Express **F** in terms of scalar components.

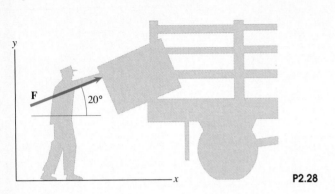

P2.28

2.29 The rocket's engine exerts a 40-kN force **F**. Express **F** in terms of scalar components.

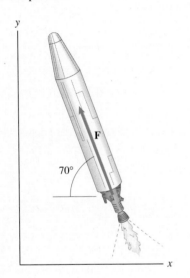

P2.29

2.30 The coordinates of two points A and B of a truss are shown. Express the position vector from point A to point B in terms of scalar components.

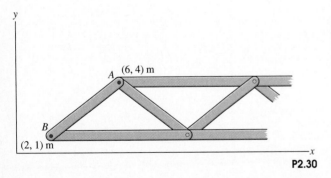

P2.30

2.31 The points A, B, ... are the joints of the hexagonal structural element. Let $\mathbf{r}_{AB}$ be the position vector from joint A to joint B,

$\mathbf{r}_{AC}$ the position vector from joint A to joint C, and so forth. Determine the scalar components of the vectors $\mathbf{r}_{AC}$ and $\mathbf{r}_{AF}$.

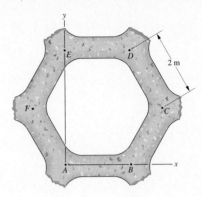

P2.31

2.32 For the hexagonal structural element in Problem 2.31, determine the scalar components of the vector $\mathbf{r}_{AB} + \mathbf{r}_{BC} + \mathbf{r}_{CD} + \mathbf{r}_{DE}$.

2.33 The position vector from point A to point B is $\mathbf{r}_{AB} = 12\mathbf{i} - 16\mathbf{j}$ (m).
(a) What is the distance from point A to point B?
(b) What is the position vector from point B to point A?

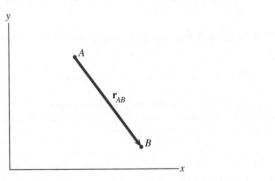

P2.33

2.34 (a) Express the position vector from point A of the front-end loader to point B in terms of scalar components.
(b) Express the position vector from point B to point C in terms of scalar components.
(c) Use the results of (a) and (b) to determine the distance from point A to point C.

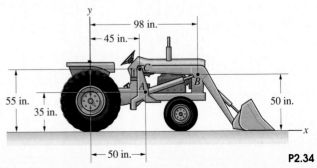

P2.34

2.35 Consider the front-end loader in Problem 2.34. To raise the bucket, the operator increases the length of the hydraulic cylinder AB. The distance between points B and C remains constant. If the length of the cylinder AB is 65 in., what is the position vector from point A to point B?

2.36 Determine the position vector $\mathbf{r}_{AB}$ in terms of its scalar components if (a) $\theta = 30°$; (b) $\theta = 225°$.

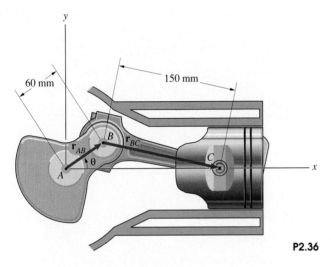

P2.36

2.37 In Problem 2.36, determine the position vector $\mathbf{r}_{BC}$ in terms of its scalar components if (a) $\theta = 30°$; (b) $\theta = 225°$.

2.38 A surveyor measures the location of point A and determines that $\mathbf{r}_{OA} = 400\mathbf{i} + 800\mathbf{j}$ (m). He wants to determine the location of a point B so that $|\mathbf{r}_{AB}| = 400$ m and $|\mathbf{r}_{OA} + \mathbf{r}_{AB}| = 1200$ m. What are the cartesian coordinates of point B?

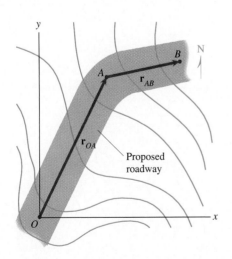

P2.38

2.39 You have designed the fixture shown to safely support a force $\mathbf{F} = F_x\mathbf{i} + F_y\mathbf{j}$ of magnitude 2000 N. If it will be subjected

to components F_x in the range -1500 N $\leq F_x \leq 1000$ N, to what range of values of F_y can it safely be subjected?

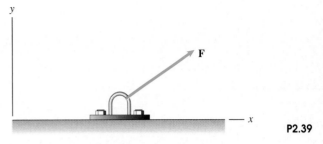

P2.39

2.40 The coordinates of two points A and B are shown. Determine a unit vector $\mathbf{e}$ that points from point A toward point B.

Strategy: Determine the position vector from point A to point B and divide it by its magnitude.

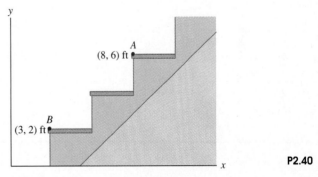

P2.40

2.41 The x and y coordinates of points A, B, and C of the sailboat are shown.
(a) Determine a unit vector that is parallel to the forestay AC and points from A toward C.
(b) Determine a unit vector that is parallel to the backstay BC and points from B toward C.

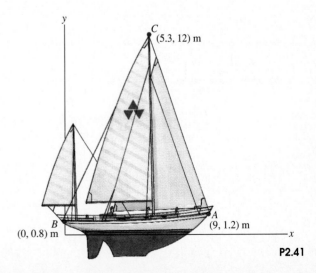

P2.41

2.42 Consider the force vector $\mathbf{F} = 3\mathbf{i} - 4\mathbf{j}$ (kN). Determine a unit vector $\mathbf{e}$ that has the same direction as $\mathbf{F}$.

2.43 The position vector from a point A to a point B is $\mathbf{r} = -8\mathbf{i} + 6\mathbf{j}$ (ft).
(a) Determine the unit vector $\mathbf{e}_{AB}$ that points from point A toward point B.
(b) Determine the unit vector $\mathbf{e}_{BA}$ that points from point B toward point A.

2.44 Two cars A and B are on a circular track of 1000-ft radius. The distance between the two cars measured along the track is 2000 ft. What is the position vector from car A to car B in terms of the coordinate system shown?

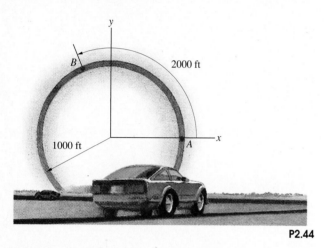

P2.44

2.45 A surveyor finds that the length of the line OA is 1500 m and the length of the line OB is 2000 m.
(a) Express the position vector from point A to point B in terms of scalar components.
(b) Use the result of (a) to determine the distance from A to B.

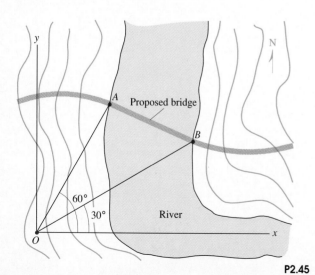

P2.45

2.46 The distance from the sun (S) to Mercury (M) is 57×10^6 km, the distance from the sun to Venus (V) is 108×10^6 km, and the distance from the sun to the earth (E) is 150×10^6 km. Assume that the planets lie in the x-y plane.
(a) Determine the components of the position vector $\mathbf{r}_M$ from the sun to Mercury, the position vector $\mathbf{r}_V$ from the sun to Venus, and the position vector $\mathbf{r}_E$ from the sun to the earth.
(b) Use the results of (a) to determine the distance from the earth to Mercury and the distance from the earth to Venus.

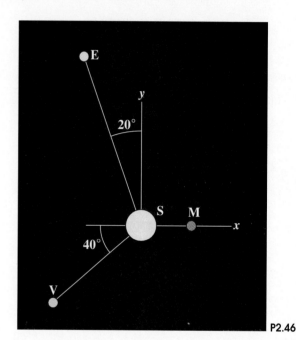

P2.46

2.47 The rope exerts forces $\mathbf{F}_A$ and $\mathbf{F}_B$ on the pulley. Their magnitudes are $|\mathbf{F}_A| = |\mathbf{F}_B| = 80$ lb. What is the magnitude of the vector sum of the forces?

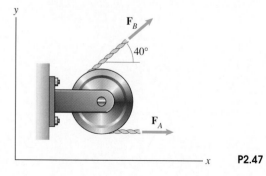

P2.47

2.48 The rope ABC exerts forces $\mathbf{F}_{BA}$ and $\mathbf{F}_{BC}$ on the block at B. Their magnitudes are $|\mathbf{F}_{BA}| = |\mathbf{F}_{BC}| = 800$ N. Determine the magnitude of the vector sum of the forces by resolving the forces into components, and compare your answer with that of Problem 2.16.

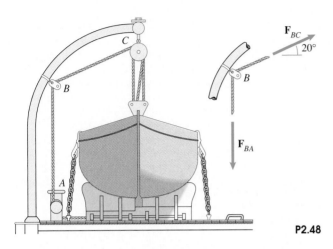

P2.48

2.49 The magnitudes of the forces are $|\mathbf{F}_1| = |\mathbf{F}_2| = |\mathbf{F}_3| = 5$ kN. What is the magnitude of the vector sum of the three forces?

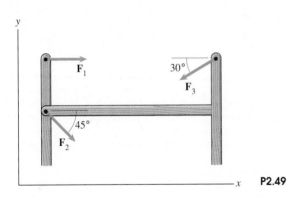

P2.49

2.50 Four groups engage in a tug-of-war. The magnitudes of the forces exerted by groups B, C, and D are $|\mathbf{F}_B| = 800$ lb, $|\mathbf{F}_C| = 1000$ lb, and $|\mathbf{F}_D| = 900$ lb. If the vector sum of the four forces equals zero, what are the magnitudes of $\mathbf{F}_A$ and the angle α?

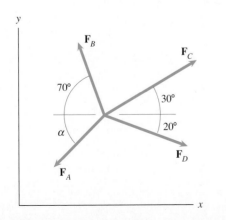

P2.50

2.51 The total thrust exerted on the booster by its main engines is 200,000 lb parallel to the y axis. Each of the two small vernier engines exerts a thrust of 5000 lb in the directions shown. Determine the magnitude and direction of the total force exerted on the booster by the main and vernier engines.

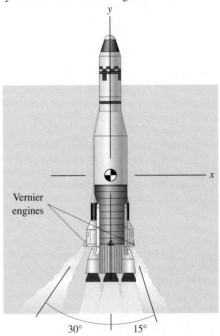

P2.51

2.52 The magnitudes of the forces acting on the support are $|\mathbf{F}_1| = |\mathbf{F}_2| = 100$ lb. The support will fail if the magnitude of the total force acting on it exceeds 150 lb. Determine the acceptable range of values of the angle α.

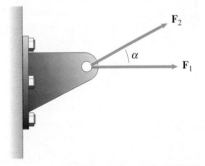

P2.52

2.53 Three forces act on the sphere. The magnitude of $\mathbf{F}_B$ is 60 lb. The vector sum of the three forces equals zero. What are the magnitudes of $\mathbf{F}_A$ and $\mathbf{F}_C$?

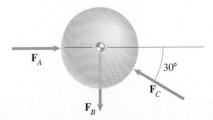

P2.53

2.54 Four forces act on a beam. The vector sum of the forces is zero. The magnitudes $|\mathbf{F}_B| = 10$ kN and $|\mathbf{F}_C| = 5$ kN. Determine the magnitudes of $\mathbf{F}_A$ and $\mathbf{F}_D$.

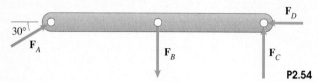

P2.54

2.55 Six forces act on a beam that forms part of the frame of a building. The vector sum of the forces is zero. The magnitudes $|\mathbf{F}_B| = |\mathbf{F}_E| = 5$ kip, $|\mathbf{F}_C| = 4$ kip, and $|\mathbf{F}_D| = 2$ kip. Determine the magnitudes of $\mathbf{F}_A$ and $\mathbf{F}_G$.

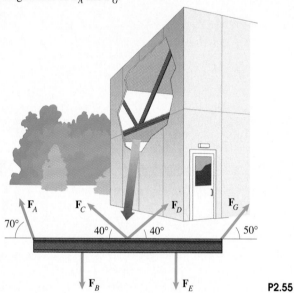

P2.55

2.56 The total weight of the man and parasail is $|\mathbf{W}| = 230$ lb. The drag force $\mathbf{D}$ is perpendicular to the lift force $\mathbf{L}$. If the vector sum of the three forces is zero, what are the magnitudes of $\mathbf{L}$ and $\mathbf{D}$?

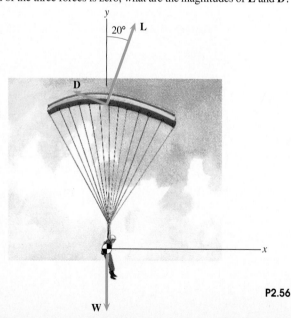

P2.56

2.57 Two cables AB and CD extend from the rocket gantry to the ground. Cable AB exerts a force of magnitude 10,000 lb on the gantry, and cable CD exerts a force of magnitude 5000 lb.

(a) Using the coordinate system shown, express each of the two forces exerted on the gantry by the cables in terms of scalar components.

(b) What is the magnitude of the total force exerted on the gantry by the two cables?

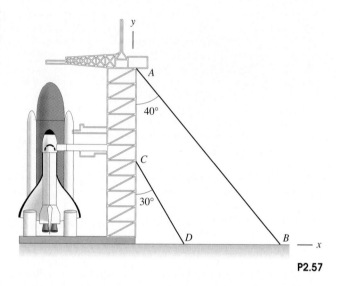

P2.57

2.58 The cables A, B, and C help support a pillar that forms part of the supports of a structure. The magnitudes of the forces exerted by the cables are equal: $|\mathbf{F}_A| = |\mathbf{F}_B| = |\mathbf{F}_C|$. The magnitude of the vector sum of the three forces is 200 kN. What is $|\mathbf{F}_A|$?

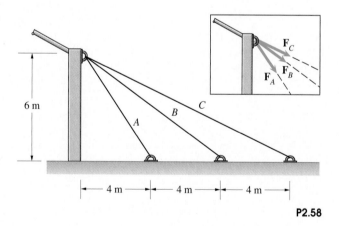

P2.58

2.59 The tension in the forestay AC of the sailboat is 300 lb. The vector sum of the forces exerted on the top of the mast C by the forestay AC and the backstay BC of the sailboat points straight downward.

(a) What is the tension in the backstay BC?
(b) What total vertical force do the two stays exert on the mast?

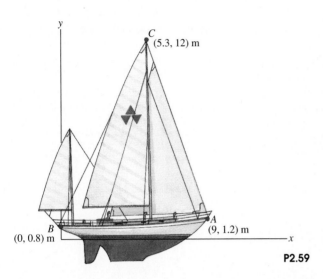

P2.59

2.60 The structure shown forms part of a truss designed by an architectural engineer to support the roof of an orchestra shell. The members AB, AC, and AD exert forces $\mathbf{F}_{AB}$, $\mathbf{F}_{AC}$, and $\mathbf{F}_{AD}$ on the joint A. The magnitude $|\mathbf{F}_{AB}| = 4$ kN. If the vector sum of the three forces equals zero, what are the magnitudes of $\mathbf{F}_{AC}$ and $\mathbf{F}_{AD}$?

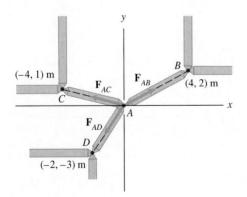

P2.60

2.61 The distance $s = 45$ in.
(a) Determine the unit vector $\mathbf{e}_{BA}$ that points from B toward A.
(b) Use the unit vector you obtained in (a) to determine the coordinates of the collar C.

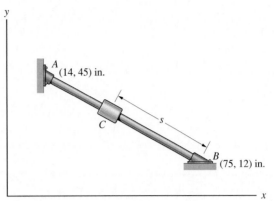

P2.61

2.62 In Problem 2.61, determine the x and y coordinates of the collar C as functions of the distance s.

2.63 The position vector $\mathbf{r}$ goes from point A to a point on the straight line between B and C. Its magnitude is $|\mathbf{r}| = 6$ ft. Express $\mathbf{r}$ in terms of scalar components.

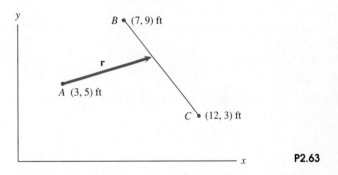

P2.63

2.64 Let $\mathbf{r}$ be the position vector from point C to the point that is a distance s meters from point A along the straight line between A and B. Express $\mathbf{r}$ in terms of scalar components. (Your answer will be in terms of s.)

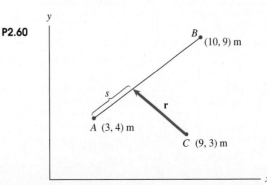

P2.64

2.4 *Components in Three Dimensions*

Many engineering applications require you to resolve vectors into components in a three-dimensional coordinate system. In this section we explain this technique and demonstrate vector operations in three dimensions.

Let's first review how to draw objects in three dimensions. Consider a three-dimensional object such as a cube. If we draw the cube as it appears when your line of sight is perpendicular to one of its faces, we obtain the diagram shown in Fig. 2.20(a). In this view the cube appears two-dimensional; you can't see the dimension perpendicular to the page. To remedy this, we can draw the cube as it appears if you move upward and to the right (Fig. 2.20b). In this oblique view you can see the third dimension. The hidden edges of the cube are shown as dashed lines.

We can use this method to draw three-dimensional coordinate systems. In Fig. 2.20(c) we align the *x*, *y*, and *z* axes of a three-dimensional cartesian coordinate system with the edges of the cube. The three-dimensional representation of the coordinate system is shown in Fig. 2.20(d).

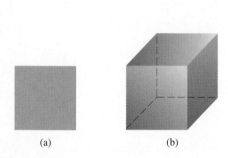

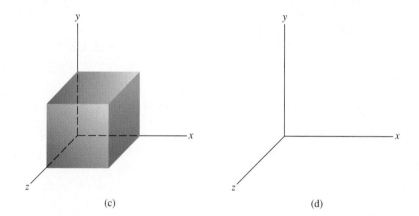

(a)   (b)

(c)   (d)

Figure 2.20

(a) A cube viewed with the line of sight perpendicular to a face.
(b) An oblique view of the cube.
(c) A cartesian coordinate system aligned with the edges of the cube.
(d) Three-dimensional representation of the coordinate system.

The coordinate system in Fig. 2.20(d) is **right-handed**. If you point the fingers of your right hand in the direction of the positive *x* axis and bend them (as in preparing to make a fist) toward the positive *y* axis, your thumb will point in the direction of the positive *z* axis (Fig. 2.21). When the positive *z* axis points in the opposite direction, the coordinate system is left-handed. For some purposes, it doesn't matter which coordinate system you use. However, some

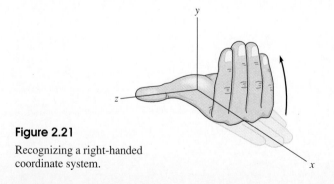

Figure 2.21

Recognizing a right-handed coordinate system.

equations we will derive do not give correct results with a left-handed coordinate system. For this reason we will use only right-handed coordinate systems.

We can resolve a vector $\mathbf{U}$ into vector components $\mathbf{U}_x$, $\mathbf{U}_y$, and $\mathbf{U}_z$ parallel to the x, y, and z axes (Fig. 2.22):

$$\mathbf{U} = \mathbf{U}_x + \mathbf{U}_y + \mathbf{U}_z. \tag{2.11}$$

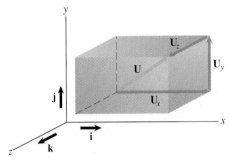

Figure 2.22

A vector $\mathbf{U}$ and its vector components.

(We have drawn a box around the vector to help you visualize the directions of the vector components.) By introducing unit vectors $\mathbf{i}$, $\mathbf{j}$, and $\mathbf{k}$ that point in the positive x, y, and z directions, we can express $\mathbf{U}$ in terms of scalar components as

$$\mathbf{U} = U_x \mathbf{i} + U_y \mathbf{j} + U_z \mathbf{k}. \tag{2.12}$$

We will refer to the scalars U_x, U_y, and U_z as the x, y, and z components of $\mathbf{U}$.

Magnitude of a Vector in Terms of Components

Consider a vector $\mathbf{U}$ and its vector components (Fig. 2.23a). From the right triangle formed by the vectors $\mathbf{U}_y$, $\mathbf{U}_z$, and their sum $\mathbf{U}_y + \mathbf{U}_z$ (Fig. 2.23b), we can see that

$$|\mathbf{U}_y + \mathbf{U}_z|^2 = |\mathbf{U}_y|^2 + |\mathbf{U}_z|^2. \tag{2.13}$$

The vector $\mathbf{U}$ is the sum of the vectors $\mathbf{U}_x$ and $\mathbf{U}_y + \mathbf{U}_z$. These three vectors form a right triangle (Fig. 2.23c), from which we obtain

$$|\mathbf{U}|^2 = |\mathbf{U}_x|^2 + |\mathbf{U}_y + \mathbf{U}_z|^2.$$

Substituting Eq. (2.13) into this result yields the equation

$$|\mathbf{U}|^2 = |\mathbf{U}_x|^2 + |\mathbf{U}_y|^2 + |\mathbf{U}_z|^2 = U_x^2 + U_y^2 + U_z^2.$$

Thus the magnitude of a vector $\mathbf{U}$ is given in terms of its components in three dimensions by

$$|\mathbf{U}| = \sqrt{U_x^2 + U_y^2 + U_z^2}. \tag{2.14}$$

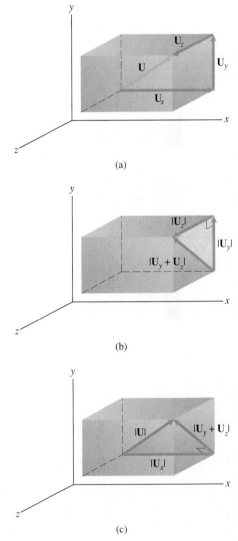

(a)

(b)

(c)

Figure 2.23

(a) A vector $\mathbf{U}$ and its vector components.
(b) The right triangle formed by the vectors $\mathbf{U}_y$, $\mathbf{U}_z$, and $\mathbf{U}_y + \mathbf{U}_z$.
(c) The right triangle formed by the vectors $\mathbf{U}$, $\mathbf{U}_x$, and $\mathbf{U}_y + \mathbf{U}_z$.

Direction Cosines

We described the direction of a vector relative to a two-dimensional cartesian coordinate system by specifying the angle between the vector and one of the coordinate axes. One of the ways we can describe the direction of a vector in three dimensions is by specifying the angles θ_x, θ_y, and θ_z between the vector and the positive coordinate axes (Fig. 2.24a).

In Figs. 2.24 (b)–(d), we demonstrate that the components of the vector $\mathbf{U}$ are given in terms of the angles θ_x, θ_y, and θ_z by

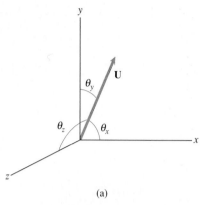

(a)

$$U_x = |\mathbf{U}|\cos\theta_x, \quad U_y = |\mathbf{U}|\cos\theta_y, \quad U_z = |\mathbf{U}|\cos\theta_z. \qquad (2.15)$$

The quantities $\cos\theta_x$, $\cos\theta_y$, and $\cos\theta_z$ are called the **direction cosines** of $\mathbf{U}$.

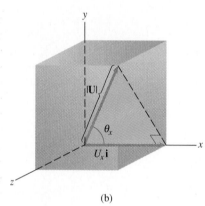

(b)

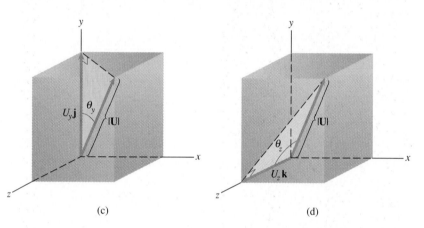

(c)   (d)

Figure 2.24

(a) A vector $\mathbf{U}$ and the angles θ_x, θ_y, and θ_z.
(b)–(d) The angles θ_x, θ_y, and θ_z and the vector components of $\mathbf{U}$.

The direction cosines of a vector are not independent. If we substitute Eqs. (2.15) into Eq. (2.14), we find that the direction cosines satisfy the relation

$$\cos^2\theta_x + \cos^2\theta_y + \cos^2\theta_z = 1. \qquad (2.16)$$

Suppose that $\mathbf{e}$ is a unit vector with the same direction as $\mathbf{U}$, so that

$$\mathbf{U} = |\mathbf{U}|\mathbf{e}.$$

In terms of components, this equation is

$$U_x\mathbf{i} + U_y\mathbf{j} + U_z\mathbf{k} = |\mathbf{U}|(e_x\mathbf{i} + e_y\mathbf{j} + e_z\mathbf{k}).$$

Thus the relations between the components of $\mathbf{U}$ and $\mathbf{e}$ are

$$U_x = |\mathbf{U}|e_x, \quad U_y = |\mathbf{U}|e_y, \quad U_z = |\mathbf{U}|e_z.$$

By comparing these equations to Eqs. (2.15), we see that

$$\cos\theta_x = e_x, \quad \cos\theta_y = e_y, \quad \cos\theta_z = e_z.$$

The direction cosines of a vector $\mathbf{U}$ are the components of a unit vector with the same direction as $\mathbf{U}$.

Position Vectors in Terms of Components

Generalizing the two-dimensional case, let's consider a point A with coordinates (x_A, y_A, z_A) and a point B with coordinates (x_B, y_B, z_B). The position vector $\mathbf{r}_{AB}$ from A to B, shown in Fig. 2.25(a), is given in terms of the coordinates of A and B by

$$\mathbf{r}_{AB} = (x_B - x_A)\,\mathbf{i} + (y_B - y_A)\,\mathbf{j} + (z_B - z_A)\,\mathbf{k}. \qquad (2.17)$$

The components are obtained by subtracting the coordinates of point A from the coordinates of point B (Fig. 2.25b).

Components of a Vector Parallel to a Given Line

In three-dimensional applications, the direction of a vector is often defined by specifying the coordinates of two points on a line that is parallel to the vector. You can use this information to determine the components of the vector.

Suppose that we know the coordinates of two points A and B on a line parallel to a vector $\mathbf{U}$ (Fig. 2.26a). We can use Eq. (2.17) to determine the position vector $\mathbf{r}_{AB}$ from A to B (Fig. 2.26b). We can divide $\mathbf{r}_{AB}$ by its magnitude to obtain a unit vector $\mathbf{e}_{AB}$ that points from A toward B (Fig. 2.26c). Since $\mathbf{e}_{AB}$ has the same direction as $\mathbf{U}$, we can determine $\mathbf{U}$ in terms of its scalar components by expressing it as the product of its magnitude and $\mathbf{e}_{AB}$.

More generally, suppose that you know the magnitude of a vector $\mathbf{U}$ and the components of any vector $\mathbf{V}$ that has the same direction as $\mathbf{U}$. Then $\mathbf{V}/|\mathbf{V}|$ is a unit vector with the same direction as $\mathbf{U}$, and you can determine the components of $\mathbf{U}$ by expressing it as $\mathbf{U} = |\mathbf{U}|(\mathbf{V}/|\mathbf{V}|)$.

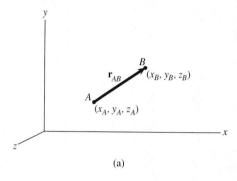
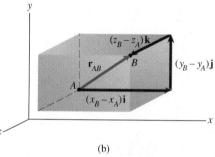

(a)

(b)

Figure 2.25

(a) The position vector from point A to point B.

(b) The components of $\mathbf{r}_{AB}$ can be determined from the coordinates of points A and B.

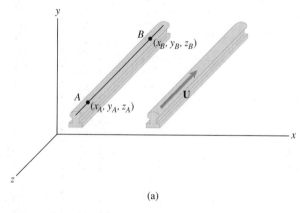

(a)

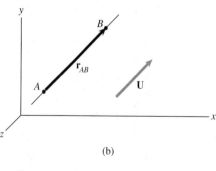

(b)

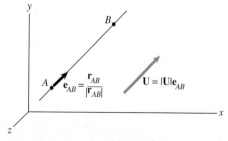

(c)

Figure 2.26

(a) Two points A and B on a line parallel to $\mathbf{U}$.

(b) The position vector from A to B.

(c) The unit vector $\mathbf{e}_{AB}$ that points from A toward B.

In the following examples we demonstrate vector operations in terms of components in three dimensions. The first example shows that vector manipulations are no more difficult in three dimensions than in two dimensions when you know the components of the vectors. However, specifying the directions of vectors and determining their components is a bit more complicated in three dimensions. We will give examples that illustrate the use of direction cosines and other methods of specifying the directions of vectors in three dimensions.

Example 2.7

An engineer designing a threshing machine determines that at a particular time the position vectors of the ends A and B of a shaft are $\mathbf{r}_A = 3\mathbf{i} - 4\mathbf{j} - 12\mathbf{k}$ (ft) and $\mathbf{r}_B = -\mathbf{i} + 7\mathbf{j} + 6\mathbf{k}$ (ft).

(a) What is the magnitude of $\mathbf{r}_A$?

(b) Determine the angles θ_x, θ_y, and θ_z between $\mathbf{r}_A$ and the positive coordinate axes.

(c) Determine the scalar components of the position vector of end B of the shaft relative to end A.

STRATEGY

(a) Since we know the components of $\mathbf{r}_A$, we can use Eq. (2.14) to determine its magnitude.

(b) We can obtain the angles θ_x, θ_y, and θ_z from Eqs. (2.15).

(c) The position vector of end B of the shaft relative to end A is $\mathbf{r}_B - \mathbf{r}_A$.

SOLUTION

(a) The magnitude of $\mathbf{r}_A$ is

$$|\mathbf{r}_A| = \sqrt{r_{Ax}^2 + r_{Ay}^2 + r_{Az}^2} = \sqrt{(3)^2 + (-4)^2 + (-12)^2} = 13 \text{ ft.}$$

(b) The direction cosines of $\mathbf{r}_A$ are

$$\cos\theta_x = \frac{r_{Ax}}{|\mathbf{r}_A|} = \frac{3}{13},$$

$$\cos\theta_y = \frac{r_{Ay}}{|\mathbf{r}_A|} = \frac{-4}{13},$$

$$\cos\theta_z = \frac{r_{Az}}{|\mathbf{r}_A|} = \frac{-12}{13}.$$

From these equations we find that the angles between $\mathbf{r}_A$ and the positive coordinate axes are $\theta_x = 76.7°$, $\theta_y = 107.9°$, and $\theta_z = 157.4°$.

(c) The position vector of end B of the shaft relative to end A is

$$\mathbf{r}_B - \mathbf{r}_A = (-\mathbf{i} + 7\mathbf{j} + 6\mathbf{k}) - (3\mathbf{i} - 4\mathbf{j} - 12\mathbf{k})$$
$$= -4\mathbf{i} + 11\mathbf{j} + 18\mathbf{k} \text{ (ft).}$$

Example 2.8

The crane in Fig. 2.27 exerts a 600-lb force $\mathbf{F}$ on the caisson. The angle between $\mathbf{F}$ and the x axis is 54°, and the angle between $\mathbf{F}$ and the y axis is 40°. The z component of $\mathbf{F}$ is positive. Express $\mathbf{F}$ in terms of scalar components.

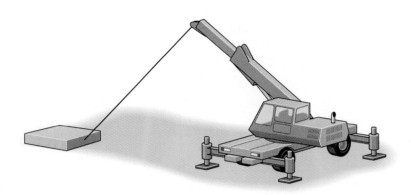

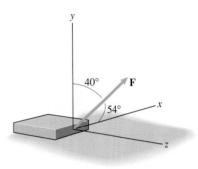

Figure 2.27

STRATEGY

Only two of the angles between the vector and the positive coordinate axes are given, but we can use Eq. (2.16) to determine the third angle. Then we can determine the components of $\mathbf{F}$ by using Eqs. (2.15).

SOLUTION

The angles between $\mathbf{F}$ and the positive coordinate axes are related by

$$\cos^2 \theta_x + \cos^2 \theta_y + \cos^2 \theta_z = (\cos 54°)^2 + (\cos 40°)^2 + \cos^2 \theta_z = 1.$$

Solving this equation for $\cos \theta_z$, we obtain the two solutions $\cos \theta_z = 0.260$ and $\cos \theta_z = -0.260$, which tells us that $\theta_z = 74.9°$ or $\theta_z = 105.1°$. The z component of the vector $\mathbf{F}$ is positive, so the angle between $\mathbf{F}$ and the positive z axis is less than 90°. Therefore $\theta_z = 74.9°$.

The components of $\mathbf{F}$ are

$$F_x = |\mathbf{F}| \cos \theta_x = 600 \cos 54° \quad = 353 \text{ lb},$$
$$F_y = |\mathbf{F}| \cos \theta_y = 600 \cos 40° \quad = 460 \text{ lb},$$
$$F_z = |\mathbf{F}| \cos \theta_z = 600 \cos 74.9° = 156 \text{ lb}.$$

Example 2.9

The tether of the balloon in Fig. 2.28 exerts an 800-N force **F** on the hook at O. The vertical line AB intersects the x-z plane at point A. The angle between the z axis and the line OA is 60°, and the angle between the line OA and **F** is 45°. Express **F** in terms of scalar components.

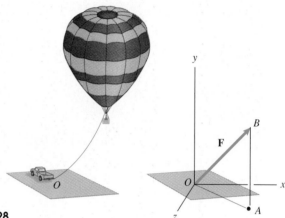

Figure 2.28

STRATEGY

We can determine the components of **F** from the given geometric information in two steps. First, we resolve **F** into two vector components parallel to the lines OA and AB. The component parallel to AB is the vector component $\mathbf{F}_y$. Then we can resolve the component parallel to OA to determine the vector components $\mathbf{F}_x$ and $\mathbf{F}_z$.

SOLUTION

In Fig. (a), we resolve **F** into its y component $\mathbf{F}_y$ and the component $\mathbf{F}_h$ parallel to OA. The magnitude of $\mathbf{F}_y$ is

$$|\mathbf{F}_y| = |\mathbf{F}| \sin 45° = 800 \sin 45° = 566 \text{ N},$$

and the magnitude of $\mathbf{F}_h$ is

$$|\mathbf{F}_h| = |\mathbf{F}| \cos 45° = 800 \cos 45° = 566 \text{ N}.$$

In Fig. (b), we resolve $\mathbf{F}_h$ into the vector components $\mathbf{F}_x$ and $\mathbf{F}_z$. The magnitude of $\mathbf{F}_x$ is

$$|\mathbf{F}_x| = |\mathbf{F}_h| \sin 60° = 566 \sin 60° = 490 \text{ N},$$

and the magnitude of $\mathbf{F}_z$ is

$$|\mathbf{F}_z| = |\mathbf{F}_h| \cos 60° = 566 \cos 60° = 283 \text{ N}.$$

The vector components $\mathbf{F}_x$, $\mathbf{F}_y$, and $\mathbf{F}_z$ all point in the positive axis directions, so the scalar components of **F** are positive:

$$\mathbf{F} = 490\mathbf{i} + 566\mathbf{j} + 283\mathbf{k} \text{ (N)}.$$

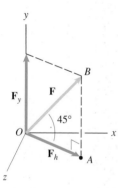

(a) Resolving **F** into vector components parallel to OA and OB.

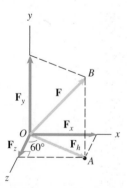

(b) Resolving $\mathbf{F}_h$ into vector components parallel to the x and z axes.

Example 2.10

The bar AB in Fig. 2.29 exerts a 140-N force $\mathbf{F}$ on its support at A. The force is parallel to the bar and points toward B. Express $\mathbf{F}$ in terms of scalar components.

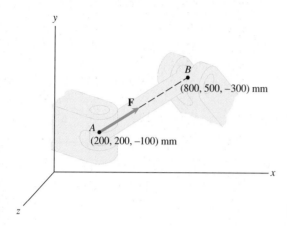

Figure 2.29

STRATEGY

Since we are given the coordinates of points A and B, we can determine the components of the position vector from A to B. By dividing the position vector by its magnitude, we can obtain a unit vector with the same direction as $\mathbf{F}$. Then by multiplying the unit vector by the magnitude of $\mathbf{F}$, we obtain $\mathbf{F}$ in terms of its components.

SOLUTION

The position vector from A to B is (Fig. a)

$$\mathbf{r}_{AB} = (x_B - x_A)\,\mathbf{i} + (y_B - y_A)\,\mathbf{j} + (z_B - z_A)\,\mathbf{k}$$
$$= [(800) - (200)]\,\mathbf{i} + [(500) - (200)]\,\mathbf{j} + [(-300) - (-100)]\,\mathbf{k}$$
$$= 600\mathbf{i} + 300\mathbf{j} - 200\mathbf{k}\ \text{mm},$$

and its magnitude is

$$|\mathbf{r}_{AB}| = \sqrt{(600)^2 + (300)^2 + (-200)^2} = 700\ \text{mm}.$$

By dividing $\mathbf{r}_{AB}$ by its magnitude, we obtain a unit vector with the same direction as $\mathbf{F}$ (Fig. b),

$$\mathbf{e}_{AB} = \frac{\mathbf{r}_{AB}}{|\mathbf{r}_{AB}|} = \frac{6}{7}\mathbf{i} + \frac{3}{7}\mathbf{j} - \frac{2}{7}\mathbf{k}.$$

Then, in terms of its scalar components, $\mathbf{F}$ is

$$\mathbf{F} = |\mathbf{F}|\mathbf{e}_{AB} = (140)\left(\frac{6}{7}\mathbf{i} + \frac{3}{7}\mathbf{j} - \frac{2}{7}\mathbf{k}\right) = 120\mathbf{i} + 60\mathbf{j} - 40\mathbf{k}\ \text{(N)}.$$

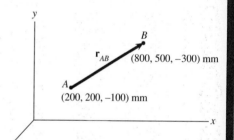

(a) The position vector $\mathbf{r}_{AB}$.

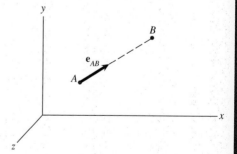

(b) The unit vector $\mathbf{e}_{AB}$ pointing from A toward B.

Example 2.11

The rope in Fig. 2.30 extends from point B through a metal loop attached to the wall at A to point C. The rope exerts forces $\mathbf{F}_{AB}$ and $\mathbf{F}_{AC}$ on the loop at A with magnitudes $|\mathbf{F}_{AB}| = |\mathbf{F}_{AC}| = 200$ lb. What is the magnitude of the total force $\mathbf{F} = \mathbf{F}_{AB} + \mathbf{F}_{AC}$ exerted on the loop by the rope?

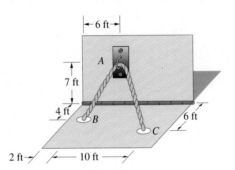

Figure 2.30

STRATEGY

The force $\mathbf{F}_{AB}$ is parallel to the line from A to B, and the force $\mathbf{F}_{AC}$ is parallel to the line from A to C. Since we can determine the coordinates of points A, B, and C from the given dimensions, we can determine the components of unit vectors that have the same directions as the two forces and use them to express the forces in terms of scalar components.

SOLUTION

Let $\mathbf{r}_{AB}$ be the position vector from point A to point B and let $\mathbf{r}_{AC}$ be the position vector from point A to point C (Fig. a). From the given dimensions, the coordinates of points A, B, and C are

$$A: (6, 7, 0) \text{ ft}, \qquad B: (2, 0, 4) \text{ ft}, \qquad C: (12, 0, 6) \text{ ft}.$$

Therefore the components of $\mathbf{r}_{AB}$ and $\mathbf{r}_{AC}$ are

$$\mathbf{r}_{AB} = (x_B - x_A)\,\mathbf{i} + (y_B - y_A)\,\mathbf{j} + (z_B - z_A)\,\mathbf{k}$$
$$= (2 - 6)\,\mathbf{i} + (0 - 7)\,\mathbf{j} + (4 - 0)\,\mathbf{k}$$
$$= -4\mathbf{i} - 7\mathbf{j} + 4\mathbf{k} \text{ (ft)}$$

and

$$\mathbf{r}_{AC} = (x_C - x_A)\,\mathbf{i} + (y_C - y_A)\,\mathbf{j} + (z_C - z_A)\,\mathbf{k}$$
$$= (12 - 6)\,\mathbf{i} + (0 - 7)\,\mathbf{j} + (6 - 0)\,\mathbf{k}$$
$$= 6\mathbf{i} - 7\mathbf{j} + 6\mathbf{k} \text{ (ft)}.$$

(a) The position vectors $\mathbf{r}_{AB}$ and $\mathbf{r}_{AC}$.

Their magnitudes are $|\mathbf{r}_{AB}| = 9$ ft and $|\mathbf{r}_{AC}| = 11$ ft. By dividing $\mathbf{r}_{AB}$ and $\mathbf{r}_{AC}$ by their magnitudes, we obtain unit vectors $\mathbf{e}_{AB}$ and $\mathbf{e}_{AC}$ that point in the directions of $\mathbf{F}_{AB}$ and $\mathbf{F}_{AC}$ (Fig. b):

$$\mathbf{e}_{AB} = \frac{\mathbf{r}_{AB}}{|\mathbf{r}_{AB}|} = -0.444\mathbf{i} - 0.778\mathbf{j} + 0.444\mathbf{k},$$

$$\mathbf{e}_{AC} = \frac{\mathbf{r}_{AC}}{|\mathbf{r}_{AC}|} = 0.545\mathbf{i} - 0.636\mathbf{j} + 0.545\mathbf{k}.$$

The forces $\mathbf{F}_{AB}$ and $\mathbf{F}_{AC}$ are

$$\mathbf{F}_{AB} = 200\,\mathbf{e}_{AB} = -88.9\mathbf{i} - 155.6\mathbf{j} + 88.9\mathbf{k} \text{ (lb)},$$

$$\mathbf{F}_{AC} = 200\,\mathbf{e}_{AC} = 109.1\mathbf{i} - 127.3\mathbf{j} + 109.1\mathbf{k} \text{ (lb)}.$$

The total force exerted on the loop by the rope is

$$\mathbf{F} = \mathbf{F}_{AB} + \mathbf{F}_{AC} = 20.2\mathbf{i} - 282.8\mathbf{j} + 198.0\mathbf{k} \text{ (lb)},$$

and its magnitude is

$$|\mathbf{F}| = \sqrt{(20.2)^2 + (-282.8)^2 + (198.0)^2} = 346 \text{ lb}.$$

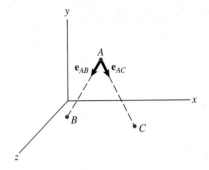

(b) The unit vectors $\mathbf{e}_{AB}$ and $\mathbf{e}_{AC}$.

Example 2.12

The cable AB in Fig. 2.31 exerts a 50-N force $\mathbf{T}$ on the collar at A. Express $\mathbf{T}$ in terms of scalar components.

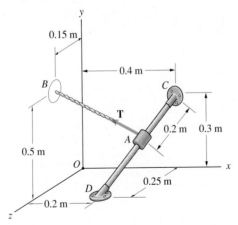

Figure 2.31

STRATEGY

Let $\mathbf{r}_{AB}$ be the position vector from A to B. We will divide $\mathbf{r}_{AB}$ by its magnitude to obtain a unit vector $\mathbf{e}_{AB}$ having the same direction as the force $\mathbf{T}$. Then we can obtain $\mathbf{T}$ in terms of scalar components by expressing it as the product of its magnitude and $\mathbf{e}_{AB}$. To begin this procedure, we must first determine the coordinates of the collar A. We will do so by obtaining a unit vector $\mathbf{e}_{CD}$ pointing from C toward D and multiplying it by 0.2 m to determine the position of the collar A relative to C.

SOLUTION

Determining the Coordinates of Point A The position vector from C to D is

$$\mathbf{r}_{CD} = (0.2 - 0.4)\,\mathbf{i} + (0 - 0.3)\,\mathbf{j} + (0.25 - 0)\,\mathbf{k}$$
$$= -0.2\mathbf{i} - 0.3\mathbf{j} + 0.25\mathbf{k}\ (\text{m}).$$

Dividing this vector by its magnitude, we obtain the unit vector $\mathbf{e}_{CD}$ (Fig. a):

$$\mathbf{e}_{CD} = \frac{\mathbf{r}_{CD}}{|\mathbf{r}_{CD}|} = \frac{-0.2\mathbf{i} - 0.3\mathbf{j} + 0.25\mathbf{k}}{\sqrt{(-0.2)^2 + (-0.3)^2 + (0.25)^2}}$$
$$= -0.456\mathbf{i} - 0.684\mathbf{j} + 0.570\mathbf{k}.$$

Using this vector, we obtain the position vector from C to A:

$$\mathbf{r}_{CA} = (0.2\ \text{m})\,\mathbf{e}_{CD} = -0.091\mathbf{i} - 0.137\mathbf{j} + 0.114\mathbf{k}\ (\text{m}).$$

The position vector from the origin of the coordinate system to C is $\mathbf{r}_{OC} = 0.4\mathbf{i} + 0.3\mathbf{j}$ (m), so the position vector from the origin to A is

$$\mathbf{r}_{OA} = \mathbf{r}_{OC} + \mathbf{r}_{CA} = (0.4\mathbf{i} + 0.3\mathbf{j}) + (-0.091\mathbf{i} - 0.137\mathbf{j} + 0.114\mathbf{k})$$
$$= 0.309\mathbf{i} + 0.163\mathbf{j} + 0.114\mathbf{k}\ (\text{m}).$$

The coordinates of A are $(0.309, 0.163, 0.114)$ m.

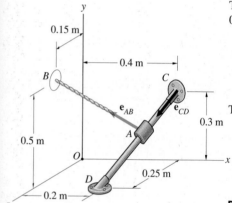

(a) The unit vectors $\mathbf{e}_{AB}$ and $\mathbf{e}_{CD}$.

Determining the Components of T Using the coordinates of point A, the position vector from A to B is

$$\mathbf{r}_{AB} = (0 - 0.309)\,\mathbf{i} + (0.5 - 0.163)\,\mathbf{j} + (0.15 - 0.114)\,\mathbf{k}$$
$$= -0.309\mathbf{i} + 0.337\mathbf{j} + 0.036\mathbf{k}\ (\text{m}).$$

Dividing this vector by its magnitude, we obtain the unit vector $\mathbf{e}_{AB}$ (Fig. a):

$$\mathbf{e}_{AB} = \frac{\mathbf{r}_{AB}}{|\mathbf{r}_{AB}|} = \frac{-0.309\mathbf{i} + 0.337\mathbf{j} + 0.036\mathbf{k}}{\sqrt{(-0.309)^2 + (0.337)^2 + (0.036)^2}}$$
$$= -0.674\mathbf{i} + 0.735\mathbf{j} + 0.079\mathbf{k}.$$

The force $\mathbf{T}$ is

$$\mathbf{T} = |\mathbf{T}|\,\mathbf{e}_{AB} = (50\ \text{N})(-0.674\mathbf{i} + 0.735\mathbf{j} + 0.079\mathbf{k})$$
$$= -33.7\mathbf{i} + 36.7\mathbf{j} + 3.9\mathbf{k}\ (\text{N}).$$

Problems

2.65 A vector $\mathbf{U} = 3\mathbf{i} - 4\mathbf{j} - 12\mathbf{k}$. What is its magnitude?

Strategy: The magnitude of a vector is given in terms of its components by Eq. (2.14).

2.66 A force vector $\mathbf{F} = 20\mathbf{i} + 60\mathbf{j} - 90\mathbf{k}$ (N). Determine its magnitude.

2.67 A force vector $\mathbf{F} = F_x \mathbf{i} - 120\mathbf{j} - 40\mathbf{k}$ (lb). Its magnitude $|\mathbf{F}| = 130$ lb. What is F_x?

2.68 A vector $\mathbf{U} = U_x \mathbf{i} + U_y \mathbf{j} + U_z \mathbf{k}$. Its magnitude $|\mathbf{U}| = 30$. Its scalar components are related by the equations $U_y = -2U_x$ and $U_z = 4U_y$. Determine the scalar components.

2.69 A vector $\mathbf{U} = 100\mathbf{i} + 200\mathbf{j} - 600\mathbf{k}$, and a vector $\mathbf{V} = -200\mathbf{i} + 450\mathbf{j} + 100\mathbf{k}$. Determine the magnitude of the vector $-2\mathbf{U} + 3\mathbf{V}$.

2.70 Two vectors $\mathbf{U} = 3\mathbf{i} - 2\mathbf{j} + 6\mathbf{k}$ and $\mathbf{V} = 4\mathbf{i} + 12\mathbf{j} - 3\mathbf{k}$.
(a) Determine the magnitudes of $\mathbf{U}$ and $\mathbf{V}$.
(b) Determine the magnitude of the vector $3\mathbf{U} + 2\mathbf{V}$.

2.71 A vector $\mathbf{U} = 40\mathbf{i} - 70\mathbf{j} - 40\mathbf{k}$.
(a) What is its magnitude?
(b) What are the angles θ_x, θ_y, and θ_z between $\mathbf{U}$ and the positive coordinate axes?

Strategy: Since you know the components of $\mathbf{U}$, you can determine the angles θ_x, θ_y, and θ_z from Eqs. (2.15).

2.72 A force $\mathbf{F} = 600\mathbf{i} - 700\mathbf{j} + 600\mathbf{k}$ (lb). What are the angles θ_x, θ_y, and θ_z between the vector $\mathbf{F}$ and the positive coordinate axes?

2.73 The cable exerts a 50-lb force $\mathbf{F}$ on the metal hook at O. The angle between $\mathbf{F}$ and the x axis is $40°$, and the angle between $\mathbf{F}$ and the y axis is $70°$. The z component of $\mathbf{F}$ is positive.
(a) Express $\mathbf{F}$ in terms of scalar components.
(b) What are the direction cosines of $\mathbf{F}$?

Strategy: Since you are given only two of the angles between $\mathbf{F}$ and the coordinate axes, you must first determine the third one. Then you can obtain the components of $\mathbf{F}$ from Eqs. (2.15).

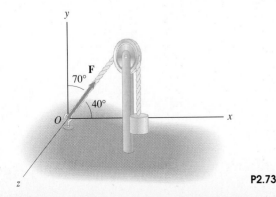

P2.73

2.74 A unit vector has direction cosines $\cos\theta_x = -0.5$ and $\cos\theta_y = 0.2$. Its z component is positive. Express it in terms of scalar components.

2.75 The airplane's engines exert a total thrust force $\mathbf{T}$ of 200-kN magnitude. The angle between $\mathbf{T}$ and the x axis is $120°$, and the angle between $\mathbf{T}$ and the y axis is $130°$. The z component of $\mathbf{T}$ is positive.
(a) What is the angle between $\mathbf{T}$ and the z axis?
(b) Express $\mathbf{T}$ in terms of scalar components.

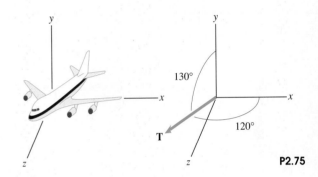

P2.75

2.76 The position vector from a point A to a point B is $3\mathbf{i} + 4\mathbf{j} - 4\mathbf{k}$ (ft). The position vector from point A to a point C is $-3\mathbf{i} + 13\mathbf{j} - 2\mathbf{k}$ (ft).
(a) What is the distance from point B to point C?
(b) What are the direction cosines of the position vector from point B to point C?

2.77 A vector $\mathbf{U} = 3\mathbf{i} - 2\mathbf{j} + 6\mathbf{k}$. Determine the components of the unit vector that has the same direction as $\mathbf{U}$.

2.78 A force vector $\mathbf{F} = 3\mathbf{i} - 4\mathbf{j} - 2\mathbf{k}$ (N).
(a) What is the magnitude of $\mathbf{F}$?
(b) Determine the components of the unit vector that has the same direction as $\mathbf{F}$.

2.79 A force vector $\mathbf{F}$ points in the same direction as the unit vector $\mathbf{e} = \frac{2}{7}\mathbf{i} - \frac{6}{7}\mathbf{j} - \frac{3}{7}\mathbf{k}$. The magnitude of $\mathbf{F}$ is 700 lb. Express $\mathbf{F}$ in terms of scalar components.

2.80 A force vector $\mathbf{F}$ points in the same direction as the position vector $\mathbf{r} = 4\mathbf{i} + 4\mathbf{j} - 7\mathbf{k}$ (m). The magnitude of $\mathbf{F}$ is 90 kN. Express $\mathbf{F}$ in terms of scalar components.

2.81 Astronauts on the space shuttle use radar to determine the magnitudes and direction cosines of the position vectors of two satellites A and B. The vector $\mathbf{r}_A$ from the shuttle to satellite A has magnitude 2 km, and direction cosines $\cos \theta_x = 0.768$, $\cos \theta_y = 0.384$, $\cos \theta_z = 0.512$. The vector $\mathbf{r}_B$ from the shuttle to satellite B has magnitude 4 km and direction cosines $\cos \theta_x = 0.743$, $\cos \theta_y = 0.557$, $\cos \theta_z = -0.371$. What is the distance between the satellites?

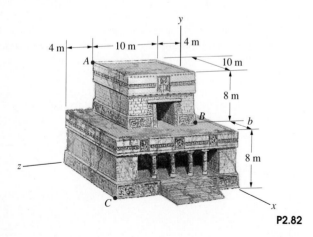

P2.81

2.82 Archaeologists measure a pre-Columbian ceremonial structure and obtain the dimensions shown. Determine (a) the magnitude and (b) the direction cosines of the position vector from point A to point B.

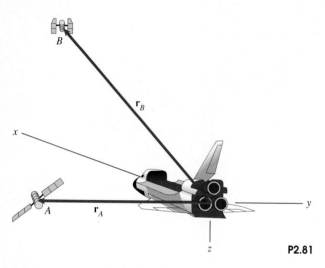

P2.82

2.83 Consider the structure described in Problem 2.82. After returning to the United States, an archaeologist discovers that he lost the notes containing the dimension b, but other notes indicate that the distance from point B to point C is 16.4 m. What are the direction cosines of the vector from B to C?

2.84 Observers at A and B use theodolites to measure the direction from their positions to a rocket in flight. If the coordinates of the rocket's position at a given instant are (4, 4, 2) km, determine the direction cosines of the vectors $\mathbf{r}_{AR}$ and $\mathbf{r}_{BR}$ that the observers would measure at that instant.

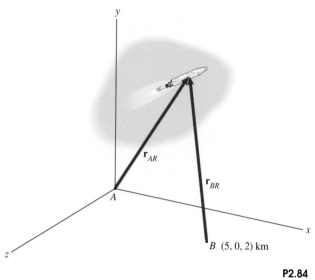

P2.84

2.85 In Problem 2.84, suppose that the coordinates of the rocket's position are unknown. At a given instant, the person at A determines that the direction cosines of $\mathbf{r}_{AR}$ are $\cos \theta_x = 0.535$, $\cos \theta_y = 0.802$, and $\cos \theta_z = 0.267$, and the person at B determines that the direction cosines of $\mathbf{r}_{BR}$ are $\cos \theta_x = -0.576$, $\cos \theta_y = 0.798$, and $\cos \theta_z = -0.177$. What are the coordinates of the rocket's position at that instant?

2.86 The height of Mount Everest was originally measured by a surveyor using the following procedure. He first measured the distance between two points A and B of equal altitude. Suppose that they are 10,000 ft above sea level and are 32,000 ft apart. He then used a theodolite to measure the direction cosines of the vectors from point A to the top of the mountain P and from point B to P. Suppose that for $\mathbf{r}_{AP}$, the direction cosines are $\cos \theta_x = 0.509$, $\cos \theta_y = 0.509$, $\cos \theta_z = 0.694$, and for $\mathbf{r}_{BP}$ they are $\cos \theta_x = -0.605$, $\cos \theta_y = 0.471$, $\cos \theta_z = 0.642$. The z axis of the coordinate system is vertical. What is the height of Mount Everest above sea level?

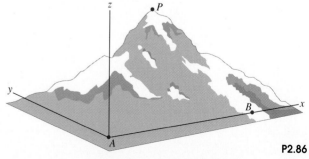

P2.86

2.87 The distance from point O to point A is 20 ft. The straight line AB is parallel to the y axis, and point B is in the x-z plane. Express the vector $\mathbf{r}_{OA}$ in terms of scalar components.

   Strategy: You can resolve $\mathbf{r}_{OA}$ into a vector from O to B and a vector from B to A. You can then resolve the vector from O to B into vector components parallel to the x and z axes. See Example 2.9.

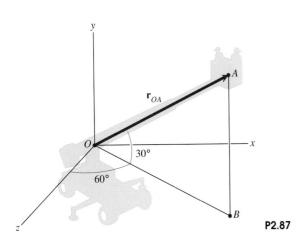

P2.87

2.88 The magnitude of $\mathbf{r}$ is 100 in. The straight line from the head of $\mathbf{r}$ to point A is parallel to the x axis, and point A is contained in the y-z plane. Express $\mathbf{r}$ in terms of scalar components.

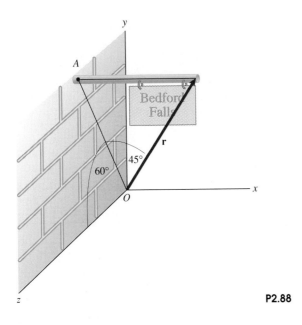

P2.88

2.89 The straight line from the head of $\mathbf{F}$ to point A is parallel to the y axis, and point A is contained in the x-z plane. The x component of $\mathbf{F}$ is $F_x = 100$ N.

(a) What is the magnitude of $\mathbf{F}$?
(b) Determine the angles θ_x, θ_y, and θ_z between $\mathbf{F}$ and the positive coordinate axes.

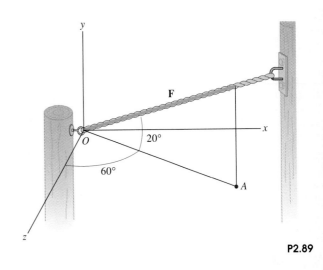

P2.89

2.90 The position of a point P on the surface of the earth is specified by the longitude λ, measured from the point G on the equator directly south of Greenwich, England, and the latitude L measured from the equator. Longitude is given as west (W) longitude or east (E) longitude, indicating whether the angle is measured west or east from point G. Latitude is given as north (N) latitude or south (S) latitude, indicating whether the angle is measured north or south from the equator. Suppose that P is at longitude 30°W and latitude 45°N. Let R_E be the radius of the earth. Using the coordinate system shown, determine the components of the position vector of P relative to the center of the earth. (Your answer will be in terms of R_E.)

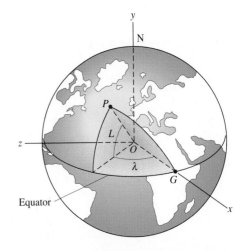

P2.90

2.91 (a) Write the position vector $\mathbf{r}_{AB}$ from point A to point B in terms of scalar components.

(b) The vector $\mathbf{F}$ has magnitude $|\mathbf{F}| = 200$ N and is parallel to the line from A to B. Write $\mathbf{F}$ in terms of scalar components.

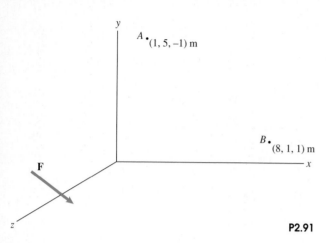

P2.91

2.92 The part of the cable from A to B exerts a 110-lb force $\mathbf{T}$ at A.

(a) Determine the unit vector that points from point A toward point B.

(b) Express the force $\mathbf{T}$ in terms of scalar components.

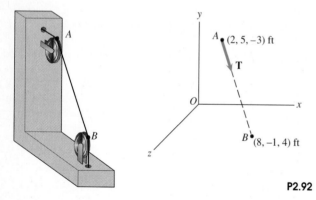

P2.92

2.93 A cable extends from point A to point B. It exerts a 1-kN force $\mathbf{F}$ at A that is directed along the line from A to B. Express $\mathbf{F}$ in terms of scalar components.

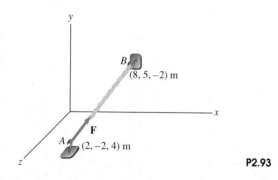

P2.93

2.94 In Problem 2.93, what are the direction cosines of the force $\mathbf{F}$?

2.95 The cable AB exerts a 200-lb force $\mathbf{F}_{AB}$ at point A that is directed along the line from A to B. Express $\mathbf{F}_{AB}$ in terms of scalar components.

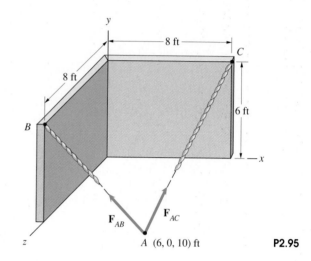

P2.95

2.96 Consider the cables and wall described in Problem 2.95. Cable AB exerts a 200-lb force $\mathbf{F}_{AB}$ at point A that is directed along the line from A to B. The cable AC exerts a 100-lb force $\mathbf{F}_{AC}$ at point A that is directed along the line from A to C. Determine the magnitude of the total force exerted at point A by the two cables.

2.97 The 70-m-tall tower is supported by three cables that exert forces $\mathbf{F}_{AB}$, $\mathbf{F}_{AC}$, and $\mathbf{F}_{AD}$ on it. The magnitude of each force is 2 kN. Express the total force exerted on the tower by the three cables in terms of scalar components.

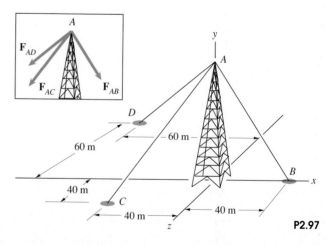

P2.97

2.98 Consider the tower described in Problem 2.97. The magnitude of the force $\mathbf{F}_{AB}$ is 2 kN. The x and z components of the vector sum of the forces exerted on the tower by the three cables are zero. What are the magnitudes of $\mathbf{F}_{AC}$ and $\mathbf{F}_{AD}$?

2.99 Express the position vector from point O to the collar at A in terms of scalar components.

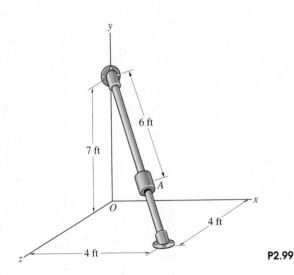

P2.99

2.100 The cable AB exerts a 32-lb force $\mathbf{T}$ on the collar at A. Express $\mathbf{T}$ in terms of scalar components.

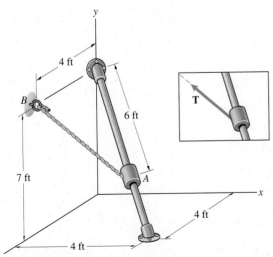

P2.100

2.101 The circular bar has a 4-m radius and lies in the x-y plane. Express the position vector from point B to the collar at A in terms of scalar components.

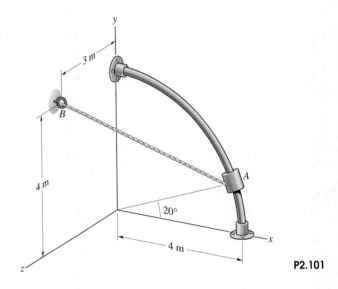

P2.101

2.102 The cable AB in Problem 2.101 exerts a 60-N force $\mathbf{T}$ on the collar at A that is directed along the line from A toward B. Express $\mathbf{T}$ in terms of scalar components.

Products of Vectors

Two kinds of products of vectors, the dot and cross products, have been found to have applications in science and engineering, especially in mechanics and electromagnetic field theory. We use both of these products in Chapter 4 to evaluate moments of forces about points and lines. We discuss them here so that you can concentrate on mechanics when we introduce moments and not be distracted by the details of vector operations.

2.5 Dot Products

The dot product of two vectors has many uses, including resolving a vector into components parallel and perpendicular to a given line and determining the angle between two lines in space.

Definition

Consider two vectors **U** and **V** (Fig. 2.32a). The **dot product** of **U** and **V**, denoted by **U · V** (hence the name "dot product"), is defined to be the product of the magnitude of **U**, the magnitude of **V**, and the cosine of the angle θ between **U** and **V** when they are placed tail to tail (Fig. 2.32b):

$$\mathbf{U} \cdot \mathbf{V} = |\mathbf{U}||\mathbf{V}| \cos \theta. \qquad (2.18)$$

Because the result of the dot product is a scalar, the dot product is sometimes called the scalar product. The units of the dot product are the product of the units of the two vectors. Notice that the dot product of two nonzero vectors is equal to zero if and only if the two vectors are perpendicular.

The dot product has the properties

$$\mathbf{U} \cdot \mathbf{V} = \mathbf{V} \cdot \mathbf{U}, \quad \text{The dot product is commutative.} \qquad (2.19)$$

$$a(\mathbf{U} \cdot \mathbf{V}) = (a\mathbf{U}) \cdot \mathbf{V} = \mathbf{U} \cdot (a\mathbf{V}), \quad \text{The dot product is associative with respect to scalar multiplication.} \qquad (2.20)$$

and

$$\mathbf{U} \cdot (\mathbf{V} + \mathbf{W}) = \mathbf{U} \cdot \mathbf{V} + \mathbf{U} \cdot \mathbf{W} \quad \text{The dot product is distributive with respect to vector addition.} \qquad (2.21)$$

for any scalar a and vectors **U**, **V**, and **W**.

Dot Products in Terms of Components

In this section we derive an equation that allows you to determine the dot product of two vectors if you know their scalar components. The derivation also results in an equation for the angle between the vectors. The first step is to

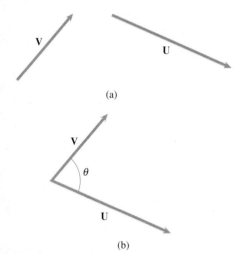

Figure 2.32

(a) The vectors **U** and **V**.
(b) The angle θ between **U** and **V** when the two vectors are placed tail to tail.

determine the dot products formed from the unit vectors **i**, **j**, and **k**. Let's evaluate the dot product **i** · **i**. The magnitude |**i**| = 1, and the angle between two identical vectors placed tail to tail is zero, so we obtain

$$\mathbf{i} \cdot \mathbf{i} = |\mathbf{i}| \, |\mathbf{i}| \, \cos(0) = (1)(1)(1) = 1.$$

The dot product of **i** and **j** is

$$\mathbf{i} \cdot \mathbf{j} = |\mathbf{i}| \, |\mathbf{j}| \, \cos(90°) = (1)(1)(0) = 0.$$

Continuing in this way, we obtain

$$\begin{aligned}
\mathbf{i} \cdot \mathbf{i} = 1, \quad & \mathbf{i} \cdot \mathbf{j} = 0, \quad \mathbf{i} \cdot \mathbf{k} = 0, \\
\mathbf{j} \cdot \mathbf{i} = 0, \quad & \mathbf{j} \cdot \mathbf{j} = 1, \quad \mathbf{j} \cdot \mathbf{k} = 0, \\
\mathbf{k} \cdot \mathbf{i} = 0, \quad & \mathbf{k} \cdot \mathbf{j} = 0, \quad \mathbf{k} \cdot \mathbf{k} = 1.
\end{aligned} \tag{2.22}$$

The dot product of two vectors **U** and **V** expressed in terms of their components is

$$\begin{aligned}
\mathbf{U} \cdot \mathbf{V} = &(U_x \mathbf{i} + U_y \mathbf{j} + U_z \mathbf{k}) \cdot (V_x \mathbf{i} + V_y \mathbf{j} + V_z \mathbf{k}) \\
= &\, U_x V_x (\mathbf{i} \cdot \mathbf{i}) + U_x V_y (\mathbf{i} \cdot \mathbf{j}) + U_x V_z (\mathbf{i} \cdot \mathbf{k}) \\
&+ U_y V_x (\mathbf{j} \cdot \mathbf{i}) + U_y V_y (\mathbf{j} \cdot \mathbf{j}) + U_y V_z (\mathbf{j} \cdot \mathbf{k}) \\
&+ U_z V_x (\mathbf{k} \cdot \mathbf{i}) + U_z V_y (\mathbf{k} \cdot \mathbf{j}) + U_z V_z (\mathbf{k} \cdot \mathbf{k}).
\end{aligned}$$

In obtaining this result, we used Eqs. (2.20) and (2.21). Substituting Eqs. (2.22) into this expression, we obtain an equation for the dot product in terms of the scalar components of the two vectors:

$$\boxed{\mathbf{U} \cdot \mathbf{V} = U_x V_x + U_y V_y + U_z V_z.} \tag{2.23}$$

To obtain an equation for the angle θ in terms of the components of the vectors, we equate the expression for the dot product given by Eq. (2.23) to the definition of the dot product, Eq. (2.18), and solve for $\cos \theta$:

$$\cos \theta = \frac{\mathbf{U} \cdot \mathbf{V}}{|\mathbf{U}| \, |\mathbf{V}|} = \frac{U_x V_x + U_y V_y + U_z V_z}{|\mathbf{U}| \, |\mathbf{V}|}. \tag{2.24}$$

Vector Components Parallel and Normal to a Line

In some engineering applications you must resolve a vector into components that are parallel and normal (perpendicular) to a given line. The component of a vector parallel to a line is called the **projection** of the vector onto the line. For example, when the vector represents a force, the projection of the force onto a line is the component of the force in the direction of the line.

We can determine the components of a vector parallel and normal to a line by using the dot product. Consider a vector **U** and a straight line L (Fig. 2.33a). We can resolve **U** into components $\mathbf{U}_p$ and $\mathbf{U}_n$ that are parallel and normal to L (Fig. 2.33b).

Figure 2.33

(a) A vector **U** and line L.
(b) Resolving **U** into components parallel and normal to L.

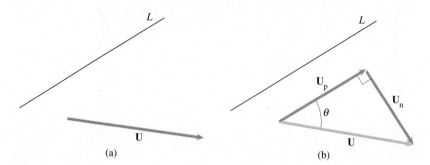

(a)   (b)

The Parallel Component In terms of the angle θ between **U** and the component $\mathbf{U}_p$, the magnitude of $\mathbf{U}_p$ is

$$|\mathbf{U}_p| = |\mathbf{U}| \cos \theta. \tag{2.25}$$

Let **e** be a unit vector parallel to L (Fig. 2.34). The dot product of **e** and **U** is

$$\mathbf{e} \cdot \mathbf{U} = |\mathbf{e}||\mathbf{U}| \cos \theta = |\mathbf{U}| \cos \theta.$$

Figure 2.34

The unit vector **e** is parallel to L.

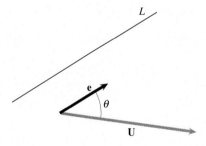

Comparing this result with Eq. (2.25), we see that the magnitude of $\mathbf{U}_p$ is

$$|\mathbf{U}_p| = \mathbf{e} \cdot \mathbf{U}.$$

Therefore the parallel component, or projection of **U** onto L, is

$$\mathbf{U}_p = (\mathbf{e} \cdot \mathbf{U}) \, \mathbf{e}. \tag{2.26}$$

(This equation holds even if **e** doesn't point in the direction of $\mathbf{U}_p$. In that case, the angle $\theta > 90°$ and $\mathbf{e} \cdot \mathbf{U}$ is negative.) When you know the components of a vector and the components of a unit vector **e** parallel to a line L, you can use Eq. (2.26) to determine the component of the vector parallel to L.

The Normal Component Once you have determined the parallel component, you can obtain the normal component from the relation $\mathbf{U} = \mathbf{U}_p + \mathbf{U}_n$:

$$\mathbf{U}_n = \mathbf{U} - \mathbf{U}_p. \tag{2.27}$$

In the following examples we show you how to evaluate and use dot products. Example 2.14 illustrates that the dot product is an easy way to determine the angle between two straight lines in three dimensions, and Example 2.15 demonstrates the use of the dot product to determine a projection of a vector.

Example 2.13

The magnitude of the force **F** in Fig. 2.35 is 100 lb. The magnitude of the vector **r** from point O to point A is 8 ft.
(a) Use the definition of the dot product to determine $\mathbf{r} \cdot \mathbf{F}$.
(b) Use Eq. (2.23) to determine $\mathbf{r} \cdot \mathbf{F}$.

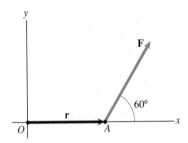

Figure 2.35

STRATEGY

(a) Since we know the magnitudes of **r** and **F** and the angle between them when they are placed tail to tail, we can determine $\mathbf{r} \cdot \mathbf{F}$ directly from the definition.
(b) We can determine the components of **r** and **F** and use Eq. (2.23) to determine their dot product.

SOLUTION

(a) Using the definition of the dot product,

$$\mathbf{r} \cdot \mathbf{F} = |\mathbf{r}||\mathbf{F}| \cos \theta = (8)(100) \cos 60° = 400 \text{ ft-lb}.$$

(b) The vector $\mathbf{r} = 8\mathbf{i}$ (ft). The vector **F** in terms of scalar components is

$$\mathbf{F} = 100 \cos 60° \, \mathbf{i} + 100 \sin 60° \, \mathbf{j} \text{ (lb)}.$$

Therefore the dot product of **r** and **F** is

$$\mathbf{r} \cdot \mathbf{F} = r_x F_x + r_y F_y + r_z F_z$$
$$= (8)(100 \cos 60°) + (0)(100 \sin 60°) + (0)(0) = 400 \text{ ft-lb}.$$

Example 2.14

What is the angle θ between the lines AB and AC in Fig. 2.36?

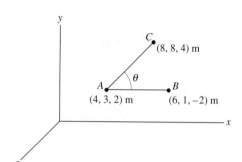

Figure 2.36

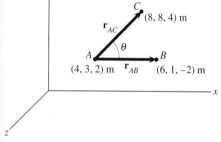

(a) The position vectors $\mathbf{r}_{AB}$ and $\mathbf{r}_{AC}$.

STRATEGY

We know the coordinates of the points A, B, and C, so we can determine the components of the vector $\mathbf{r}_{AB}$ from A to B and the vector $\mathbf{r}_{AC}$ from A to C (Fig. a). Then we can use Eq. (2.24) to determine θ.

SOLUTION

The vectors $\mathbf{r}_{AB}$ and $\mathbf{r}_{AC}$ are

$$\mathbf{r}_{AB} = (6 - 4)\,\mathbf{i} + (1 - 3)\,\mathbf{j} + (-2 - 2)\,\mathbf{k} = 2\mathbf{i} - 2\mathbf{j} - 4\mathbf{k}\ (\text{m}),$$

$$\mathbf{r}_{AC} = (8 - 4)\,\mathbf{i} + (8 - 3)\,\mathbf{j} + (4 - 2)\,\mathbf{k} = 4\mathbf{i} + 5\mathbf{j} + 2\mathbf{k}\ (\text{m}).$$

Their magnitudes are

$$|\mathbf{r}_{AB}| = \sqrt{(2)^2 + (-2)^2 + (-4)^2} = 4.90\ \text{m},$$

$$|\mathbf{r}_{AC}| = \sqrt{(4)^2 + (5)^2 + (2)^2} = 6.71\ \text{m}.$$

The dot product of $\mathbf{r}_{AB}$ and $\mathbf{r}_{AC}$ is

$$\mathbf{r}_{AB} \cdot \mathbf{r}_{AC} = (2)(4) + (-2)(5) + (-4)(2) = -10\ \text{m}^2.$$

Therefore

$$\cos \theta = \frac{\mathbf{r}_{AB} \cdot \mathbf{r}_{AC}}{|\mathbf{r}_{AB}||\mathbf{r}_{AC}|} = \frac{-10}{(4.90)(6.71)} = -0.304.$$

The angle $\theta = \arccos(-0.304) = 107.7°$.

Example 2.15

Suppose that you pull on the cable OA in Fig. 2.37, exerting a 50-N force $\mathbf{F}$ at O. What are the components of $\mathbf{F}$ parallel and normal to the cable OB?

STRATEGY

Resolving $\mathbf{F}$ into components parallel and normal to OB (Fig. a), we can determine the components by using Eqs. (2.26) and (2.27). But to apply them, we must first express $\mathbf{F}$ in terms of scalar components and determine the components of a unit vector parallel to OB. We can obtain the components of $\mathbf{F}$ by determining the components of the unit vector pointing from O toward A and multiplying them by $|\mathbf{F}|$.

SOLUTION

The position vectors from O to A and from O to B are (Fig. b)

$$\mathbf{r}_{OA} = 6\mathbf{i} + 6\mathbf{j} - 3\mathbf{k} \text{ (m)},$$

$$\mathbf{r}_{OB} = 10\mathbf{i} - 2\mathbf{j} + 3\mathbf{k} \text{ (m)}.$$

Their magnitudes are $|\mathbf{r}_{OA}| = 9$ m and $|\mathbf{r}_{OB}| = 10.6$ m. Dividing these vectors by their magnitudes, we obtain unit vectors that point from the origin toward A and B (Fig. c):

$$\mathbf{e}_{OA} = \frac{\mathbf{r}_{OA}}{|\mathbf{r}_{OA}|} = \frac{6\mathbf{i} + 6\mathbf{j} - 3\mathbf{k}}{9} = 0.667\mathbf{i} + 0.667\mathbf{j} - 0.333\mathbf{k},$$

$$\mathbf{e}_{OB} = \frac{\mathbf{r}_{OB}}{|\mathbf{r}_{OB}|} = \frac{10\mathbf{i} - 2\mathbf{j} + 3\mathbf{k}}{10.6} = 0.941\mathbf{i} - 0.188\mathbf{j} + 0.282\mathbf{k}.$$

The force $\mathbf{F}$ in terms of scalar components is

$$\mathbf{F} = |\mathbf{F}|\mathbf{e}_{OA} = (50)(0.667\mathbf{i} + 0.667\mathbf{j} - 0.333\mathbf{k})$$

$$= 33.3\mathbf{i} + 33.3\mathbf{j} - 16.7\mathbf{k} \text{ (N)}.$$

Taking the dot product of $\mathbf{e}_{OB}$ and $\mathbf{F}$, we obtain

$$\mathbf{e}_{OB} \cdot \mathbf{F} = (0.941)(33.3) + (-0.188)(33.3) + (0.282)(-16.7)$$

$$= 20.4 \text{ N}.$$

The parallel component of $\mathbf{F}$ is

$$\mathbf{F}_p = (\mathbf{e}_{OB} \cdot \mathbf{F})\,\mathbf{e}_{OB} = (20.4)(0.941\mathbf{i} - 0.188\mathbf{j} + 0.282\mathbf{k})$$

$$= 19.2\mathbf{i} - 3.8\mathbf{j} + 5.8\mathbf{k} \text{ (N)},$$

and the normal component is

$$\mathbf{F}_n = \mathbf{F} - \mathbf{F}_p = 14.2\mathbf{i} + 37.2\mathbf{j} - 22.4\mathbf{k} \text{ (N)}.$$

DISCUSSION

You can confirm that two vectors are perpendicular by making sure their dot product is zero. In this example,

$$\mathbf{F}_p \cdot \mathbf{F}_n = (19.2)(14.2) + (-3.8)(37.2) + (5.8)(-22.4) = 0.$$

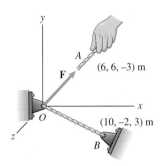

Figure 2.37

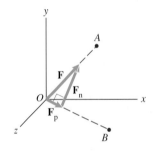

(a) The components of $\mathbf{F}$ parallel and normal to OB.

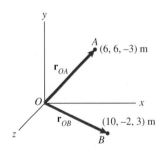

(b) The position vectors $\mathbf{r}_{OA}$ and $\mathbf{r}_{OB}$.

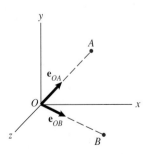

(c) The unit vectors $\mathbf{e}_{OA}$ and $\mathbf{e}_{OB}$.

Problems

2.103   Determine the dot product $\mathbf{U} \cdot \mathbf{V}$ of the vectors $\mathbf{U} = 2\mathbf{i} - 4\mathbf{j} + 3\mathbf{k}$ and $\mathbf{V} = -3\mathbf{i} + 6\mathbf{j} + 3\mathbf{k}$.

Strategy: Since you know the components of the vectors, you can use Eq. (2.23) to determine their dot product.

2.104   Determine the dot product $\mathbf{U} \cdot \mathbf{V}$ of the vectors $\mathbf{U} = 40\mathbf{i} + 20\mathbf{j} + 60\mathbf{k}$ and $\mathbf{V} = -30\mathbf{i} + 15\mathbf{k}$.

2.105   What is the dot product of the position vector $\mathbf{r} = -10\mathbf{i} + 25\mathbf{j}$ (m) and the force $\mathbf{F} = 300\mathbf{i} + 250\mathbf{j} + 300\mathbf{k}$ (N)?

2.106   What is the dot product of the position vector $\mathbf{r} = 4\mathbf{i} - 12\mathbf{j} - 3\mathbf{k}$ (ft) and the force $\mathbf{F} = 20\mathbf{i} + 30\mathbf{j} - 10\mathbf{k}$ (lb)?

2.107   Two vectors $\mathbf{U} = -6\mathbf{i} + \mathbf{j} + 8\mathbf{k}$ and $\mathbf{V} = 3\mathbf{i} + 2\mathbf{j} + 2\mathbf{k}$.
(a) Determine the dot product of $\mathbf{U}$ and $\mathbf{V}$.
(b) What can you conclude about $\mathbf{U}$ and $\mathbf{V}$ from the result of (a)?

2.108   Two vectors $\mathbf{U} = U_x \mathbf{i} - 4\mathbf{j}$ and $\mathbf{V} = -2\mathbf{i} + 6\mathbf{j}$ are perpendicular. What is U_x?

2.109   The magnitudes $|\mathbf{U}| = 10$ and $|\mathbf{V}| = 20$.
(a) Use the definition of the dot product to determine $\mathbf{U} \cdot \mathbf{V}$.
(b) Use Eq. (2.23) to determine $\mathbf{U} \cdot \mathbf{V}$.

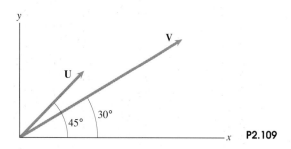

P2.109

2.110   By evaluating the dot product $\mathbf{U} \cdot \mathbf{V}$, prove the identity $\cos(\theta_1 - \theta_2) = \cos\theta_1 \cos\theta_2 + \sin\theta_1 \sin\theta_2$.

Strategy: Evaluate the dot product both by using the definition and by using Eq. (2.23).

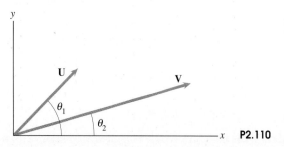

P2.110

2.111   Two vectors $\mathbf{U} = U_x \mathbf{i} - 6\mathbf{j} + \mathbf{k}$ and $\mathbf{V} = -3\mathbf{i} + V_y \mathbf{j} + \mathbf{k}$. Their dot product $\mathbf{U} \cdot \mathbf{V} = -35$, and the magnitude of their sum is $|\mathbf{U} + \mathbf{V}| = 3$. What are the components U_x and V_y?

2.112   What is the angle θ between the straight lines AB and AC?

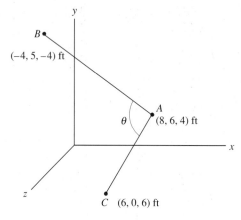

P2.112

2.113   The ship O measures the positions of the ship A and the airplane B and obtains the coordinates shown. What is the angle θ between the lines of sight OA and OB?

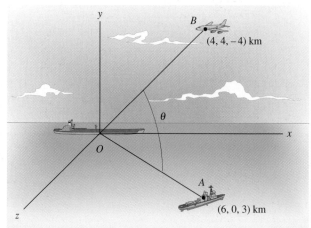

P2.113

2.114   Astronauts on the space shuttle use radar to determine the magnitudes and direction cosines of the position vectors of two satellites A and B. The vector $\mathbf{r}_A$ from the shuttle to satellite A has magnitude 2 km and direction cosines $\cos\theta_x = 0.768$, $\cos\theta_y = 0.384$, $\cos\theta_z = 0.512$. The vector $\mathbf{r}_B$ from the shuttle to satellite B has magnitude 4 km and direction cosines $\cos\theta_x = 0.743$,

$\cos \theta_y = 0.557$, $\cos \theta_z = -0.371$. What is the angle θ between the vectors $\mathbf{r}_A$ and $\mathbf{r}_B$?

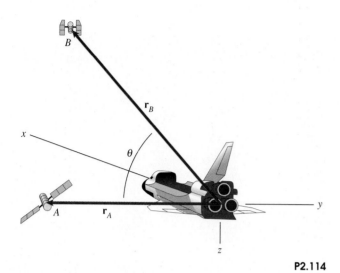

P2.114

2.115 The force vector $\mathbf{F} = F_x\mathbf{i} + F_y\mathbf{j} + F_z\mathbf{k}$.

(a) Use Eq. (2.26) to determine the vector component of $\mathbf{F}$ parallel to the x axis.
(b) Determine the vector component of $\mathbf{F}$ parallel to the line L.

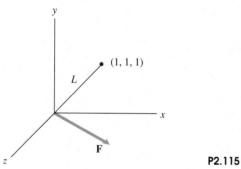

P2.115

2.116 The force $\mathbf{F} = 21\mathbf{i} + 14\mathbf{j}$ (kN). Resolve it into vector components parallel and normal to the line OA.

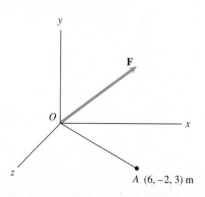

P2.116

2.117 At the instant shown, the Harrier's thrust vector is $\mathbf{T} = 3800\mathbf{i} + 15{,}300\mathbf{j} - 1800\mathbf{k}$ (lb), and its velocity vector is $\mathbf{v} = 24\mathbf{i} + 6\mathbf{j} - 2\mathbf{k}$ (ft/s). Resolve $\mathbf{T}$ into vector components parallel and normal to $\mathbf{v}$. (These are the components of the airplane's thrust parallel and normal to the direction of its motion.)

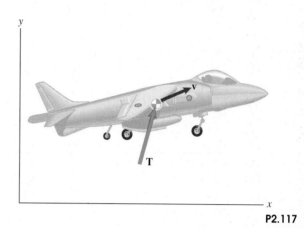

P2.117

2.118 Cables extend from A to B and from A to C. The cable AC exerts a 1000-lb force $\mathbf{F}$ at A.

(a) What is the angle between the cables AB and AC?
(b) Determine the vector component of $\mathbf{F}$ parallel to the cable AB.

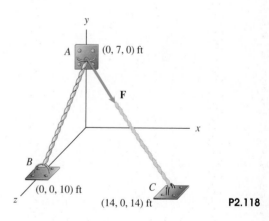

P2.118

2.119 Consider the cables AB and AC shown in Problem 2.118. Let $\mathbf{r}_{AB}$ be the position vector from point A to point B. Determine the vector component of $\mathbf{r}_{AB}$ parallel to the cable AC.

2.120 Consider the cables AB and AC shown in Problem 2.118. Use the dot product to determine the perpendicular distance from point B to the cable AC. That is, determine the length of the shortest straight line from point B to a point on the cable AC.

2.121 The rope AB exerts a 50-N force $\mathbf{T}$ on collar A. Determine the vector component of $\mathbf{T}$ parallel to the bar CD.

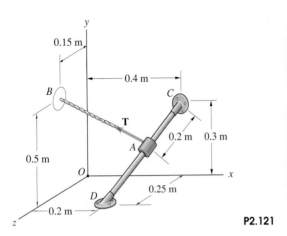

P2.121

2.122 In Problem 2.121, determine the vector component of $\mathbf{T}$ normal to the bar CD.

2.123 The circular bar has a 4-m radius and lies in the x-y plane. The cable AB exerts a 60-N force $\mathbf{T}$ on the collar at A that is directed along the line from A toward B. Determine the vector component of $\mathbf{T}$ in the direction tangent to the axis of the circular bar at A.

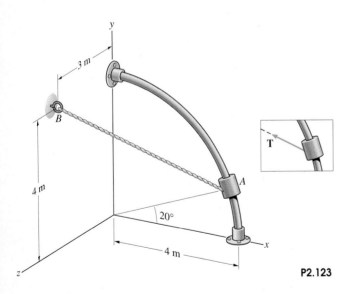

P2.123

2.124 In Problem 2.123, determine the vector component of $\mathbf{T}$ in the direction normal to the axis of the circular bar at A.

2.125 An astronaut in a maneuvering unit approaches a space station. At the present instant, the station informs him that his position relative to the origin of the station's coordinate system is $\mathbf{r}_G = 50\mathbf{i} + 80\mathbf{j} + 180\mathbf{k}$ (m) and his velocity is $\mathbf{v} = -2.2\mathbf{j} - 3.6\mathbf{k}$ (m/s).

The position of an airlock is $\mathbf{r}_A = -12\mathbf{i} + 20\mathbf{k}$ (m). Determine the angle between his velocity vector and the line from his position to the airlock's position.

P2.125

2.126 In Problem 2.125, determine the vector component of the astronaut's velocity parallel to the line from his position to the airlock's position.

2.127 Point P is at longitude 30°W and latitude 45°N on the Atlantic Ocean between Nova Scotia and France. (See Problem 2.90.) Point Q is at longitude 60°E and latitude 20°N in the Arabian Sea. Use the dot product to determine the shortest distance along the surface of the earth from P to Q in terms of the radius of the earth R_E.

Strategy: Use the dot product to determine the angle between the lines OP and OQ; then use the definition of an angle in radians to determine the distance along the surface of the earth from P to Q.

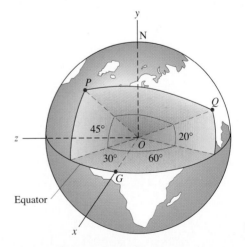

P2.127

2.6 *Cross Products*

Like the dot product, the cross product of two vectors has many applications, including determining the rate of rotation of a fluid particle and calculating the force exerted on a charged particle by a magnetic field. Because of its usefulness for determining moments of forces, the cross product is an indispensable tool in mechanics. In this section we show you how to evaluate cross products and give examples of simple applications.

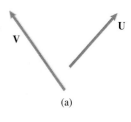

(a)

Definition

Consider two vectors **U** and **V** (Fig. 2.38a). The **cross product** of **U** and **V**, denoted **U** × **V**, is defined by

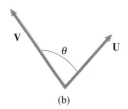

(b)

$$\mathbf{U} \times \mathbf{V} = |\mathbf{U}||\mathbf{V}| \sin \theta \, \mathbf{e}. \tag{2.28}$$

The angle θ is the angle between **U** and **V** when they are placed tail to tail (Fig. 2.38b). The vector **e** is a unit vector defined to be perpendicular to both **U** and **V**. Since this leaves two possibilities for the direction of **e**, the vectors **U**, **V**, and **e** are defined to be a right-handed system. The **right-hand rule** for determining the direction of **e** is shown in Fig. 2.38(c). When you point the four fingers of your right hand in the direction of the vector **U** (the first vector in the cross product) and close your fingers toward the vector **V** (the second vector in the cross product), your thumb points in the direction of **e**.

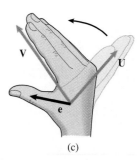

(c)

Figure 2.38

(a) The vectors **U** and **V**.
(b) The angle θ between the vectors when they are placed tail to tail.
(c) Determining the direction of **e** by the right-hand rule.

Because the result of the cross product is a vector, it is sometimes called the vector product. The units of the cross product are the product of the units of the two vectors. Notice that the cross product of two nonzero vectors is equal to zero if and only if the two vectors are parallel.

An interesting property of the cross product is that it is *not* commutative. Eq. (2.28) implies that the magnitude of the vector **U** × **V** is equal to the magnitude of the vector **V** × **U**, but the right-hand rule indicates that they are opposite in direction (Fig. 2.39). That is,

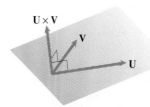

$$\mathbf{U} \times \mathbf{V} = -\mathbf{V} \times \mathbf{U}. \qquad \text{\textbf{The cross product is} \textit{not} \textbf{commutative.}} \tag{2.29}$$

The cross product also satisfies the relations

$$a(\mathbf{U} \times \mathbf{V}) = (a\mathbf{U}) \times \mathbf{V} = \mathbf{U} \times (a\mathbf{V}) \qquad \text{\textbf{The cross product is associative with respect to scalar multiplication.}} \tag{2.30}$$

and

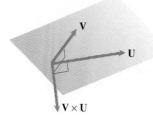

$$\mathbf{U} \times (\mathbf{V} + \mathbf{W}) = (\mathbf{U} \times \mathbf{V}) + (\mathbf{U} \times \mathbf{W}) \qquad \text{\textbf{The cross product is distributive with respect to vector addition.}} \tag{2.31}$$

for any scalar a and vectors **U**, **V**, and **W**.

Figure 2.39

Directions of **U** × **V** and **V** × **U**.

Cross Products in Terms of Components

To obtain an equation for the cross product of two vectors in terms of their components, we must determine the cross products formed from the unit vectors $\mathbf{i}$, $\mathbf{j}$, and $\mathbf{k}$. Since the angle between two identical vectors placed tail to tail is zero,

$$\mathbf{i} \times \mathbf{i} = |\mathbf{i}|\,|\mathbf{i}|\,\sin(0)\,\mathbf{e} = \mathbf{0}.$$

The cross product $\mathbf{i} \times \mathbf{j}$ is

$$\mathbf{i} \times \mathbf{j} = |\mathbf{i}|\,|\mathbf{j}|\,\sin(90)^\circ\,\mathbf{e} = \mathbf{e},$$

where $\mathbf{e}$ is a unit vector perpendicular to $\mathbf{i}$ and $\mathbf{j}$. Either $\mathbf{e} = \mathbf{k}$ or $\mathbf{e} = -\mathbf{k}$. Applying the right-hand rule, we find that $\mathbf{e} = \mathbf{k}$ (Fig. 2.40). Therefore

$$\mathbf{i} \times \mathbf{j} = \mathbf{k}.$$

Continuing in this way, we obtain

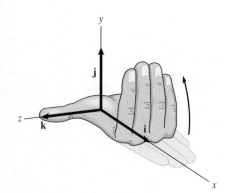

Figure 2.40

The right-hand rule indicates that $\mathbf{i} \times \mathbf{j} = \mathbf{k}$.

$$
\begin{aligned}
\mathbf{i} \times \mathbf{i} &= \ 0, & \mathbf{i} \times \mathbf{j} &= \ \mathbf{k}, & \mathbf{i} \times \mathbf{k} &= -\mathbf{j}, \\
\mathbf{j} \times \mathbf{i} &= -\mathbf{k}, & \mathbf{j} \times \mathbf{j} &= \ 0, & \mathbf{j} \times \mathbf{k} &= \ \mathbf{i}, \\
\mathbf{k} \times \mathbf{i} &= \ \mathbf{j}, & \mathbf{k} \times \mathbf{j} &= -\mathbf{i}, & \mathbf{k} \times \mathbf{k} &= \ 0.
\end{aligned}
\tag{2.32}
$$

You can remember these results easily by arranging the unit vectors in a circle, as shown in Fig. 2.41(a). The cross product of adjacent vectors is equal to the third vector with a positive sign if the order of the vectors in the cross product is the order indicated by the arrows and a negative sign otherwise. For example, in Fig. 2.41(b) you can see that $\mathbf{i} \times \mathbf{j} = \mathbf{k}$, but $\mathbf{i} \times \mathbf{k} = -\mathbf{j}$.

The cross product of two vectors $\mathbf{U}$ and $\mathbf{V}$ expressed in terms of their components is

$$
\begin{aligned}
\mathbf{U} \times \mathbf{V} = {}& (U_x\mathbf{i} + U_y\mathbf{j} + U_z\mathbf{k}) \times (V_x\mathbf{i} + V_y\mathbf{j} + V_z\mathbf{k}) \\
= {}& U_x V_x\,(\mathbf{i} \times \mathbf{i}) + U_x V_y\,(\mathbf{i} \times \mathbf{j}) + U_x V_z\,(\mathbf{i} \times \mathbf{k}) \\
& + U_y V_x\,(\mathbf{j} \times \mathbf{i}) + U_y V_y\,(\mathbf{j} \times \mathbf{j}) + U_y V_z\,(\mathbf{j} \times \mathbf{k}) \\
& + U_z V_x\,(\mathbf{k} \times \mathbf{i}) + U_z V_y\,(\mathbf{k} \times \mathbf{j}) + U_z V_z\,(\mathbf{k} \times \mathbf{k}).
\end{aligned}
$$

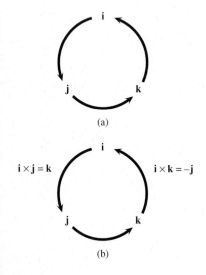

Figure 2.41

(a) Arrange the unit vectors in a circle with arrows to indicate their order.
(b) You can use the circle to determine their cross products.

By substituting Eqs. (2.32) into this expression, we obtain the equation

$$
\begin{aligned}
\mathbf{U} \times \mathbf{V} = {}& (U_y V_z - U_z V_y)\,\mathbf{i} - (U_x V_z - U_z V_x)\,\mathbf{j} \\
& + (U_x V_y - U_y V_x)\,\mathbf{k}.
\end{aligned}
\tag{2.33}
$$

This result can be compactly written as the determinant

$$\tag{2.34}$$

This equation is based on Eqs. (2.32), which we obtained using a right-handed coordinate system. It gives the correct result for the cross product only if you use a right-handed coordinate system to determine the components of **U** and **V**.

Evaluating a 3 × 3 Determinant

You can evaluate a 3 × 3 determinant by repeating its first two columns as shown and evaluating the products of the terms along the six diagonal lines.

$$
\begin{array}{ccc|cc}
\mathbf{i} & \mathbf{j} & \mathbf{k} & \mathbf{i} & \mathbf{j} \\
U_x & U_y & U_z & U_x & U_y \\
V_x & V_y & V_z & V_x & V_y
\end{array}
$$

$$
(-)\quad(-)\quad(-)\quad(+)\quad(+)\quad(+)
$$

Adding the terms obtained from the diagonals that run downward to the right (blue arrows) and subtracting the terms obtained from the diagonals that run downward to the left (red arrows) gives the value of the determinant:

$$
\begin{vmatrix}
\mathbf{i} & \mathbf{j} & \mathbf{k} \\
U_x & U_y & U_z \\
V_x & V_y & V_z
\end{vmatrix}
=
U_y V_z\,\mathbf{i} + U_z V_x\,\mathbf{j} + U_x V_y\,\mathbf{k} \\
- U_y V_x\,\mathbf{k} - U_z V_y\,\mathbf{i} - U_x V_z\,\mathbf{j}.
$$

You can also evaluate a 3 × 3 determinant by expressing it as

$$
\begin{vmatrix}
\mathbf{i} & \mathbf{j} & \mathbf{k} \\
U_x & U_y & U_z \\
V_x & V_y & V_z
\end{vmatrix}
=
\mathbf{i}\begin{vmatrix} U_y & U_z \\ V_y & V_z \end{vmatrix}
- \mathbf{j}\begin{vmatrix} U_x & U_z \\ V_x & V_z \end{vmatrix}
+ \mathbf{k}\begin{vmatrix} U_x & U_y \\ V_x & V_y \end{vmatrix}.
$$

The terms on the right are obtained by multiplying each element of the first row of the 3 × 3 determinant by the 2 × 2 determinant obtained by crossing out that element's row and column. For example, the first element of the first row, **i**, is multiplied by the 2 × 2 determinant

$$
\begin{vmatrix}
\mathbf{i} & \mathbf{j} & \mathbf{k} \\
U_x & U_y & U_z \\
V_x & V_y & V_z
\end{vmatrix}
$$

Be sure to remember that the second term is subtracted. Expanding the 2 × 2 determinants, we obtain the value of the determinant:

$$
\begin{vmatrix}
\mathbf{i} & \mathbf{j} & \mathbf{k} \\
U_x & U_y & U_z \\
V_x & V_y & V_z
\end{vmatrix}
=
(U_y V_z - U_z V_y)\,\mathbf{i} - (U_x V_z - U_z V_x)\,\mathbf{j} \\
+ (U_x V_y - U_y V_x)\,\mathbf{k}.
$$

2.7 Mixed Triple Products

In Chapter 4, when we discuss the moment of a force about a line, we will use an operation called the **mixed triple product**, defined by

$$\mathbf{U} \cdot (\mathbf{V} \times \mathbf{W}). \tag{2.35}$$

In terms of the scalar components of the vectors,

$$\mathbf{U} \cdot (\mathbf{V} \times \mathbf{W}) = (U_x\mathbf{i} + U_y\mathbf{j} + U_z\mathbf{k}) \cdot \begin{vmatrix} \mathbf{i} & \mathbf{j} & \mathbf{k} \\ V_x & V_y & V_z \\ W_x & W_y & W_z \end{vmatrix}$$

$$= (U_x\mathbf{i} + U_y\mathbf{j} + U_z\mathbf{k}) \cdot [(V_yW_z - V_zW_y)\mathbf{i}$$
$$- (V_xW_z - V_zW_x)\mathbf{j} + (V_xW_y - V_yW_x)\mathbf{k}]$$

$$= U_x(V_yW_z - V_zW_y) - U_y(V_xW_z - V_zW_x)$$
$$+ U_z(V_xW_y - V_yW_x).$$

This result can be expressed as the determinant

$$\mathbf{U} \cdot (\mathbf{V} \times \mathbf{W}) = \begin{vmatrix} U_x & U_y & U_z \\ V_x & V_y & V_z \\ W_x & W_y & W_z \end{vmatrix}. \tag{2.36}$$

Interchanging any two of the vectors in the mixed triple product changes the sign but not the absolute value of the result. For example,

$$\mathbf{U} \cdot (\mathbf{V} \times \mathbf{W}) = -\mathbf{W} \cdot (\mathbf{V} \times \mathbf{U}).$$

If the vectors **U**, **V**, and **W** in Fig. 2.42 form a right-handed system, it can be shown that the volume of the parallelepiped equals $\mathbf{U} \cdot (\mathbf{V} \times \mathbf{W})$.

Figure 2.42
Parallelepiped defined by the vectors **U**, **V**, and **W**.

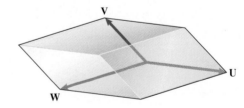

In the following examples we show you how to evaluate cross products and demonstrate simple applications. Example 2.18 demonstrates the use of the cross product to determine a unit vector that is perpendicular to two straight lines and to determine the minimum distance from a point to a straight line.

Example 2.16

Determine the cross product $\mathbf{U} \times \mathbf{V}$ of the vectors $\mathbf{U} = -2\mathbf{i} + \mathbf{j}$ and $\mathbf{V} = 3\mathbf{i} - 4\mathbf{k}$.

STRATEGY

We can evaluate the cross product of the vectors in two ways: by evaluating the cross products of their components term by term and by using Eq. (2.34).

SOLUTION

$$\mathbf{U} \times \mathbf{V} = (-2\mathbf{i} + \mathbf{j}) \times (3\mathbf{i} - 4\mathbf{k})$$

$$= (-2)(3)(\mathbf{i} \times \mathbf{i}) + (-2)(-4)(\mathbf{i} \times \mathbf{k}) + (1)(3)(\mathbf{j} \times \mathbf{i})$$

$$+ (1)(-4)(\mathbf{j} \times \mathbf{k})$$

$$= (-6)(0) + (8)(-\mathbf{j}) + (3)(-\mathbf{k}) + (-4)(\mathbf{i})$$

$$= -4\mathbf{i} - 8\mathbf{j} - 3\mathbf{k}.$$

Using Eq. (2.34), we obtain

$$\mathbf{U} \times \mathbf{V} = \begin{vmatrix} \mathbf{i} & \mathbf{j} & \mathbf{k} \\ U_x & U_y & U_z \\ V_x & V_y & V_z \end{vmatrix} = \begin{vmatrix} \mathbf{i} & \mathbf{j} & \mathbf{k} \\ -2 & 1 & 0 \\ 3 & 0 & -4 \end{vmatrix} = -4\mathbf{i} - 8\mathbf{j} - 3\mathbf{k}.$$

Example 2.17

The magnitude of the force $\mathbf{F}$ in Fig. 2.43 is 100 lb. The magnitude of the vector $\mathbf{r}$ from point O to point A is 8 ft.
(a) Use the definition of the cross product to determine $\mathbf{r} \times \mathbf{F}$.
(b) Use Eq. (2.34) to determine $\mathbf{r} \times \mathbf{F}$.

Figure 2.43

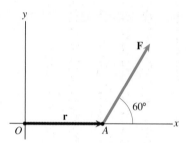

STRATEGY

(a) We know the magnitudes of $\mathbf{r}$ and $\mathbf{F}$ and the angle between them when they are placed tail to tail. Since both vectors lie in the x-y plane, the unit vector $\mathbf{k}$ is perpendicular to both $\mathbf{r}$ and $\mathbf{F}$. We therefore have all the information we need to determine $\mathbf{r} \times \mathbf{F}$ directly from the definition.
(b) We can determine the components of $\mathbf{r}$ and $\mathbf{F}$ and use Eq. (2.34) to determine $\mathbf{r} \times \mathbf{F}$.

SOLUTION

(a) Using the definition of the cross product,

$$\mathbf{r} \times \mathbf{F} = |\mathbf{r}|\,|\mathbf{F}|\,\sin\theta\,\mathbf{e} = (8)(100)\sin 60°\,\mathbf{e} = 693\,\mathbf{e}\ \text{(ft-lb)}.$$

Since $\mathbf{e}$ is defined to be perpendicular to $\mathbf{r}$ and $\mathbf{F}$, either $\mathbf{e} = \mathbf{k}$ or $\mathbf{e} = -\mathbf{k}$. Pointing the fingers of the right hand in the direction of $\mathbf{r}$ and closing them toward $\mathbf{F}$, the right-hand rule indicates that $\mathbf{e} = \mathbf{k}$. Therefore

$$\mathbf{r} \times \mathbf{F} = 693\mathbf{k}\ \text{(ft-lb)}.$$

(b) The vector $\mathbf{r} = 8\mathbf{i}$ (ft). The vector $\mathbf{F}$ in terms of scalar components is

$$\mathbf{F} = 100\cos 60°\,\mathbf{i} + 100\sin 60°\,\mathbf{j}\ \text{(lb)}.$$

From Eq. (2.34),

$$\mathbf{r} \times \mathbf{F} = \begin{vmatrix} \mathbf{i} & \mathbf{j} & \mathbf{k} \\ r_x & r_y & r_z \\ F_x & F_y & F_z \end{vmatrix} = \begin{vmatrix} \mathbf{i} & \mathbf{j} & \mathbf{k} \\ 8 & 0 & 0 \\ 100\cos 60° & 100\sin 60° & 0 \end{vmatrix}$$

$$= (8)(100\sin 60°)\,\mathbf{k} = 693\mathbf{k}\ \text{(ft-lb)}.$$

Example 2.18

Consider the straight lines OA and OB in Fig. 2.44.
(a) Determine the components of a unit vector that is perpendicular to both OA and OB.
(b) What is the minimum distance from point A to the line OB?

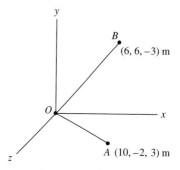

Figure 2.44

STRATEGY

(a) Let $\mathbf{r}_{OA}$ and $\mathbf{r}_{OB}$ be the position vectors from O to A and from O to B (Fig. a). Since the cross product $\mathbf{r}_{OA} \times \mathbf{r}_{OB}$ is perpendicular to $\mathbf{r}_{OA}$ and $\mathbf{r}_{OB}$, we will determine it and divide it by its magnitude to obtain a unit vector perpendicular to the lines OA and OB.

(b) The minimum distance from A to the line OB is the length d of the straight line from A to OB that is perpendicular to OB (Fig. b). We can see that $d = |\mathbf{r}_{OA}| \sin\theta$, where θ is the angle between $\mathbf{r}_{OA}$ and $\mathbf{r}_{OB}$. From the definition of the cross product, the magnitude of $\mathbf{r}_{OA} \times \mathbf{r}_{OB}$ is $|\mathbf{r}_{OA}||\mathbf{r}_{OB}| \sin\theta$, so we can determine d by dividing the magnitude of $\mathbf{r}_{OA} \times \mathbf{r}_{OB}$ by the magnitude of $\mathbf{r}_{OB}$.

SOLUTION

(a) The components of $\mathbf{r}_{OA}$ and $\mathbf{r}_{OB}$ are

$$\mathbf{r}_{OA} = 10\mathbf{i} - 2\mathbf{j} + 3\mathbf{k} \ (\text{m}),$$

$$\mathbf{r}_{OB} = 6\mathbf{i} + 6\mathbf{j} - 3\mathbf{k} \ (\text{m}).$$

By using Eq. (2.34), we obtain $\mathbf{r}_{OA} \times \mathbf{r}_{OB}$:

$$\mathbf{r}_{OA} \times \mathbf{r}_{OB} = \begin{vmatrix} \mathbf{i} & \mathbf{j} & \mathbf{k} \\ 10 & -2 & 3 \\ 6 & 6 & -3 \end{vmatrix} = -12\mathbf{i} + 48\mathbf{j} + 72\mathbf{k} \ (\text{m}^2).$$

This vector is perpendicular to $\mathbf{r}_{OA}$ and $\mathbf{r}_{OB}$. Dividing it by its magnitude, we obtain a unit vector $\mathbf{e}$ that is perpendicular to the lines OA and OB:

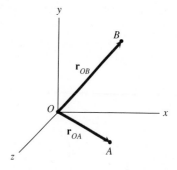

(a) The vectors $\mathbf{r}_{OA}$ and $\mathbf{r}_{OB}$.

$$\mathbf{e} = \frac{\mathbf{r}_{OA} \times \mathbf{r}_{OB}}{|\mathbf{r}_{OA} \times \mathbf{r}_{OB}|} = \frac{-12\mathbf{i} + 48\mathbf{j} + 72\mathbf{k}}{\sqrt{(-12)^2 + (48)^2 + (72)^2}}$$

$$= -0.137\mathbf{i} + 0.549\mathbf{j} + 0.824\mathbf{k}.$$

(b) From Fig. (b), the minimum distance d is

$$d = |\mathbf{r}_{OA}| \sin\theta.$$

The magnitude of $\mathbf{r}_{OA} \times \mathbf{r}_{OB}$ is

$$|\mathbf{r}_{OA} \times \mathbf{r}_{OB}| = |\mathbf{r}_{OA}||\mathbf{r}_{OB}| \sin\theta.$$

Solving this equation for $\sin\theta$, the distance d is

$$d = |\mathbf{r}_{OA}| \left(\frac{|\mathbf{r}_{OA} \times \mathbf{r}_{OB}|}{|\mathbf{r}_{OA}||\mathbf{r}_{OB}|} \right) = \frac{|\mathbf{r}_{OA} \times \mathbf{r}_{OB}|}{|\mathbf{r}_{OB}|}$$

$$= \frac{\sqrt{(-12)^2 + (48)^2 + (72)^2}}{\sqrt{(6)^2 + (6)^2 + (-3)^2}} = 9.71 \ \text{m}.$$

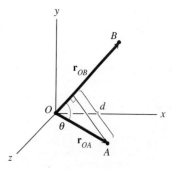

(b) The minimum distance d from A to the line OB.

Example 2.19

The rope CE in Fig. 2.45 exerts a 500-N force $\mathbf{T}$ on the door $ABCD$. What is the magnitude of the component of $\mathbf{T}$ perpendicular to the door?

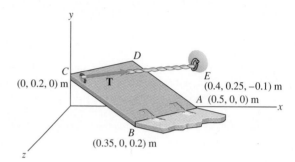

Figure 2.45

STRATEGY

We are given the coordinates of the corners A, B, and C of the door. By taking the cross product of the position vector $\mathbf{r}_{CB}$ from C to B and the position vector $\mathbf{r}_{CA}$ from C to A, we will obtain a vector that is perpendicular to the door. We can divide the resulting vector by its magnitude to obtain a unit vector perpendicular to the door and then apply Eq. (2.26) to determine the component of $\mathbf{T}$ perpendicular to the door.

SOLUTION

The components of $\mathbf{r}_{CB}$ and $\mathbf{r}_{CA}$ are

$$\mathbf{r}_{CB} = 0.35\mathbf{i} - 0.2\mathbf{j} + 0.2\mathbf{k} \text{ (m)},$$

$$\mathbf{r}_{CA} = 0.5\mathbf{i} - 0.2\mathbf{j} \text{ (m)}.$$

Their cross product is

$$\mathbf{r}_{CB} \times \mathbf{r}_{CA} = \begin{vmatrix} \mathbf{i} & \mathbf{j} & \mathbf{k} \\ 0.35 & -0.2 & 0.2 \\ 0.5 & -0.2 & 0 \end{vmatrix} = 0.04\mathbf{i} + 0.1\mathbf{j} + 0.03\mathbf{k} \text{ (m}^2).$$

Dividing this vector by its magnitude, we obtain a unit vector $\mathbf{e}$ that is perpendicular to the door (Fig. a):

$$\mathbf{e} = \frac{\mathbf{r}_{CB} \times \mathbf{r}_{CA}}{|\mathbf{r}_{CB} \times \mathbf{r}_{CA}|} = \frac{0.04\mathbf{i} + 0.1\mathbf{j} + 0.03\mathbf{k}}{\sqrt{(0.04)^2 + (0.1)^2 + (0.03)^2}}$$

$$= 0.358\mathbf{i} + 0.894\mathbf{j} + 0.268\mathbf{k}.$$

To use Eq. (2.26), we must express $\mathbf{T}$ in terms of its scalar components. The position vector from C to E is

$$\mathbf{r}_{CE} = 0.4\mathbf{i} + 0.05\mathbf{j} - 0.1\mathbf{k} \text{ (m)},$$

so we can express the force $\mathbf{T}$ as

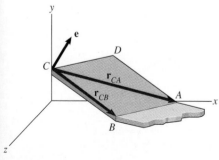

(a) Determining a unit vector perpendicular to the door.

$$\mathbf{T} = |\mathbf{T}|\frac{\mathbf{r}_{CE}}{|\mathbf{r}_{CE}|} = (500)\ \frac{0.4\mathbf{i} + 0.05\mathbf{j} - 0.1\mathbf{k}}{\sqrt{(0.4)^2 + (0.05)^2 + (-0.1)^2}}$$

$$= 481.5\mathbf{i} + 60.2\mathbf{j} - 120.4\mathbf{k}\ (\text{N}).$$

The component of **T** parallel to the unit vector **e**, which is the component perpendicular to the door, is

$$\mathbf{T}_\mathrm{p} = (\mathbf{e} \cdot \mathbf{T})\,\mathbf{e} = [(0.358)(481.5) + (0.894)(60.2) + (0.268)(-120.4)]\,\mathbf{e}$$

$$= 194\mathbf{e}\ (\text{N}).$$

The magnitude of $\mathbf{T}_\mathrm{p}$ is 194 N.

Problems

2.128 Determine the cross product $\mathbf{U} \times \mathbf{V}$ of the vectors $\mathbf{U} = 3\mathbf{i} - 10\mathbf{j}$ and $\mathbf{V} = -6\mathbf{j} + 2\mathbf{k}$.

  Strategy: You can evaluate the cross product of the two vectors either term by term or by using Eq. (2.34). See Example 2.16.

2.129 Two vectors $\mathbf{U} = 3\mathbf{i} + 2\mathbf{j}$ and $\mathbf{V} = 2\mathbf{i} + 4\mathbf{j}$.
(a) What is the cross product $\mathbf{U} \times \mathbf{V}$?
(b) What is the cross product $\mathbf{V} \times \mathbf{U}$?

2.130 What is the cross product $\mathbf{r} \times \mathbf{F}$ of the position vector $\mathbf{r} = 2\mathbf{i} + 2\mathbf{j} + 2\mathbf{k}$ (m) and the force $\mathbf{F} = 20\mathbf{i} - 40\mathbf{k}$ (N)?

2.131 What is the cross product $\mathbf{r} \times \mathbf{F}$ of the position vector $\mathbf{r} = 4\mathbf{i} - 12\mathbf{j} - 3\mathbf{k}$ (ft) and the force $\mathbf{F} = 20\mathbf{i} + 30\mathbf{j} - 10\mathbf{k}$ (lb)?

2.132 Consider the vectors $\mathbf{U} = 6\mathbf{i} - 2\mathbf{j} - 3\mathbf{k}$ and $\mathbf{V} = -12\mathbf{i} + 4\mathbf{j} + 6\mathbf{k}$.
(a) Determine the cross product $\mathbf{U} \times \mathbf{V}$.
(b) What can you conclude about **U** and **V** from the result of (a)?

2.133 The vectors $\mathbf{U} = U_x\,\mathbf{i} - 6\mathbf{j} + U_z\,\mathbf{k}$ and $\mathbf{V} = 2\mathbf{i} - 3\mathbf{j} + \mathbf{k}$ are parallel. Use the cross product to determine U_x and U_z.

2.134 The magnitudes $|\mathbf{U}| = 10$ and $|\mathbf{V}| = 20$.
(a) Use the definition of the cross product to determine $\mathbf{U} \times \mathbf{V}$.
(b) Use the definition of the cross product to determine $\mathbf{V} \times \mathbf{U}$.

(c) Use Eq. (2.34) to determine $\mathbf{U} \times \mathbf{V}$.
(d) Use Eq. (2.34) to determine $\mathbf{V} \times \mathbf{U}$.

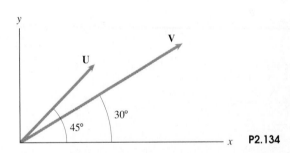

P2.134

2.135 The force $\mathbf{F} = 10\mathbf{i} - 4\mathbf{j}$ (N). Determine the cross product $\mathbf{r}_{AB} \times \mathbf{F}$.

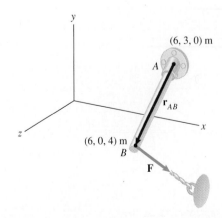

P2.135

2.136 By evaluating the cross product $\mathbf{U} \times \mathbf{V}$, prove the identity $\sin(\theta_1 - \theta_2) = \sin\theta_1 \cos\theta_2 - \cos\theta_1 \sin\theta_2$.

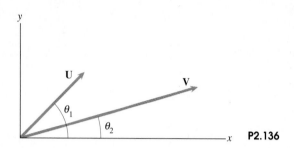

P2.136

2.137 Use the cross product to determine the components of a unit vector $\mathbf{e}$ that is perpendicular to the vectors $\mathbf{U} = 3\mathbf{i} - 10\mathbf{j}$ and $\mathbf{V} = -6\mathbf{j} + 2\mathbf{k}$.

2.138 (a) What is the cross product $\mathbf{r}_{OA} \times \mathbf{r}_{OB}$?
(b) Determine a unit vector $\mathbf{e}$ that is perpendicular to $\mathbf{r}_{OA}$ and $\mathbf{r}_{OB}$.

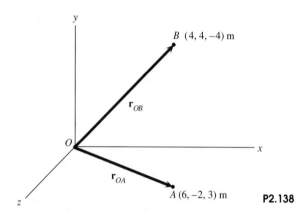

P2.138

2.139 Consider the points O, A, and B shown in Problem 2.138. Use the cross product to determine the minimum distance from point A to the line OB.

2.140 The cable BC exerts a 1000-lb force $\mathbf{F}$ on the hook at B. Determine $\mathbf{r}_{AB} \times \mathbf{F}$.

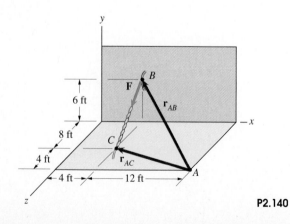

P2.140

2.141 The cable BC shown in Problem 2.140 exerts a 300-lb force $\mathbf{F}$ on the hook at B.
(a) Determine $\mathbf{r}_{AB} \times \mathbf{F}$ and $\mathbf{r}_{AC} \times \mathbf{F}$.
(b) Use the definition of the cross product to explain why the results of (a) are equal.

2.142 The rope AB exerts a 50-N force $\mathbf{T}$ on the collar at A. Let $\mathbf{r}_{CA}$ be the position vector from point C to point A. Determine the cross product $\mathbf{r}_{CA} \times \mathbf{T}$.

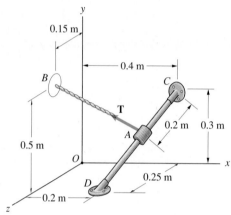

P2.142

2.143 In Problem 2.142, let $\mathbf{r}_{OA}$ and $\mathbf{r}_{OB}$ be the position vectors from the origin O to points A and B. Determine the cross products $\mathbf{r}_{OA} \times \mathbf{T}$ and $\mathbf{r}_{OB} \times \mathbf{T}$.

2.144 The bar AB is 6 m long and is perpendicular to the bars AC and AD. Use the cross product to determine the coordinates x_B, y_B, z_B of point B.

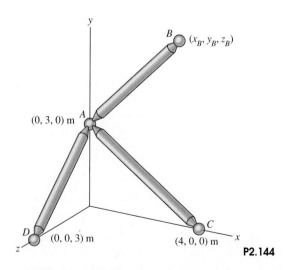

P2.144

2.145 Determine the minimum distance from point P to the plane defined by the three points A, B, and C.

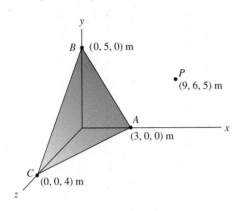

P2.145

2.146 Consider vectors $\mathbf{U} = 3\mathbf{i} - 10\mathbf{j}$, $\mathbf{V} = -6\mathbf{j} + 2\mathbf{k}$, and $\mathbf{W} = 2\mathbf{i} + 6\mathbf{j} - 4\mathbf{k}$.
(a) Determine the value of the mixed triple product $\mathbf{U} \cdot (\mathbf{V} \times \mathbf{W})$ by first evaluating the cross product $\mathbf{V} \times \mathbf{W}$ and then taking the dot product of the result with the vector $\mathbf{U}$.
(b) Determine the value of the mixed triple product $\mathbf{U} \cdot (\mathbf{V} \times \mathbf{W})$ by using Eq. (2.36).

2.147 For the vectors $\mathbf{U} = 6\mathbf{i} + 2\mathbf{j} - 4\mathbf{k}$, $\mathbf{V} = 2\mathbf{i} + 7\mathbf{j}$, and $\mathbf{W} = 3\mathbf{i} + 2\mathbf{k}$, evaluate the following mixed triple products: (a) $\mathbf{U} \cdot (\mathbf{V} \times \mathbf{W})$; (b) $\mathbf{W} \cdot (\mathbf{V} \times \mathbf{U})$; (c) $\mathbf{V} \cdot (\mathbf{W} \times \mathbf{U})$.

2.148 Use the mixed triple product to calculate the volume of the parallelepiped.

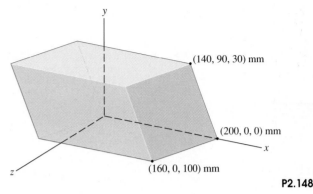

P2.148

2.149 By using Eqs. (2.23) and (2.34), show that

$$\mathbf{U} \cdot (\mathbf{V} \times \mathbf{W}) = \begin{vmatrix} U_x & U_y & U_z \\ V_x & V_y & V_z \\ W_x & W_y & W_z \end{vmatrix}.$$

2.150 The vectors $\mathbf{U} = \mathbf{i} + U_y\mathbf{j} + 4\mathbf{k}$, $\mathbf{V} = 2\mathbf{i} + \mathbf{j} - 2\mathbf{k}$, and $\mathbf{W} = -3\mathbf{i} + \mathbf{j} - 2\mathbf{k}$ are coplanar (they lie in the same plane). What is the component U_y?

Chapter Summary

In this chapter we have defined scalars, vectors, and vector operations. We showed how to express vectors in terms of cartesian components and carry out vector operations in terms of components. We introduced the definitions of the dot and cross products and the mixed triple product and demonstrated some applications of these operations, particularly the use of the dot product to resolve a vector into components parallel and perpendicular to a given direction. In Chapter 3 we will use vector operations to analyze forces acting on objects in equilibrium.

A physical quantity completely described by a real number is a **scalar**. A **vector** has both **magnitude** and **direction**. A vector is represented graphically by an arrow whose length is defined to be proportional to its magnitude.

Rules for Manipulating Vectors

The sum of two vectors is defined by the **triangle rule** (Fig. a) or the equivalent **parallelogram rule** (Fig. b).

The product of a scalar a and a vector $\mathbf{U}$ is a vector $a\mathbf{U}$ with magnitude $|a||\mathbf{U}|$. Its direction is the same as $\mathbf{U}$ when a is positive and opposite to $\mathbf{U}$ when a is negative. The product $(-1)\mathbf{U}$ is written $-\mathbf{U}$ and is called the negative of $\mathbf{U}$. The division of $\mathbf{U}$ by a is the product $(1/a)\mathbf{U}$.

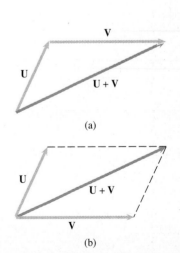

A **unit vector** is a vector whose magnitude is 1. A unit vector specifies a direction. Any vector **U** can be expressed as $|\mathbf{U}|\mathbf{e}$, where **e** is a unit vector with the same direction as **U**. Dividing any vector by its magnitude yields a unit vector with the same direction as the vector.

Cartesian Components

A vector **U** is expressed in terms of **scalar components** as

$$\mathbf{U} = U_x\mathbf{i} + U_y\mathbf{j} + U_z\mathbf{k} \qquad \text{Eq. (2.12)}$$

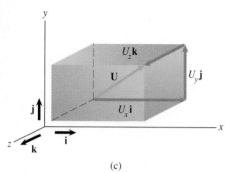

(c)

(Fig. c). The coordinate system is **right-handed** (Fig. d): If the fingers of the right hand are pointed in the positive x direction and then closed toward the positive y direction, the thumb points in the z direction. The magnitude of **U** is

$$|\mathbf{U}| = \sqrt{U_x^2 + U_y^2 + U_z^2}. \qquad \text{Eq. (2.14)}$$

Let θ_x, θ_y, and θ_z be the angles between **U** and the positive coordinate axes (Fig. e). Then the scalar components of **U** are

$$U_x = |\mathbf{U}|\cos\theta_x, \quad U_y = |\mathbf{U}|\cos\theta_y, \quad U_z = |\mathbf{U}|\cos\theta_z. \qquad \text{Eq. (2.15)}$$

The quantities $\cos\theta_x$, $\cos\theta_y$, and $\cos\theta_z$ are the **direction cosines** of **U**. They satisfy the relation

$$\cos^2\theta_x + \cos^2\theta_y + \cos^2\theta_z = 1. \qquad \text{Eq. (2.16)}$$

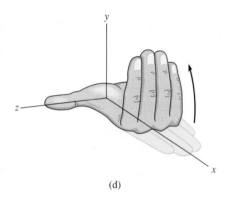

(d)

The **position vector** $\mathbf{r}_{AB}$ from a point A with coordinates (x_A, y_A, z_A) to a point B with coordinates (x_B, y_B, z_B) is given by

$$\mathbf{r}_{AB} = (x_B - x_A)\mathbf{i} + (y_B - y_A)\mathbf{j} + (z_B - z_A)\mathbf{k}. \qquad \text{Eq. (2.17)}$$

Dot Products

The dot product of two vectors **U** and **V** is

$$\mathbf{U} \cdot \mathbf{V} = |\mathbf{U}||\mathbf{V}|\cos\theta, \qquad \text{Eq. (2.18)}$$

where θ is the angle between the vectors when they are placed tail to tail. The dot product of two nonzero vectors is equal to zero if and only if the two vectors are perpendicular.

In terms of scalar components,

$$\mathbf{U} \cdot \mathbf{V} = U_xV_x + U_yV_y + U_zV_z. \qquad \text{Eq. (2.23)}$$

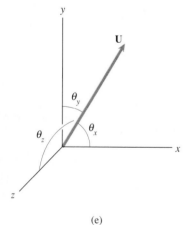

(e)

A vector **U** can be resolved into vector components $\mathbf{U}_p$ and $\mathbf{U}_n$ parallel and normal to a straight line L. In terms of a unit vector **e** that is parallel to L,

$$\mathbf{U}_p = (\mathbf{e} \cdot \mathbf{U})\mathbf{e} \qquad \text{Eq. (2.26)}$$

and

$$\mathbf{U}_n = \mathbf{U} - \mathbf{U}_p. \qquad \text{Eq. (2.27)}$$

Cross Products

The cross product of two vectors **U** and **V** is

$$\mathbf{U} \times \mathbf{V} = |\mathbf{U}||\mathbf{V}| \sin \theta\, \mathbf{e}, \qquad \textbf{Eq. (2.28)}$$

where θ is the angle between the vectors **U** and **V** when they are placed tail to tail and **e** is a unit vector perpendicular to **U** and **V**. The direction of **e** is specified by the **right-hand rule**: When the fingers of the right hand are pointed in the direction of **U** (the first vector in the cross product) and closed toward **V** (the second vector in the cross product), the thumb points in the direction of **e**. The cross product of two nonzero vectors is equal to zero if and only if the two vectors are parallel.

In terms of scalar components,

$$\mathbf{U} \times \mathbf{V} = \begin{vmatrix} \mathbf{i} & \mathbf{j} & \mathbf{k} \\ U_x & U_y & U_z \\ V_x & V_y & V_z \end{vmatrix}. \qquad \textbf{Eq. (2.34)}$$

Mixed Triple Products

The **mixed triple product** is the operation

$$\mathbf{U} \cdot (\mathbf{V} \times \mathbf{W}). \qquad \textbf{Eq. (2.35)}$$

In terms of scalar components,

$$\mathbf{U} \cdot (\mathbf{V} \times \mathbf{W}) = \begin{vmatrix} U_x & U_y & U_z \\ V_x & V_y & V_z \\ W_x & W_y & W_z \end{vmatrix} \qquad \textbf{Eq. (2.36)}$$

Review Problems

2.151 The magnitude of **F** is 8 kN. Express **F** in terms of scalar components.

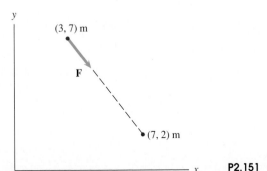

P2.151

2.152 The magnitude of the vertical force **W** is 600 lb, and the magnitude of the force **B** is 1500 lb. Given that $\mathbf{A} + \mathbf{B} + \mathbf{W} = \mathbf{0}$, determine the magnitude of the force **A** and the angle α.

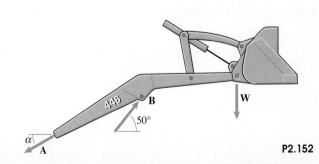

P2.152

2.153 The magnitude of the vertical force vector **A** is 200 lb. If **A** + **B** + **C** = **0**, what are the magnitudes of the force vectors **B** and **C**?

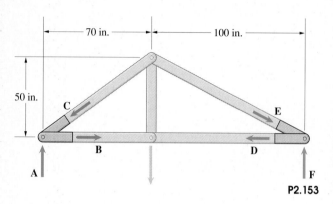

P2.153

2.154 The magnitude of the horizontal force vector **D** in Problem 2.153 is 280 lb. If **D** + **E** + **F** = **0**, what are the magnitudes of the force vectors **E** and **F**?

Refer to the following diagram when solving Problems 2.155 through 2.160.

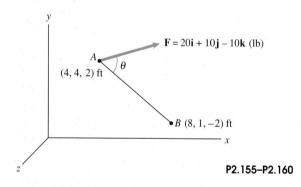

P2.155–P2.160

2.155 What are the direction cosines of **F**?

2.156 Determine the scalar components of a unit vector parallel to line *AB* that points from *A* toward *B*.

2.157 What is the angle θ between the line *AB* and the force **F**?

2.158 Determine the vector component of **F** that is parallel to the line *AB*.

2.159 Determine the vector component of **F** that is normal to the line *AB*.

2.160 Determine the vector $\mathbf{r}_{BA} \times \mathbf{F}$, where $\mathbf{r}_{BA}$ is the position vector from *B* to *A*.

2.161 An engineer calculates that the magnitude of the axial force in one of the beams of a geodesic dome is $|\mathbf{P}| = 7.65$ kN. The cartesian coordinates of the endpoints *A* and *B* of the straight beam are $(-12.4, 22.0, -18.4)$ m and $(-9.2, 24.4, -15.6)$ m, respectively. Express the force **P** in terms of scalar components.

P2.161

2.162 The rope exerts a force of magnitude $|\mathbf{F}| = 200$ lb on the top of the pole at *B*.
(a) Determine the vector $\mathbf{r}_{AB} \times \mathbf{F}$, where $\mathbf{r}_{AB}$ is the position vector from *A* to *B*.
(b) Determine the vector $\mathbf{r}_{AC} \times \mathbf{F}$, where $\mathbf{r}_{AC}$ is the position vector from *A* to *C*.

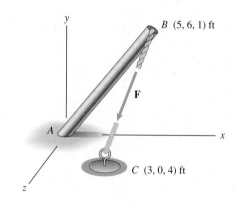

P2.162

2.163 The magnitudes $|\mathbf{F}_A| = 600$ N and $|\mathbf{F}_B| = 400$ N. Determine $\mathbf{F}_A + \mathbf{F}_B$.

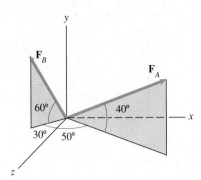

P2.163

2.164 Suppose that the forces $\mathbf{F}_A$ and $\mathbf{F}_B$ shown in Problem 2.163 have the same magnitude and $\mathbf{F}_A \cdot \mathbf{F}_B = 600$ N^2. What are $\mathbf{F}_A$ and $\mathbf{F}_B$?

2.165 The magnitude of the force vector $\mathbf{F}_B$ is 2 kN. Express it in terms of scalar components.

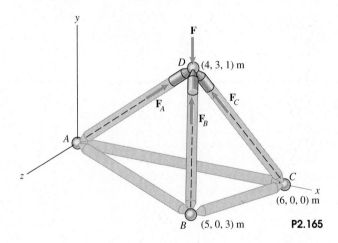

P2.165

2.166 The magnitude of the vertical force vector $\mathbf{F}$ in Problem 2.165 is 6 kN. Determine the vector components of $\mathbf{F}$ parallel and normal to the line from B to D.

2.167 The magnitude of the vertical force vector $\mathbf{F}$ in Problem 2.165 is 6 kN. Given that $\mathbf{F} + \mathbf{F}_A + \mathbf{F}_B + \mathbf{F}_C = \mathbf{0}$, what are the magnitudes of $\mathbf{F}_A$, $\mathbf{F}_B$, and $\mathbf{F}_C$?

2.168 The magnitude of the vertical force $\mathbf{W}$ is 160 N. The direction cosines of the position vector from A to B are $\cos\theta_x = 0.500$, $\cos\theta_y = 0.866$, and $\cos\theta_z = 0$, and the direction cosines of the position vector from B to C are $\cos\theta_x = 0.707$, $\cos\theta_y = 0.619$, and $\cos\theta_z = -0.342$. Point G is the midpoint of the line from B to

C. Determine the vector $\mathbf{r}_{AG} \times \mathbf{W}$, where $\mathbf{r}_{AG}$ is the position vector from A to G.

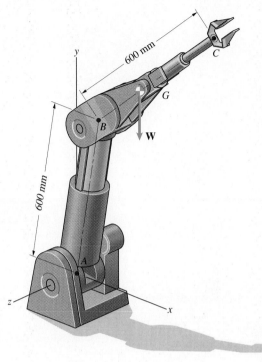

P2.168

2.169 The rope CE exerts a 500-N force $\mathbf{T}$ on the door $ABCD$. Determine the vector component of $\mathbf{T}$ in the direction parallel to the line from point A to point B.

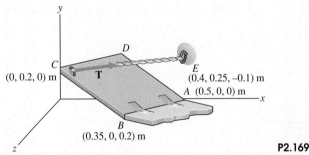

P2.169

2.170 In Problem 2.169, let $\mathbf{r}_{BC}$ be the position vector from point B to point C. Determine the cross product $\mathbf{r}_{BC} \times \mathbf{T}$.

2.171 In Problem 2.169, let $\mathbf{r}_{BC}$ be the position vector from point B to point C, and let $\mathbf{e}_{AB}$ be a unit vector that points from point A toward point B. Evaluate the mixed triple product $\mathbf{e}_{AB} \cdot (\mathbf{r}_{BC} \times \mathbf{T})$.

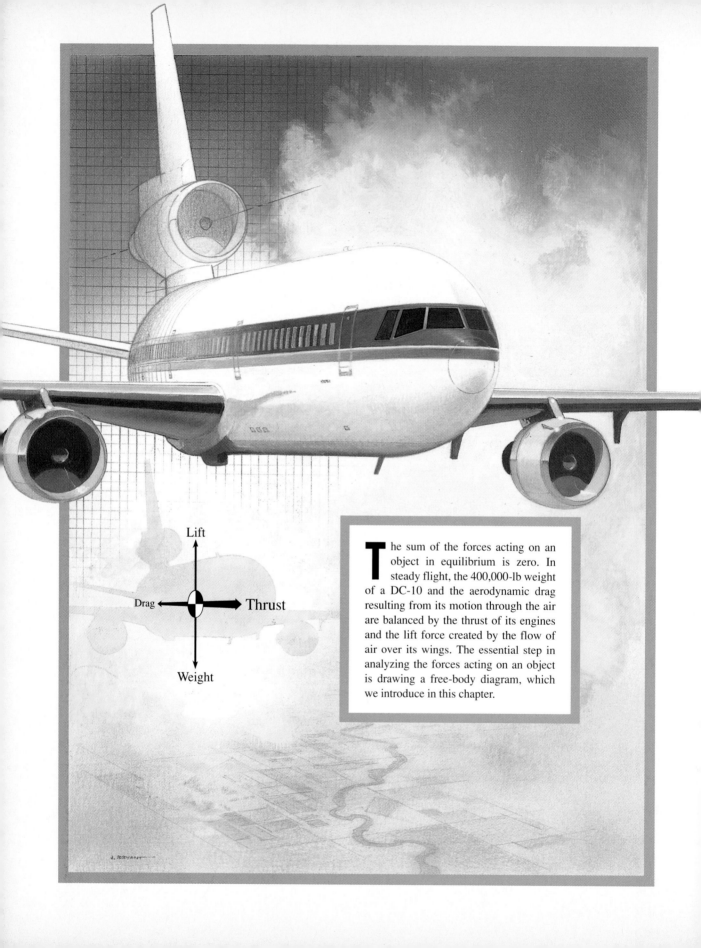

Lift

Drag ← | → **Thrust**

Weight

The sum of the forces acting on an object in equilibrium is zero. In steady flight, the 400,000-lb weight of a DC-10 and the aerodynamic drag resulting from its motion through the air are balanced by the thrust of its engines and the lift force created by the flow of air over its wings. The essential step in analyzing the forces acting on an object is drawing a free-body diagram, which we introduce in this chapter.

Chapter 3

Forces

ENGINEERS design devices to exert and control forces. In ancient times, engineers designed catapults to hurl stones and fortress walls to resist them. Modern engineers design hydraulic cylinders and jet engines to exert forces and structures to resist them. Your first step toward understanding how to work with forces will be learning to determine forces acting on objects in equilibrium.

In Chapter 2 we represented forces by vectors and used vector addition to sum forces. In this chapter we discuss forces in more detail and introduce two of the most important concepts in mechanics, equilibrium and the free-body diagram. We will use free-body diagrams to identify the forces on objects and use equilibrium to determine unknown forces.

3.1 *Types of Forces*

Force is a familiar concept, as is evident from the words *push, pull,* and *lift* used in everyday conversation. In engineering you must deal with different types of forces having a large range of magnitudes (Table 3.1). In this section we introduce some terms used to describe forces and discuss particular forces that occur frequently in engineering applications.

Magnetic tape drives are used to store and retrieve data.

Table 3.1 Magnitudes of some forces

Tension of the tape in a magnetic tape drive	2.2 N (0.5 lb)	
Force of the atmosphere on a 1-m^2 surface at sea level	1.0×10^5 N	$(2.2 \times 10^4$ lb)
Maximum pulling force of a locomotive	9.0×10^5 N	$(2.0 \times 10^5$ lb)
Thrust of the *Energia* rocket	3.9×10^7 N	$(8.8 \times 10^6$ lb)
Pull of the main cables of the Verrazano-Narrows Bridge (New York)	1.1×10^9 N	$(2.5 \times 10^8$ lb)

Terminology

Line of Action   When a force is represented by a vector, the straight line collinear with the vector is called the **line of action** of the force (Fig. 3.1).

The *Energia* booster is one of the components of the joint U.S.-Russian space station program.

Figure 3.1

A force **F** and its line of action.

Systems of Forces   A **system of forces** is simply a particular set of forces. A system of forces is **coplanar**, or **two-dimensional**, if the lines of action of the forces lie in a plane. Otherwise, it is **three-dimensional**. A system of forces is **concurrent** if the lines of action of the forces intersect at a point (Fig. 3.2a) and **parallel** if the lines of action are parallel (Fig. 3.2b).

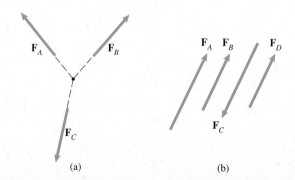

Figure 3.2

(a) Concurrent forces.
(b) Parallel forces.

External and Internal Forces   We say that a given object is subjected to an **external force** if the force is exerted by a different object. When one part of a given object is subjected to a force by another part of the same object, we say it is subjected to an **internal force**. These definitions require that you clearly define the object you are considering. For example, suppose that you are the object. When you are standing, the floor—a different object—exerts an external force on your feet. If you press your hands together, your left hand exerts an internal force on your right hand. However, if your right hand is the object you are considering, the force exerted by your left hand is an external force.

Body and Surface Forces   A force acting on an object is called a **body force** if it acts on the volume of the object and a **surface force** if it acts on its surface. The gravitational force on an object is a body force. A surface force can be exerted on an object by contact with another object. Both body and surface forces can result from electromagnetic effects.

Gravitational Forces

You are aware of the force exerted on an object by the earth's gravity whenever you pick up something heavy. We can represent the gravitational force, or weight, of an object by a vector (Fig. 3.3).

The magnitude of an object's weight is related to its mass m by

$$|\mathbf{W}| = mg,$$

where g is the acceleration due to gravity at sea level. We will use the values $g = 9.81 \text{ m/s}^2$ in SI units and $g = 32.2 \text{ ft/s}^2$ in U.S. Customary units.

Gravitational forces, and also electromagnetic forces, act at a distance. The objects they act on are not necessarily in contact with the objects exerting the forces. In the next section we discuss forces resulting from contacts between objects.

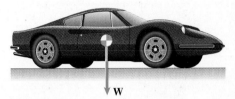

Figure 3.3

Representing an object's weight by a vector.

Contact Forces

Contact forces are the forces that result from contacts between objects. For example, you exert a contact force when you push on a wall (Fig. 3.4a). The surface of your hand exerts a force on the surface of the wall that can be represented by a vector $\mathbf{F}$ (Fig. 3.4b). The wall exerts an equal and opposite force $-\mathbf{F}$ on your hand (Fig. 3.4c). (Recall Newton's third law: The forces exerted on each other by any two particles are equal in magnitude and opposite in direction. If you have any doubt that the wall exerts a force on your hand, try pushing on the wall while standing on roller skates.)

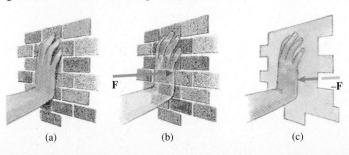

(a)   (b)   (c)

Figure 3.4

(a) Exerting a contact force on a wall by pushing on it.
(b) The vector $\mathbf{F}$ represents the force you exert on the wall.
(c) The wall exerts a force $-\mathbf{F}$ on your hand.

We will be concerned with contact forces exerted on objects by contact with the surfaces of other objects and by ropes, cables, and springs.

Surfaces Consider two plane surfaces in contact (Fig. 3.5a). We represent the force exerted on the right surface by the left surface by the vector **F** in Fig. 3.5(b). We can resolve **F** into a component **N** that is normal to the surface and a component **f** that is parallel to the surface (Fig. 3.5c). The component **N** is called the **normal force**, and the component **f** is called the **friction force**. We sometimes assume that the friction force between two surfaces is negligible in comparison to the normal force, a condition we describe by saying that the surfaces are **smooth**. In this case we show only the normal force (Fig. 3.5d). When the friction force cannot be neglected, we say the surfaces are **rough**.

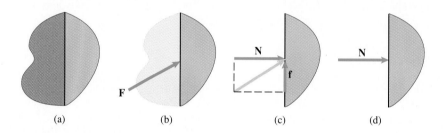

(a)   (b)   (c)   (d)

Figure 3.5

(a) Two plane surfaces in contact.
(b) The force **F** exerted on the right surface.
(c) The force **F** resolved into components normal and parallel to the surface.
(d) Only the normal force is shown when friction is neglected.

If the contacting surfaces are curved (Fig. 3.6a), the normal force and the friction force are perpendicular and parallel to the plane tangent to the surfaces at their point of contact (Fig. 3.6b).

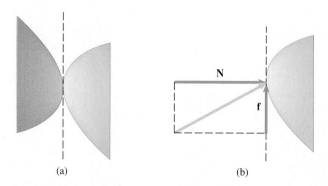

(a)   (b)

Figure 3.6

(a) Curved contacting surfaces. The dashed line indicates the plane tangent to the surfaces at their point of contact.
(b) The normal force and friction force on the right surface.

Ropes and Cables You can exert a contact force on an object by attaching a rope or cable to the object and pulling on it. In Fig. 3.7(a), the crane's cable is attached to a container of building materials. We can represent the force the cable exerts on the container by a vector **T** (Fig. 3.7b). The magnitude of **T** is called the **tension** in the cable, and the line of action of **T** is collinear with the cable. The cable exerts an equal and opposite force −**T** on the crane (Fig. 3.7c).

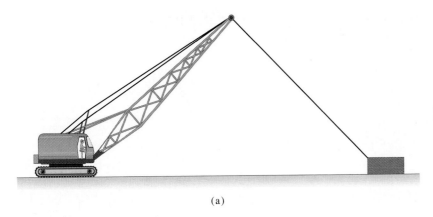

(a)

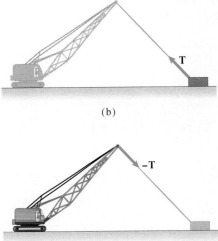

(b)

(c)

Figure 3.7
(a) A crane with its cable attached to a container.
(b) The force **T** exerted on the container by the cable.
(c) The force −**T** exerted on the crane by the cable.

Notice that we have assumed that the cable is straight and that the tension where the cable is connected to the container equals the tension near the crane. This is approximately true if the weight of the cable is small compared to the tension. Otherwise, the cable will sag significantly and the tension will vary along its length. In Chapter 9 we will discuss ropes and cables whose weights are not small in comparison to their tensions. For now, you should assume that ropes and cables are straight and that their tensions are constant along their lengths.

A **pulley** is a wheel with a grooved rim that can be used to change the direction of a rope or cable (Fig. 3.8a). For now, we assume that the tension is the same on both sides of a pulley (Fig. 3.8b). This is true, or at least approximately true, when the pulley can turn freely and the rope or cable either is stationary or turns the pulley at a constant rate.

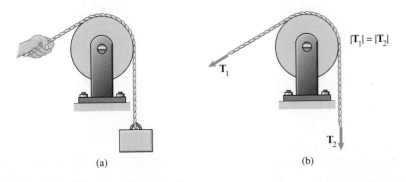

(a)

(b)

Figure 3.8
(a) A pulley changes the direction of a rope or cable.
(b) For now, you should assume that the tensions on each side of the pulley are equal.

Springs Springs are used to exert contact forces in mechanical devices, for example, in the suspensions of cars (Fig. 3.9). Let's consider a coil spring whose unstretched length, the length of the spring when its ends are free, is L_0 (Fig. 3.10a). When the spring is stretched to a length L greater than L_0 (Fig. 3.10b), it pulls on the object to which it is attached with a force $\mathbf{F}$ (Fig. 3.10c). The object exerts an equal and opposite force $-\mathbf{F}$ on the spring (Fig. 3.10d).

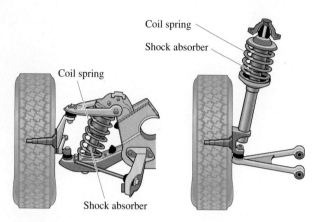

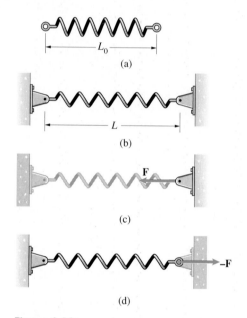

Figure 3.9

Coil springs in car suspensions. The arrangement on the right is called a MacPherson strut.

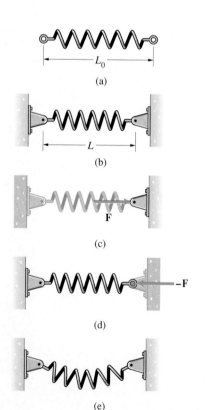

Figure 3.10

(a) A spring of unstretched length L_0.
(b) The spring stretched to a length $L > L_0$.
(c, d) The force $\mathbf{F}$ exerted by the spring and the force $-\mathbf{F}$ on the spring.

When the spring is compressed to a length L less than L_0 (Figs. 3.11a, b), the spring pushes on the object with a force $\mathbf{F}$ and the object exerts an equal and opposite force $-\mathbf{F}$ on the spring (Figs. 3.11c, d). If a spring is compressed too much, it may buckle (Fig. 3.11e). A spring designed to exert a force by being compressed is often provided with lateral support to prevent buckling, for example, by enclosing it in a cylindrical sleeve. In the car suspensions shown in Fig. 3.9, the shock absorbers within the coils prevent the springs from buckling.

Figure 3.11

(a) A spring of length L_0.
(b) The spring compressed to a length $L < L_0$.
(c, d) The spring pushes on an object with a force $\mathbf{F}$, and the object exerts a force $-\mathbf{F}$ on the spring.
(e) A coil spring will buckle if it is compressed too much.

The magnitude of the force exerted by a spring depends on the material it is made of, its design, and how much it is stretched or compressed relative to its unstretched length. When the change in length is not too large compared to the unstretched length, the coil springs commonly used in mechanical devices exert a force approximately proportional to the change in length:

$$|\mathbf{F}| = k\,|L - L_0|. \tag{3.1}$$

Because the force is a linear function of the change in length (Fig. 3.12), a spring that satisfies this relation is called a **linear spring**. The value of the **spring constant** k depends on the material and design of the spring. Its dimensions are (force)/(length). Notice from Eq. (3.1) that k equals the magnitude of the force required to stretch or compress the spring a unit of length.

Suppose that the unstretched length of a spring is $L_0 = 1$ m and $k = 3000$ N/m. If the spring is stretched to a length $L = 1.2$ m, the magnitude of the pull it exerts is

$$k\,|L - L_0| = 3000(1.2 - 1) = 600 \text{ N}.$$

Although coil springs are commonly used in mechanical devices, we are interested in them for a much more general reason. Springs can be used to *model* situations in which forces depend on displacements. For example, the force necessary to bend the steel beam in Fig. 3.13(a) is a linear function of the displacement δ,

$$|\mathbf{F}| = k\delta,$$

if δ is not too large. Therefore we can model the force-deflection behavior of the beam with a linear spring (Fig. 3.13b). This simple example foreshadows a powerful technique: analyzing complicated structures by modeling them as assemblages of small elements connected by linear springs.

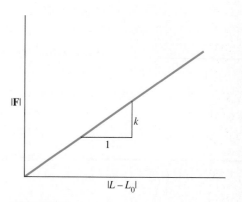

Figure 3.12

The graph of the force exerted by a linear spring as a function of its stretch or compression is a straight line with slope k.

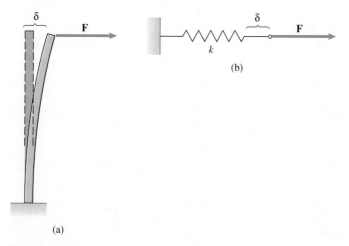

(b)

Figure 3.13

(a) A steel beam deflected by a force.
(b) Modeling the beam's behavior with a linear spring.

3.2 Equilibrium and Free-Body Diagrams

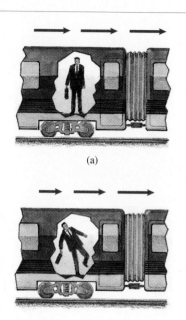

(a)

Figure 3.14

(a) While the train moves at a constant speed, a person standing in the aisle is in equilibrium.

(b) If the train starts to speed up, the person is no longer in equilibrium.

Statics is the study of objects in equilibrium. In everyday conversation, *equilibrium* means an unchanging state—a state of balance. Before we explain precisely what this term means in mechanics, let's consider some examples. Pieces of furniture sitting at rest in a room and a person standing stationary in the room are in equilibrium. If a train travels at constant speed on a straight track, objects that are at rest relative to the train, such as a person standing in the aisle, are in equilibrium (Fig. 3.14a). The person standing in the room and the person standing in the aisle of the train are not accelerating. If the train should start to increase or decrease its speed, however, the person standing in the aisle would no longer be in equilibrium and might lose his balance (Fig. 3.14b).

We say that an object is in **equilibrium** only if each point of the object has the same constant velocity, which is called **steady translation**. The velocity must be measured relative to a frame of reference in which Newton's laws are valid, which is called an **inertial reference frame**. In most engineering applications, the velocity can be measured relative to the earth.

The vector sum of the external forces acting on an object in equilibrium is zero. We will use the symbol $\Sigma \mathbf{F}$ to denote the sum of the external forces. Thus when an object is in equilibrium,

$$\boxed{\Sigma \mathbf{F} = \mathbf{0}.} \tag{3.2}$$

In some situations you can use this **equilibrium equation** to determine unknown forces acting on an object in equilibrium. Your first step will be to draw a free-body diagram of the object to identify the external forces acting on it. The **free-body diagram** is an essential tool in mechanics. It focuses your attention on the object of interest and helps you identify the external forces acting on it. Although in statics we will be concerned only with objects in equilibrium, free-body diagrams are also used in dynamics to analyze the motions of objects.

The free-body diagram is a simple concept. It is a drawing of an object and the external forces acting on it. Otherwise, *nothing* other than the object of interest is included. The drawing shows the object isolated, or freed, from its surroundings. Drawing a free-body diagram involves three steps:

1. *Identify the object you want to isolate.* As the following examples show, your choice is often dictated by particular forces you want to determine.

2. *Draw a sketch of the object isolated from its surroundings, and show relevant dimensions and angles.* Your drawing should be reasonably accurate, but it can omit irrelevant details.

3. *Draw vectors representing all of the external forces acting on the isolated object, and label them.* Don't forget to include the gravitational force if you are not intentionally neglecting it.

You will also need to choose a coordinate system so that you can express the forces on the isolated object in terms of components. Often you will find it convenient to choose the coordinate system before drawing the free-body diagram, but in some situations the best choice of coordinate system will not be apparent until after you have drawn it.

A simple example demonstrates how you can choose free-body diagrams to determine particular forces and also that you must distinguish carefully between external and internal forces. Two stationary blocks of equal weight W are suspended by cables in Fig. 3.15. The system is in equilibrium. Suppose that we want to determine the tensions in the two cables.

To determine the tension in cable AB, we first isolate an "object" consisting of the lower block and part of cable AB (Fig. 3.16a). We then ask ourselves what forces can be exerted on our isolated object by objects not included in the diagram. The earth exerts a gravitational force of magnitude W on the block. Also, where we "cut" cable AB, the cable is subjected to a contact force equal to the tension in the cable (Fig. 3.16b). The arrows in this figure indicate the directions of the forces. The scalar W is the weight of the block and T_{AB} is the tension in cable AB. We assume that the weight of the part of cable AB included in the free-body diagram can be neglected in comparison to the weight of the block.

Since the free-body diagram is in equilibrium, the sum of the external forces equals zero. In terms of a coordinate system with the y axis upward (Fig. 3.16c), we obtain the equilibrium equation

$$\Sigma\,\mathbf{F} = T_{AB}\,\mathbf{j} - W\,\mathbf{j} = (T_{AB} - W)\,\mathbf{j} = \mathbf{0}.$$

Thus the tension in cable AB is $T_{AB} = W$.

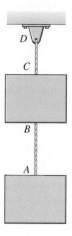

Figure 3.15

Stationary blocks suspended by cables.

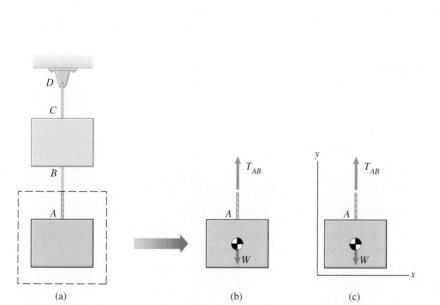

(a) (b) (c)

Figure 3.16

(a) Isolating the lower block and part of cable AB.
(b) Indicating the external forces completes the free-body diagram.
(c) Introducing a coordinate system.

We can determine the tension in cable *CD* by isolating the upper block (Fig. 3.17a). The external forces are the weight of the upper block and the tensions in the two cables (Fig. 3.17b). In this case we obtain the equilibrium equation

$$\Sigma\mathbf{F} = T_{CD}\mathbf{j} - T_{AB}\mathbf{j} - W\mathbf{j} = (T_{CD} - T_{AB} - W)\mathbf{j} = \mathbf{0}.$$

Since $T_{AB} = W$, we find that $T_{CD} = 2W$.

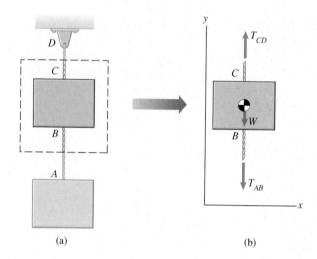

Figure 3.17

(a) Isolating the upper block to determine the tension in cable *CD*.
(b) Free-body diagram of the upper block.

(a)   (b)

We could also have determined the tension in cable *CD* by treating the two blocks and the cable *AB* as a single object (Figs. 3.18a, b). The equilibrium equation is

$$\Sigma\mathbf{F} = T_{CD}\mathbf{j} - W\mathbf{j} - W\mathbf{j} = (T_{CD} - 2W)\mathbf{j} = \mathbf{0},$$

and we again obtain $T_{CD} = 2W$.

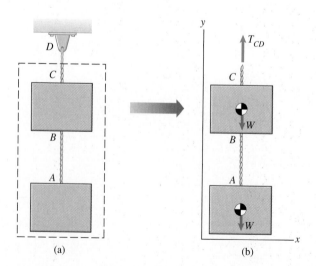

Figure 3.18

(a) An alternative choice for determining the tension in cable *CD*.
(b) Free-body diagram including both blocks and cable *AB*.

(a)   (b)

Why doesn't the tension in cable *AB* appear on the free-body diagram in Fig. 3.18(b)? Remember that only external forces are shown on free-body diagrams. Since cable *AB* is part of the free-body diagram in this case, the forces it exerts on the upper and lower blocks are internal forces.

We have described the procedure for drawing free-body diagrams. In the next section we will draw free-body diagrams of objects subjected to two-dimensional systems of forces and use them to determine unknown forces acting on objects in equilibrium.

3.3 *Two-Dimensional Force Systems*

Suppose that the system of external forces acting on an object in equilibrium is two-dimensional (coplanar). By orienting a coordinate system so that the forces lie in the *x-y* plane, we can express the sum of the external forces as

$$\Sigma \mathbf{F} = (\Sigma F_x)\, \mathbf{i} + (\Sigma F_y)\, \mathbf{j} = \mathbf{0},$$

where ΣF_x and ΣF_y are the sums of the *x* and *y* components of the forces. Since a vector is zero only if each of its components is zero, we obtain two scalar equilibrium equations:

$$\boxed{\Sigma F_x = 0, \qquad \Sigma F_y = 0.} \qquad\qquad (3.3)$$

The sums of the *x* and *y* components of the external forces acting on an object in equilibrium must each equal zero.

The following examples illustrate situations in which you can use Eqs. (3.3) to determine unknown forces acting on objects in equilibrium. Two steps are typically required:

- Draw a free-body diagram—*You must choose an object to isolate that results in a free-body diagram including both known forces and forces you want to determine.*

- Apply the equilibrium equations—*Introduce a coordinate system and use Eqs. (3.3) to obtain equations relating the known and unknown forces.*

Example 3.1

For display at an automobile show, the 1440-kg car in Fig. 3.19 is held in place on the inclined surface by the horizontal cable from A to B. Determine the tension that the cable (and the fixture to which it is connected at B) must support. The car's brakes are not engaged, so the tires exert only normal forces on the inclined surface.

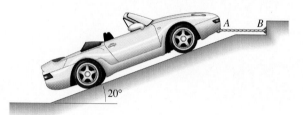

Figure 3.19

STRATEGY

Since the car is in equilibrium, we can draw its free-body diagram and use Eqs. (3.3) to determine the forces exerted on the car by the cable and the inclined surface.

SOLUTION

Draw the Free-Body Diagram We first draw a diagram of the car isolated from its surrounding (Fig. a) and then complete the free-body diagram by showing the force exerted by the car's weight, the force T exerted by the cable, and the normal force N exerted by the inclined surface (Fig. b).

Apply the Equilibrium Equations In Fig. (c), we introduce a coordinate system and resolve the normal force into x and y components. The equilibrium equations are

$$\Sigma F_x = T - N \sin 20° = \mathbf{0},$$

$$\Sigma F_y = N \cos 20° - mg = \mathbf{0}.$$

We can solve the second equilibrium equation for N,

$$N = \frac{mg}{\cos 20°} = \frac{(1440)(9.81)}{\cos 20°} = 1.50 \text{ kN},$$

and then solve the first equilibrium equation for the tension T:

$$T = N \sin 20° = 5.14 \text{ kN}.$$

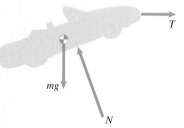

(a) Isolating the car.

(b) The completed free-body diagram shows the known and unknown external forces.

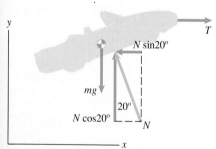

(c) Introducing a coordinate system and resolving N into its components.

Example 3.2

The 3000-lb crate in Fig. 3.20(a) is held in place in the hold of a ship by straps. When the ship has a list angle α (Fig. 3.20b), the supporting straps behave like a linear spring with constant k. The slack straps on the opposite side exert no force on the crate, and friction between the crate and the floor of the hold is negligible. Determine the spring constant k necessary for the straps to hold the crate in equilibrium when $b = 4$ ft and $\alpha = 20°$. (Assume that the straps are unstretched when the crate is centered in the hold.)

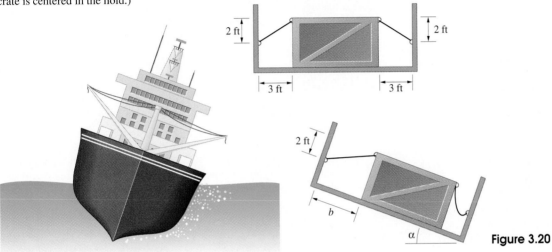

2 ft   2 ft

3 ft   3 ft

2 ft

b

α

Figure 3.20

STRATEGY

We can draw the free-body diagram of the crate and apply the equilibrium equations to determine the force the supporting straps must exert. With the given dimensions, we can calculate the unstretched and stretched lengths of the supporting straps, then use Eq. (3.1) to determine k.

SOLUTION

Draw the Free-Body Diagram   We isolate the crate from its surroundings and complete the free-body diagram by showing its weight W, the force T exerted by the straps, and the normal force N exerted by the floor (Fig. a). The angle $\beta = \arctan(2/4) = 26.6°$.

Apply the Equilibrium Equations   Introducing the coordinate system in Fig. (b) and resolving T and W into x and y components, the sum of the forces in the x direction is

$$\Sigma F_x = W \sin 20° - T \cos 26.6° = 0.$$

Solving, the force the supporting straps must exert is

$$T = \frac{W \sin 20°}{\cos 26.6°} = \frac{(3000) \sin 20°}{\cos 26.6°} = 1147 \text{ lb}.$$

The unstretched length of the straps is $L_0 = \sqrt{2^2 + 3^2} = 3.606$ ft, and the stretched length is $L = \sqrt{2^2 + 4^2} = 4.472$ ft. From the relation

$$T = k(L - L_0): \qquad 1147 = k(4.472 - 3.606),$$

we determine that the necessary spring constant is $k = 1324$ lb/ft.

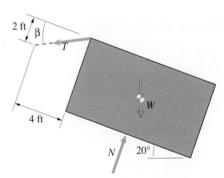

2 ft β

T

W

N 20°

(a) Free-body diagram of the tilted crate.

4 ft

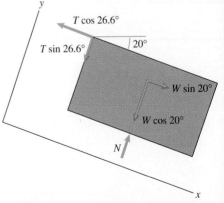

y

$T \cos 26.6°$

$T \sin 26.6°$

20°

$W \sin 20°$

$W \cos 20°$

N

x

(b) Introducing a coordinate system and resolving T and W into their components.

Example 3.3

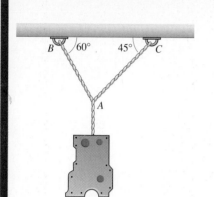

Figure 3.21

The automobile engine block in Fig. 3.21 is suspended by a system of cables. The mass of the block is 200 kg. What are the tensions in cables AB and AC?

STRATEGY

We need a free-body diagram that is subjected to the forces we want to determine. By isolating part of the cable system near point A where the cables are joined, we can obtain a free-body diagram that is subjected to the weight of the block and the unknown tensions in cables AB and AC.

SOLUTION

Draw the Free-Body Diagram Isolating part of the cable system near point A (Fig. a), we obtain a free-body diagram subjected to the weight of the block $W = mg = (200 \text{ kg})(9.81 \text{ m/s}^2) = 1962$ N and the tensions in cables AB and AC (Fig. b).

Apply the Equilibrium Equations We select the coordinate system shown in Fig. (c) and resolve the cable tensions into x and y components. The resulting equilibrium equations are

$$\Sigma F_x = T_{AC} \cos 45° - T_{AB} \cos 60° = 0,$$

$$\Sigma F_y = T_{AC} \sin 45° + T_{AB} \sin 60° - 1962 = 0.$$

Solving these equations, we find that the tensions in the cables are $T_{AB} = 1436$ N and $T_{AC} = 1016$ N.

Alternative Solution: We can determine the tensions in the cables in another way that will also help you visualize the conditions for equilibrium. Since the sum of the three forces acting on our free-body diagram is zero, the vectors form a closed polygon when placed head to tail (Fig. d). You can see that the sum of the vertical components of the tensions supports the weight and that the horizontal components of the tensions must balance each other. The angle of the triangle opposite the weight W is $180° - 30° - 45° = 105°$. By applying the law of sines,

$$\frac{\sin 45°}{T_{AB}} = \frac{\sin 30°}{T_{AC}} = \frac{\sin 105°}{1962},$$

we obtain $T_{AB} = 1436$ N and $T_{AC} = 1016$ N.

DISCUSSION

How were we able to choose a free-body diagram that permitted us to determine the unknown tensions in the cables? There are no definite rules for choosing free-body diagrams. You will learn what to do in many cases from the examples we present, but you will also encounter new situations. It may be necessary to try several free-body diagrams before finding one that provides the information you need. Remember that forces you want to determine should appear as external forces on your free-body diagram, and your objective is to obtain a number of equilibrium equations equal to the number of unknown forces.

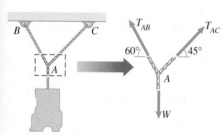

(a) Isolating part of the cable system.   **(b)** The completed free-body diagram.

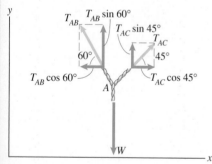

(c) Selecting a coordinate system and resolving the forces into components.

(d) The triangle formed by the sum of the three forces.

Example 3.4

The mass of each pulley of the system in Fig. 3.22 is m, and the mass of the suspended object A is m_A. Determine the force T necessary for the system to be in equilibrium.

Figure 3.22

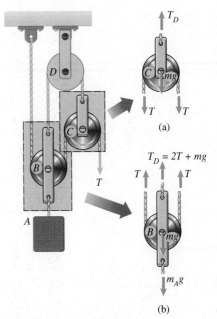

STRATEGY

By drawing free-body diagrams of the individual pulleys and applying equilibrium, we can relate the force T to the weights of the pulleys and the object A.

SOLUTION

We first draw a free-body diagram of the pulley C to which the force T is applied (Fig. a). Notice that we assume the tension in the cable supported by the pulley to equal T on both sides (see Fig. 3.8). From the equilibrium equation

$$T_D - T - T - mg = 0,$$

we determine that the tension in the cable supported by pulley D is

$$T_D = 2T + mg.$$

We now know the tensions in the cables extending from pulleys C and D to pulley B in terms of T. Drawing the free-body diagram of pulley B (Fig. b), we obtain the equilibrium equation

$$T + T + 2T + mg - mg - m_Ag = 0.$$

Solving, we obtain $T = m_Ag/4$.

(a) Free-body diagram of pulley C.
(b) Free-body diagram of pulley B.

Example 3.5

Application to Engineering

Steady Flight

The forces acting on the airplane in Fig. 3.23 are its weight W, the thrust T exerted by its engines, and aerodynamic forces. The dashed line indicates the path along which the airplane is moving. The aerodynamic forces are resolved into a component perpendicular to the path, the lift L, and a component parallel to the path, the drag D. The angle γ between the horizontal and the path is called the flight path angle, and α is the angle of attack. If the airplane remains in equilibrium for an interval of time, it is said to be in steady flight. If $\gamma = 6°$, $D = 125$ kN, $L = 680$ kN, and the mass of the airplane is 72 Mg (megagrams), what values of T and α are necessary to maintain steady flight?

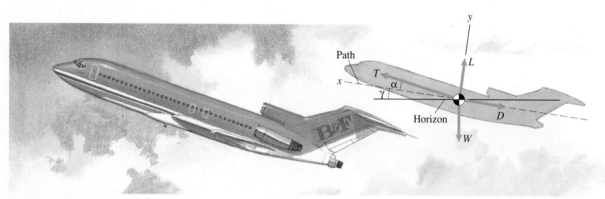

Figure 3.23
External forces on an airplane in flight.

SOLUTION

In terms of the coordinate system in Fig. 3.23, the equilibrium equations are

$$\Sigma F_x = T \cos \alpha - D - W \sin \gamma = 0, \qquad (3.4)$$

$$\Sigma F_y = T \sin \alpha + L - W \cos \gamma = 0. \qquad (3.5)$$

We solve Eq. (3.5) for $\sin \alpha$, solve Eq. (3.4) for $\cos \alpha$, and divide to obtain an equation for $\tan \alpha$:

$$\tan \alpha = \frac{\sin \alpha}{\cos \alpha} = \frac{W \cos \gamma - L}{W \sin \gamma + D}$$

$$= \frac{(72,000)(9.81) \cos 6° - 680,000}{(72,000)(9.81) \sin 6° + 125,000} = 0.113.$$

The angle of attack $\alpha = \arctan (0.113) = 6.44°$. Now we use Eq. (3.4) to determine the thrust:

$$T = \frac{W \sin \gamma + D}{\cos \alpha} = \frac{(72,000)(9.81) \sin 6° + 125,000}{\cos 6.44°} = 200,094 \text{ N.}$$

Notice that the thrust necessary for steady flight is 28 percent of the airplane's weight.

DESIGN ISSUES

In the examples we have considered so far, the values of certain forces acting on an object in equilibrium were given, and our goal was simply to determine the unknown forces by setting the sum of the forces equal to zero. In many situations in engineering, an object in equilibrium is subjected to forces that have different values under different conditions, and this has a profound effect on its design.

When an airplane cruises at constant altitude ($\gamma = 0$), Eqs. (3.4) and (3.5) reduce to

$$T \cos \alpha = D,$$

$$T \sin \alpha + L = W.$$

The horizontal component of the thrust must equal the drag, and the sum of the vertical component of the thrust and the lift must equal the weight. For a fixed value of α, the lift and drag increase as the speed of the airplane increases. A principal design concern is to minimize D at cruising speed in order to minimize the thrust (and consequently the fuel consumption) needed to satisfy the first equilibrium equation. Much of the research on airplane design, including both theoretical analyses and model tests in wind tunnels (Fig. 3.24), is devoted to developing airplane shapes that minimize drag.

Figure 3.25
The Concorde must land with a large angle of attack to generate sufficient lift.

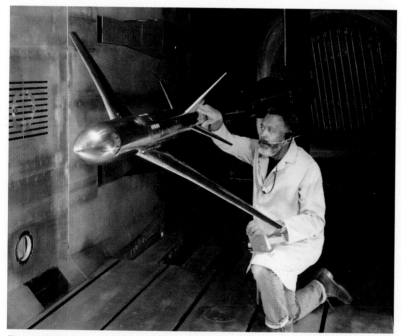

Figure 3.24
Wind tunnels are used to measure the aerodynamic forces on airplane models.

When an airplane is on its landing approach, its speed is much lower and drag is of less concern in design. The lift is also smaller and satisfying Eq. (3.5) becomes the principal design concern. Within limits, the pilot can increase both the terms $T \sin \alpha$ and L by increasing the angle of attack. This is especially evident in the large angle of attack during landing of the Concorde (Fig. 3.25), an airplane for which the engineers had to sacrifice good aerodynamic characteristics during landing to achieve sufficiently low drag at very high cruising speeds. In the case of the F-14 (Fig. 3.26), the engineers used wings with variable sweep to obtain both good lift characteristics at low speed and low drag at high speed.

Figure 3.26
The F-14 with its wings in the takeoff and landing configuration and in the high-speed configuration.

Problems

3.1 The external forces acting on an object in equilibrium are shown. If $F_1 = 75$ N, what are F_2 and F_3?

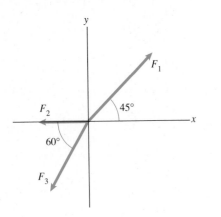

P3.1

3.2 The force $F_1 = 100$ lb and the angle $\alpha = 70°$. The weight of the ring is negligible. Determine the forces F_2 and F_3.

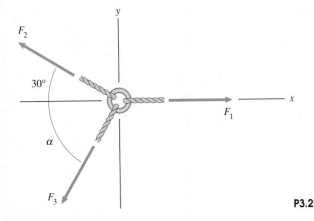

P3.2

3.3 In Problem 3.2, the force $F_1 = 100$ lb. Choose the angle α so that the magnitude of the force F_3 is a minimum. What is the resulting magnitude of F_3?

Strategy: Draw a vector diagram of the sum of the three forces.

3.4 The beam is in equilibrium. If $B = 260$ kN, $C = 100$ kN, and the beam's weight is negligible, what are the forces A_x and A_y?

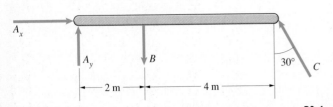

P3.4

3.5 The mass of the beam in Problem 3.4 is 12 kg and it is in equilibrium. The force A_y points upward. If $B = 83$ N and $C = 100$ N, what are the forces A_x and A_y?

3.6 A zoologist estimates that the jaw of a predator, *Martes*, is subjected to a force P as large as 800 N. What forces T and M must be exerted by the temporalis and masseter muscles to support this value of P?

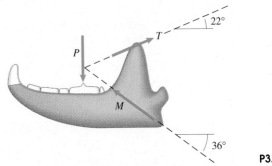

P3.6

3.7 The two springs are identical, with unstretched lengths 250 mm and spring constants $k = 1200$ N/m.
(a) Draw the free-body diagram of block A.
(b) Draw the free-body diagram of block B.
(c) What are the masses of the two blocks?

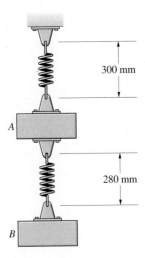

P3.7

3.8 The two springs in Problem 3.7 are identical, with unstretched lengths 250 mm and spring constants k. The sum of the masses of blocks A and B is 10 kg. Determine the value of k and the masses of the two blocks.

3.9 The 200-lb horizontal bar is suspended by the springs A, B, and C. The unstretched lengths of the springs are equal. The spring constants are $k_A = k_C = 400$ lb/ft and $k_B = 300$ lb/ft. What are the tensions in the springs?

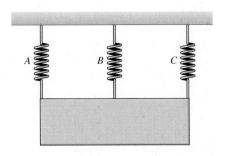

P3.9

3.10 The mass of the crane is 20 Mg (megagrams), and the tension in its cable is 1 kN. The crane's cable is attached to a caisson whose mass is 400 kg. Determine the magnitudes of the normal and friction forces exerted on the crane by the level ground.

Strategy: Draw the free-body diagram of the crane and the part of its cable within the dashed line.

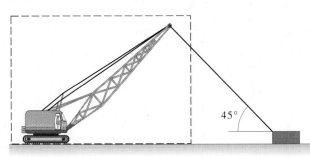

P3.10

3.11 The 2400-lb car is parked on the sloping road with its brakes locked. If $\alpha = 25°$, what are the magnitudes of the total normal and friction forces exerted on the car by the road?

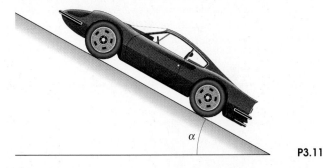

P3.11

3.12 Suppose that the 2400-lb car in Problem 3.11 will remain in equilibrium on the sloping road only if the friction force exerted on the car by the road is not greater than 0.6 times the normal force. What is the largest angle α for which the car will remain in equilibrium?

3.13 The mass of the crate is 40 kg. The inclined surface is rough. The length of the spring is 180 mm, its unstretched length is 200 mm, and the spring constant is $k = 2500$ N/m. What is the

magnitude of the friction force exerted on the crate by the rough surface?

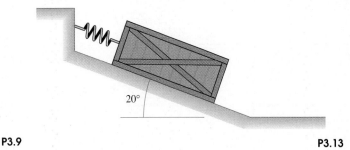

P3.13

3.14 The 600-lb box is held in place on the smooth bed of the dump truck by the rope *AB*. (Remember that "smooth" means that friction is negligible.)
(a) If $\alpha = 25°$, what is the tension in the rope?
(b) If the rope will safely support a tension of 400 lb, what is the maximum allowable value of α?

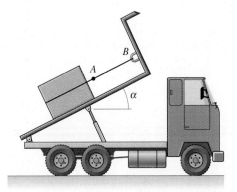

P3.14

3.15 The cable *AB* prevents the 5-Mg (megagram) crate from sliding across the smooth floor of the listing ship's hold. If the cable will safely support a tension of 40 kN, what is the largest safe value of the list angle α?

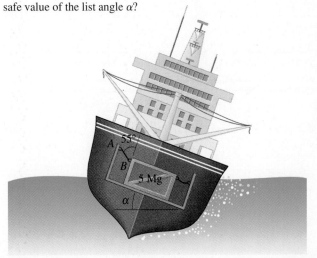

P3.15

3.16 The weights of the two blocks are $W_1 = 200$ lb and $W_2 = 50$ lb. Neglecting friction, determine the force the man must exert to hold the blocks in place.

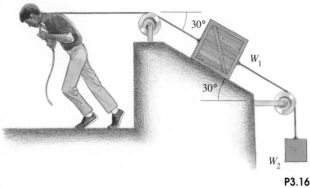

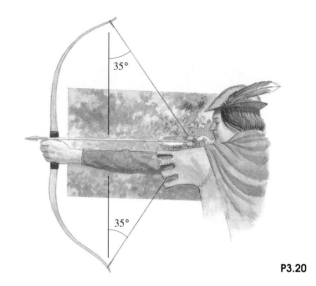

P3.16

3.17 The two springs have the same unstretched length, and the inclined surface is smooth. Show that the magnitudes of the forces exerted by the two springs are

$$F_1 = \frac{W \sin \alpha}{1 + k_2/k_1}, \quad F_2 = \frac{W \sin \alpha}{1 + k_1/k_2}.$$

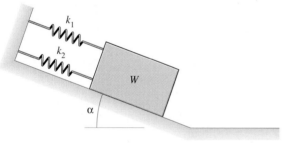

P3.17

3.18 A 10-kg painting is suspended by a wire. If $\alpha = 25°$, what is the tension in the wire?

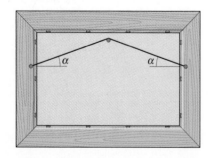

P3.18

3.19 If the wire supporting the suspended painting in Problem 3.18 breaks when the tension exceeds 150 N and you want a 100 percent safety factor (that is, you want the wire to be able to sup-

port twice the actual weight of the painting), what is the smallest value of α you can use?

3.20 A 50-lb pull is required to draw the bow to the position shown. Determine the tension in the bowstring (a) by drawing a free-body diagram of the bowstring and (b) by drawing a free-body diagram of the bow.

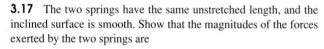

P3.20

3.21 The rocket is suspended by two cables. The mass of the rocket is 45 Mg (megagrams).
(a) What is the tension in the cables when the rocket is in the position shown?
(b) If the rocket is raised by reeling in the two cables by equal amounts, what tension must the cables be able to support (based on the tension required to support the stationary rocket) to raise the rocket 2 m above the position shown?

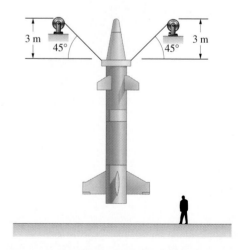

P3.21

3.22 A construction worker holds a 500-lb crate in the position shown. What force must she exert on the cable?

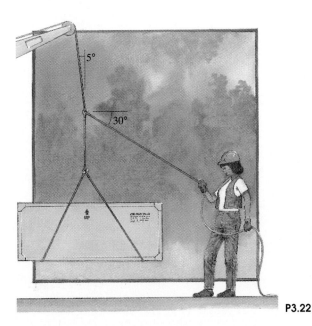

P3.22

3.23 A construction worker on the moon (acceleration due to gravity = 5.32 ft/s^2) holds the same crate described in Problem 3.22 in the position shown. What force must she exert on the cable?

P3.23

3.24 A student on his summer job needs to pull a crate across the floor. Pulling as shown in Fig. (a), he can exert a tension of 60 lb. He finds that the crate doesn't move, so he tries the arrangement in Fig. (b), exerting a vertical force of 60 lb on the rope. What is the

magnitude of the horizontal force he exerts on the crate in each case?

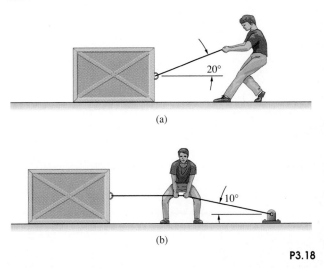

(a)

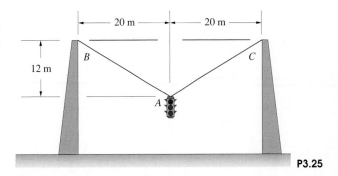

(b)

P3.18

3.25 The 140-kg traffic light is suspended above the street by two cables. What is the tension in the cables?

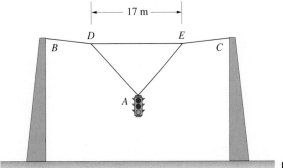

P3.25

3.26 Consider the suspended traffic light in Problem 3.25. To raise the light temporarily during a parade, an engineer wants to connect the 17-m length of cable *DE* to the midpoints of cables *AB* and *AC* as shown. However, for safety considerations, he doesn't want to subject any of the cables to a tension larger than 4 kN. Can he do it?

P3.26

3.27 The mass of the suspended crate is 5 kg. What are the tensions in cables AB and AC?

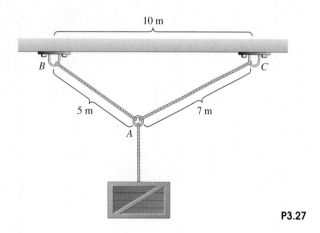

P3.27

3.28 What are the tensions in the upper and lower cables? (Your answers will be in terms of W. Neglect the weight of the pulley.)

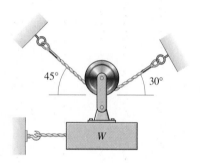

P3.28

3.29 The mass $m_A = 30$ kg. Determine the mass of the pulley B.

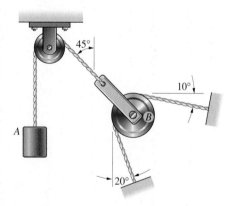

P3.29

3.30 An astronaut candidate conducts experiments on an air-bearing platform. While he carries out calibrations, the platform is held in place by the horizontal tethers AB, AC, and AD. The forces exerted by the tethers are the only horizontal forces acting on the

platform. If the tension in tether AC is 2 N, what are the tensions in the other two tethers?

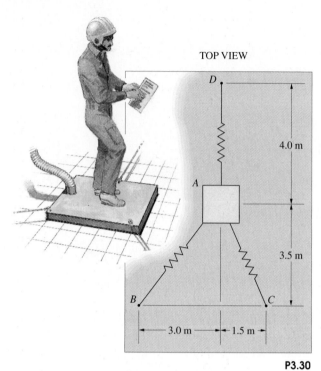

P3.30

3.31 The post anchors a cable that helps support an oil derrick. If $\alpha = 35°$ and $\beta = 50°$, what are the tensions in cables AB and AC? (Your answers will be in terms of the tension T.)

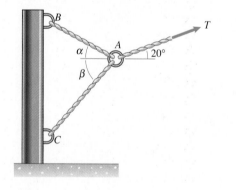

P3.31

3.32 Consider the system in Problem 3.31. The angle $\alpha = 35°$.
(a) Determine the value of the angle β that minimizes the tension in cable AC.
(b) When β has the value determined in (a), what are the tensions in cables AB and AC? (Your answers will be in terms of the tension T.)

3.33 The unstretched length of the spring AB is 660 mm, and the spring constant $k = 1000$ N/m. What is the mass of the suspended object?

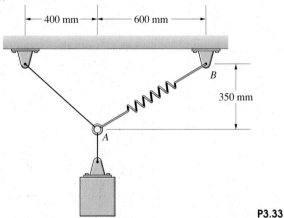

P3.33

3.34 If the system described in Problem 3.33 is on Mars (acceleration due to gravity = 4.02 m/s^2), what is the mass of the suspended object?

3.35 The collar A slides on the smooth vertical bar. The masses $m_A = 20$ kg and $m_B = 10$ kg. When $h = 0.1$ m, the spring is unstretched. When the system is in equilibrium, $h = 0.3$ m. Determine the spring constant k.

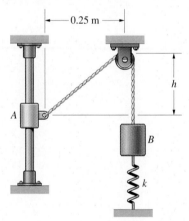

P3.35

3.36 The breeches buoy is used to transfer the person B between two ships. The person is attached to a pulley that rolls on the overhead cable. The total weight of the person and the buoy is 250 lb. What tension in the horizontal line AB is necessary to hold the person in equilibrium in the position shown?

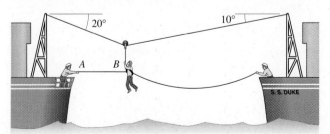

P3.36

3.37 The system of cables suspends a 1000-lb bank of lights above a movie set. Determine the tensions in cables AB, CD, and CE.

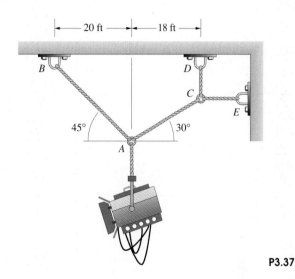

P3.37

3.38 Consider the 1000-lb bank of lights in Problem 3.37. A technician changes the position of the lights by removing the cable CE. What is the tension in cable AB after the change?

3.39 While working on another exhibit, a curator at the Smithsonian Institution pulls the suspended *Voyager* aircraft to one side by attaching three horizontal cables as shown. The mass of the aircraft is 1250 kg. Determine the tensions in the cable segments AB, BC, and CD.

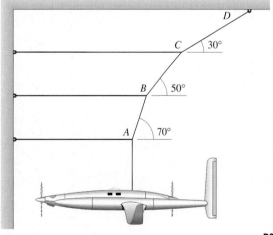

P3.39

3.40 A truck dealer wants to suspend a 4-Mg (megagram) truck as shown for advertising. The distance $b = 15$ m, and the sum of the lengths of the cables AB and BC is 42 m. What are the tensions in the cables?

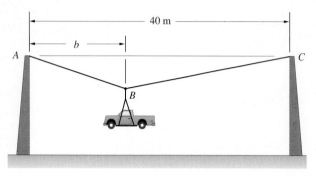

P3.40

3.41 The distance $h = 12$ in., and the tension in cable AD is 200 lb. What are the tensions in cables AB and AC?

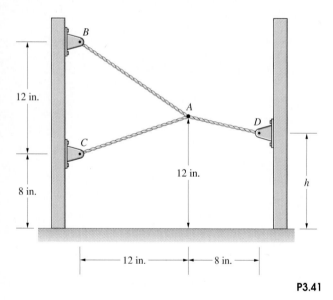

P3.41

3.42 You are designing a cable system to support a suspended object of weight W. Because your design requires points A and B to be placed as shown, you have no control over the angle α, but you can choose the angle β by placing point C wherever you wish. Show that to minimize the tensions in cables AB and BC, you must choose $\beta = \alpha$ if the angle $\alpha \geq 45°$.

Strategy: Draw a diagram of the sum of the forces exerted by the three cables at A.

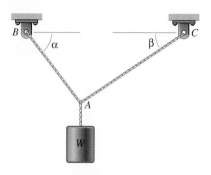

P3.42

3.43 In Problem 3.42, suppose that you have no control over the angle α and you want to design the cable system so that the tension in cable AC is a minimum. What is the required angle β?

3.44 The mass of the box on the left is 30 kg, and the mass of the box on the right is 40 kg. The surfaces are smooth.
(a) What is the tension in the cable?
(b) What is the angle α?

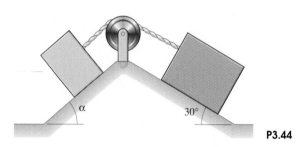

P3.44

3.45 Consider the stationary system in Problem 3.44. The mass of the box on the right is 40 kg and the angle α is 45°.
(a) What is the tension in the cable?
(b) What is the mass of the box on the left?

3.46 The 3000-lb car and the 4600-lb tow truck are stationary. The muddy surface on which the car rests exerts a negligible friction force on the car. What is the tension in the tow cable?

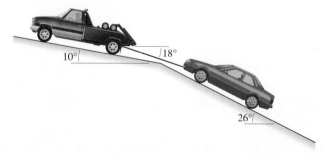

P3.46

3.47 In Problem 3.46, determine the magnitudes of the total normal and friction forces exerted on the truck's tires.

3.48 The 50-lb cylinder rests on two smooth surfaces.
(a) Draw the free-body diagram of the cylinder.
(b) If $\alpha = 30°$, what are the magnitudes of the forces exerted on the cylinder by the left and right surfaces?

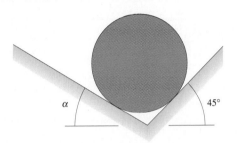

P3.48

3.49 For the 50-lb cylinder in Problem 3.48, obtain an equation for the force exerted on the cylinder by the left surface in terms of the angle α in two ways: (a) using a coordinate system with the y axis vertical, (b) using a coordinate system with the y axis parallel to the right surface.

3.50 The 50-kg sphere is at rest on the smooth horizontal surface. The horizontal force $F = 500$ N. What is the normal force exerted on the sphere by the surface?

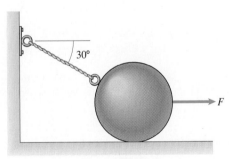

P3.50

3.51 Consider the stationary sphere in Problem 3.50.
(a) Draw a graph of the normal force exerted on the sphere by the surface as a function of the force F from $F = 0$ to $F = 1$ kN.
(b) In the result of (a), notice that the normal force decreases to zero and becomes negative as F increases. What does that mean?

3.52 The 3175-lb car is moving with constant speed on a road with the slope shown. The aerodynamic forces on the car are the drag $D = 120$ lb, which is parallel to the road, and the lift $L = 80$ lb, which is perpendicular to the road. Determine the magnitudes

of the total normal and friction forces exerted on the car by the road.

P3.52

3.53 The device shown is towed beneath a ship to measure water temperature and salinity. The mass of the device is 130 kg. The angle $\alpha = 20°$. The motion of the water relative to the device causes a horizontal drag force D. The hydrostatic pressure distribution in the water exerts a vertical "buoyancy" force B. The magnitude of the buoyancy force is equal to the product of the volume of the device, $V = 0.075$ m^3, and the weight density of the water, $\gamma = 9500$ N/m^3. Determine the drag force D and the tension in the cable.

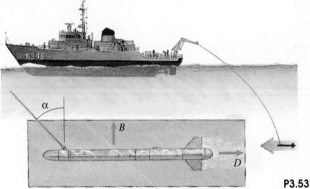

P3.53

3.54 The mass of each pulley of the system is m and the mass of the suspended object A is m_A. Determine the force T necessary for the system to be in equilibrium.

P3.54

3.55 The mass of each pulley of the system is m and the mass of the suspended object A is m_A. Determine the force T necessary for the system to be in equilibrium.

A

T

P3.55

3.56 The system is in equilibrium. What are the coordinates of point A?

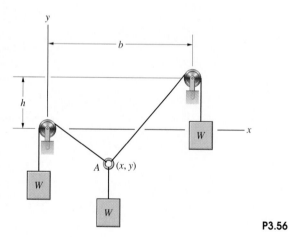

P3.56

3.57 The light fixture of weight W is suspended from a circular arch by a large number N of equally spaced cables. The tension T in each cable is the same. Show that

$$T = \frac{\pi W}{2N}.$$

Strategy: Consider an element of the arch defined by an angle $d\theta$ measured from the point where the cables join:

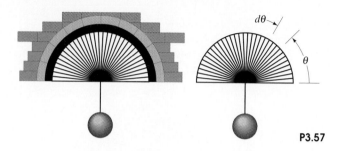

P3.57

Since the total angle described by the arch is π radians, the number of cables attached to the element is $(N/\pi)d\theta$. You can use this result to write the equilibrium equations for the part of the cable system where the cables join.

3.58 The solution to Problem 3.57 is an "asymptotic" result whose accuracy increases as N increases. Determine the exact tension T_{exact} for $N = 3, 5, 9$, and 17, and confirm the numbers in the following table. (For example, for $N = 3$, the cables are attached at $\theta = 0$, $\theta = 90°$, and $\theta = 180°$.)

N	3	5	9	17
$\dfrac{T_{exact}}{\pi W/2N}$	1.91	1.32	1.14	1.07

3.59 The system in Fig. (a) provides lateral support for a load resting on the smooth bed of a truck. The spring constant $k = 100$ lb/ft, and the unstretched length of each spring is 2 ft. When the load is subjected to an effective lateral load F (Fig. b), the distance from the original position of the load to its equilibrium position is $\delta = 1$ ft. What is F?

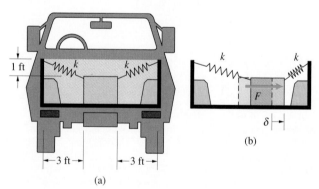

(a)

(b)

P3.59

Problems 3.60–3.62 are related to Example 3.5.

3.60 An airplane is in steady level flight ($\gamma = 0$). Its weight $W = 30{,}000$ lb, thrust $T = 8000$ lb, and angle of attack $\alpha = 10°$. What are the drag D and lift L?

3.61 An airplane is in steady flight, the angle of attack $\alpha = 0$, the thrust-to-drag ratio $T/D = 2$, and the lift-to-drag ratio $L/D = 4$. What is the flight path angle γ?

3.62 An airplane glides in steady flight ($T = 0$), and its lift-to-drag ratio is $L/D = 4$.
(a) What is the flight path angle γ?
(b) If the airplane glides from an altitude of 1000 m to zero altitude, what horizontal distance does it travel?

3.4 *Three-Dimensional Force Systems*

The equilibrium situations we have considered so far have involved only coplanar forces. When the system of external forces acting on an object in equilibrium is three-dimensional, we can express the sum of the external forces as

$$\Sigma \mathbf{F} = (\Sigma F_x)\, \mathbf{i} + (\Sigma F_y)\, \mathbf{j} + (\Sigma F_z)\, \mathbf{k} = \mathbf{0}.$$

Each component of this equation must equal zero, resulting in three scalar equilibrium equations:

$$\boxed{\Sigma F_x = 0, \qquad \Sigma F_y = 0, \qquad \Sigma F_z = 0.} \tag{3.6}$$

The sums of the x, y, and z components of the external forces acting on an object in equilibrium must each equal zero.

In the following examples we discuss situations in which you can use Eqs. (3.6) to determine unknown forces acting on objects in equilibrium.

Example 3.6

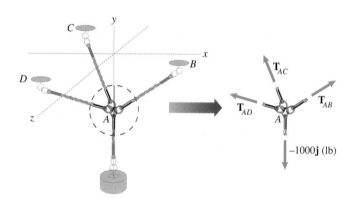

The 1000-lb cylinder in Fig. 3.27 is suspended from the ceiling by cables attached at points B, C, and D. What are the tensions in cables AB, AC, and AD?

Figure 3.27

STRATEGY

We can determine the tensions by the approach we used for similar two-dimensional problems. By isolating part of the cable system near point A, we can obtain a free-body diagram subjected to forces due to the tensions in the cables. Since the sums of the x, y, and z components of the external forces must each equal zero, we obtain three equations for the three unknown tensions.

SOLUTION

Draw the Free-Body Diagram   We isolate part of the cable system near point A (Fig. a) and complete the free-body diagram by showing the forces exerted by the tensions in the cables (Fig. b). The magnitudes of the vectors $\mathbf{T}_{AB}$, $\mathbf{T}_{AC}$, and $\mathbf{T}_{AD}$ are the tensions in cables AB, AC, and AD, respectively.

(a) Isolating part of the cable system.

(b) The completed free-body diagram showing the forces exerted by the tensions in the cables.

Apply the Equilibrium Equations   The sum of the external forces acting on the free-body diagram is

$$\Sigma \mathbf{F} = \mathbf{T}_{AB} + \mathbf{T}_{AC} + \mathbf{T}_{AD} - 1000\mathbf{j} = \mathbf{0}.$$

To solve this equation for the tensions in the cables, we need to express the vectors $\mathbf{T}_{AB}$, $\mathbf{T}_{AC}$, and $\mathbf{T}_{AD}$ in terms of their components.

We first determine the components of a unit vector that points in the direction of the vector $\mathbf{T}_{AB}$. Let $\mathbf{r}_{AB}$ be the position vector from point A to point B (Fig. c):

$$\mathbf{r}_{AB} = (x_B - x_A)\,\mathbf{i} + (y_B - y_A)\,\mathbf{j} + (z_B - z_A)\,\mathbf{k} = 4\mathbf{i} + 4\mathbf{j} + 2\mathbf{k} \text{ (ft)}.$$

Dividing $\mathbf{r}_{AB}$ by its magnitude, we obtain a unit vector that has the same direction as $\mathbf{T}_{AB}$:

$$\mathbf{e}_{AB} = \frac{\mathbf{r}_{AB}}{|\mathbf{r}_{AB}|} = 0.667\mathbf{i} + 0.667\mathbf{j} + 0.333\mathbf{k}.$$

Now we can write the vector $\mathbf{T}_{AB}$ as the product of the tension T_{AB} in cable AB and $\mathbf{e}_{AB}$:

$$\mathbf{T}_{AB} = T_{AB}\mathbf{e}_{AB} = T_{AB}(0.667\mathbf{i} + 0.667\mathbf{j} + 0.333\mathbf{k}).$$

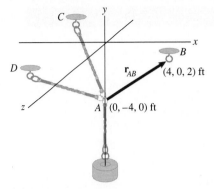

(c) The position vector $\mathbf{r}_{AB}$.

We now express the force vectors $\mathbf{T}_{AC}$ and $\mathbf{T}_{AD}$ in terms of the tensions T_{AC} and T_{AD} in cables AC and AD in the same way. The results are

$$\mathbf{T}_{AC} = T_{AC}(-0.408\mathbf{i} + 0.816\mathbf{j} - 0.408\mathbf{k}),$$
$$\mathbf{T}_{AD} = T_{AD}(-0.514\mathbf{i} + 0.686\mathbf{j} + 0.514\mathbf{k}).$$

We use these expressions to write the sum of the external forces in terms of the tensions T_{AB}, T_{AC}, and T_{AD}:

$$\begin{aligned}
\Sigma \mathbf{F} &= \mathbf{T}_{AB} + \mathbf{T}_{AC} + \mathbf{T}_{AD} - 1000\mathbf{j} \\
&= (0.667T_{AB} - 0.408T_{AC} - 0.514T_{AD})\,\mathbf{i} \\
&\quad + (0.667T_{AB} + 0.816T_{AC} + 0.686T_{AD} - 1000)\,\mathbf{j} \\
&\quad + (0.333T_{AB} - 0.408T_{AC} + 0.514T_{AD})\,\mathbf{k} \\
&= \mathbf{0}.
\end{aligned}$$

The sums of the forces in the x, y, and z directions must each equal zero:

$$\Sigma F_x = 0.667T_{AB} - 0.408T_{AC} - 0.514T_{AD} = 0,$$
$$\Sigma F_y = 0.667T_{AB} + 0.816T_{AC} + 0.686T_{AD} - 1000 = 0,$$
$$\Sigma F_z = 0.333T_{AB} - 0.408T_{AC} + 0.514T_{AD} = 0.$$

Solving these equations, we find that the tensions are $T_{AB} = 529$ lb, $T_{AC} = 648$ lb, and $T_{AD} = 171$ lb.

DISCUSSION

Notice that this example required several of the techniques we covered in Chapter 2. In particular, we had to determine the components of a position vector, divide the position vector by its magnitude to obtain a unit vector with the same direction as a particular force, and express the force in terms of its components by writing it as the product of the unit vector and the magnitude of the force.

Example 3.7

The 100-lb "slider" C in Fig. 3.28 is held in place on the smooth bar by the cable AC. Determine the tension in the cable and the force exerted on the slider by the bar.

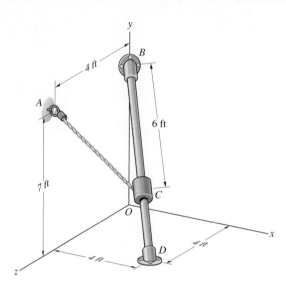

Figure 3.28

STRATEGY

Since we want to determine forces that act on the slider, we need to draw its free-body diagram. The external forces acting on the slider are its weight and the forces exerted on it by the cable and the bar. If we approached this example as we did the previous one, our next step would be to express the forces in terms of their components. However, we don't know the direction of the force exerted on the slider by the bar. Since the smooth bar exerts negligible friction force, we do know that the force exerted by the bar is normal to its axis. Therefore we can eliminate this force from the equation $\Sigma \mathbf{F} = \mathbf{0}$ by taking the dot product of the equation with a unit vector that is parallel to the bar.

SOLUTION

Draw the Free-Body Diagram   We isolate the slider (Fig. a) and complete the free-body diagram by showing the weight of the slider, the force $\mathbf{T}$ exerted by the tension in the cable, and the normal force $\mathbf{N}$ exerted by the bar (Fig. b).

Apply the Equilibrium Equations   The sum of the external forces acting on the free-body diagram is

$$\Sigma \mathbf{F} = \mathbf{T} + \mathbf{N} - 100\mathbf{j} = \mathbf{0}. \tag{3.7}$$

Let $\mathbf{e}_{BD}$ be the unit vector pointing from point B toward point D. Since $\mathbf{N}$ is perpendicular to the bar, $\mathbf{e}_{BD} \cdot \mathbf{N} = 0$. Therefore

$$\mathbf{e}_{BD} \cdot (\Sigma \mathbf{F}) = \mathbf{e}_{BD} \cdot (\mathbf{T} - 100\mathbf{j}) = 0. \tag{3.8}$$

This equation has a simple interpretation: The component of the slider's weight parallel to the bar is balanced by the component of $\mathbf{T}$ parallel to the bar.

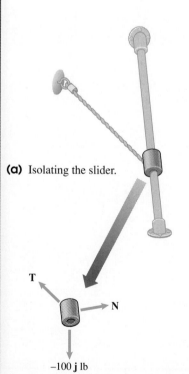

(a) Isolating the slider.

(b) Free-body diagram of the slider showing the forces exerted by its weight, the cable, and the bar.

Determining $\mathbf{e}_{BD}$: We determine the vector from point B to point D,

$$\mathbf{r}_{BD} = (4 - 0)\,\mathbf{i} + (0 - 7)\,\mathbf{j} + (4 - 0)\,\mathbf{k} = 4\mathbf{i} - 7\mathbf{j} + 4\mathbf{k} \text{ (ft)},$$

and divide it by its magnitude to obtain the unit vector $\mathbf{e}_{BD}$:

$$\mathbf{e}_{BD} = \frac{\mathbf{r}_{BD}}{|\mathbf{r}_{BD}|} = \frac{4}{9}\mathbf{i} - \frac{7}{9}\mathbf{j} + \frac{4}{9}\mathbf{k}.$$

Resolving $\mathbf{T}$ *into components:* To express $\mathbf{T}$ in terms of its components, we need to determine the coordinates of the slider C. We can write the vector from B to C in terms of the unit vector $\mathbf{e}_{BD}$,

$$\mathbf{r}_{BC} = 6\mathbf{e}_{BD} = 2.67\mathbf{i} - 4.67\mathbf{j} + 2.67\mathbf{k} \text{ (ft)},$$

and then add it to the vector from the origin O to B to obtain the vector from O to C:

$$\mathbf{r}_{OC} = \mathbf{r}_{OB} + \mathbf{r}_{BC} = 7\mathbf{j} + (2.67\mathbf{i} - 4.67\mathbf{j} + 2.67\mathbf{k})$$
$$= 2.67\mathbf{i} + 2.33\mathbf{j} + 2.67\mathbf{k} \text{ (ft)}.$$

The components of this vector are the coordinates of point C.

Now we can determine a unit vector with the same direction as $\mathbf{T}$. The vector from C to A is

$$\mathbf{r}_{CA} = (0 - 2.67)\,\mathbf{i} + (7 - 2.33)\,\mathbf{j} + (4 - 2.67)\,\mathbf{k}$$
$$= -2.67\mathbf{i} + 4.67\mathbf{j} + 1.33\mathbf{k} \text{ (ft)},$$

and the unit vector that points from point C toward point A is

$$\mathbf{e}_{CA} = \frac{\mathbf{r}_{CA}}{|\mathbf{r}_{CA}|} = -0.482\mathbf{i} + 0.843\mathbf{j} + 0.241\mathbf{k}.$$

Let T be the tension in the cable AC. Then we can write the vector $\mathbf{T}$ as

$$\mathbf{T} = T\mathbf{e}_{CA} = T(-0.482\mathbf{i} + 0.843\mathbf{j} + 0.241\mathbf{k}).$$

Determining $\mathbf{T}$ *and* $\mathbf{N}$: Substituting our expressions for $\mathbf{e}_{BD}$ and $\mathbf{T}$ in terms of their components into Eq. (3.8),

$$\mathbf{e}_{BD} \cdot (\mathbf{T} - 100\mathbf{j})$$

$$= \left[\frac{4}{9}\mathbf{i} - \frac{7}{9}\mathbf{j} + \frac{4}{9}\mathbf{k}\right] \cdot [-0.482T\mathbf{i} + (0.843T - 100)\,\mathbf{j} + 0.241T\mathbf{k}]$$

$$= -0.762T + 77.8 = 0,$$

we obtain the tension $T = 102$ lb.

Now we can determine the force exerted on the slider by the bar by using Eq. (3.7):

$$\mathbf{N} = -\mathbf{T} + 100\mathbf{j} = -102(-0.482\mathbf{i} + 0.843\mathbf{j} + 0.241\mathbf{k}) + 100\mathbf{j}$$
$$= 49.1\mathbf{i} + 14.0\mathbf{j} - 24.6\mathbf{k} \text{ (lb)}.$$

Problems

3.63 The external forces acting on an object in equilibrium are $\mathbf{F}_1 = 120\mathbf{i} - 30\mathbf{j} + 20\mathbf{k}$ (lb), $\mathbf{F}_2 = -100\mathbf{i} + 40\mathbf{k}$ (lb), $\mathbf{F}_3$, and $\mathbf{F}_4$. What is the magnitude of $\mathbf{F}_3 + \mathbf{F}_4$?

3.64 The force $\mathbf{F} = 5\mathbf{i}$ (kN) acts on point A where the cables AB, AC, and AD are joined. What are the tensions in the three cables?

Strategy: Isolate part of the cable system near point A. See Example 3.6.

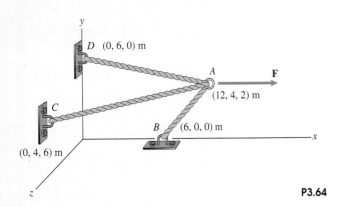

P3.64

3.65 Consider the system of cables in Problem 3.64. The force $\mathbf{F} = F\mathbf{i}$ acts on point A where the cables AB, AC, and AD are joined. The tension in cable AB is 10 kN. Determine the tensions in cables AC and AD and the force F.

3.66 To support the tent, the tension in the rope AB must be 40 lb. What are the tensions in the ropes AC, AD, and AE?

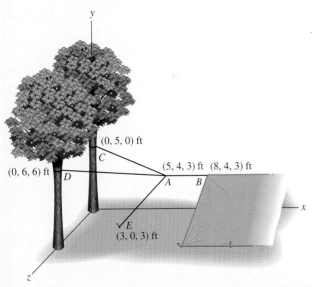

P3.66

3.67 Cable AB is attached to the top of the vertical 3-m post, and its tension is 50 kN. What are the tensions in cables AO, AC, and AD?

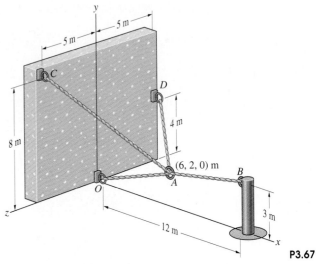

P3.67

3.68 A weather balloon is tethered by the lines AB, AC, and AD. The mass of the balloon and the gas it contains is 80 kg, and the (upward) buoyancy force on the balloon is 1000 N. What are the tensions in the cables?

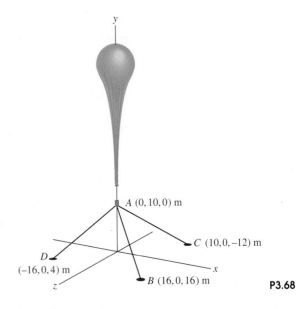

P3.68

3.69 Consider the tethered balloon in Problem 3.68. The mass of the balloon and the gas it contains is 80 kg. The tension in the line AB is measured and determined to be 150 N. What is the buoyancy force?

3.70 The weight of the horizontal wall section is $W = 20,000$ lb. Determine the tensions in the cables AB, AC, and AD.

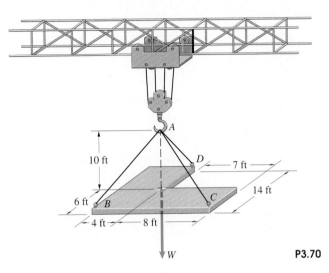

P3.70

3.71 In Problem 3.70, each cable will safely support a tension of 40,000 lb. Based on this criterion, what is the largest safe value of the weight W?

3.72 The climber A is being helped up an icy slope by two friends. His mass is 80 kg, and the direction cosines of the force exerted on him by the slope are $\cos \theta_x = -0.286$, $\cos \theta_y = 0.429$, and $\cos \theta_z = 0.857$. The y axis is vertical. If the climber is in equilibrium in the position shown, what are the tensions in the ropes AB and AC and the magnitude of the force exerted on him by the slope?

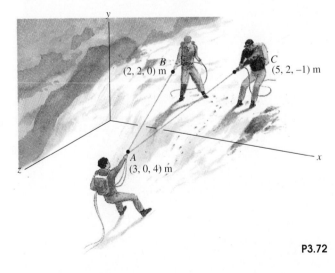

P3.72

3.73 Consider the climber A being helped by his friends in Problem 3.72. To try to make the tensions in the ropes more equal, the friend at B moves to the position $(4, 2, 0)$ m. What are the new tensions in the ropes AB and AC and the magnitude of the force exerted on the climber by the slope?

3.74 The small sphere A weighs 20 lb, and its coordinates are $(4, 0, 6)$ ft. It is supported by two smooth flat plates labeled 1 and 2 and the cable AB. The unit vector $\mathbf{e}_1 = \frac{4}{9}\mathbf{i} + \frac{7}{9}\mathbf{j} + \frac{4}{9}\mathbf{k}$ is perpendicular to plate 1, and the unit vector $\mathbf{e}_2 = -\frac{9}{11}\mathbf{i} + \frac{2}{11}\mathbf{j} + \frac{6}{11}\mathbf{k}$ is perpendicular to plate 2. What is the tension in the cable?

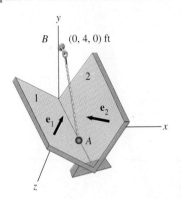

P3.74

3.75 The 1350-kg car is at rest on a plane surface. The unit vector $\mathbf{e}_n = 0.231\mathbf{i} + 0.923\mathbf{j} + 0.308\mathbf{k}$ is perpendicular to the surface. The y axis points upward. Determine the magnitudes of the normal and friction forces the car's wheels exert on the surface.

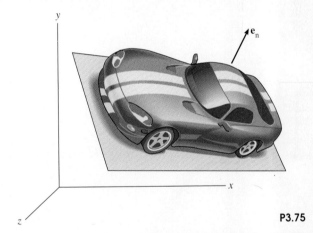

P3.75

3.76 The system shown anchors a stanchion of a cable-suspended roof. If the tension in cable AB is 900 kN, what are the tensions in cables EF and EG?

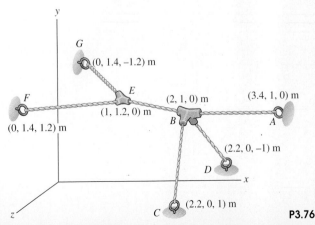

P3.76

3.77 The cables of the system in Problem 3.76 will each safely support a tension of 1500 kN. Based on this criterion, what is the largest safe value of the tension in cable *AB*?

3.78 The 200-kg slider at *A* is held in place on the smooth vertical bar by the cable *AB*.
(a) Determine the tension in the cable.
(b) Determine the force exerted on the slider by the bar.

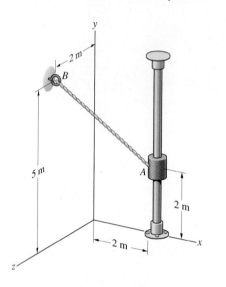

P3.78

3.79 The 100-lb slider at *A* is held in place on the smooth circular bar by the cable *AB*. The circular bar is contained in the *x*-*y* plane.
(a) Determine the tension in the cable.
(b) Determine the normal force exerted on the slider by the bar.

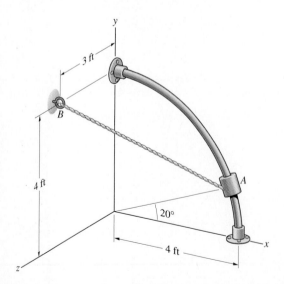

P3.79

3.80 The cable *AB* keeps the 8-kg collar *A* in place on the smooth bar *CD*. The *y* axis points upward. What is the tension in the cable?

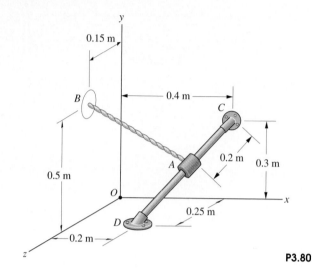

P3.80

3.81 In Problem 3.80, determine the magnitude of the normal force exerted on the collar *A* by the smooth bar.

3.82 The 10-kg collar *A* and 20-kg collar *B* are held in place on the smooth bars by the 3-m cable from *A* to *B* and the force *F* acting on *A*. The force *F* is parallel to the bar. Determine *F*.

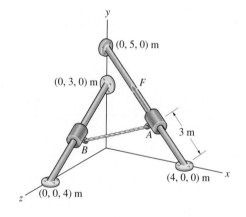

P3.82

Computational Mechanics

The following examples and problems are designed for the use of a programmable calculator or computer. Example 3.8 is similar to previous examples and problems except that the solution must be calculated for a range of input quantities. Example 3.9 leads to an algebraic equation that must be solved numerically.

Example 3.8

The system of cables in Fig. 3.29 is designed to suspend a load with a mass of 1 Mg (megagram). The dimension $b = 2$ m, and the length of cable AB is 1 m. The height of the load can be adjusted by changing the length of cable AC.
(a) Graph the tensions in cables AB and AC for values of the length of cable AC from 1.2 m to 2.2 m.
(b) Cables AB and AC can each safely support a tension equal to the weight of the load. Use the results of (a) to estimate the allowable range of the length of cable AC.

Figure 3.29

STRATEGY

By drawing the free-body diagram of the part of the cable system where the cables join, we can determine the tensions in the cables in terms of the length of cable AC.

SOLUTION

(a) Let the lengths of the cables be $L_{AB} = 1$ m and L_{AC}. We can apply the law of cosines to the triangle in Fig. (a) to determine α in terms of L_{AC}:

$$\alpha = \arccos\left(\frac{b^2 + L_{AB}^2 - L_{AC}^2}{2bL_{AB}}\right).$$

Then we can use the law of sines to determine β:

$$\beta = \arcsin\left(\frac{L_{AB}\sin\alpha}{L_{AC}}\right).$$

Draw the Free-Body Diagram We draw the free-body diagram of the part of the cable system where the cables join in Fig. (b), where T_{AB} and T_{AC} are the tensions in the cables.

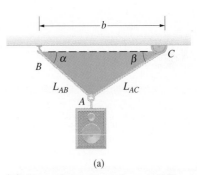

(a) Determining the angles α and β.

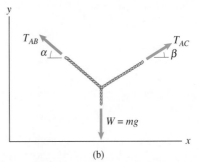

(b) Free-body diagram of part of the cable system.

Apply the Equilibrium Equations Selecting the coordinate system shown in Fig. (b), the equilibrium equations are

$$\Sigma F_x = -T_{AB} \cos \alpha + T_{AC} \cos \beta = 0,$$

$$\Sigma F_y = T_{AB} \sin \alpha + T_{AC} \sin \beta - W = 0.$$

Solving these equations for the cable tensions, we obtain

$$T_{AB} = \frac{W \cos \beta}{\sin \alpha \cos \beta + \cos \alpha \sin \beta},$$

$$T_{AC} = \frac{W \cos \alpha}{\sin \alpha \cos \beta + \cos \alpha \sin \beta}.$$

To compute the results, we input a value of the length L_{AC} and calculate the angle α, then the angle β, and then the tensions T_{AB} and T_{AC}. The resulting values of T_{AB}/W and T_{AC}/W are plotted as functions of L_{AC} in Fig. 3.30.

(b) The allowable range of the length of cable AC is the range over which the tensions in both cables are less than W. From Fig. 3.30 we can see that the tension T_{AB} exceeds W for values of L_{AC} less than about 1.35 m, so the safe range is $L_{AC} > 1.35$ m.

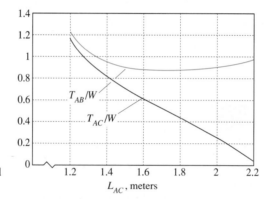

Figure 3.30

Ratios of the cable tensions to the suspended weight as functions of L_{AC}.

Example 3.9

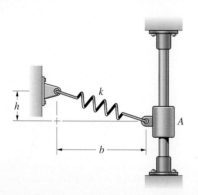

The 12-lb collar A in Fig. 3.31 is held in equilibrium on the smooth vertical bar by the spring. The spring constant $k = 300$ lb/ft, the unstretched length of the spring is $L_0 = b$, and the distance $b = 1$ ft. What is the distance h?

STRATEGY

Both the direction and the magnitude of the force exerted on the collar by the spring depend on h. By drawing the free-body diagram of the collar and applying the equilibrium equations, we can obtain an equation for h.

Figure 3.31

SOLUTION

Draw the Free-Body Diagram   We isolate the collar (Fig. a) and complete the free-body diagram by showing its weight $W = 12$ lb, the force F exerted by the spring, and the normal force N exerted by the bar (Fig. b).

Apply the Equilibrium Equations   Selecting the coordinate system shown in Fig. (b), we obtain the equilibrium equations

$$\Sigma F_x = N - \left(\frac{b}{\sqrt{h^2 + b^2}}\right) F = 0,$$

$$\Sigma F_y = \left(\frac{h}{\sqrt{h^2 + b^2}}\right) F - W = 0.$$

In terms of the length of the spring $L = \sqrt{h^2 + b^2}$, the force exerted by the spring is

$$F = k(L - L_0) = k\left(\sqrt{h^2 + b^2} - b\right).$$

Substituting this expression into the second equilibrium equation, we obtain the equation

$$\left(\frac{h}{\sqrt{h^2 + b^2}}\right) k \left(\sqrt{h^2 + b^2} - b\right) - W = 0.$$

Inserting the values of k, b, and W, we find that the distance h is a root of the equation

$$f(h) = \left(\frac{300h}{\sqrt{h^2 + 1}}\right)\left(\sqrt{h^2 + 1} - 1\right) - 12 = 0. \tag{3.9}$$

How can we solve this nonlinear algebraic equation for h? Some calculators and software are designed to obtain roots of such equations. Another approach is to calculate the value of $f(h)$ for a range of values of h and plot the results, as we have done in Fig. 3.32. From the graph we see that the solution is approximately $h = 0.45$ ft. By examining the computed results near $h = 0.45$ ft,

h (ft)	$f(h)$
0.449	−0.1818
0.450	−0.1094
0.451	−0.0368
0.452	0.0361
0.453	0.1092
0.454	0.1826

we see that the solution (to three significant digits) is $h = 0.452$ ft.

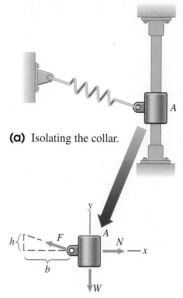

(a) Isolating the collar.

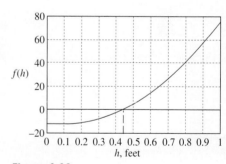

(b) The free-body diagram.

Figure 3.32

Graph of the function $f(h)$.

Problems

3.83 (a) Graph the tensions in cables AB and AC for values of d from $d = 0$ to $d = 1.8$ m.
(b) Each cable will safely support a tension of 1 kN. Use your graph to estimate the acceptable range of values of d.

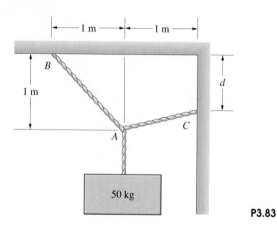

P3.83

3.84 The suspended traffic light weighs 100 lb. The cables AB, BC, AD, and DE are each 11 ft long. Determine the smallest permissible length of the cable BD if the tensions in the cables must not exceed 1000 lb.

Strategy: Graph the tensions in the cables for a range of lengths of the cable BD.

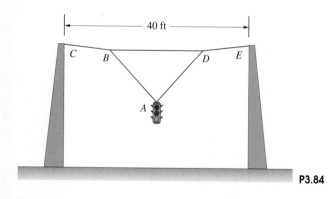

P3.84

3.85 The 2000-lb scoreboard A is suspended above a sports arena by the cables AB and AC. Each cable is 160 ft long. Suppose you want to move the scoreboard out of the way for a tennis match by shortening cable AB while keeping the length of cable AC constant.
(a) Graph the tension in cable AB as a function of its length for values of the length from 142 ft to 160 ft.

(b) Use your graph to estimate how much you can raise the scoreboard relative to its original position if you don't want to subject cable AB to a tension greater than 6000 lb.

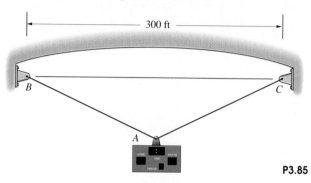

P3.85

3.86 Consider the suspended 4-Mg truck in Problem 3.40. The sum of the lengths of the cables AB and BC is 42 m.
(a) Graph the tensions in cables AB and BC for values of b from zero to 20 m.
(b) Each cable will safely support a tension of 60 kN. Use the results of (a) to estimate the allowable range of the distance b.

3.87 Consider the system in Problem 3.41. The tension in cable AD is 200 lb.
(a) Graph the tensions in cables AB and AC for values of h from zero to 20 in.
(b) For what range of values of h are the tensions in cables AB and AC both positive? What happens when h is not in that range?

3.88 Assigned to design a scale to measure small weights, a group of engineering students devises the system shown. A weight is placed in the pan, and the angle α is measured using the protractor at B. The length of the string AB is 500 mm. The unstretched length of the spring AC is 660 mm, and the spring constant $k = 20$ N/m.
(a) Graph the angle α for values of the suspended weight (neglecting the weight of the pan) from zero to 20 N.
(b) Use the result of (a) to estimate the angle α corresponding to a weight of 10 N.

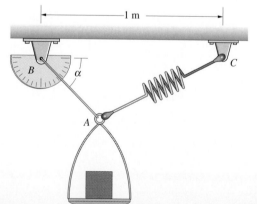

P3.88

3.89   The system in Problem 3.59 provides lateral support for a load resting on the smooth bed of a truck. When the load is subjected to an effective lateral load F (Fig. b), the distance from the original position of the load to its equilibrium position is δ. The unstretched length of each spring is 1 ft. Suppose that the load is subjected to an effective lateral load $F = 200$ lb.
(a) Graph the spring constant k for values of δ from 0.5 ft to 3 ft.
(b) Use the results of (a) to estimate the values of k for $\delta = 1$ ft and $\delta = 2$ ft.

3.90   Consider the tethered balloon in Problem 3.68. The mass of the balloon and the gas it contains is 80 kg, and the buoyancy force is 1000 N. If the lines AB, AC, and AD will each safely support a tension of 500 N and the coordinates of point A are $(0, h, 0)$, what is the smallest height h at which the balloon can be tethered?

3.91   The collar A slides on the smooth vertical bar. The masses $m_A = 20$ kg and $m_B = 10$ kg, and the spring constant $k = 360$ N/m. When $h = 0.2$ m, the spring is unstretched. Determine the value of h when the system is in equilibrium.

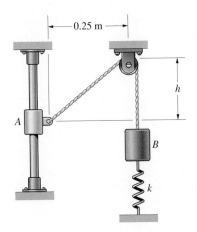

P3.91

3.92   The cable AB keeps the 8-kg collar A in place on the smooth bar CD. The y axis points upward. Determine the distance s from C to the collar A for which the tension in the cable is 150 N.

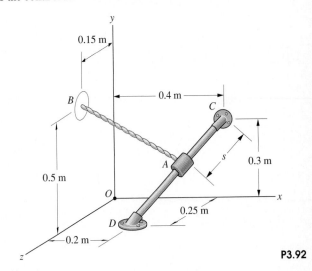

P3.92

3.93   In Problem 3.92, determine the distance s from C to the collar A for which the magnitude of the normal force exerted on the collar A by the smooth bar is 50 N.

3.94   The 10-kg collar A and 20-kg collar B slide on the smooth bars. The cable from A to B is 3 m in length. Determine the value of the distance s in the range $1 \leq s \leq 5$ m for which the system is in equilibrium.

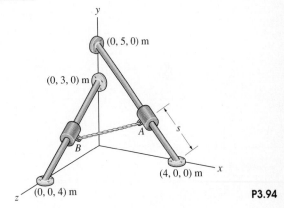

P3.94

Chapter Summary

In this chapter we discussed the forces that occur frequently in engineering applications and introduced two of the most important concepts in mechanics: the free-body diagram and equilibrium. By drawing free-body diagrams and applying the vector techniques developed in Chapter 2, we showed how unknown forces acting on objects in equilibrium can be determined from the condition that the sum of the external forces must equal zero. The sum of the moments of the external forces on an object in equilibrium must also equal zero, and this condition can be used to obtain additional information about unknown forces on objects. We will discuss moments of forces in Chapter 4.

We will then apply equilibrium to individual objects in Chapter 5 and to structures in Chapter 6.

The straight line coincident with a force vector is called the **line of action** of the force. A system of forces is **coplanar**, or **two-dimensional**, if the lines of action of the forces lie in a plane. Otherwise, it is **three-dimensional**. A system of forces is **concurrent** if the lines of action of the forces intersect at a point and **parallel** if the lines of action are parallel.

An object is subjected to an **external force** if the force is exerted by a different object. When one part of an object is subjected to a force by another part of the same object, the force is **internal**.

A **body force** acts on the volume of an object, and a **surface** or **contact force** acts on its surface.

Gravitational Forces

The weight of an object is related to its mass by $W = mg$, where $g = 9.81$ m/s^2 in SI units and $g = 32.2$ ft/s^2 in U.S. Customary units.

Surfaces

Two surfaces in contact exert forces on each other that are equal in magnitude and opposite in direction. Each force can be resolved into the **normal force** and the **friction force**. If the friction force is negligible in comparison to the normal force, the surfaces are said to be **smooth**. Otherwise, they are **rough**.

Ropes and Cables

A rope or cable attached to an object exerts a force on the object whose magnitude is equal to the tension and whose line of action is parallel to the rope or cable at the point of attachment.

A **pulley** is a wheel with a grooved rim that can be used to change the direction of a rope or cable. When a pulley can turn freely and the rope or cable either is stationary or turns the pulley at a constant rate, the tension is approximately the same on both sides of the pulley.

1. Choose an object to isolate.

Springs

The force exerted by a **linear spring** is

$$|\mathbf{F}| = k|L - L_0|, \qquad \textbf{Eq. (3.1)}$$

where k is the **spring constant**, L is the length of the spring, and L_0 is its unstretched length.

2. Draw the isolated object.

Free-Body Diagrams

A free-body diagram is a drawing of an object in which the object is isolated from its surroundings and the external forces acting on the object are shown. Drawing a free-body diagram requires the steps shown in Figs. 1–3. A coordinate system must be chosen to express the forces on the isolated object in terms of components.

3. Show the external forces.

Equilibrium

If an object is in equilibrium, the sum of the external forces acting on it is zero:

$$\Sigma \mathbf{F} = \mathbf{0}. \qquad \textbf{Eq. (3.2)}$$

Expressing the sum of the forces in terms of components,

$$\Sigma \mathbf{F} = (\Sigma F_x)\,\mathbf{i} + (\Sigma F_y)\,\mathbf{j} + (\Sigma F_z)\,\mathbf{k} = \mathbf{0},$$

the sums of the external forces in the x, y, and z directions each equal zero:

$$\Sigma F_x = 0, \qquad \Sigma F_y = 0, \qquad \Sigma F_z = 0. \qquad \textbf{Eqs. (3.6)}$$

Review Problems

3.95 The 100-lb crate is held in place on the smooth surface by the rope AB. Determine the tension in the rope and the magnitude of the normal force exerted on the crate by the surface.

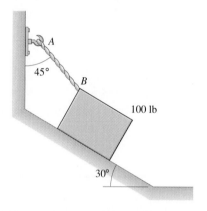

P3.95

3.96 The system shown is called Russell's traction. If the sum of the downward forces exerted at A and B by the patient's leg is 32.2 lb, what is the weight W?

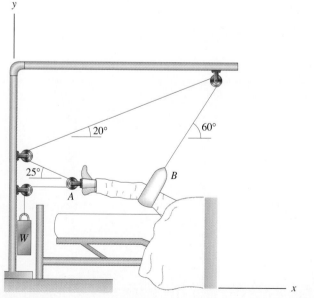

P3.96

3.97 A heavy rope used as a hawser for a cruise ship sags as shown. If it weighs 200 lb, what are the tensions in the rope at A and B?

P3.97

3.98 The cable AB is horizontal, and the box on the right weighs 100 lb. The surfaces are smooth.
(a) What is the tension in the cable?
(b) What is the weight of the box on the left?

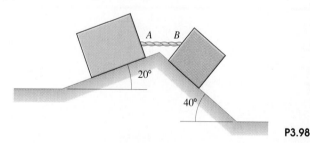

P3.98

3.99 You are designing a cable system to support a suspended object of weight W. The two wires must be identical, and the dimension b is fixed. The ratio of the tension T in each wire to its cross-sectional area A must equal a specified value $T/A = \sigma$. The "cost" of your design is the total volume of material in the two wires, $V = 2A\sqrt{b^2 + h^2}$. Determine the value of h that minimizes the cost.

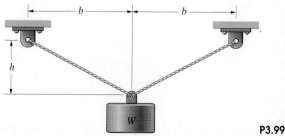

P3.99

3.100 The mass of the suspended object A is m_A and the masses of the pulleys are negligible. Determine the force T necessary for the system to be in equilibrium.

P3.100

3.101 The assembly A, including the pulley, weighs 60 lb. What force F is necessary for the system to be in equilibrium?

P3.101

3.102 The mass of block A is 42 kg, and the mass of block B is 50 kg. The surfaces are smooth. If the blocks are in equilibrium, what is the force F?

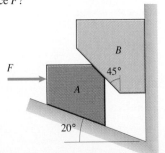

P3.102

3.103 The 680-kg load suspended from the helicopter is in equilibrium. The aerodynamic drag force on the load is horizontal. The y axis is vertical, and cable OA lies in the x-y plane. Determine the magnitude of the drag force and the tension in cable OA.

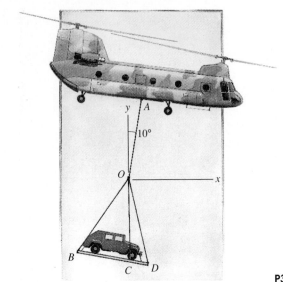

P3.103

3.104 In Problem 3.103, the coordinates of the three cable attachment points B, C, and D are $(-3.3, -4.5, 0)$ m, $(1.1, -5.3, 1)$ m, and $(1.6, -5.4, -1)$ m, respectively. What are the tensions in cables OB, OC, and OD?

3.105 A climber helps his friend up an icy slope. His friend is hauling a box of supplies. If the mass of the friend is 90 kg and the mass of the supplies is 22 kg, what are the tensions in the ropes AB and CD? Assume that the slope is smooth.

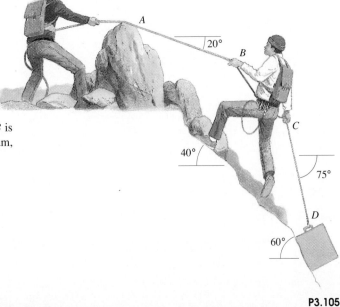

P3.105

3.106 The small sphere of mass m is attached to a string of length L and rests on the smooth surface of a sphere of radius R. Determine the tension in the string in terms of m, L, h, and R.

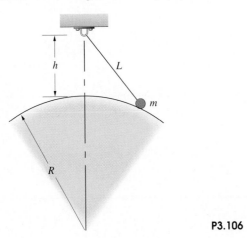

P3.106

3.107 An engineer doing preliminary design studies for a new radio telescope envisions a triangular receiving platform suspended by cables from three equally spaced 40-m towers. The receiving platform has a mass of 20 Mg (megagrams) and is 10 m below the tops of the towers. What tension would the cables be subjected to?

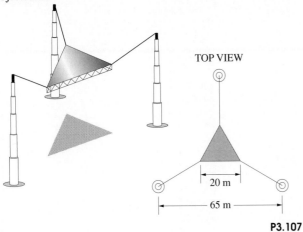

TOP VIEW

P3.107

3.108 The metal disk A weighs 10 lb. It is held in place at the center of the smooth inclined surface by the strings AB and AC. What are the tensions in the strings?

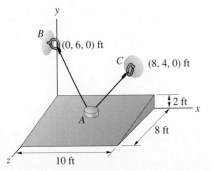

P3.108

3.109 The bulldozer exerts a force $\mathbf{F} = 2\mathbf{i}$ (kip) at A. What are the tensions in cables AB, AC, and AD?

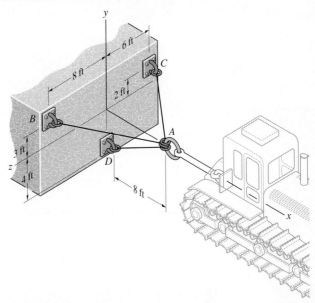

3.110 The 1350-kg car is at rest on a plane surface with its brakes locked. The unit vector $\mathbf{e}_n = 0.231\mathbf{i} + 0.923\mathbf{j} + 0.308\mathbf{k}$ is perpendicular to the surface. The y axis points upward. The direction cosines of the cable from A to B are $\cos\theta_x = -0.816$, $\cos\theta_y = 0.408$, $\cos\theta_z = -0.408$, and the tension in the cable is 1.2 kN. Determine the magnitudes of the normal and friction forces the car's wheels exert on the surface.

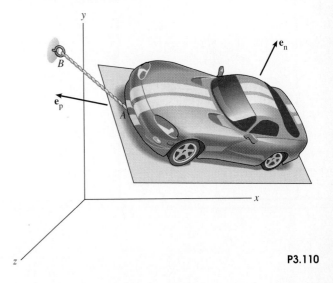

P3.110

3.111 The brakes of the car in Problem 3.110 are released, and the car is held in place on the plane surface by the cable AB. The car's front wheels are aligned so that the tires exert no friction forces parallel to the car's longitudinal axis. The unit vector $\mathbf{e}_p = -0.941\mathbf{i} + 0.131\mathbf{j} + 0.314\mathbf{k}$ is parallel to the plane surface and aligned with the car's longitudinal axis. What is the tension in the cable?

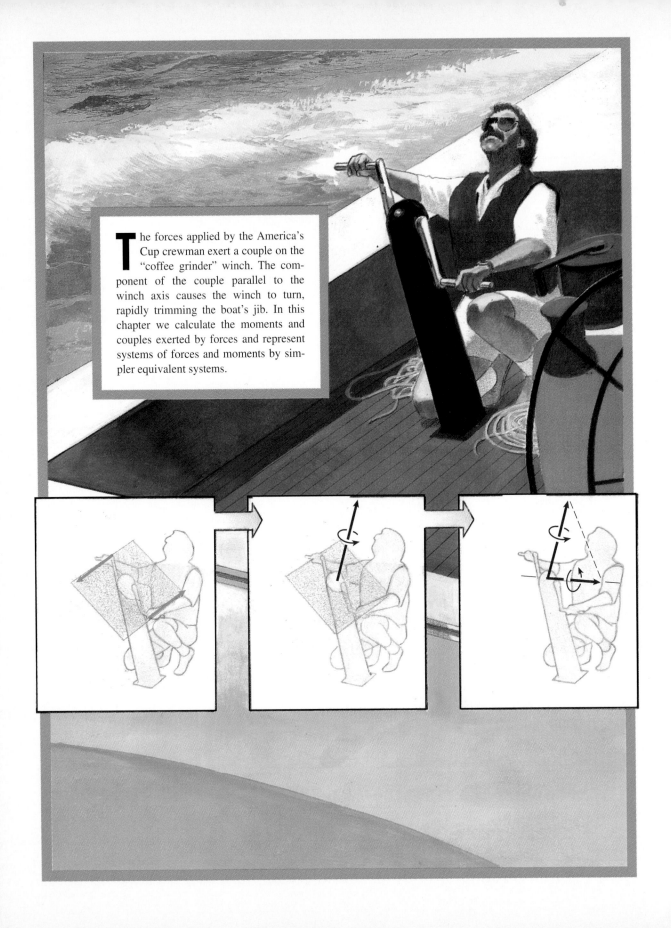

The forces applied by the America's Cup crewman exert a couple on the "coffee grinder" winch. The component of the couple parallel to the winch axis causes the winch to turn, rapidly trimming the boat's jib. In this chapter we calculate the moments and couples exerted by forces and represent systems of forces and moments by simpler equivalent systems.

Chapter 4

Systems of Forces and Moments

THE effects of forces depend not only on their magnitudes and directions but also on the moments, or torques, they exert. The moment of a force is the measure of its tendency to cause rotation. Moments cause the rotation of machinery such as the winch on an America's Cup sailboat, the wheels of a vehicle, crankshafts, and turbines. If an object is in equilibrium, the sum of the moments about any point due to the external forces acting on it is zero. Before continuing our study of the free-body diagram and equilibrium, we must discuss moments and the concept of equivalent systems of forces and moments.

4.1 Two-Dimensional Description of the Moment

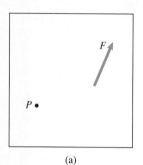

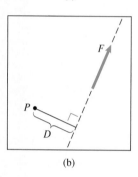

Figure 4.1

(a) The force and point P.
(b) The perpendicular distance D from point P to the line of action of F.
(c) The sense of the moment is counterclockwise.

Consider a force of magnitude F and a point P, and let's view them in the direction perpendicular to the plane containing the force vector and the point (Fig. 4.1a). The **magnitude of the moment** of the force about P is DF, where D is the perpendicular distance from P to the line of action of the force (Fig. 4.1b). In this example, the force would tend to cause counterclockwise rotation about point P. That is, if you imagine the force acts on an object that can rotate about point P, the force would tend to cause counterclockwise rotation (Fig. 4.1c). We say that the **sense of the moment** is counterclockwise. *We define counterclockwise moments to be positive and clockwise moments to be negative.* (This is the usual convention, although we occasionally encounter situations in which it is more convenient to define clockwise moments to be positive.) Thus the moment of the force about P is

$$M_P = DF. \tag{4.1}$$

Notice that if the line of action of F passes through P, the perpendicular distance $D = 0$ and the moment of F about P is zero.

The dimensions of the moment are (distance) $\times$ (force). For example, moments can be expressed in newton-meters in SI units and in foot-pounds in U.S. Customary units.

Suppose that you want to place a television set on a shelf, and you aren't certain the attachment of the shelf to the wall is strong enough to support it. Instinctively, you place it near the wall (Fig. 4.2a), knowing that the attachment is more likely to fail if you place it away from the wall (Fig. 4.2b). What is the difference in the two cases? The magnitude and direction of the force exerted on the shelf by the weight of the television are the same in each case, but the moments exerted on the attachment are different. The moment exerted about P by its weight when it is near the wall, $M_P = -D_1 W$, is smaller in magnitude than the moment about P when it is placed away from the wall, $M_P = -D_2 W$.

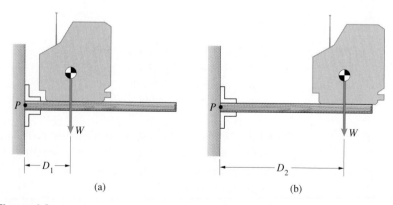

(a)   (b)

Figure 4.2

It is better to place the television near the wall (a) instead of away from it (b) because the moment exerted on the support at P is smaller.

You can use the method we describe in this section to determine the sum of the moments of a system of forces about a point if the forces are two-dimensional (coplanar) and the point lies in the same plane. For example, consider the construction crane shown in Fig. 4.3. The sum of the moments exerted about point P by the load W_1 and the counterweight W_2 is

$$\Sigma M_P = D_1 W_1 - D_2 W_2.$$

This moment tends to cause the top of the vertical tower to rotate and could cause it to collapse. If the distance D_2 is adjusted so that $D_1 W_1 = D_2 W_2$, the moment about point P due to the load and the counterweight is zero.

If you resolve a force into components, the moment of the force about a point P is equal to the sum of the moments of its components about P. We prove this very useful result in the next section.

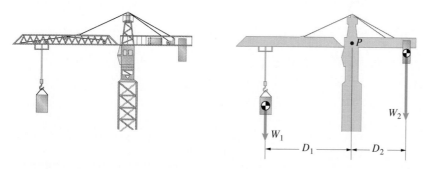

Figure 4.3

A tower crane used in the construction of high-rise buildings.

The following examples illustrate some situations in which you can use the two-dimensional description to calculate moments. Determining the moment of a force about a point P requires three steps:

1. Determine the perpendicular distance—*You must determine the perpendicular distance from P to the line of action of the force. If this appears difficult, you may find it easier to resolve the force into components and determine the perpendicular distances to the components.*

2. Calculate the magnitude of the moment—*The magnitude of the moment is the product of the force and the perpendicular distance.*

3. Determine the sign—*The moment is defined to be positive if the sense of the moment (the direction the force would tend to cause rotation about P) is counterclockwise.*

Example 4.1

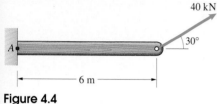

Figure 4.4

What is the moment of the 40-kN force in Fig. 4.4 about point A?

STRATEGY

We can calculate the moment in two ways: by determining the perpendicular distance from point A to the line of action of the force or by resolving the force into components and determining the sum of the moments of the components about A.

SOLUTION

First Method   From Fig. (a), the perpendicular distance from A to the line of action of the force is

$$D = 6 \sin 30° = 3 \text{ m}.$$

The magnitude of the moment of the force about A is (3 m)(40 kN) = 120 kN-m, and the sense of the moment about A is counterclockwise. Therefore the moment is

$$M_A = 120 \text{ kN-m}.$$

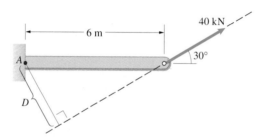

(a) Determining the perpendicular distance D.

Second Method   In Fig. (b), we resolve the force into horizontal and vertical components. The perpendicular distance from A to the line of action of the horizontal component is zero, so the horizontal component exerts no moment about A. The magnitude of the moment of the vertical component about A is (6 m)(40 sin 30° kN) = 120 kN-m, and the sense of its moment about A is counterclockwise. The moment is

$$M_A = 120 \text{ kN-m}.$$

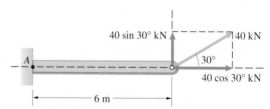

(b) Resolving the force into components.

Example 4.2

Four forces act on the machine part in Fig. 4.5. What is the sum of the moments of the forces about the origin O?

STRATEGY

We can determine the moments of the forces about point O directly from the given information except for the 4-kN force. We will determine its moment by resolving it into components and summing the moments of the components.

SOLUTION

Moment of the 3-kN Force The line of action of the 3-kN force passes through O. It exerts no moment about O.

Moment of the 5-kN Force The line of action of the 5-kN force also passes through O. It too exerts no moment about O.

Moment of the 2-kN Force The perpendicular distance from O to the line of action of the 2-kN force is 0.3 m, and the sense of the moment about O is clockwise. The moment of the 2-kN force about O is

$$-(0.3 \text{ m})(2 \text{ kN}) = -0.600 \text{ kN-m.}$$

(Notice that we converted the perpendicular distance from millimeters into meters to express the result in terms of kilonewton-meters.)

Moment of the 4-kN Force In Fig. (a), we introduce a coordinate system and resolve the 4-kN force into x and y components. The perpendicular distance from O to the line of action of the x component is 0.3 m, and the sense of the moment about O is clockwise. The moment of the x component about O is

$$-(0.3 \text{ m})(4 \cos 30° \text{ kN}) = -1.039 \text{ kN-m.}$$

The perpendicular distance from point O to the line of action of the y component is 0.7 m, and the sense of the moment about O is counterclockwise. The moment of the y component about O is

$$(0.7 \text{ m})(4 \sin 30° \text{ kN}) = 1.400 \text{ kN-m.}$$

The sum of the moments of the four forces about point O is

$$\Sigma M_0 = -0.600 - 1.039 + 1.400 = -0.239 \text{ kN-m.}$$

The four forces exert a 0.239 kN-m clockwise moment about point O.

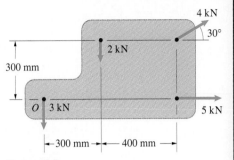

Figure 4.5

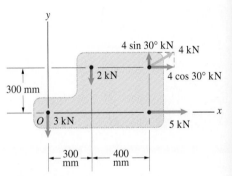

(a) Resolving the 4-kN force into components.

Example 4.3

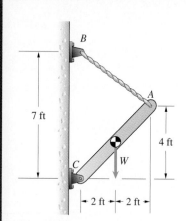

Figure 4.6

The weight $W = 300$ lb (Fig. 4.6). The sum of the moments about C due to the weight W and the force exerted on the bar CA by the cable AB is zero. What is the tension in the cable?

STRATEGY

Let T be the tension in cable AB. Using the given dimensions, we can express the horizontal and vertical components of the force exerted on the bar by the cable in terms of T. Then by setting the sum of the moments about C due to the weight of the bar and the force exerted by the cable equal to zero, we can obtain an equation for T.

SOLUTION

Using similar triangles, we resolve the force exerted on the bar by the cable into horizontal and vertical components (Fig. a). The sum of the moments about C due to the weight of the bar and the force exerted by the cable AB is

$$\Sigma M_C = 4\left(\frac{4}{5}T\right) + 4\left(\frac{3}{5}T\right) - 2W = 0.$$

Solving for T, we obtain

$$T = 0.357W = 107.1 \text{ lb.}$$

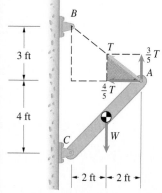

(a) Resolving the force exerted by the cable into horizontal and vertical components.

Problems

4.1 Determine the moment of the 50-N force about (a) point A, (b) point B.

4.2 Determine the moment of the 50-lb force about (a) point A, (b) point B, (c) point C.

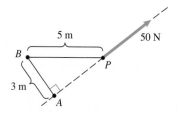

P4.1

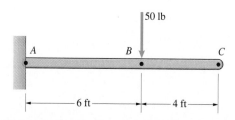

P4.2

4.3 The wheels of the overhead crane exert downward forces on the horizontal I-beam at B and C. If the force at B is 40 kip and the force at C is 44 kip, determine the sum of the moments of the forces on the beam about (a) point A, (b) point D.

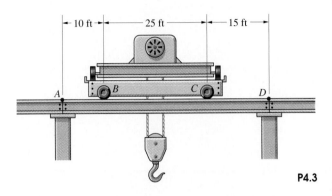

P4.3

4.4 If you exert a 90-N force on the wrench in the direction shown, what moment do you exert about the center of the nut?

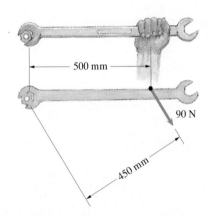

P4.4

4.5 If you exert a force F on the wrench in the direction shown and a 50 N-m moment is required to loosen the nut, what force F must you apply?

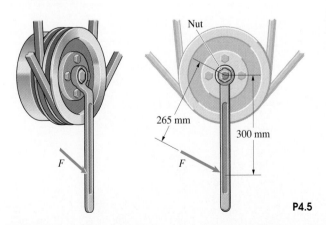

P4.5

4.6 The 20-N force exerts a 20 N-m counterclockwise moment about P.
(a) What is the perpendicular distance from P to the line of action of the force?
(b) What is the angle α?

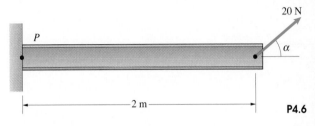

P4.6

4.7 The gears exert 200-N forces on each other at their point of contact.
(a) Determine the moment about A due to the force exerted on the left gear.
(b) Determine the moment about B due to the force exerted on the right gear.

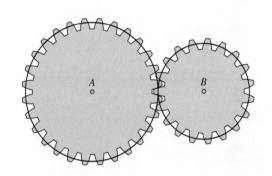

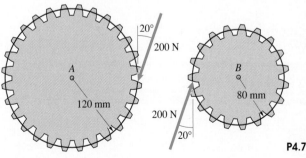

P4.7

4.8 The 5-ft beam *AB* will fail at *A* if the moment of the force *F* about *A* exceeds 10 ft-kip. Based on this criterion, how large can the force *F* be?

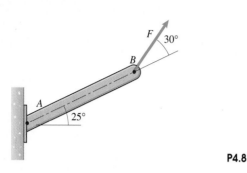

P4.8

4.9 Determine the moment of the 80-lb force about *P*.

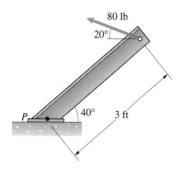

P4.9

4.10 The 20-N force *F* exerts a 20 N-m counterclockwise moment about *P*.
(a) What is the perpendicular distance from *P* to the line of action of *F*?
(b) What is the angle α?

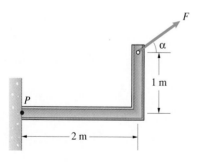

P4.10

4.11 Consider the beam and force shown in Problem 4.10. If *F* is the minimum force that will exert a 20 N-m counterclockwise moment about *P*, what are *F* and α?

4.12 Two students attempt to loosen a lug nut with a lug wrench. One of the students exerts the two 60-lb forces; the other, having to

reach around his friend, can only exert the two 30-lb forces. What torque (moment) do they exert on the nut?

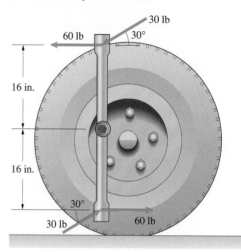

P4.12

4.13 The two students described in Problem 4.12, having failed to loosen the lug nut, try a different tactic. One of them stands on the lug wrench, exerting a 150-lb force on it. The other pulls on the wrench with the force *F*. If a torque of 245 ft-lb is required to loosen the lug nut, what force *F* must the student exert?

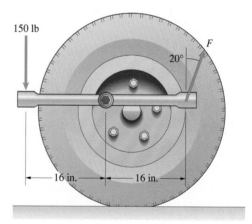

P4.13

4.14 Determine the moment exerted by the 30-lb weight (a) about *E*, (b) about *S*.

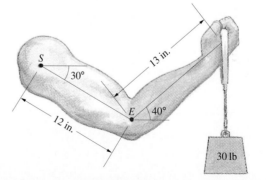

P4.14

4.15   Three forces act on the square plate. Determine the sum of the moments of the forces (a) about A, (b) about B, (c) about C.

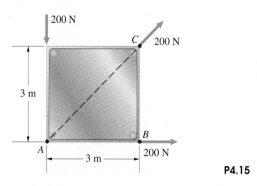

P4.15

4.16   Determine the sum of the moments of the three forces about (a) point A, (b) point B, (c) point C.

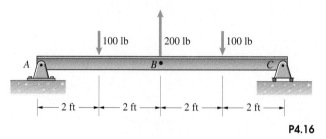

P4.16

4.17   Determine the sum of the moments of the five forces acting on the Howe truss about point A.

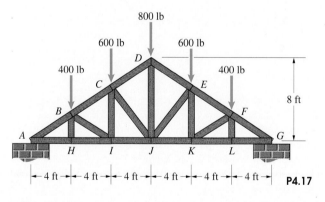

P4.17

4.18   In Problem 4.17, determine the sum of the moments of the five forces about point K.

4.19   The forces $F_1 = 50$ N and $F_2 = 100$ N. Determine the sum of the moments of the two forces (a) about A, (b) about B.

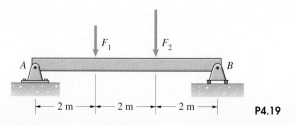

P4.19

4.20   Consider the beam shown in Problem 4.19. If the two forces exert a 140 kN-m clockwise moment about A and a 20 kN-m clockwise moment about B, what are F_1 and F_2?

4.21   The force $F = 140$ lb. The vector sum of the forces acting on the beam is zero, and the sum of the moments about the left end of the beam is zero.
(a) What are the forces A_x, A_y, and B?
(b) What is the sum of the moments about the right end of the beam?

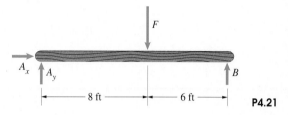

P4.21

4.22   The vector sum of the three forces is zero, and the sum of the moments of the three forces about A is zero.
(a) What are F_A and F_B?
(b) What is the sum of the moments of the three forces about B?

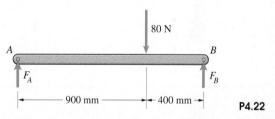

P4.22

4.23   The force F exerts a 240 N-m clockwise moment about A and a 120 N-m counterclockwise moment about B. Determine F and the distance from A to its point of application.

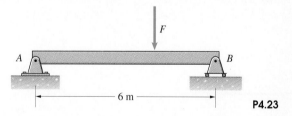

P4.23

4.24   Determine the sum of the moments of the three forces about (a) point A, (b) point B.

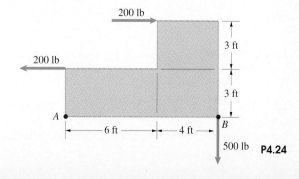

P4.24

4.25 The 160-N weights of the arms AB and BC of the robotic manipulator act at their midpoints. Determine the sum of the moments of the three weights about A.

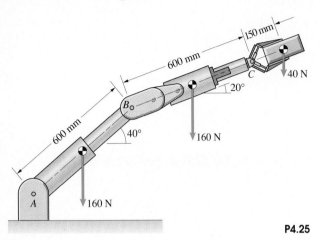

P4.25

4.26 The space shuttle's attitude thrusters exert two forces of magnitude $F = 1740$ lb. What moment do the thrusters exert about G?

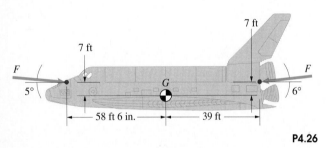

P4.26

4.27 The force F exerts a 200 ft-lb counterclockwise moment about A and a 100 ft-lb clockwise moment about B. What are F and θ?

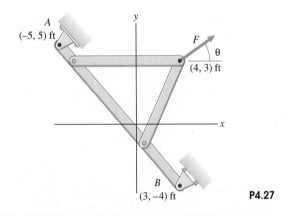

P4.27

4.28 Five forces act on a link in the gear-shifting mechanism of a lawn mower. The vector sum of the five forces on the bar is zero. The sum of their moments about the point where the forces A_x and A_y act is zero.

(a) Determine the forces A_x, A_y, and B.
(b) Determine the sum of the moments of the forces about the point where the force B acts.

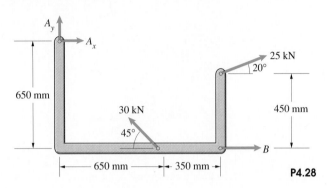

P4.28

4.29 Five forces act on a model truss built by a civil engineering student as part of a design project. The dimensions are $b = 300$ mm and $h = 400$ mm; $F = 100$ N. The sum of the moments of the forces about the point where A_x and A_y act is zero. If the weight of the truss is negligible, what is the force B?

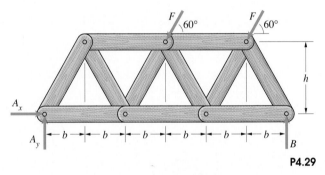

P4.29

4.30 Consider the truss shown in Problem 4.29. The dimensions are $b = 3$ ft and $h = 4$ ft; $F = 300$ lb. The vector sum of the forces acting on the truss is zero, and the sum of the moments of the forces about the point where A_x and A_y act is zero.
(a) Determine the forces A_x, A_y, and B.
(b) Determine the sum of the moments of the forces about the point where the force B acts.

4.31 The mass $m = 70$ kg. What is the moment about A due to the force exerted on the beam at B by the cable?

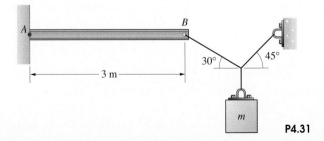

P4.31

4.32 Consider the system shown in Problem 4.31. The beam will collapse at A if the magnitude of the moment about A due to the force exerted on the beam at B by the cable exceeds 2 kN-m. What is the largest mass m that can be suspended?

4.33 The bar AB exerts a force at B that helps support the vertical retaining wall. The force is parallel to the bar. The civil engineer wants the bar to exert a 38 kN-m moment about O. What is the magnitude of the force the bar must exert?

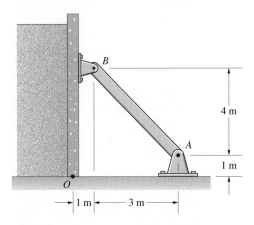

P4.33

4.34 A contestant in a fly-casting contest snags his line in some grass. If the tension in the line is 5 lb, what moment does the force exerted on the rod by the line exert about point H, where he holds the rod?

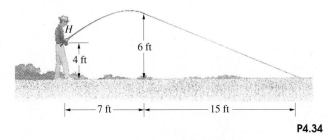

P4.34

4.35 The cables AB and AC help support the tower. The tension in cable AB is 5 kN. The points $A, B, C,$ and O are contained in the same vertical plane.
(a) What is the moment about O due to the force exerted on the tower by cable AB?
(b) If the sum of the moments about O due to the forces exerted on the tower by the two cables is zero, what is the tension in cable AC?

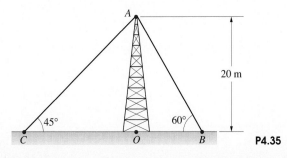

P4.35

4.36 Consider the tower shown in Problem 4.35. If the magnitude of the vertical component of the total force exerted on the tower by cables AB and AC is 15 kN and the sum of the moments about O due to the forces is zero, what are the tensions in the cables?

4.37 The tension in each cable is the same. The forces exerted on the beam by the three cables exert a 1.2 kN-m counterclockwise moment about O. What is the tension in the cables?

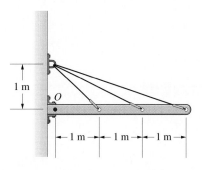

P4.37

4.38 The tension in cable AB is 300 lb. The sum of the moments about O due to the forces exerted on the beam by the two cables is zero. What is the magnitude of the sum of the forces exerted on the beam by the two cables?

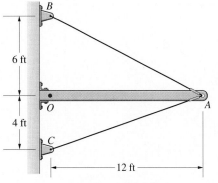

P4.38

4.39 The beam shown in Problem 4.38 will safely support the forces exerted by the two cables at A if the magnitude of the horizontal component of the total force exerted at A does not exceed 1000 lb and the sum of the moments about O due to the forces exerted by the cables equals zero. Based on these criteria, what are the maximum permissible tensions in the two cables?

4.40 The hydraulic cylinder BC exerts a 300-kN force on the boom of the crane at C. The force is parallel to the cylinder. What is the moment of the force about A?

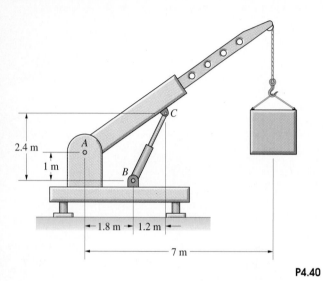

P4.40

4.41 The hydraulic cylinder BC exerts a 2200-lb force on the boom of the crane at C. The force is parallel to the cylinder. The angle $\alpha = 40°$. What is the moment of the force about A?

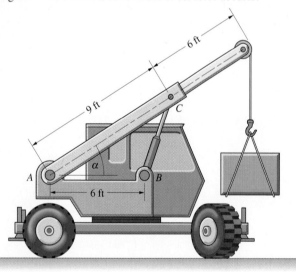

P4.41

4.42 The hydraulic cylinder BC in Problem 4.41 exerts a 2200-lb force on the boom of the crane at C. The force is parallel to the cylinder. The cable supporting the suspended crate exerts a downward force at the end of the boom equal to the weight of the crate. The angle $\alpha = 35°$. If the sum of the moments about A due to the two forces exerted on the boom is zero, what is the weight of the crate?

4.43 The unstretched length of the spring is 1 m, and the spring constant is $k = 20$ N/m. If $\alpha = 30°$, what is the moment about A due to the force exerted by the spring on the circular bar at B?

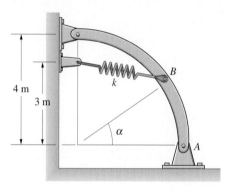

P4.43

4.44 The exercise equipment shown is used by resting the elbow on the fixed pad and rotating the forearm to stretch the elastic cord AB. The cord behaves like a linear spring. Its unstretched length is 1 ft, and its spring constant is 100 lb/ft. If $\alpha = 45°$, what moment is exerted about the elbow joint E by the force exerted by the elastic cord at B?

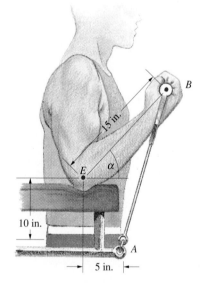

P4.44

4.2 *The Moment Vector*

The moment of a force about a point is a vector. In this section we define the moment and explain how it is evaluated. We then show that when you use the two-dimensional description of the moment described in Section 4.1, you are specifying the magnitude and direction of the moment vector.

Consider a force vector $\mathbf{F}$ and point P (Fig. 4.7a). The **moment** of $\mathbf{F}$ about P is the vector

$$\boxed{\mathbf{M}_P = \mathbf{r} \times \mathbf{F},} \qquad (4.2)$$

where $\mathbf{r}$ is a position vector from P to *any* point on the line of action of $\mathbf{F}$ (Fig. 4.7b).

Magnitude of the Moment

The magnitude of $\mathbf{M}_P$ is

$$|\mathbf{M}_P| = |\mathbf{r}|\,|\mathbf{F}|\sin\theta,$$

where θ is the angle between the vectors $\mathbf{r}$ and $\mathbf{F}$ when they are placed tail to tail. The perpendicular distance from P to the line of action of $\mathbf{F}$ is $D = |\mathbf{r}|\sin\theta$ (Fig. 4.7c). Therefore the magnitude of the moment $\mathbf{M}_P$ equals the product of the perpendicular distance from P to the line of action of $\mathbf{F}$ and the magnitude of $\mathbf{F}$:

$$|\mathbf{M}_P| = D\,|\mathbf{F}|. \qquad (4.3)$$

Notice that if you know the vectors $\mathbf{M}_P$ and $\mathbf{F}$, you can solve this equation for the perpendicular distance D.

Sense of the Moment

We know from the definition of the cross product that $\mathbf{M}_P$ is perpendicular to both $\mathbf{r}$ and $\mathbf{F}$. That means that $\mathbf{M}_P$ is perpendicular to the plane containing P and $\mathbf{F}$ (Fig. 4.8a). Notice in this figure that we denote a moment by a circular arrow around the vector.

The direction of $\mathbf{M}_P$ also indicates the sense of the moment: If you point the thumb of your right hand in the direction of $\mathbf{M}_P$, the "arc" of your fingers indicates the sense of the rotation that $\mathbf{F}$ tends to cause about P (Fig. 4.8b).

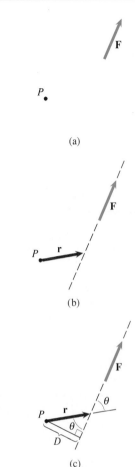

Figure 4.7

(a) The force $\mathbf{F}$ and point P.
(b) A vector $\mathbf{r}$ from P to a point on the line of action of $\mathbf{F}$.
(c) The angle θ and the perpendicular distance D.

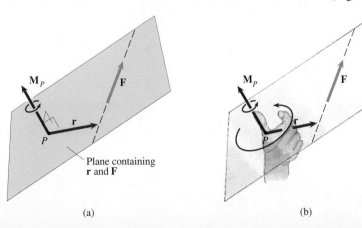

Figure 4.8

(a) $\mathbf{M}_P$ is perpendicular to the plane containing P and $\mathbf{F}$.
(b) The direction of $\mathbf{M}_P$ indicates the sense of the moment.

The result you obtain with Eq. (4.2) doesn't depend on where the vector **r** intersects the line of action of **F**. Instead of using the vector **r** in Fig. 4.9(a), we could use the vector **r′** in Fig. 4.9(b). The vector **r** = **r′** + **u**, where **u** is parallel to **F** (Fig. 4.9c). Therefore

$$\mathbf{r} \times \mathbf{F} = (\mathbf{r}' + \mathbf{u}) \times \mathbf{F} = \mathbf{r}' \times \mathbf{F}$$

because the cross product of the parallel vectors **u** and **F** is zero.

Figure 4.9

(a) A vector **r** from P to the line of action of **F**.
(b) A different vector **r′**.
(c) **r** = **r′** + **u**.

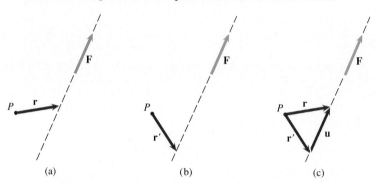

(a) (b) (c)

In summary, the moment of a force **F** about a point P has three properties:

1. The magnitude of $\mathbf{M}_P$ is equal to the product of the magnitude of **F** and the perpendicular distance from P to the line of action of **F**. If the line of action of **F** passes through P, $\mathbf{M}_P = \mathbf{0}$.

2. $\mathbf{M}_P$ is perpendicular to the plane containing P and **F**.

3. The direction of $\mathbf{M}_P$ indicates the sense of the moment through a right-hand rule (Fig. 4.8b). Since the cross product is not commutative, you must be careful to maintain the correct sequence of the vectors in the equation $\mathbf{M}_P = \mathbf{r} \times \mathbf{F}$.

Let's determine the moment of the force **F** in Fig. 4.10(a) about the point P. Since the vector **r** in Eq. (4.2) can be a position vector to any point on the line of action of **F**, we can use the vector from P to the point of application of **F** (Fig. 4.10b):

$$\mathbf{r} = (12 - 3)\,\mathbf{i} + (6 - 4)\,\mathbf{j} + (-5 - 1)\,\mathbf{k} = 9\mathbf{i} + 2\mathbf{j} - 6\mathbf{k} \text{ (ft).}$$

The moment is

$$\mathbf{M}_p = \mathbf{r} \times \mathbf{F} = \begin{vmatrix} \mathbf{i} & \mathbf{j} & \mathbf{k} \\ 9 & 2 & -6 \\ 4 & 4 & 7 \end{vmatrix} = 38\mathbf{i} - 87\mathbf{j} + 28\mathbf{k} \text{ (ft-lb).}$$

The magnitude of $\mathbf{M}_P$,

$$|\mathbf{M}_p| = \sqrt{(38)^2 + (-87)^2 + (28)^2} = 99.0 \text{ ft-lb,}$$

equals the product of the magnitude of **F** and the perpendicular distance D from point P to the line of action of **F**. Therefore

$$D = \frac{|\mathbf{M}_p|}{|\mathbf{F}|} = \frac{99.0 \text{ ft-lb}}{9 \text{ lb}} = 11.0 \text{ ft.}$$

The direction of $\mathbf{M}_P$ tells us both the orientation of the plane containing P and $\mathbf{F}$ and the sense of the moment (Fig. 4.10c).

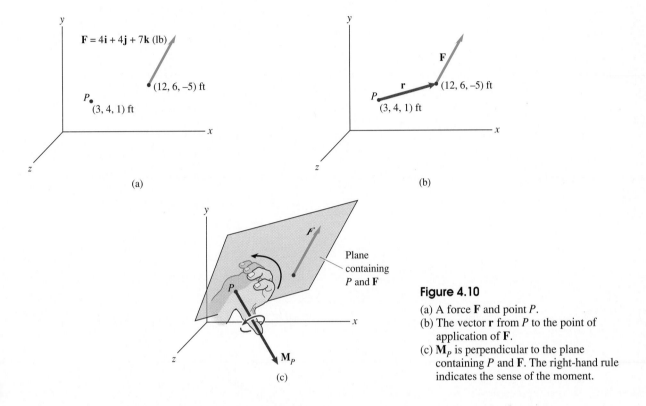

(a)

(b)

(c)

Figure 4.10

(a) A force $\mathbf{F}$ and point P.
(b) The vector $\mathbf{r}$ from P to the point of application of $\mathbf{F}$.
(c) $\mathbf{M}_P$ is perpendicular to the plane containing P and $\mathbf{F}$. The right-hand rule indicates the sense of the moment.

Relation to the Two-Dimensional Description

If our view is perpendicular to the plane containing the point P and the force $\mathbf{F}$, the two-dimensional description of the moment we used in Section 4.1 specifies both the magnitude and direction of $\mathbf{M}_P$. In this situation, $\mathbf{M}_P$ is perpendicular to the page, and the right-hand rule indicates whether it points out of or into the page.

For example, in Fig. 4.11(a), the view is perpendicular to the x-y plane and the 10-N force is contained in the x-y plane. Suppose that we want to determine the moment of the force about the origin O. The perpendicular distance from O to the line of action of the force is 4 m. The two-dimensional description of the moment of the force about O is that its magnitude is $(4 \text{ m})(10 \text{ N}) = 40$ N-m and its sense is counterclockwise, or

$$M_O = 40 \text{ N-m}.$$

That tells us that the magnitude of the vector $\mathbf{M}_O$ is 40 N-m, and the right-hand rule (Fig. 4.11b) indicates that it points out of the page. Therefore

$$\mathbf{M}_O = 40\mathbf{k} \text{ (N-m)}.$$

We can confirm this result by using Eq. (4.2). If we let **r** be the vector from O to the point of application of the force (Fig. 4.11c),

$$\mathbf{M}_O = \mathbf{r} \times \mathbf{F} = (4\mathbf{i} + 2\mathbf{j}) \times 10\mathbf{j} = 40\mathbf{k} \text{ (N-m)}.$$

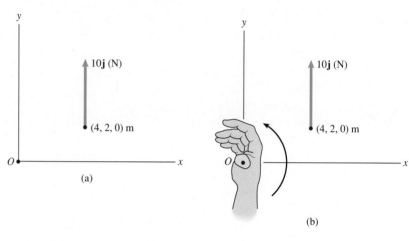

Figure 4.11

(a) The force is contained in the x-y plane.
(b) The sense of the moment indicates that $\mathbf{M}_O$ points out of the page.
(c) The vector **r** from O to the point of application of **F**.

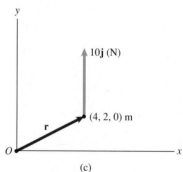

As this example illustrates, the two-dimensional description of the moment determines the moment vector. The converse is also true. The magnitude of $\mathbf{M}_O$ equals the product of the magnitude of the force and the perpendicular distance from O to the line of action of the force, 40 N-m, and the direction of $\mathbf{M}_O$ indicates that the sense of the moment is counterclockwise (Fig. 4.11b).

Varignon's Theorem

Let $\mathbf{F}_1, \mathbf{F}_2, \ldots, \mathbf{F}_N$ be a concurrent system of forces whose lines of action intersect at a point Q. The moment of the system about a point P is

$$(\mathbf{r}_{PQ} \times \mathbf{F}_1) + (\mathbf{r}_{PQ} \times \mathbf{F}_2) + \cdots + (\mathbf{r}_{PQ} \times \mathbf{F}_N)$$
$$= \mathbf{r}_{PQ} \times (\mathbf{F}_1 + \mathbf{F}_2 + \cdots + \mathbf{F}_N),$$

where $\mathbf{r}_{PQ}$ is the vector from P to Q (Fig. 4.12). This result, known as **Varignon's theorem**, follows from the distributive property of the cross product, Eq. (2.31). It confirms that the moment of a force about a point P is equal to the sum of the moments of its components about P.

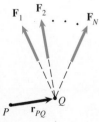

Figure 4.12

A system of concurrent forces and a point P.

The following examples demonstrate how to determine the moments of forces about points. When the two-dimensional description is applicable, we compare it to the vector expression for the moment. We also include examples in which you must use the equation $M_P = r \times F$ to determine the moment. This typically requires two steps:

1. Choose the vector **r**—*You must choose a position vector from P to any point on the line of action of* **F** *and determine its components.*

2. Evaluate **r** × **F**—*Determine the components of* **F**, *if necessary, and evaluate the cross product to determine the moment.*

Example 4.4

Determine the moment of the 400-N force in Fig. 4.13 about O.
(a) What is the two-dimensional description of the moment?
(b) Express the moment as a vector without using Eq. (4.2).
(c) Use Eq. (4.2) to determine the moment.

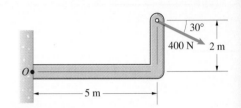

Figure 4.13

SOLUTION

(a) Resolving the force into horizontal and vertical components (Fig. a), the two-dimensional description of the moment is

$$M_O = -(2 \text{ m})(400 \cos 30° \text{ N}) - (5 \text{ m})(400 \sin 30° \text{ N})$$

$$= -1.69 \text{ kN-m}.$$

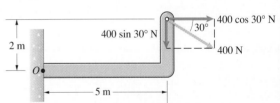

(a) Resolving the force into components.

(b) To express the moment as a vector, we introduce the coordinate system shown in Fig. (b). The magnitude of the moment is 1.69 kN-m, and its sense is clockwise. Pointing the arc of the fingers of the right hand clockwise, the thumb points into the page. Therefore

$$\mathbf{M}_O = -1.69\mathbf{k} \text{ (kN-m)}.$$

(c) We apply Eq. (4.2):

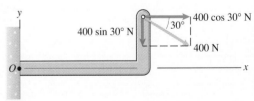

(b) Introducing a coordinate system.

Choose the Vector r   We can let **r** be the vector from O to the point of application of the force (Fig. c):

$$\mathbf{r} = 5\mathbf{i} + 2\mathbf{j} \text{ (m)}.$$

Evaluate r × F   The moment is

$$\mathbf{M}_O = \mathbf{r} \times \mathbf{F} = (5\mathbf{i} + 2\mathbf{j}) \times (400 \cos 30° \, \mathbf{i} - 400 \sin 30° \, \mathbf{j})$$

$$= -1.69\mathbf{k} \text{ (kN-m)}.$$

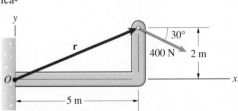

(c) The vector **r** from O to the point of application of the force.

Example 4.5

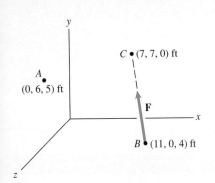

Figure 4.14

The line of action of the 90-lb force **F** in Fig. 4.14 passes through points B and C.
(a) What is the moment of **F** about point A?
(b) What is the perpendicular distance from point A to the line of action of **F**?

STRATEGY

(a) We must use Eq. (4.2) to determine the moment. Since **r** is a vector from A to any point on the line of action of **F**, we can use either the vector from A to B or the vector from A to C. To demonstrate that we obtain the same result, we will determine the moment using both.
(b) Since the magnitude of the moment is equal to the product of the magnitude of **F** and the perpendicular distance from A to the line of action of **F**, we can use the result of (a) to determine the perpendicular distance.

SOLUTION

(a) To evaluate the cross product in Eq. (4.2), we need the components of **F**. The vector from B to C is

$$(7 - 11)\,\mathbf{i} + (7 - 0)\,\mathbf{j} + (0 - 4)\,\mathbf{k} = -4\mathbf{i} + 7\mathbf{j} - 4\mathbf{k}\ (\text{ft}).$$

Dividing this vector by its magnitude, we obtain a unit vector $\mathbf{e}_{BC}$ that has the same direction as **F** (Fig. a):

$$\mathbf{e}_{BC} = -\frac{4}{9}\mathbf{i} + \frac{7}{9}\mathbf{j} - \frac{4}{9}\mathbf{k}.$$

Now we express **F** as the product of its magnitude and $\mathbf{e}_{BC}$:

$$\mathbf{F} = 90\mathbf{e}_{BC} = -40\mathbf{i} + 70\mathbf{j} - 40\mathbf{k}\ (\text{lb}).$$

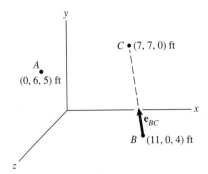

(a) The unit vector $\mathbf{e}_{BC}$.

Choose the Vector r   The position vector from A to B (Fig. b) is

$$\mathbf{r}_{AB} = (11 - 0)\,\mathbf{i} + (0 - 6)\,\mathbf{j} + (4 - 5)\,\mathbf{k} = 11\mathbf{i} - 6\mathbf{j} - \mathbf{k}\ (\text{ft}).$$

Evaluate r × F   The moment of $\mathbf{F}$ about A is

$$\mathbf{M}_A = \mathbf{r}_{AB} \times \mathbf{F} = \begin{vmatrix} \mathbf{i} & \mathbf{j} & \mathbf{k} \\ 11 & -6 & -1 \\ -40 & 70 & -40 \end{vmatrix}$$

$$= 310\mathbf{i} + 480\mathbf{j} + 530\mathbf{k}\ (\text{ft-lb}).$$

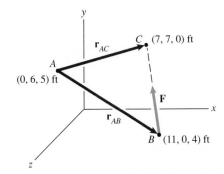

(b) The moment can be determined using either $\mathbf{r}_{AB}$ or $\mathbf{r}_{AC}$.

Alternative Choice of Position Vector   If we use the vector from A to C instead,

$$\mathbf{r}_{AC} = (7 - 0)\,\mathbf{i} + (7 - 6)\,\mathbf{j} + (0 - 5)\,\mathbf{k} = 7\mathbf{i} + \mathbf{j} - 5\mathbf{k}\ (\text{ft}),$$

we obtain the same result:

$$\mathbf{M}_A = \mathbf{r}_{AC} \times \mathbf{F} = \begin{vmatrix} \mathbf{i} & \mathbf{j} & \mathbf{k} \\ 7 & 1 & -5 \\ -40 & 70 & -40 \end{vmatrix}$$

$$= 310\mathbf{i} + 480\mathbf{j} + 530\mathbf{k}\ (\text{ft-lb}).$$

(b) The perpendicular distance is

$$\frac{|\mathbf{M}_A|}{|\mathbf{F}|} = \frac{\sqrt{(310)^2 + (480)^2 + (530)^2}}{} = 8.66\ \text{ft}.$$

Example 4.6

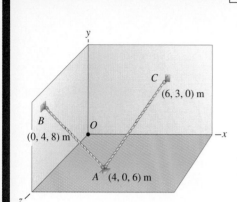

Figure 4.15

The cables AB and AC in Fig. 4.15 extend from an attachment point A on the floor to attachment points B and C in the walls. The tension in cable AB is 10 kN, and the tension in cable AC is 20 kN. What is the sum of the moments about O due to the forces exerted on A by the two cables?

SOLUTION

Let $\mathbf{F}_{AB}$ and $\mathbf{F}_{AC}$ be the forces exerted on the attachment point A by the two cables (Fig. a). To express $\mathbf{F}_{AB}$ in terms of its components, we determine the position vector from A to B,

$$(0 - 4)\,\mathbf{i} + (4 - 0)\,\mathbf{j} + (8 - 6)\,\mathbf{k} = -4\mathbf{i} + 4\mathbf{j} + 2\mathbf{k}\ (\text{m}),$$

and divide it by its magnitude to obtain a unit vector $\mathbf{e}_{AB}$ with the same direction as $\mathbf{F}_{AB}$ (Fig. b):

$$\mathbf{e}_{AB} = \frac{-4\mathbf{i} + 4\mathbf{j} + 2\mathbf{k}}{6} = -\frac{2}{3}\mathbf{i} + \frac{2}{3}\mathbf{j} + \frac{1}{3}\mathbf{k}.$$

Now we write $\mathbf{F}_{AB}$ as

$$\mathbf{F}_{AB} = 10\mathbf{e}_{AB} = -6.67\mathbf{i} + 6.67\mathbf{j} + 3.33\mathbf{k}\ (\text{kN}).$$

We express the force $\mathbf{F}_{AC}$ in terms of its components in the same way:

$$\mathbf{F}_{AC} = 5.71\mathbf{i} + 8.57\mathbf{j} - 17.14\mathbf{k}\ (\text{kN}).$$

Choose the Vector r   Since the lines of action of both forces pass through point A, we can use the vector from O to A to determine the moments of both forces about point O (Fig. a):

$$\mathbf{r} = 4\mathbf{i} + 6\mathbf{k}\ (\text{m}).$$

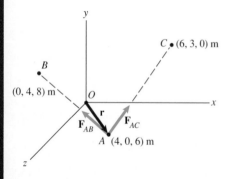

(a) The forces $\mathbf{F}_{AB}$ and $\mathbf{F}_{AC}$ exerted at A by the cables.

Evaluate r × F   The sum of the moments is

$$\Sigma\,\mathbf{M}_O = (\mathbf{r} \times \mathbf{F}_{AB}) + (\mathbf{r} \times \mathbf{F}_{AC})$$

$$= \begin{vmatrix} \mathbf{i} & \mathbf{j} & \mathbf{k} \\ 4 & 0 & 6 \\ -6.67 & 6.67 & 3.33 \end{vmatrix} + \begin{vmatrix} \mathbf{i} & \mathbf{j} & \mathbf{k} \\ 4 & 0 & 6 \\ 5.71 & 8.57 & -17.14 \end{vmatrix}$$

$$= -91.4\mathbf{i} + 49.5\mathbf{j} + 61.0\mathbf{k}\ (\text{kN-m}).$$

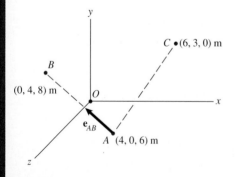

(b) The unit vector $\mathbf{e}_{AB}$ has the same direction as $\mathbf{F}_{AB}$.

Problems

4.45 Determine the moment of the 50-lb force about the origin O.
(a) What is the two-dimensional description of the moment?
(b) Express the moment as a vector without using the cross product.
(c) Use the cross product to determine the moment.

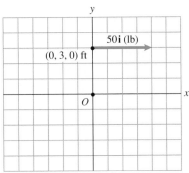

P4.45

4.46 Determine the moment of the 80-N force about the origin O.
(a) What is the two-dimensional description of the moment?
(b) Express the moment as a vector without using the cross product.
(c) Use the cross product to determine the moment, letting **r** be the vector from O to A.
(d) Use the cross product to determine the moment, letting **r** be the vector from O to B.

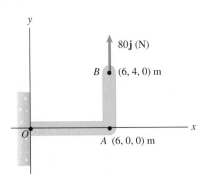

P4.46

4.47 A bioengineer studying an injury sustained in throwing the javelin estimates that the athlete's shoulder was subjected to an 1800 in-lb moment and that the magnitude of the force exerted was $|\mathbf{F}| = 80$ lb. The vector **F** and the point O are contained in the x-y plane.
(a) What was the perpendicular distance from O to the line of action of **F**?

(b) Express the moment of **F** about O as a vector.

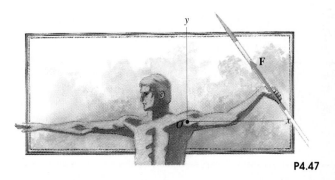

P4.47

4.48 Use the cross product to determine the moment of the 100-kN force (a) about A, (b) about B.

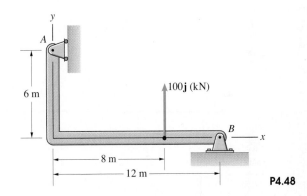

P4.48

4.49 The line of action of the 100-lb force is contained in the x-y plane.
(a) Use the cross product to determine the moment of the force about the origin O.
(b) Use the result of (a) to determine the perpendicular distance from O to the line of action of the force.

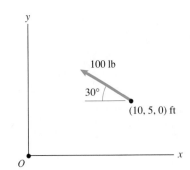

P4.49

4.50 The line of action of **F** is contained in the *x-y* plane. The moment of **F** about *O* is 140**k** (N-m), and the moment of **F** about *A* is 280**k** (N-m). What are the components of **F**?

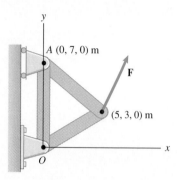

P4.50

4.51 To test the bending stiffness of a light composite beam, engineering students subject it to the vertical forces shown.
(a) What is the two-dimensional description of the moment of the 200-lb force about *A*?
(b) Express the moment determined in (a) as a vector.

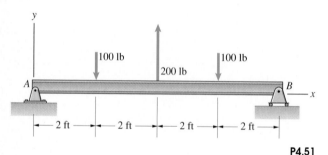

P4.51

4.52 Consider the beam and forces shown in Problem 4.51. Use the cross product to determine the sum of the moments of the three forces (a) about *A*, (b) about *B*.

4.53 Three forces are applied to the plate. Use the cross product to determine the sum of the moments of the three forces about the origin *O*.

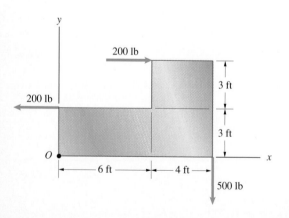

P4.53

4.54 (a) Determine the magnitude of the moment of the 150-N force about *A* by calculating the perpendicular distance from *A* to the line of action of the force.
(b) Use the cross product to determine the moment of the 150-N force about *A*.
(c) Use the result of (b) to determine the magnitude of the moment.

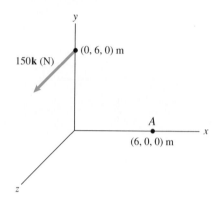

P4.54

4.55 A force **F** = 4**i** + 6**j** − 2**k** (kN) is applied at the point (8, 4, 4) m. What is the moment of **F** about the point (2, 2, 2) m?

4.56 A force **F** = 20**i** − 30**j** + 60**k** (lb) is applied at the point (2, 3, 6) ft. What is the moment of **F** about the point (−2, −1, −1) ft?

4.57 A force **F** = 20**i** − 30**j** + 60**k** (lb). The moment of **F** about a point *P* is **M**$_p$ = 450**i** − 100**j** − 200**k** (ft-lb). What is the perpendicular distance from point *P* to the line of action of **F**?

4.58 A force **F** is applied at the point (8, 6, 13) m. Its magnitude is |**F**| = 90 N, and the moment of **F** about the point (4, 2, 6) is zero. What are the components of **F**?

4.59 The force **F** = 30**i** + 20**j** + 10**k** (N).
(a) Determine the moment of **F** about *A*.
(b) What is the product of the magnitude of **F** and the perpendicular distance from point *A* to the line of action of **F**?

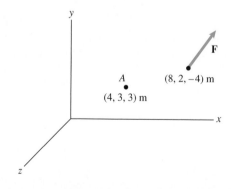

P4.59

4.60   The direction cosines of the force **F** are cos $\theta_x = 0.818$, cos $\theta_y = 0.182$, and cos $\theta_z = -0.545$. The support of the beam at O will fail if the magnitude of the moment of **F** about O exceeds 100 kN-m. Determine the magnitude of the largest force **F** that can safely be applied to the beam.

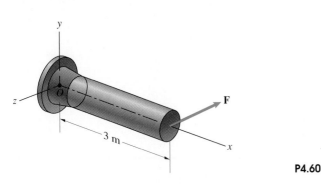

P4.60

4.61   The force **F** exerted on a grip of an exercise machine points in the direction of the unit vector $\mathbf{e} = \frac{2}{3}\mathbf{i} - \frac{2}{3}\mathbf{j} - \frac{1}{3}\mathbf{k}$, and its magnitude is 30 lb.
(a) What is the moment of **F** about the origin O?
(b) What is the product of the magnitude of **F** and the perpendicular distance from O to the line of action of **F**?

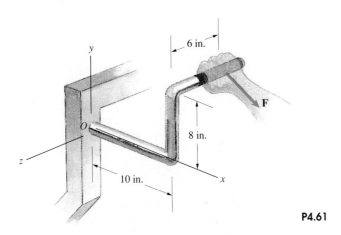

P4.61

4.62   In Problem 4.61, the force **F** points in the direction of the unit vector $\mathbf{e} = \frac{2}{3}\mathbf{i} - \frac{2}{3}\mathbf{j} - \frac{1}{3}\mathbf{k}$. The support at O will safely support a moment of magnitude 5000 in-lb. Based on this criterion, what is the largest safe magnitude of **F**?

4.63   An engineer estimates that under the most adverse expected weather conditions, the total force on the highway sign will be **F** = $\pm 1.4\mathbf{i} - 2.0\mathbf{j}$ (kN). What moment does this force exert about the base O?

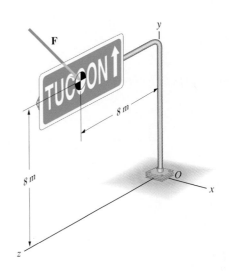

P4.63

4.64   After executing a "slam dunk" shot, a basketball player hangs on the rim of the basket, exerting the force shown. Determine the moment he exerts about the origin O in terms of the dimensions h and r and the angle α.

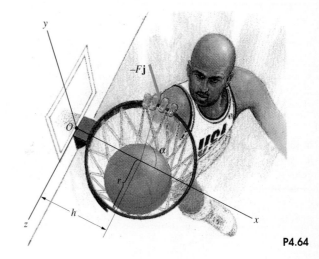

P4.64

4.65   The tension in cable AC is 100 lb. Determine the moment about the origin O due to the force exerted at A by cable AC. Use the cross product, letting **r** be the vector (a) from O to A, (b) from O to C.

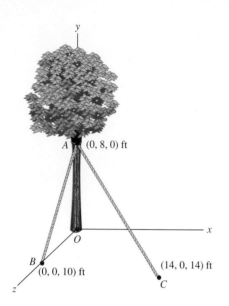

A   (0, 8, 0) ft

O

x

B
(0, 0, 10) ft

(14, 0, 14) ft

C

z

P4.65

4.66   Consider the tree in Problem 4.65. The tension in cable AB is 100 lb, and the tension in cable AC is 140 lb. Determine the magnitude of the sum of the moments about O due to the forces exerted at A by the two cables.

4.67   Consider the tree in Problem 4.65. Cables AB and AC have the same tension T. The tree will collapse if the magnitude of the moment about O due to the forces exerted on the tree by the cables exceeds 5000 ft-lb. Based on this criterion, what is the maximum allowable value of T?

4.68   The tension in cable OA is 500 N. Determine the moment about B due to the force exerted by the cable (a) on point O, (b) on point A. (c) Determine the perpendicular distance from B to the cable OA.

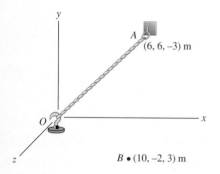

y

A
(6, 6, –3) m

O

x

z

B • (10, –2, 3) m

P4.68

4.69   The tower is 70 m tall. The tensions in cables AB, AC, and AD are 4 kN, 2 kN, and 2 kN, respectively. Determine the sum of the moments about the origin O due to the forces exerted by the cables at point A.

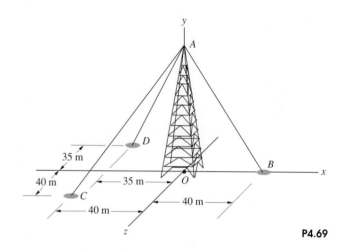

y

A

35 m

D

40 m

35 m

O

x

C

B

40 m

40 m

z

P4.69

4.70   Consider the 70-m tower in Problem 4.69. Suppose that the tension in cable AB is 4 kN, and you want to adjust the tensions in cables AC and AD so that the sum of the moments about the origin O due to the forces exerted by the cables at point A is zero. Determine the tensions.

4.71   The tension in cable AB is 150 N. The tension in cable AC is 100 N. Determine the sum of the moments about D due to the forces exerted on the wall by the cables.

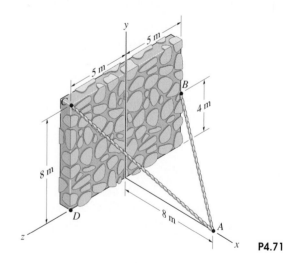

y

5 m

5 m

C

B

5 m

4 m

8 m

D

A

8 m

z

x

P4.71

4.72   Consider the wall shown in Problem 4.71. The total force exerted by the two cables in the direction perpendicular to the wall is 2 kN. The magnitude of the sum of the moments about D due to the forces exerted on the wall by the cables is 18 kN-m. What are the tensions in the cables?

4.73 The force $F = 800$ lb. The sum of the moments about O due to the force F and the forces exerted at A by the cables AB and AC is zero. What are the tensions in the cables?

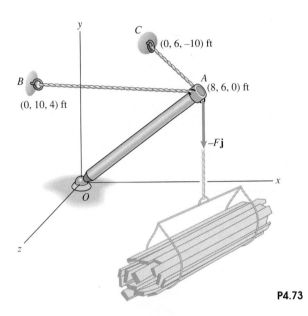

P4.73

4.74 In Problem 4.73, the sum of the moments about O due to the force F and the forces exerted at A by the cables AB and AC is zero. Each cable will safely support a tension of 2000 lb. Based on this criterion, what is the largest safe value of the force F?

4.75 Two forces are exerted on the crankshaft by the connecting rods. The direction cosines of $\mathbf{F}_A$ are $\cos\theta_x = -0.182$, $\cos\theta_y = 0.818$, and $\cos\theta_z = 0.545$, and its magnitude is 4 kN. The direction cosines of $\mathbf{F}_B$ are $\cos\theta_x = 0.182$, $\cos\theta_y = 0.818$, and $\cos\theta_z = -0.545$, and its magnitude is 2 kN. What is the sum of the moments about O due to the two forces?

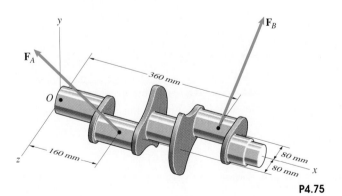

P4.75

4.76 The weights of the arms OA and AB of the robotic manipulator act at their midpoints. The direction cosines of the centerline of arm OA are $\cos\theta_x = 0.500$, $\cos\theta_y = 0.866$, and $\cos\theta_z = 0$, and the direction cosines of the centerline of arm AB are $\cos\theta_x = 0.707$, $\cos\theta_y = 0.619$, and $\cos\theta_z = -0.342$. What is the sum of the moments about O due to the two forces?

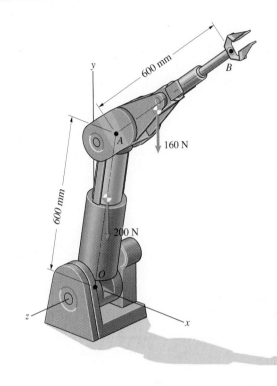

P4.76

4.3 Moment of a Force About a Line

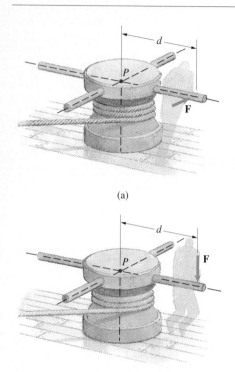

(a)

(b)

Figure 4.16

(a) Turning a capstan.
(b) A vertical force does not turn the capstan.

Figure 4.17

Applying a force to a turbine with axis of rotation L.

The device in Fig. 4.16, called a *capstan*, was used in the days of square-rigged sailing ships. Crewmen turned it by pushing on the handles as shown in Fig. 4.16(a), providing power for such tasks as raising anchors and hoisting yards. You know that a vertical force **F** applied to one of the handles as shown in Fig. 4.16(b) does not cause the capstan to turn, even though the magnitude of the moment about point P is $d|\mathbf{F}|$ in both cases.

The measure of the tendency of a force to cause rotation about a line, or axis, is called the moment of the force about the line. Suppose that a force **F** acts on an object such as a turbine that rotates about an axis L, and we resolve **F** into components in terms of the coordinate system shown in Fig. 4.17. You can see that the components F_x and F_z do not tend to rotate the turbine, just as the force parallel to the axis of the capstan did not cause it to turn. It is the component F_y that tends to cause rotation, by exerting a moment of magnitude aF_y about the turbine's axis. In this example we can determine the moment of **F** about L easily because the coordinate system is conveniently placed. We now introduce an expression you can use to determine the moment of a force about any line.

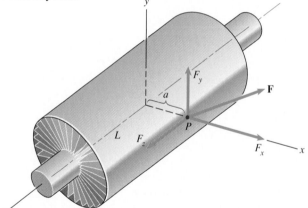

Definition

Consider a line L and force **F** (Fig. 4.18a). Let $\mathbf{M}_P$ be the moment of **F** about an arbitrary point P on L (Fig. 4.18b). The moment of **F** about L is the component of $\mathbf{M}_P$ parallel to L, which we denote by $\mathbf{M}_L$ (Fig. 4.18c). The magnitude of the moment of **F** about L is $|\mathbf{M}_L|$, and when the thumb of the right hand is pointed in the direction of $\mathbf{M}_L$, the arc of the fingers indicates the sense of the moment about L.

In terms of a unit vector **e** along L (Fig. 4.18d), $\mathbf{M}_L$ is given by

$$\mathbf{M}_L = (\mathbf{e} \cdot \mathbf{M}_P)\,\mathbf{e}. \qquad (4.4)$$

(The unit vector **e** can point in either direction. See our discussion of vector components parallel and normal to a line in Section 2.5.) The moment $\mathbf{M}_P = \mathbf{r} \times \mathbf{F}$, so we can also express $\mathbf{M}_L$ as

$$\mathbf{M}_L = [\mathbf{e} \cdot (\mathbf{r} \times \mathbf{F})]\,\mathbf{e}. \qquad (4.5)$$

The mixed triple product in this expression is given in terms of the components of the three vectors by

$$\mathbf{e} \cdot (\mathbf{r} \times \mathbf{F}) = \begin{vmatrix} e_x & e_y & e_z \\ r_x & r_y & r_z \\ F_x & F_y & F_z \end{vmatrix}. \tag{4.6}$$

Notice that the value of the scalar $\mathbf{e} \cdot \mathbf{M}_P = \mathbf{e} \cdot (\mathbf{r} \times \mathbf{F})$ tells you both the magnitude and direction of $\mathbf{M}_L$. The absolute value of $\mathbf{e} \cdot \mathbf{M}_P$ is the magnitude of $\mathbf{M}_L$. If $\mathbf{e} \cdot \mathbf{M}_P$ is positive, $\mathbf{M}_L$ points in the direction of $\mathbf{e}$, and if $\mathbf{e} \cdot \mathbf{M}_P$ is negative, $\mathbf{M}_L$ points in the direction opposite to $\mathbf{e}$.

The result you obtain with Eq. (4.4) or (4.5) doesn't depend on which point on L you choose to determine $\mathbf{M}_P = \mathbf{r} \times \mathbf{F}$. If we use point P in Fig. 4.19 to determine the moment of $\mathbf{F}$ about L, we get the result given by Eq. (4.5). If we use P' instead, we obtain the same result,

$$[\mathbf{e} \cdot (\mathbf{r}' \times \mathbf{F})] \, \mathbf{e} = \{\mathbf{e} \cdot [(\mathbf{r} + \mathbf{u}) \times \mathbf{F}]\} \, \mathbf{e}$$

$$= [\mathbf{e} \cdot (\mathbf{r} \times \mathbf{F}) + \mathbf{e} \cdot (\mathbf{u} \times \mathbf{F})] \, \mathbf{e}$$

$$= [\mathbf{e} \cdot (\mathbf{r} \times \mathbf{F})] \, \mathbf{e},$$

because $\mathbf{u} \times \mathbf{F}$ is perpendicular to $\mathbf{e}$.

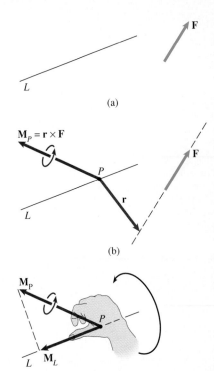

(a)

(b)

(c)

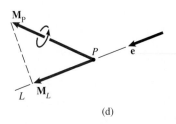

(d)

Figure 4.18

(a) The line L and force $\mathbf{F}$.
(b) $\mathbf{M}_P$ is the moment of $\mathbf{F}$ about any point P on L.
(c) The component $\mathbf{M}_L$ is the moment of $\mathbf{F}$ about L.
(d) A unit vector $\mathbf{e}$ along L.

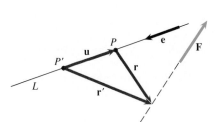

Figure 4.19

Using different points P and P' to determine the moment of $\mathbf{F}$ about L.

Applying the Definition

To demonstrate that $\mathbf{M}_L$ is the measure of the tendency of $\mathbf{F}$ to cause rotation about L, we return to the turbine in Fig. 4.17. Let Q be a point on L at an arbitrary distance b from the origin (Fig. 4.20a). The vector $\mathbf{r}$ from Q to P is $\mathbf{r} = a\mathbf{i} - b\mathbf{k}$, so the moment of $\mathbf{F}$ about Q is

$$\mathbf{M}_Q = \mathbf{r} \times \mathbf{F} = \begin{vmatrix} \mathbf{i} & \mathbf{j} & \mathbf{k} \\ a & 0 & -b \\ F_x & F_y & F_z \end{vmatrix} = bF_y\mathbf{i} - (aF_z + bF_x)\mathbf{j} + aF_y\mathbf{k}.$$

Since the z axis is coincident with L, the unit vector $\mathbf{k}$ is along L. Therefore the moment of $\mathbf{F}$ about L is

$$\mathbf{M}_L = (\mathbf{k} \cdot \mathbf{M}_Q) \, \mathbf{k} = aF_y\mathbf{k}.$$

The components F_x and F_z exert no moment about L. If we assume that F_y is positive, it exerts a moment of magnitude aF_y about the turbine's axis in the direction shown in Fig. 4.20(b).

Figure 4.20

(a) An arbitrary point Q on L and the vector $\mathbf{r}$ from Q to P.

(b) $\mathbf{M}_L$ and the sense of the moment about L.

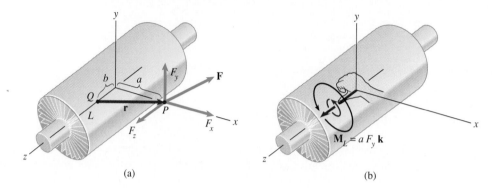

(a)   (b)

Now let's determine the moment of a force about an arbitrary line L (Fig. 4.21a). The first step is to choose a point on the line. If we choose point A (Fig. 4.21b), the vector $\mathbf{r}$ from A to the point of application of $\mathbf{F}$ is

$$\mathbf{r} = (8 - 2)\,\mathbf{i} + (6 - 0)\,\mathbf{j} + (4 - 4)\,\mathbf{k} = 6\mathbf{i} + 6\mathbf{j} \text{ (m).}$$

The moment of $\mathbf{F}$ about A is

$$\mathbf{M}_A = \mathbf{r} \times \mathbf{F} = \begin{vmatrix} \mathbf{i} & \mathbf{j} & \mathbf{k} \\ 6 & 6 & 0 \\ 10 & 60 & -20 \end{vmatrix}$$

$$= -120\mathbf{i} + 120\mathbf{j} + 300\mathbf{k} \text{ (N-m).}$$

The next step is to determine a unit vector along L. The vector from A to B is

$$(-7 - 2)\,\mathbf{i} + (6 - 0)\,\mathbf{j} + (2 - 4)\,\mathbf{k} = -9\mathbf{i} + 6\mathbf{j} - 2\mathbf{k} \text{ (m).}$$

Dividing this vector by its magnitude, we obtain a unit vector $\mathbf{e}_{AB}$ that points from A toward B (Fig. 4.21c):

$$\mathbf{e}_{AB} = -\frac{9}{11}\mathbf{i} + \frac{6}{11}\mathbf{j} - \frac{2}{11}\mathbf{k}.$$

The moment of $\mathbf{F}$ about L is

$$\mathbf{M}_L = (\mathbf{e}_{AB} \cdot \mathbf{M}_A)\,\mathbf{e}_{AB}$$

$$= \left[\left(-\frac{9}{11} \right)(-120) + \left(\frac{6}{11} \right)(120) + \left(-\frac{2}{11} \right)(300) \right] \mathbf{e}_{AB}$$

$$= 109\,\mathbf{e}_{AB} \text{ (N-m).}$$

The magnitude of $\mathbf{M}_L$ is 109 N-m; pointing the thumb of the right hand in the direction of $\mathbf{e}_{AB}$ indicates the direction.

If we calculate $\mathbf{M}_L$ using the unit vector $\mathbf{e}_{BA}$ that points from B toward A instead, we obtain

$$\mathbf{M}_L = -109\mathbf{e}_{BA} \text{ (N-m).}$$

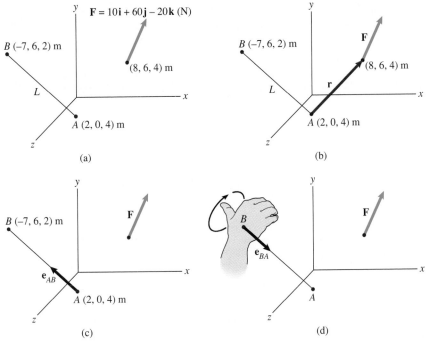

Figure 4.21
(a) A force **F** and line L.
(b) The vector **r** from A to the point of application of **F**.
(c) $\mathbf{e}_{AB}$ points from A toward B.
(d) The right-hand rule indicates the sense of the moment.

We obtain the same magnitude, and the minus sign indicates that $\mathbf{M}_L$ points in the direction opposite to $\mathbf{e}_{BA}$, so the direction of $\mathbf{M}_L$ is the same. Therefore the right-hand rule indicates the same sense (Fig. 4.21d).

Special Cases

The preceding examples demonstrate three important results that we can state in more general terms:

1. When the line of action of **F** is perpendicular to a plane containing L (Fig. 4.22a), the magnitude of the moment of **F** about L is equal to the product of the magnitude of **F** and the perpendicular distance D from L to the point where the line of action intersects the plane: $|\mathbf{M}_L| = |\mathbf{F}|D$.

2. When the line of action of **F** is parallel to L (Fig. 4.22b), the moment of **F** about L is zero: $\mathbf{M}_L = 0$. Since $\mathbf{M}_P = \mathbf{r} \times \mathbf{F}$ is perpendicular to **F**, $\mathbf{M}_P$ is perpendicular to L and the vector component of $\mathbf{M}_P$ parallel to L is zero.

3. When the line of action of **F** intersects L (Fig. 4.22c), the moment of **F** about L is zero. Since we can choose any point on L to evaluate $\mathbf{M}_P$, we can use the point where the line of action of **F** intersects L. The moment $\mathbf{M}_P$ about that point is zero, so its vector component parallel to L is zero.

The following examples demonstrate how to determine the moment of a force about a line. Using Eqs. (4.4)–(4.6) requires three steps:

1. Determine a vector **r**—*Choose any point P on L, and determine the components of a vector* **r** *from P to any point on the line of action of* **F**.

2. Determine a vector **e**—*Determine the components of a unit vector along L. It doesn't matter in which direction along L it points.*

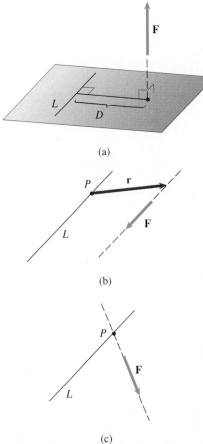

Figure 4.22
(a) **F** is perpendicular to a plane containing L.
(b) **F** is parallel to L.
(c) The line of action of **F** intersects L at P.

3. Evaluate $\mathbf{M}_L$—*You can calculate* $\mathbf{M}_P = \mathbf{r} \times \mathbf{F}$ *and determine* $\mathbf{M}_L$ *by using Eq. (4.4), or you can use Eq. (4.6) to evaluate the mixed triple product and substitute the result into Eq. (4.5).*

Example 4.7

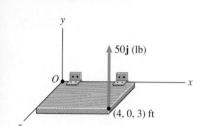

Figure 4.23

What is the moment of the 50-lb force in Fig. 4.23 about the x axis?

STRATEGY

We can determine the moment in two ways.

First Method   We can use Eqs. (4.5) and (4.6). Since $\mathbf{r}$ can extend from any point on the x axis to the line of action of the force, we can use the vector from O to the point of application of the force. The vector $\mathbf{e}$ must be a unit vector along the x axis, so we can use either $\mathbf{i}$ or $-\mathbf{i}$.

Second Method   This example is the first of the special cases we just discussed, because the 50-lb force is perpendicular to the x-z plane. We can determine the magnitude and direction of the moment directly from the given information.

SOLUTION

First Method   *Determine a vector* $\mathbf{r}$. The vector from O to the point of application of the force is (Fig. a)

$$\mathbf{r} = 4\mathbf{i} + 3\mathbf{k} \text{ (ft).}$$

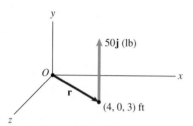

(a) The vector $\mathbf{r}$ from O to the point of application of the force.

Determine a vector $\mathbf{e}$. We can use the unit vector $\mathbf{i}$.
Evaluate $\mathbf{M}_L$. Using Eq. (4.6), the mixed triple product is

$$\mathbf{i} \cdot (\mathbf{r} \times \mathbf{F}) = \begin{vmatrix} 1 & 0 & 0 \\ 4 & 0 & 3 \\ 0 & 50 & 0 \end{vmatrix} = -150 \text{ ft-lb.}$$

Then from Eq. (4.5), the moment of the force about the x axis is

$$\mathbf{M}_{(x\text{ axis})} = [\mathbf{i} \cdot (\mathbf{r} \times \mathbf{F})]\,\mathbf{i} = -150\mathbf{i} \text{ (ft-lb).}$$

The magnitude of the moment is 150 ft-lb, and its sense is as shown in Fig. (b).

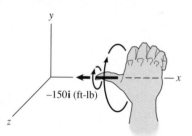

(b) The sense of the moment.

Second Method   Since the 50-lb force is perpendicular to a plane (the x-z plane) containing the x axis, the magnitude of the moment about the x axis is equal to the perpendicular distance from the x axis to the point where the line of action of the force intersects the x-z plane (Fig. c):

$$|\mathbf{M}_{(x\text{ axis})}| = (3 \text{ ft})(50 \text{ lb}) = 150 \text{ ft-lb.}$$

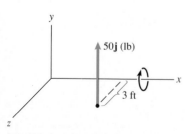

(c) The distance from the x axis to the point where the line of action of the force intersects the x-z plane is 3 ft. The arrow indicates the sense of the moment about the x axis.

Pointing the arc of the fingers in the direction of the sense of the moment about the x axis (Fig. c), we find that the right-hand rule indicates that $\mathbf{M}_{(x\text{ axis})}$ points in the negative x-axis direction. Therefore

$$\mathbf{M}_{(x\text{ axis})} = -150\mathbf{i} \text{ (ft-lb).}$$

Example 4.8

What is the moment of the force **F** in Fig. 4.24 about the bar BC?

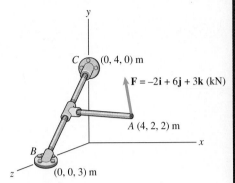

Figure 4.24

STRATEGY

We can use Eqs. (4.5) and (4.6) to determine the moment. Since we know the coordinates of points B and C, we can determine the components of a vector **r** that extends either from B to the point of application of the force or from C to the point of application. We can also use the coordinates of points B and C to determine a unit vector along the line BC.

SOLUTION

Determine a Vector r   We need a vector from any point on the line BC to any point on the line of action of the force. We can let **r** be the vector from B to the point of application of **F** (Fig. a):

$$\mathbf{r} = (4 - 0)\,\mathbf{i} + (2 - 0)\,\mathbf{j} + (2 - 3)\,\mathbf{k} = 4\mathbf{i} + 2\mathbf{j} - \mathbf{k}\ (\text{m}).$$

Determine a Vector e   To obtain a unit vector along the bar BC, we determine the vector from B to C,

$$(0 - 0)\,\mathbf{i} + (4 - 0)\,\mathbf{j} + (0 - 3)\,\mathbf{k} = 4\mathbf{j} - 3\mathbf{k}\ (\text{m}),$$

and divide it by its magnitude (Fig. a):

$$\mathbf{e}_{BC} = \frac{4\mathbf{j} - 3\mathbf{k}}{5} = 0.8\mathbf{j} - 0.6\mathbf{k}.$$

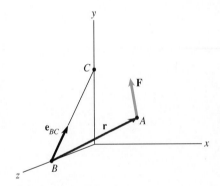

(a) The vectors **r** and $\mathbf{e}_{BC}$.

Evaluate M$_L$ Using Eq. (4.6), the mixed triple product is

$$\mathbf{e}_{BC} \cdot (\mathbf{r} \times \mathbf{F}) = \begin{vmatrix} 0 & 0.8 & -0.6 \\ 4 & 2 & -1 \\ -2 & 6 & 3 \end{vmatrix} = -24.8 \text{ kN-m}.$$

Substituting this result into Eq. (4.5), the moment of **F** about the bar BC is

$$\mathbf{M}_{BC} = [\mathbf{e}_{BC} \cdot (\mathbf{r} \times \mathbf{F})]\,\mathbf{e}_{BC} = -24.8\mathbf{e}_{BC}\ (\text{kN-m}).$$

The magnitude of $\mathbf{M}_{BC}$ is 24.8 kN-m, and its direction is opposite to that of $\mathbf{e}_{BC}$. The sense of the moment is shown in Fig. (b).

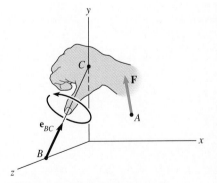

(b) The right-hand rule indicates the sense of the moment about BC.

Example 4.9

Application to Engineering

Rotating Machines

The crewman in Fig. 4.25 exerts the forces shown on the handles of the coffee grinder winch, where $\mathbf{F} = 4\mathbf{j} + 32\mathbf{k}$ N. Determine the total moment he exerts (a) about point O, (b) about the axis of the winch, which coincides with the x axis.

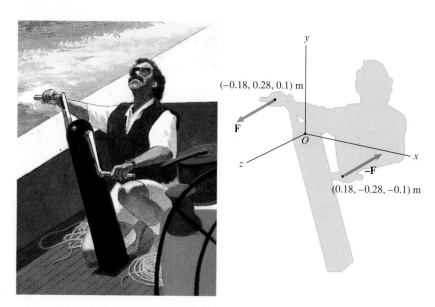

Figure 4.25

STRATEGY

(a) To obtain the total moment about point O, we must sum the moments of the two forces about O. Let the sum be denoted by $\Sigma\,\mathbf{M}_O$. (b) Because point O is on the x axis, the total moment about the x axis is the component of $\Sigma\,\mathbf{M}_O$ parallel to the x axis, which is the x component of $\Sigma\,\mathbf{M}_O$.

SOLUTION

(a) The total moment about point O is

$$\Sigma\,\mathbf{M}_O = \begin{vmatrix} \mathbf{i} & \mathbf{j} & \mathbf{k} \\ -0.18 & 0.28 & 0.1 \\ 0 & 4 & 32 \end{vmatrix} + \begin{vmatrix} \mathbf{i} & \mathbf{j} & \mathbf{k} \\ 0.18 & -0.28 & -0.1 \\ 0 & & -32 \end{vmatrix}$$

$$= 17.3\mathbf{i} + 11.5\mathbf{j} - 1.1\mathbf{k} \text{ (N-m)}.$$

(b) The total moment about the x axis is the x component of $\Sigma\,\mathbf{M}_O$ (Fig. a):

$$\Sigma\,\mathbf{M}_{(x\text{ axis})} = 17.3\mathbf{i} \text{ (N-m)}.$$

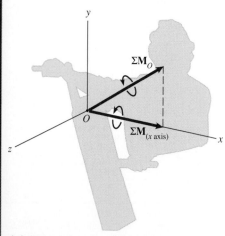

(a) The total moment about the x axis.

Notice that this is the result given by Eq. (4.4): Since **i** is a unit vector parallel to the x axis,

$$\Sigma \, \mathbf{M}_{(x \text{ axis})} = (\mathbf{i} \cdot \Sigma \, \mathbf{M}_O) \, \mathbf{i} = 17.3\mathbf{i} \text{ (N-m)}.$$

DESIGN ISSUES

The winch in this example is a simple representative of a class of rotating machines that includes hydrodynamic and aerodynamic power turbines, propellers, jet engines, and electric motors and generators. The ancestors of hydrodynamic and aerodynamic power turbines—water wheels and windmills—were among the earliest machines. These devices illustrate the importance of the concept of the moment of a force about a line. Their common feature is a part designed to rotate and perform some function when it is subjected to a moment about its axis of rotation. In the case of the winch, the forces exerted on the handles by the crewman exert a moment about the axis of rotation, causing the winch to rotate and wind a rope onto a drum, trimming the boat's sails. A hydrodynamic power turbine (Fig. 4.26) has turbine blades that are subjected to forces by flowing water, exerting a moment about the axis of rotation. This moment rotates the shaft to which the blades are attached, turning an electric generator that is connected to the same shaft.

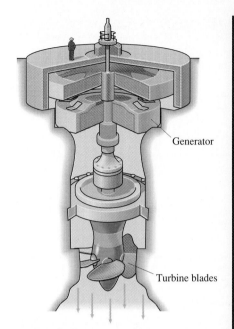

Figure 4.26

A hydroelectric turbine. Water flowing through the turbine blades exerts a moment about the axis of the shaft, turning the generator.

Problems

4.77 Determine the moment of the 40-lb force about the z axis (a) without using Eqs. (4.4)–(4.6), (b) by using Eqs. (4.5) and (4.6). (See Example 4.7.)

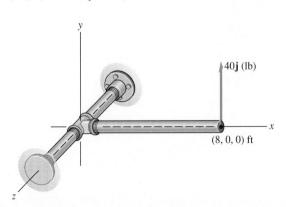

P4.77

4.78 Determine the moment of the 20-N force about (a) the x axis, (b) the y axis, (c) the z axis.

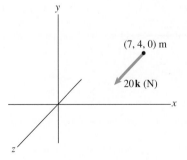

P4.78

4.79   Three forces parallel to the y axis act on the rectangular plate. Determine the sum of the moments of the forces about the x axis (a) without using Eqs. (4.4)–(4.6), (b) by using Eqs. (4.5) and (4.6).

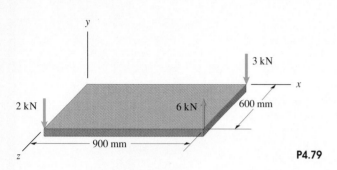

P4.79

4.80   Consider the rectangular plate shown in Problem 4.79. The three forces are parallel to the y axis. Determine the sum of the moments of the forces (a) about the y axis, (b) about the z axis.

4.81   A person exerts a force $\mathbf{F} = -2\mathbf{i} - 4\mathbf{j} - 12\mathbf{k}$ (lb) on the gate at C. What moment does the force exert about the hinge line AB of the gate?

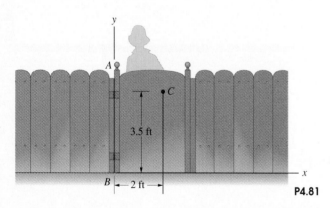

P4.81

4.82   Three forces parallel to the y axis act on the rectangular plate. The vector sum of the forces is zero, and the magnitude of the sum of the moments of the forces about O is 600 ft-lb. What is the sum of the moments of the three forces about the x axis?

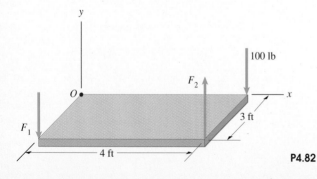

P4.82

4.83   The force $\mathbf{F} = 100\mathbf{i} + 60\mathbf{j} - 40\mathbf{k}$ (lb). What is the moment of $\mathbf{F}$ about the y axis? Draw a sketch to indicate the sense of the moment.

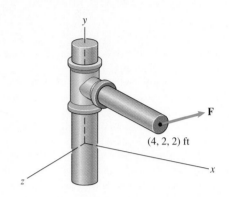

P4.83

4.84   Suppose that the moment of the force $\mathbf{F}$ shown in Problem 4.83 about the x axis is $-80\mathbf{i}$ (ft-lb), the moment about the y axis is zero, and the moment about the z axis is $160\mathbf{k}$ (ft-lb). If $F_y = 80$ lb, what are F_x and F_z?

4.85   The robotic manipulator is stationary. The weights of the arms AB and BC act at their midpoints. The direction cosines of the centerline of arm AB are $\cos\theta_x = 0.500$, $\cos\theta_y = 0.866$, $\cos\theta_z = 0$, and the direction cosines of the centerline of arm BC are $\cos\theta_x = 0.707$, $\cos\theta_y = 0.619$, $\cos\theta_z = -0.342$. What total moment is exerted about the z axis by the weights of the arms?

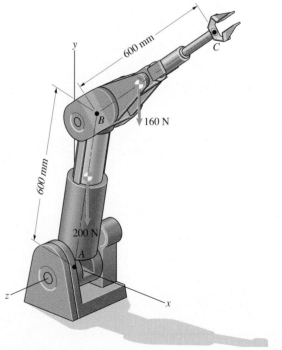

P4.85

4.86   In Problem 4.85, what total moment is exerted about the x axis by the weights of the arms?

4.87 Two forces are exerted on the crankshaft by the connecting rods. The direction cosines of $\mathbf{F}_A$ are cos $\theta_x = -0.182$, cos $\theta_y = 0.818$, and cos $\theta_z = 0.545$, and its magnitude is 4 kN. The direction cosines of $\mathbf{F}_B$ are cos $\theta_x = 0.182$, cos $\theta_y = 0.818$, and cos $\theta_z = -0.545$, and its magnitude is 2 kN. What is the sum of the moments of the two forces about the x axis? (This is the moment that causes the crankshaft to rotate.)

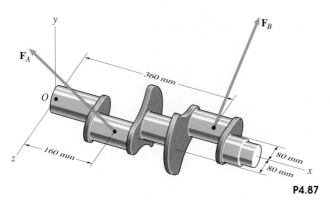

P4.87

4.88 Determine the moment of the 20-N force about the line AB. Draw a sketch to indicate the sense of the moment.

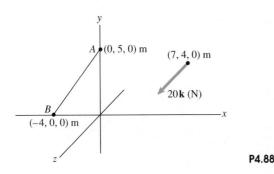

P4.88

4.89 The force $\mathbf{F} = -10\mathbf{i} + 5\mathbf{j} - 5\mathbf{k}$ (kip). Determine the moment of $\mathbf{F}$ about the line AB. Draw a sketch to indicate the sense of the moment.

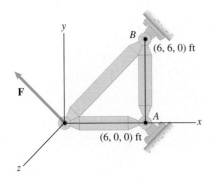

P4.89

4.90 The force $\mathbf{F} = 10\mathbf{i} + 12\mathbf{j} - 6\mathbf{k}$ (N). What is the moment of $\mathbf{F}$ about the line AO? Draw a sketch to indicate the sense of the moment.

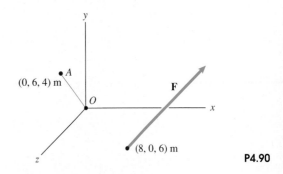

P4.90

4.91 The tension in the cable AB is 1 kN. Determine the moment about the x axis due to the force exerted on the hatch by the cable at point B. Draw a sketch to indicate the sense of the moment.

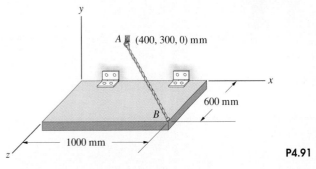

P4.91

4.92 Determine the moment of the force applied at D about the straight line through the hinges A and B.

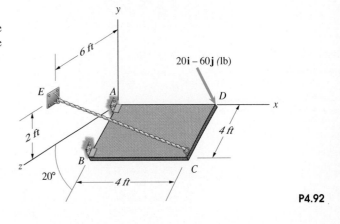

P4.92

4.93 In Problem 4.92, the tension in the cable CE is 160 lb. Determine the moment of the force exerted by the cable on the hatch at C about the straight line through the hinges A and B.

4.94 The coordinates of A are $(-2.4, 0, -0.6)$ m, and the coordinates of B are $(-2.2, 0.7, -1.2)$ m. The force exerted at B by the sailboat's main sheet AB is 130 N. Determine the moment of the force about the centerline of the mast (the y axis). Draw a sketch to indicate the sense of the moment.

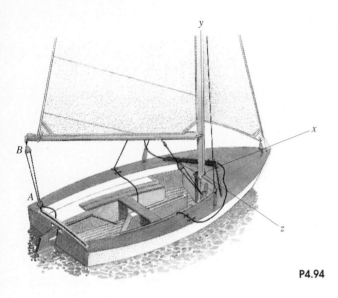

P4.94

4.95 The tension in cable AB is 200 lb. Determine the moments about each of the coordinate axes due to the force exerted on point B by the cable. Draw sketches to indicate the senses of the moments.

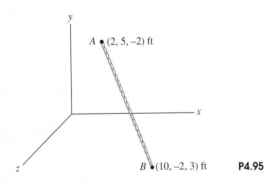

P4.95

4.96 The total force exerted on the blades of the turbine by the steam nozzle is $\mathbf{F} = 20\mathbf{i} - 120\mathbf{j} + 100\mathbf{k}$ (N), and it effectively acts at the point (100, 80, 300) mm. What moment is exerted about the axis of the turbine (the x axis)?

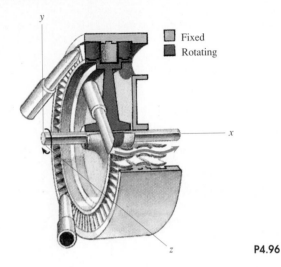

P4.96

4.97 The tension in cable AB is 50 N. Determine the moment about the line OC due to the force exerted by the cable at B. Draw a sketch to indicate the sense of the moment.

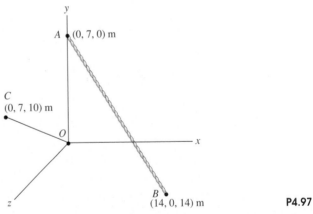

P4.97

4.98 The tension in cable AB is 80 lb. What is the moment about the line CD due to the force exerted by the cable on the wall at B?

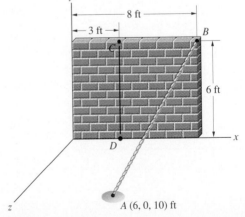

P4.98

4.99 The universal joint is connected to the drive shaft at A and A'. The coordinates of A are $(0, 40, 0)$ mm, and the coordinates of A' are $(0, -40, 0)$ mm. The forces exerted on the drive shaft by the universal joint are $-30\mathbf{j} + 400\mathbf{k}$ (N) at A and $30\mathbf{j} - 400\mathbf{k}$ (N) at A'. What is the magnitude of the torque (moment) exerted by the universal joint on the drive shaft about the shaft axis $O\text{-}O'$?

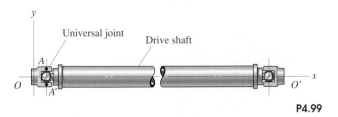

P4.99

4.102 The force $\mathbf{F} = -100\mathbf{i} + 200\mathbf{j} + 200\mathbf{k}$ (N). The moment of $\mathbf{F}$ about the line from O to A is $470\mathbf{i} + 315\mathbf{j} + 160\mathbf{k}$ (N-m). What are the direction cosines of the position vector $\mathbf{r}_{OA}$ from O to A?

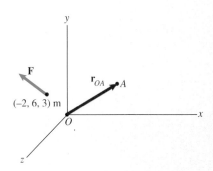

P4.102

4.100 A motorist applies the two forces shown to loosen a lug nut. The direction cosines of $\mathbf{F}$ are $\cos\theta_x = \frac{4}{13}$, $\cos\theta_y = \frac{12}{13}$, and $\cos\theta_z = \frac{3}{13}$. If the magnitude of the moment about the x axis must be 32 ft-lb to loosen the nut, what is the magnitude of the forces the motorist must apply?

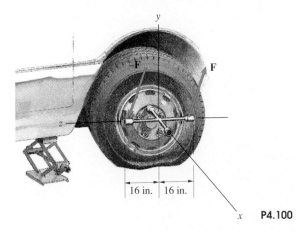

P4.100

4.101 The tension in cable AB is 2 kN. What is the magnitude of the moment about the shaft CD due to the force exerted by the cable at A? Draw a sketch to indicate the sense of the moment about the shaft.

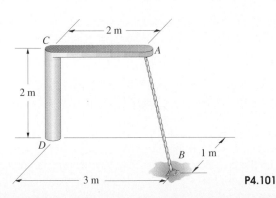

P4.101

4.103 The direction cosines of the centerline OA are $\cos\theta_x = 0.500$, $\cos\theta_y = 0.866$, and $\cos\theta_z = 0$, and the direction cosines of the line AG are $\cos\theta_x = 0.707$, $\cos\theta_y = 0.619$, and $\cos\theta_z = -0.342$. What is the moment about OA due to the 250-N weight? Draw a sketch to indicate the sense of the moment about the shaft.

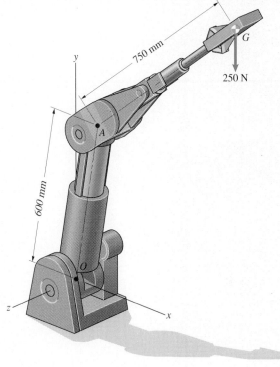

P4.103

4.104 The radius of the steering wheel is 200 mm. The distance from O to C is 1 m. The center C of the steering wheel lies in the x-y plane. The driver exerts a force $\mathbf{F} = 10\mathbf{i} + 10\mathbf{j} - 5\mathbf{k}$ (N) on the wheel at A. If the angle $\alpha = 0$, what is the magnitude of the moment about the shaft OC? Draw a sketch to indicate the sense of the moment about the shaft.

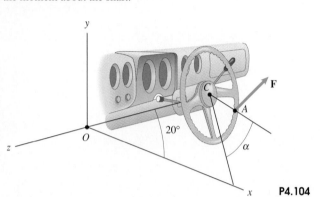

P4.104

4.105 Consider the steering wheel in Problem 4.104. Determine the moment of $\mathbf{F}$ about the shaft OC of the steering wheel if $\alpha = 30°$. Draw a sketch to indicate the sense of the moment about the shaft.

4.106 The weight W causes a tension of 100 lb in cable CD. If $d = 2$ ft, what is the moment about the z axis due to the force exerted by the cable CD at point C?

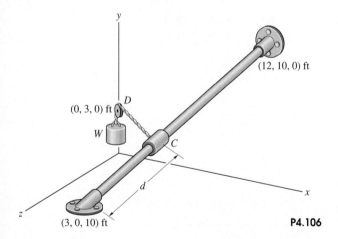

P4.106

4.107 The rod AB supports the open hood of the car. The force exerted by the rod on the hood at B is parallel to the rod. If the rod must exert a moment of 100 ft-lb magnitude about the x axis to support the hood and the distance $d = 2$ ft, what is the magnitude of the force the rod must exert on the hood?

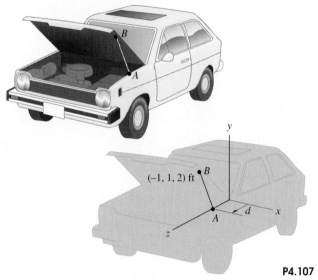

P4.107

4.4 Couples

Now that you've seen how to calculate the moment due to a force, consider this question: Is it possible to exert a moment on an object without subjecting it to a net force? The answer is yes, and it occurs when you play a compact disk, start the rotor of an electric motor, or tighten a screw with a screwdriver. Forces are exerted on these objects, but in such a way that the net force is zero while the net moment is not zero.

Two forces that have equal magnitudes, opposite directions, and different lines of action are called a **couple** (Fig. 4.27a). A couple tends to cause rotation of an object even though the vector sum of the forces is zero, and it has the remarkable property that *the moment it exerts is the same about any point.*

The moment of a couple is simply the sum of the moments of the forces about a point P (Fig. 4.27b):

$$\mathbf{M} = [\mathbf{r}_1 \times \mathbf{F}] + [\mathbf{r}_2 \times (-\mathbf{F})] = (\mathbf{r}_1 - \mathbf{r}_2) \times \mathbf{F}.$$

The vector $\mathbf{r}_1 - \mathbf{r}_2$ is equal to the vector $\mathbf{r}$ shown in Fig. 4.27(c), so we can express the moment as

$$\mathbf{M} = \mathbf{r} \times \mathbf{F}.$$

Since $\mathbf{r}$ doesn't depend on the position of P, the moment $\mathbf{M}$ is the same for *any* point P.

Because a couple exerts a moment but the sum of the forces is zero, it is often represented in diagrams simply by showing the moment (Fig. 4.27d). Like the Cheshire cat in *Alice's Adventures in Wonderland*, which vanished except for its grin, the forces don't appear; you see only the moment they exert. But we recognize the origin of the moment by referring to it as a **moment of a couple**, or simply a **couple**.

Notice in Fig. 4.27(c) that $\mathbf{M} = \mathbf{r} \times \mathbf{F}$ is the moment of $\mathbf{F}$ about a point on the line of action of the force $-\mathbf{F}$. The magnitude of the moment of a force about a point equals the product of the magnitude of the force and the perpendicular distance from the point to the line of action of the force, so $|\mathbf{M}| = D|\mathbf{F}|$, where D is the perpendicular distance between the lines of action of the two forces (Fig. 4.27e). The cross product $\mathbf{r} \times \mathbf{F}$ is perpendicular to $\mathbf{r}$ and $\mathbf{F}$, which means that $\mathbf{M}$ is perpendicular to the plane containing $\mathbf{F}$ and $-\mathbf{F}$ (Fig. 4.27f). Pointing the thumb of the right hand in the direction of $\mathbf{M}$, the arc of the fingers indicates the sense of the moment.

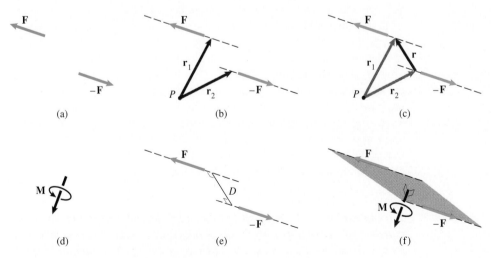

(a)

(b)

(c)

(d)

(e)

(f)

Figure 4.27
(a) A couple.
(b) Determining the moment about P.
(c) The vector $\mathbf{r} = \mathbf{r}_1 - \mathbf{r}_2$.
(d) Representing the moment of the couple.
(e) The distance D between the lines of action.
(f) $\mathbf{M}$ is perpendicular to the plane containing $\mathbf{F}$ and $-\mathbf{F}$.

In Fig. 4.28(a), our view is perpendicular to the plane containing the two forces. The distance between the lines of action of the forces is 4 m, so the magnitude of the moment of the couple is $|\mathbf{M}| = (4 \text{ m})(2 \text{ kN}) = 8$ kN-m. The

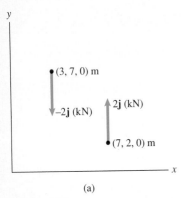

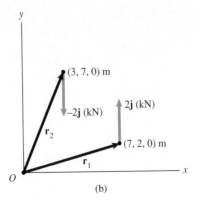

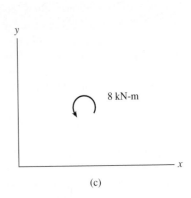

(a)

(b)

(c)

Figure 4.28

(a) A couple consisting of 2-kN forces.

(b) Determining the sum of the moments of the forces about O.

(c) Representing a couple in two dimensions.

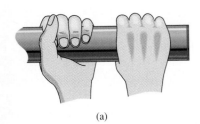

(a)

(b)

(c)

Figure 4.29

(a) Twisting a bar.

(b) The moment about the axis of the bar.

(c) The same effect is obtained by applying two equal and opposite forces.

moment $\mathbf{M}$ is perpendicular to the plane containing the two forces. Pointing the arc of the fingers of the right hand counterclockwise, we find that the right-hand rule indicates that $\mathbf{M}$ points out of the page. Therefore the moment of the couple is

$$\mathbf{M} = 8\mathbf{k} \text{ (kN-m)}.$$

We can also determine the moment of the couple by calculating the sum of the moments of the two forces about *any* point. The sum of the moments of the forces about the origin O is (Fig. 4.28b)

$$\mathbf{M} = [\mathbf{r}_1 \times (2\mathbf{j})] + [\mathbf{r}_2 \times (-2\mathbf{j})]$$
$$= [(7\mathbf{i} + 2\mathbf{j}) \times (2\mathbf{j})] + [(3\mathbf{i} + 7\mathbf{j}) \times (-2\mathbf{j})]$$
$$= 8\mathbf{k} \text{ (kN-m)}.$$

In a two-dimensional situation like this example, it isn't convenient to represent a couple by showing the moment vector, because the vector is perpendicular to the page. Instead, we represent the couple by showing its magnitude and a circular arrow that indicates its sense (Fig. 4.28c).

If you grasp a bar and twist it (Fig. 4.29a), you can exert a moment about its axis (Fig. 4.29b). Although the system of forces you exert on the bar is distributed over its surface in a complicated way, the effect is the same as if you exerted two equal and opposite forces (Fig. 4.29c). When we represent a couple as in Fig. 4.29(b), or by showing the moment vector $\mathbf{M}$, we imply that some system of forces exerts that moment. The system of forces (such as the forces you exert in twisting the bar, or the forces on the crankshaft that exert a moment on the drive shaft of a car) is nearly always more complicated than two equal and opposite forces, but the effect is the same. For this reason, we can *model* the actual system as a simple system of two forces.

You can determine the moment $\mathbf{M}$ of a couple by calculating the sum of the moments of its two forces about any point. In some cases you can also determine it from two of its properties: The magnitude of $\mathbf{M}$ is equal to the product of the magnitude of one of the forces and the perpendicular distance between the lines of action of the forces, and the direction of $\mathbf{M}$ is perpendicular to the plane containing the lines of action. The following examples demonstrate both approaches.

Example 4.10

The force **F** in Fig. 4.30 is $10\mathbf{i} - 4\mathbf{j}$ (N). Determine the moment of the couple and represent it as shown in Fig. 4.29(b).

STRATEGY

We can determine the moment in two ways: We can calculate the sum of the moments of the forces about a point, or we can sum the moments of the two couples formed by the x and y components of the forces.

SOLUTION

First Method If we calculate the sum of the moments of the forces about a point on the line of action of one of the forces, the moment of that force is zero and we only need to calculate the moment of the other force. Choosing the point of application of **F** (Fig. a), we calculate the moment as

$$\mathbf{M} = \mathbf{r} \times (-\mathbf{F}) = (-2\mathbf{i} + 3\mathbf{j}) \times (-10\mathbf{i} + 4\mathbf{j}) = 22\mathbf{k} \text{ (N-m)}.$$

We would obtain the same result by calculating the sum of the moments about any point. For example, the sum of the moments about the point P in Fig. (b) is

$$\mathbf{M} = [\mathbf{r}_1 \times \mathbf{F}] + [\mathbf{r}_2 \times (-\mathbf{F})]$$

$$= \begin{vmatrix} \mathbf{i} & \mathbf{j} & \mathbf{k} \\ -2 & -4 & -3 \\ 10 & -4 & 0 \end{vmatrix} + \begin{vmatrix} \mathbf{i} & \mathbf{j} & \mathbf{k} \\ -4 & -1 & -3 \\ -10 & 4 & 0 \end{vmatrix}$$

$$= 22\mathbf{k} \text{ (N-m)}.$$

Second Method The x and y components of the forces form two couples (Fig. c). We determine the moment of the original couple by summing the moments of the couples formed by the components.

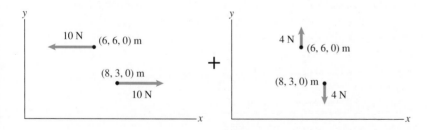

Consider the 10-N couple. The magnitude of its moment is $(3 \text{ m})(10 \text{ N}) = 30$ N-m, and its sense is counterclockwise, indicating that the moment vector points out of the page. Therefore the moment is $30\mathbf{k}$ N-m.

The 4-N couple causes a moment of magnitude $(2 \text{ m})(4 \text{ N}) = 8$ N-m and its sense is clockwise, so the moment is $-8\mathbf{k}$ N-m. The moment of the original couple is

$$\mathbf{M} = 30\mathbf{k} - 8\mathbf{k} = 22\mathbf{k} \text{ (N-m)}.$$

Its magnitude is 22 N-m, and its sense is counterclockwise (Fig. d).

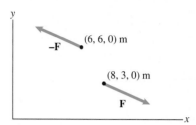

Figure 4.30

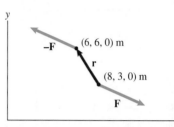

(a) Determining the moment about the point of application of **F**.

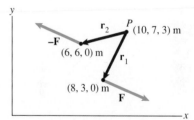

(b) Determining the moment about P.

(c) The x and y components form two couples.

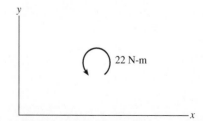

(d) Representing the moment.

Example 4.11

Two forces A and B and a 200 ft-lb couple act on the beam in Fig. 4.31. The sum of the forces is zero, and the sum of the moments about the left end of the beam is zero. What are the forces A and B?

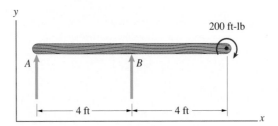

Figure 4.31

SOLUTION

The sum of the forces is

$$\Sigma F_y = A + B = 0.$$

The moment of the couple (200 ft-lb clockwise) is the same about any point, so the sum of the moments about the left end of the beam is

$$\Sigma M_{(\text{left end})} = 4B - 200 = 0.$$

The forces are $B = 50$ lb and $A = -50$ lb.

DISCUSSION

Notice that A and B form a couple (Fig. a). It causes a moment of magnitude (4 ft)(50 lb) = 200 ft-lb, and its sense is counterclockwise, so the sum of the moments of the couple formed by A and B and the 200 ft-lb clockwise couple is zero.

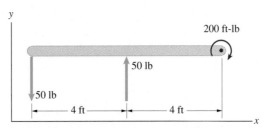

(a) The forces on the beam form a couple.

Example 4.12

Determine the sum of the moments exerted on the pipe in Fig. 4.32 by the two couples.

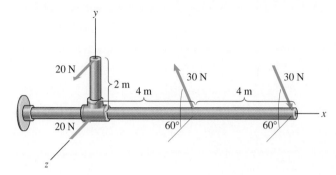

Figure 4.32

SOLUTION

Consider the 20-N couple. The magnitude of the moment of the couple is $(2 \text{ m})(20 \text{ N}) = 40$ N-m. The direction of the moment vector is perpendicular to the y-z plane, and the right-hand rule indicates that it points in the positive x-axis direction. The moment of the 20-N couple is $40\mathbf{i}$ (N-m).

By resolving the 30-N forces into y and z components, we obtain the two couples in Fig. (a). The moment of the couple formed by the y components is $-(30 \sin 60°)(4) \mathbf{k}$ (N-m), and the moment of the couple formed by the z components is $(30 \cos 60°)(4) \mathbf{j}$ (N-m).

The sum of the moments is

$$\Sigma \mathbf{M} = 40\mathbf{i} + (30 \cos 60°)(4) \mathbf{j} - (30 \sin 60°)(4) \mathbf{k}$$

$$= 40\mathbf{i} + 60\mathbf{j} - 103.9\mathbf{k} \text{ (N-m)}.$$

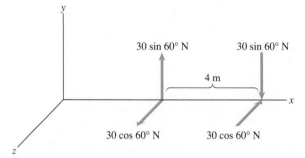

(a) Resolving the 30-N forces into y and z components.

DISCUSSION

Although the method we used in this example helps you recognize the contributions of the individual couples to the sum of the moments, it is convenient only when the orientations of the forces and their points of application relative to the coordinate system are fairly simple. When that is not the case, you can determine the sum of the moments by choosing any point and calculating the sum of the moments of the forces about that point.

Example 4.13

The force **F** in Fig. 4.33 is $-20\mathbf{i} + 20\mathbf{j} + 10\mathbf{k}$ (lb).
(a) What moment does the couple exert on the bracket?
(b) What is the perpendicular distance D between the lines of action of the two forces?

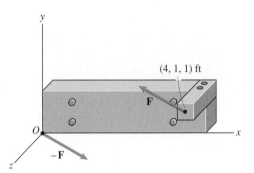

Figure 4.33

STRATEGY

(a) We can choose a point and determine the sum of the moments of the forces about that point.
(b) The magnitude of the moment of the couple equals $D|\mathbf{F}|$, so we can use the result of (a) to determine D.

SOLUTION

(a) If we determine the sum of the moments of the forces about the origin O, the moment of the force $-\mathbf{F}$ is zero. The moment of the couple is (Fig. a)

$$\mathbf{M} = \mathbf{r} \times \mathbf{F} = \begin{vmatrix} \mathbf{i} & \mathbf{j} & \mathbf{k} \\ 4 & 1 & 1 \\ -20 & 20 & 10 \end{vmatrix} = -10\mathbf{i} - 60\mathbf{j} + 100\mathbf{k} \text{ (ft-lb)}.$$

(b) The perpendicular distance is

$$D = \frac{|\mathbf{M}|}{|\mathbf{F}|} = \frac{\sqrt{(-10)^2 + (-60)^2 + (100)^2}}{} = 3.90 \text{ ft}.$$

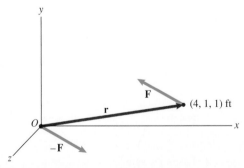

(a) Determining the sum of the moments about O.

Problems

4.108 (a) Determine the moment of the couple.
(b) What is the sum of the moments of the two forces about the point $(-4, 6, -2)$ ft?
(c) Represent the couple as shown in Fig. 4.28(c).

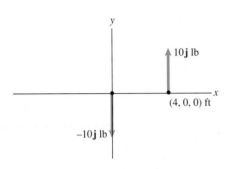

P4.108

4.109 The forces are contained in the x-y plane.
(a) Determine the moment of the couple.
(b) What is the sum of the moments of the two forces about the point $(10, -40, 20)$ ft?
(c) Represent the couple as shown in Fig. 4.28(c).

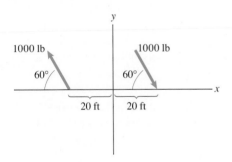

P4.109

4.110 The forces are contained in the x-y plane, and the moment of the couple is $-90\mathbf{k}$ N-m.
(a) What is the distance b?
(b) What is the sum of the moments of the two forces about the point $(3, 3, -2)$ m?

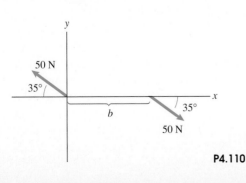

P4.110

4.111 Point P is contained in the x-y plane, $|\mathbf{F}| = 100$ N, and the moment of the couple is $-500\mathbf{k}$ (N-m). What are the coordinates of P?

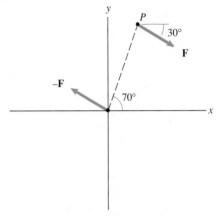

P4.111

4.112 The forces are contained in the x-y plane.
(a) Determine the sum of the moments of the two couples.
(b) What is the sum of the moments of the four forces about the point $(-6, -6, 2)$ m?
(c) Represent the result of (a) as shown in Fig. 4.28(c).

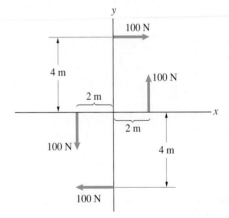

P4.112

4.113 The moment of a couple is 40 kN-m counterclockwise.
(a) Express the moment of the couple as a vector.
(b) Draw a sketch showing two equal and opposite forces that exert the given moment.

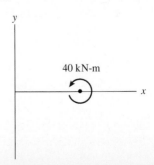

P4.113

4.114 The moments of two couples are shown. What is the sum of the moments about point P?

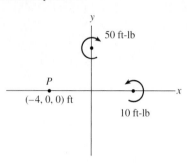

P4.114

4.115 Determine the sum of the moments exerted on the plate by the two couples.

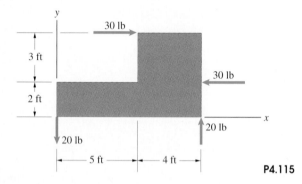

P4.115

4.116 Determine the sum of the moments exerted about A by the couple and the two forces.

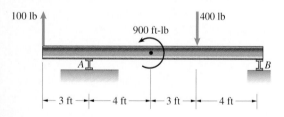

P4.116

4.117 In Problem 4.116, determine the sum of the moments exerted about B by the couple and the two forces.

4.118 What is the sum of the moments exerted on the object?

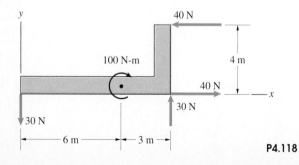

P4.118

4.119 Four forces and a couple act on the beam. The vector sum of the forces is zero, and the sum of the moments about the left end of the beam is zero. What are the forces A_x, A_y, and B?

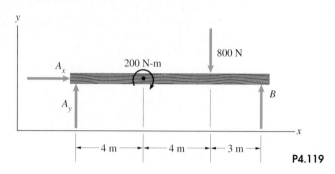

P4.119

4.120 The force $\mathbf{F} = 3\mathbf{i} + 2\mathbf{j} - \mathbf{k}$ (kN).
(a) What is the moment of the couple?
(b) What is the perpendicular distance between the lines of action of the two forces?

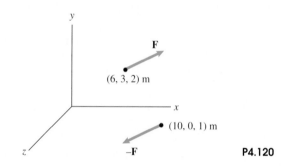

P4.120

4.121 Determine the sum of the moments exerted on the plate by the three couples. (The 80-lb forces are contained in the x-z plane.)

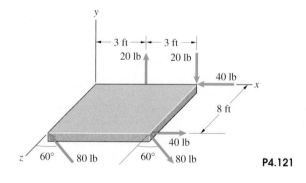

P4.121

4.122 What is the magnitude of the sum of the moments exerted on the T-shaped structure by the two couples?

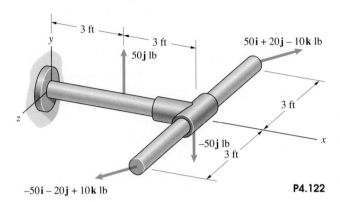

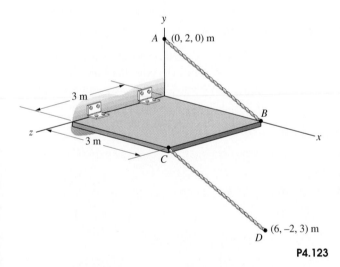

P4.122

4.123 The tension in cables AB and CD is 500 N.
(a) Show that the two forces exerted by the cables on the rectangular hatch at B and C form a couple.
(b) What is the moment exerted on the plate by the cables?

P4.123

4.124 Consider the system shown in Problem 4.123. The tensions in cables AB and CD are equal, and the magnitude of the moment exerted on the plate by the cables is 6 kN-m. What is the tension in the cables?

4.125 The bar is loaded by the forces

$$\mathbf{F}_B = 2\mathbf{i} + 6\mathbf{j} + 3\mathbf{k} \text{ (kN)},$$

$$\mathbf{F}_C = \mathbf{i} - 2\mathbf{j} + 2\mathbf{k} \text{ (kN)},$$

and the couple

$$\mathbf{M}_C = 2\mathbf{i} + \mathbf{j} - 2\mathbf{k} \text{ (kN-m)}.$$

Determine the sum of the moments of the two forces and the couple about A.

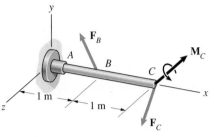

P4.125

4.126 In Problem 4.125, the forces

$$\mathbf{F}_B = 2\mathbf{i} + 6\mathbf{j} + 3\mathbf{k} \text{ (kN)},$$

$$\mathbf{F}_C = \mathbf{i} - 2\mathbf{j} + 2\mathbf{k} \text{ (kN)},$$

and the couple

$$\mathbf{M}_C = M_{Cy}\mathbf{j} + M_{Cz}\mathbf{k} \text{ (kN-m)}.$$

Determine the values of M_{Cy} and M_{Cz} so that the sum of the moments of the two forces and the couple about A is zero.

4.127 Two wrenches are used to tighten an elbow fitting. The force $\mathbf{F} = 10\mathbf{k}$ (lb) on the right wrench is applied at $(6, -5, -3)$ in., and the force $-\mathbf{F}$ on the left wrench is applied at $(4, -5, 3)$ in.
(a) Determine the moment about the x axis due to the force exerted on the right wrench.
(b) Determine the moment of the couple formed by the forces exerted on the two wrenches.
(c) Based on the results of (a) and (b), explain why two wrenches are used.

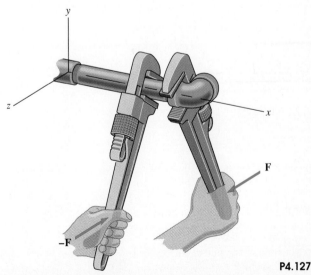

P4.127

4.5 *Equivalent Systems*

A **system of forces and moments** is simply a particular set of forces and moments of couples. The systems of forces and moments you have to deal with in engineering can be complicated. This is especially true in the case of distributed forces, such as the pressure forces exerted by water on a dam. Fortunately, if you are concerned only with the total force and moment exerted, you can represent a complicated system of forces and moments by a much simpler system.

Conditions for Equivalence

We define two systems of forces and moments, designated as system 1 and system 2, **equivalent** if the sums of the forces are equal,

$$(\Sigma \mathbf{F})_1 = (\Sigma \mathbf{F})_2, \tag{4.7}$$

and the sums of the moments about a point P are equal,

$$(\Sigma \mathbf{M}_P)_1 = (\Sigma \mathbf{M}_P)_2. \tag{4.8}$$

Demonstration of Equivalence

To see what the conditions for equivalence mean, consider the systems of forces and moments in Fig. 4.34(a). In system 1, an object is subjected to two forces $\mathbf{F}_A$ and $\mathbf{F}_B$ and a couple $\mathbf{M}_C$. In system 2, the object is subjected to a force $\mathbf{F}_D$ and two couples $\mathbf{M}_E$ and $\mathbf{M}_F$. The first condition for equivalence is

$$(\Sigma \mathbf{F})_1 = (\Sigma \mathbf{F})_2:$$
$$\mathbf{F}_A + \mathbf{F}_B = \mathbf{F}_D. \tag{4.9}$$

If we determine the sums of the moments about the point P in Fig. 4.34(b), the second condition for equivalence is

$$(\Sigma \mathbf{M}_P)_1 = (\Sigma \mathbf{M}_P)_2:$$
$$(\mathbf{r}_A \times \mathbf{F}_A) + (\mathbf{r}_B \times \mathbf{F}_B) + \mathbf{M}_C = (\mathbf{r}_D \times \mathbf{F}_D) + \mathbf{M}_E + \mathbf{M}_F. \tag{4.10}$$

If these conditions are satisfied, systems 1 and 2 are equivalent.

System 1

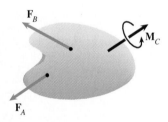

System 2

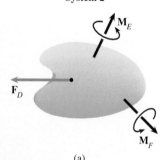

(a)

Figure 4.34

(a) Different systems of forces and moments applied to an object.
(b) Determining the sum of the moments about a point P for each system.

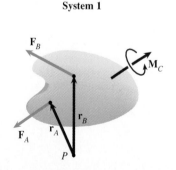

(b)

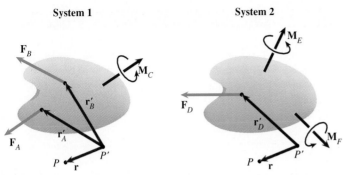

Figure 4.35

Determining the sum of the moments about a different point P' for each system.

We will use this example to demonstrate that if the sums of the forces are equal for two systems of forces and moments and the sums of the moments about one point P are equal, then the sums of the moments about any point are equal. Suppose that Eq. (4.9) is satisfied, and Eq. (4.10) is satisfied for the point P in Fig. 4.34(b). For a different point P' (Fig. 4.35), we will show that

$$(\Sigma \, \mathbf{M}_{P'})_1 = (\Sigma \, \mathbf{M}_{P'})_2:$$

$$(\mathbf{r}'_A \times \mathbf{F}_A) + (\mathbf{r}'_B \times \mathbf{F}_B) + \mathbf{M}_C = (\mathbf{r}'_D \times \mathbf{F}_D) + \mathbf{M}_E + \mathbf{M}_F. \quad (4.11)$$

In terms of the vector $\mathbf{r}$ from P' to P, the relations between the vectors $\mathbf{r}'_A$, $\mathbf{r}'_B$, and $\mathbf{r}'_D$ in Fig. 4.35 and the vectors $\mathbf{r}_A$, $\mathbf{r}_B$, and $\mathbf{r}_D$ in Fig. 4.34(b) are

$$\mathbf{r}'_A = \mathbf{r} + \mathbf{r}_A, \qquad \mathbf{r}'_B = \mathbf{r} + \mathbf{r}_B, \qquad \mathbf{r}'_D = \mathbf{r} + \mathbf{r}_D.$$

Substituting these expressions into Eq. (4.11), we obtain

$$[(\mathbf{r} + \mathbf{r}_A) \times \mathbf{F}_A] + [(\mathbf{r} + \mathbf{r}_B) \times \mathbf{F}_B] + \mathbf{M}_C$$
$$= [(\mathbf{r} + \mathbf{r}_D) \times \mathbf{F}_D] + \mathbf{M}_E + \mathbf{M}_F.$$

Rearranging terms, we can write this equation as

$$[\mathbf{r} \times (\Sigma \, \mathbf{F})_1] + (\Sigma \, \mathbf{M}_P)_1 = [\mathbf{r} \times (\Sigma \, \mathbf{F})_2] + (\Sigma \, \mathbf{M}_P)_2,$$

which holds in view of Eqs. (4.9) and (4.10). The sums of the moments of the two systems about any point are equal.

The following examples demonstrate how you can determine whether given systems of forces and moments are equivalent. You must check the two conditions for equivalence:

1. Are the sums of the forces equal?—*You must determine the vector sums of the forces in the two systems to see whether they are equal.*

2. Are the sums of the moments about an arbitrary point equal?—*You can choose any convenient point and determine the sums of the moments of the two systems about that point to see whether they are equal.*

Example 4.14

Three systems of forces and moments act on the beam in Fig. 4.36. Are they equivalent?

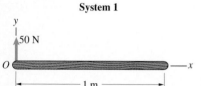

Figure 4.36

SOLUTION

Are the Sums of the Forces Equal?   The sums of the forces are

$$(\Sigma \mathbf{F})_1 = 50\mathbf{j} \ (N),$$

$$(\Sigma \mathbf{F})_2 = 50\mathbf{j} \ (N),$$

$$(\Sigma \mathbf{F})_3 = 50\mathbf{j} \ (N).$$

Are the Sums of the Moments About an Arbitrary Point Equal?   The sums of the moments about the origin O are

$$(\Sigma M_O)_1 = 0,$$

$$(\Sigma M_O)_2 = (50 \ N)(0.5 \ m) - (50 \ N\text{-}m) = -25 \ N\text{-}m,$$

$$(\Sigma M_O)_3 = (50 \ N)(1 \ m) - (50 \ N\text{-}m) = 0.$$

Systems 1 and 3 are equivalent.

DISCUSSION

Remember that you can choose any convenient point to determine whether the sums of the moments are equal. For example, the sums of the moments about the right end of the beam are

$$(\Sigma M_{\text{right end}})_1 = -(50 \ N)(1 \ m) = -50 \ N\text{-}m,$$

$$(\Sigma M_{\text{right end}})_2 = -(50 \ N)(0.5 \ m) - (50 \ N\text{-}m) = -75 \ N\text{-}m,$$

$$(\Sigma M_{\text{right end}})_3 = -50 \ N\text{-}m.$$

Example 4.15

Two systems of forces and moments act on the rectangular plate in Fig. 4.37. Are they equivalent?

System 1

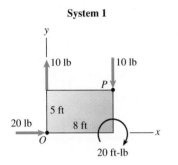

System 2

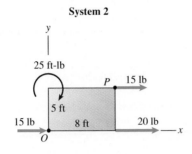

Figure 4.37

SOLUTION

Are the Sums of the Forces Equal?   The sums of the forces are

$$(\Sigma \, \mathbf{F})_1 = 20\mathbf{i} + 10\mathbf{j} - 10\mathbf{j} = 20\mathbf{i} \ (\text{lb}),$$

$$(\Sigma \, \mathbf{F})_2 = 20\mathbf{i} + 15\mathbf{i} - 15\mathbf{i} = 20\mathbf{i} \ (\text{lb}).$$

Are the Sums of the Moments About an Arbitrary Point Equal?   The sums of the moments about the origin O are

$$(\Sigma \, M_O)_1 = -(8 \text{ ft})(10 \text{ lb}) - (20 \text{ ft-lb}) = -100 \text{ ft-lb},$$

$$(\Sigma \, M_O)_2 = -(5 \text{ ft})(15 \text{ lb}) - (25 \text{ ft-lb}) = -100 \text{ ft-lb}.$$

The systems are equivalent.

DISCUSSION

Let's confirm that the sums of the moments of the two systems about a different point are equal. The sums of the moments about P are

$$(\Sigma \, M_P)_1 = -(8 \text{ ft})(10 \text{ lb}) + (5 \text{ ft})(20 \text{ lb}) - (20 \text{ ft-lb}) = 0,$$

$$(\Sigma \, M_P)_2 = -(5 \text{ ft})(15 \text{ lb}) + (5 \text{ ft})(20 \text{ lb}) - (25 \text{ ft-lb}) = 0.$$

Example 4.16

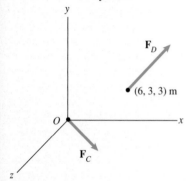

System 1

System 2

Figure 4.38

Two systems of forces and moments are shown in Fig. 4.38, where

$$\mathbf{F}_A = -10\mathbf{i} + 10\mathbf{j} - 15\mathbf{k} \text{ (kN)},$$

$$\mathbf{F}_B = 30\mathbf{i} + 5\mathbf{j} + 10\mathbf{k} \text{ (kN)},$$

$$\mathbf{M} = -90\mathbf{i} + 150\mathbf{j} + 60\mathbf{k} \text{ (kN-m)},$$

$$\mathbf{F}_C = 10\mathbf{i} - 5\mathbf{j} + 5\mathbf{k} \text{ (kN)},$$

$$\mathbf{F}_D = 10\mathbf{i} + 20\mathbf{j} - 10\mathbf{k} \text{ (kN)}.$$

Are they equivalent?

SOLUTION

Are the Sums of the Forces Equal? The sums of the forces are

$$(\Sigma\, \mathbf{F})_1 = \mathbf{F}_A + \mathbf{F}_B = 20\mathbf{i} + 15\mathbf{j} - 5\mathbf{k} \text{ (kN)},$$

$$(\Sigma\, \mathbf{F})_2 = \mathbf{F}_C + \mathbf{F}_D = 20\mathbf{i} + 15\mathbf{j} - 5\mathbf{k} \text{ (kN)}.$$

Are the Sums of the Moments About an Arbitrary Point Equal? The sum of the moments about the origin O in system 1 is

$$(\Sigma\, \mathbf{M}_O)_1 = (6\mathbf{i} \times \mathbf{F}_B) + \mathbf{M}$$

$$= \begin{vmatrix} \mathbf{i} & \mathbf{j} & \mathbf{k} \\ 6 & 0 & 0 \\ 30 & 5 & 10 \end{vmatrix} + (-90\mathbf{i} + 150\mathbf{j} + 60\mathbf{k})$$

$$= -90\mathbf{i} + 90\mathbf{j} + 90\mathbf{k} \text{ (kN-m)}.$$

The sum of the moments about O in system 2 is

$$(\Sigma\, \mathbf{M}_O)_2 = (6\mathbf{i} + 3\mathbf{j} + 3\mathbf{k}) \times \mathbf{F}_D = \begin{vmatrix} \mathbf{i} & \mathbf{j} & \mathbf{k} \\ 6 & 3 & 3 \\ 10 & 20 & -10 \end{vmatrix}$$

$$= -90\mathbf{i} + 90\mathbf{j} + 90\mathbf{k} \text{ (kN-m)}.$$

The systems are equivalent.

4.6 *Representing Systems by Equivalent Systems*

If you are concerned only with the total force and total moment exerted on an object by a given system of forces and moments, you can **represent** the system by an equivalent one. By this we mean that instead of showing the actual forces and couples acting on an object, you would show a different system that exerts the same total force and moment. In this way, you can replace a given system by a less complicated one to simplify the analysis of the forces and moments acting on an object and to gain a better intuitive understanding of their effects on the object.

Representing a System by a Force and a Couple

Let's consider an arbitrary system of forces and moments and a point P (system 1 in Fig. 4.39). We can represent this system by one consisting of a single force acting at P and a single couple (system 2). The conditions for equivalence are

$$(\Sigma\, \mathbf{F})_2 = (\Sigma\, \mathbf{F})_1:$$
$$\mathbf{F} = (\Sigma\, \mathbf{F})_1$$

and

$$(\Sigma\, \mathbf{M}_P)_2 = (\Sigma\, \mathbf{M}_P)_1:$$
$$\mathbf{M} = (\Sigma\, \mathbf{M}_P)_1.$$

These conditions are satisfied if $\mathbf{F}$ equals the sum of the forces in system 1 and $\mathbf{M}$ equals the sum of the moments about P in system 1.

Thus no matter how complicated a system of forces and moments may be, you can represent it by a single force acting at a given point and a single couple. Three particular cases occur frequently in practice.

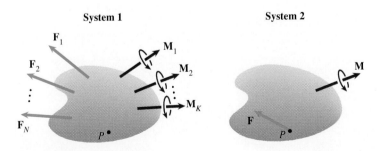

Figure 4.39

(a) An arbitrary system of forces and moments.
(b) A force acting at P and a couple.

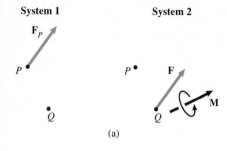

(a)

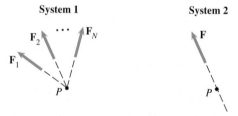

(b)

Figure 4.40

(a) System 1 is a force $\mathbf{F}_P$ acting at point P. System 2 consists of a force $\mathbf{F}$ acting at point Q and a couple $\mathbf{M}$.
(b) Determining the moment of system 1 about point Q.

Representing a Force by a Force and a Couple You can represent a force $\mathbf{F}_P$ acting at a point P (system 1 in Fig. 4.40a) by a force $\mathbf{F}$ acting at a different point Q and a couple $\mathbf{M}$ (system 2). The moment of system 1 about point Q is $\mathbf{r} \times \mathbf{F}_P$, where $\mathbf{r}$ is the vector from Q to P (Fig. 4.40b). The conditions for equivalence are

$$(\Sigma\,\mathbf{F})_2 = (\Sigma\,\mathbf{F})_1:$$
$$\mathbf{F} = \mathbf{F}_P$$

and

$$(\Sigma\,\mathbf{M}_Q)_2 = (\Sigma\,\mathbf{M}_Q)_1:$$
$$\mathbf{M} = \mathbf{r} \times \mathbf{F}_P.$$

The systems are equivalent if the force $\mathbf{F}$ equals the force $\mathbf{F}_P$ and the couple $\mathbf{M}$ equals the moment of $\mathbf{F}_P$ about Q.

Concurrent Forces Represented by a Force You can represent a system of concurrent forces whose lines of action intersect at a point P (system 1 in Fig. 4.41) by a single force whose line of action intersects P (system 2). The sums of the forces in the two systems are equal if

$$\mathbf{F} = \mathbf{F}_1 + \mathbf{F}_2 + \cdots + \mathbf{F}_N.$$

The sum of the moments about P equals zero for each system, so the systems are equivalent if the force $\mathbf{F}$ equals the sum of the forces in system 1.

Figure 4.41

A system of concurrent forces and a system consisting of a single force $\mathbf{F}$.

Parallel Forces Represented by a Force You can represent a system of parallel forces whose sum is not zero by a single force $\mathbf{F}$ (Fig. 4.42). We demonstrate this result in Example 4.20.

Figure 4.42

A system of parallel forces and a system consisting of a single force $\mathbf{F}$.

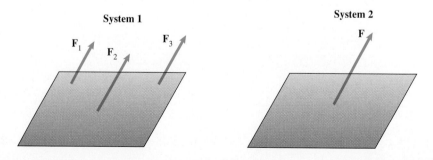

The following examples demonstrate how to represent systems of forces and moments by other systems. Given a system of forces and moments, your objective will be to represent it by a new system of a particular kind. You must determine the new system by using the conditions for equivalence:

1. The sums of the forces must be equal.

2. The sums of the moments about an arbitrary point must be equal.

Example 4.17

System 1 in Fig. 4.43 consists of a force $\mathbf{F}_A = 10\mathbf{i} + 4\mathbf{j} - 3\mathbf{k}$ (lb) acting at A. Represent it by a force acting at B and a couple.

STRATEGY

We want to represent the force $\mathbf{F}_A$ by a force $\mathbf{F}$ acting at B and a couple $\mathbf{M}$ (system 2 in Fig. a). We can determine $\mathbf{F}$ and $\mathbf{M}$ by using the two conditions for equivalence.

SOLUTION

The Sums of the Forces Must Be Equal

$$(\Sigma\,\mathbf{F})_2 = (\Sigma\,\mathbf{F})_1:$$
$$\mathbf{F} = \mathbf{F}_A = 10\mathbf{i} + 4\mathbf{j} - 3\mathbf{k}\ (\text{lb}).$$

The Sums of the Moments About an Arbitrary Point Must Be Equal   The vector from B to A is

$$\mathbf{r}_{BA} = (4 - 8)\,\mathbf{i} + (4 - 0)\,\mathbf{j} + (2 - 6)\,\mathbf{k} = -4\mathbf{i} + 4\mathbf{j} - 4\mathbf{k}\ (\text{ft}),$$

so the moment about B in system 1 is

$$\mathbf{r}_{BA} \times \mathbf{F}_A = \begin{vmatrix} \mathbf{i} & \mathbf{j} & \mathbf{k} \\ -4 & 4 & -4 \\ 10 & 4 & -3 \end{vmatrix} = 4\mathbf{i} - 52\mathbf{j} - 56\mathbf{k}\ (\text{ft-lb}).$$

The sums of the moments about B must be equal:

$$(\mathbf{M}_B)_2 = (\mathbf{M}_B)_1:$$
$$\mathbf{M} = 4\mathbf{i} - 52\mathbf{j} - 56\mathbf{k}\ (\text{ft-lb}).$$

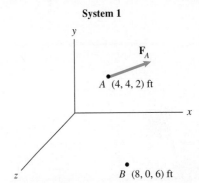

System 1

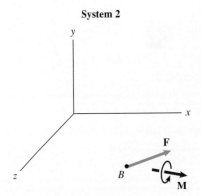

System 2

(a) A force acting at B and a couple.

Figure 4.43

Example 4.18

System 1

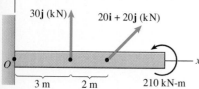

Figure 4.44

System 1 in Fig. 4.44 consists of two forces and a couple acting on a pipe. Represent system 1 by (a) a single force acting at the origin O of the coordinate system and a single couple and (b) a single force.

STRATEGY

(a) We can represent system 1 by a force $\mathbf{F}$ acting at the origin and a couple M (system 2 in Fig. a) and use the conditions for equivalence to determine $\mathbf{F}$ and M.

(b) Suppose that we place the force $\mathbf{F}$ with its point of application a distance D along the x axis (system 3 in Fig. b). The sums of the forces in systems 2 and 3 are equal. If we can choose the distance D so that the moment about O in system 3 equals M, system 3 will be equivalent to system 2 and therefore equivalent to system 1.

System 2

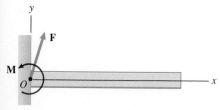

(a) A force $\mathbf{F}$ acting at O and a couple M.

SOLUTION

(a) The conditions for equivalence are

$$(\Sigma \mathbf{F})_2 = (\Sigma \mathbf{F})_1:$$

$$\mathbf{F} = 30\mathbf{j} + (20\mathbf{i} + 20\mathbf{j}) = 20\mathbf{i} + 50\mathbf{j} \text{ (kN)},$$

and

$$(\Sigma M_O)_2 = (\Sigma M_O)_1:$$

$$M = (30 \text{ kN})(3 \text{ m}) + (20 \text{ kN})(5 \text{ m}) + 210 \text{ kN-m}$$

$$= 400 \text{ kN-m}.$$

System 3

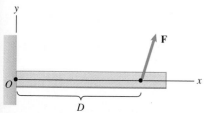

(b) A system consisting of the force $\mathbf{F}$ acting at a point on the x axis.

(b) The sums of the forces in systems 2 and 3 are equal. Equating the sums of the moments about O,

$$(\Sigma M_O)_3 = (\Sigma M_O)_2:$$

$$(50 \text{ kN})D = 400 \text{ kN-m},$$

we find that system 3 is equivalent to system 2 if $D = 8$ m.

DISCUSSION

To represent the system by a single force in (b), we needed to place the line of action of the force so that the force exerted a 400 kN-m counterclockwise moment about O. Placing the point of application of the force a distance D along the x axis was simply a convenient way to accomplish that.

Example 4.19

System 1 in Fig. 4.45 consists of the following forces and couple:

$$\mathbf{F}_A = -10\mathbf{i} + 10\mathbf{j} - 15\mathbf{k} \text{ (kN)},$$

$$\mathbf{F}_B = 30\mathbf{i} + 5\mathbf{j} + 10\mathbf{k} \text{ (kN)},$$

$$\mathbf{M}_C = -90\mathbf{i} + 150\mathbf{j} + 60\mathbf{k} \text{ (kN-m)}.$$

Suppose you want to represent it by a force **F** acting at P and a couple **M** (system 2). Determine **F** and **M**.

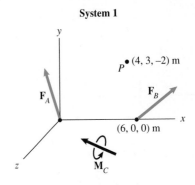

System 1

SOLUTION

The Sums of the Forces Must Be Equal

$$(\Sigma \mathbf{F})_2 = (\Sigma \mathbf{F})_1:$$

$$\mathbf{F} = \mathbf{F}_A + \mathbf{F}_B = 20\mathbf{i} + 15\mathbf{j} - 5\mathbf{k} \text{ (kN)}.$$

The Sums of the Moments About an Arbitrary Point Must Be Equal   The sums of the moments about point P must be equal:

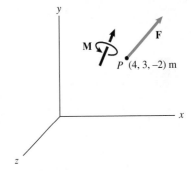

System 2

$$(\Sigma \mathbf{M}_P)_2 = (\Sigma \mathbf{M}_P)_1:$$

$$\mathbf{M} = \begin{vmatrix} \mathbf{i} & \mathbf{j} & \mathbf{k} \\ -4 & -3 & 2 \\ -10 & 10 & -15 \end{vmatrix} + \begin{vmatrix} \mathbf{i} & \mathbf{j} & \mathbf{k} \\ 2 & -3 & 2 \\ 30 & 5 & 10 \end{vmatrix}$$

$$+ (-90\mathbf{i} + 150\mathbf{j} + 60\mathbf{k})$$

$$= -105\mathbf{i} + 110\mathbf{j} + 90\mathbf{k} \text{ (kN-m)}.$$

Figure 4.45

Example 4.20

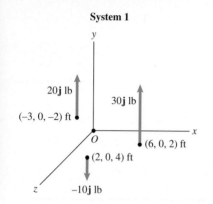

System 1

20**j** lb

30**j** lb

(−3, 0, −2) ft

O

(6, 0, 2) ft

(2, 0, 4) ft

−10**j** lb

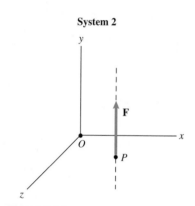

System 2

F

O

P

Figure 4.46

System 1 in Fig. 4.46 consists of parallel forces. Suppose you want to represent it by a force **F** (system 2). What is **F**, and where does its line of action intersect the *x-z* plane?

STRATEGY

We can determine **F** from the condition that the sums of the forces in the two systems must be equal. For the two systems to be equivalent, we must choose the point of application *P* so that the sums of the moments about a point are equal. This condition will tell us where the line of action intersects the *x-z* plane.

SOLUTION

The Sums of the Forces Must Be Equal

$$(\Sigma\,\mathbf{F})_2 = (\Sigma\,\mathbf{F})_1:$$

$$\mathbf{F} = 30\mathbf{j} + 20\mathbf{j} - 10\mathbf{j} = 40\mathbf{j}\ (\text{lb}).$$

The Sums of the Moments About an Arbitrary Point Must Be Equal   Let the coordinates of point *P* be (x, y, z). The sums of the moments about the origin *O* must be equal.

$$(\Sigma\,\mathbf{M}_O)_2 = (\Sigma\,\mathbf{M}_O)_1:$$

$$\begin{vmatrix} \mathbf{i} & \mathbf{j} & \mathbf{k} \\ x & y & z \\ 0 & 40 & 0 \end{vmatrix} = \begin{vmatrix} \mathbf{i} & \mathbf{j} & \mathbf{k} \\ 6 & 0 & 2 \\ 0 & 30 & 0 \end{vmatrix} + \begin{vmatrix} \mathbf{i} & \mathbf{j} & \mathbf{k} \\ 2 & 0 & 4 \\ 0 & -10 & 0 \end{vmatrix}$$

$$+ \begin{vmatrix} \mathbf{i} & \mathbf{j} & \mathbf{k} \\ -3 & 0 & -2 \\ 0 & 20 & 0 \end{vmatrix}.$$

Expanding the determinants, we obtain

$$(20 + 40z)\,\mathbf{i} + (100 - 40x)\,\mathbf{k} = \mathbf{0}.$$

The sums of the moments about the origin are equal if

$$x = 2.5\ \text{ft},$$

$$z = -0.5\ \text{ft}.$$

The systems are equivalent if $\mathbf{F} = 40\mathbf{j}$ (lb) and its line of action intersects the *x-z* plane at $x = 2.5$ ft and $z = -0.5$ ft. Notice that we did not obtain an equation for the *y* coordinate of *P*. The systems are equivalent if **F** is applied at any point along the line of action.

DISCUSSION

We could have determined the *x* and *z* coordinates of point *P* in a simpler way. Since the sums of the moments about any point must be equal for the systems to be equivalent, the sums of the moments about any *line* must also be equal. Equating the sums of the moments about the *x* axis,

$$(\Sigma M_{x\,\text{axis}})_2 = (\Sigma M_{x\,\text{axis}})_1:$$

$$-40z = -(30)(2) + (10)(4) + (20)(2),$$

we obtain $z = -0.5$ ft, and equating the sums of the moments about the z axis,

$$(\Sigma M_{z\,\text{axis}})_2 = (\Sigma M_{z\,\text{axis}})_1:$$

$$40x = (30)(6) - (10)(2) - (20)(3),$$

we obtain $x = 2.5$ ft.

Representing a System by a Wrench

We have shown that any system of forces and moments can be represented by a single force acting at a given point and a single couple. This raises an interesting question: What is the *simplest* system that can be equivalent to any system of forces and moments?

To consider this question, let's begin with an arbitrary force **F** acting at a point P and an arbitrary couple **M** (system 1 in Fig. 4.47a) and see whether we can represent this system by a simpler one. For example, can we represent it by the force **F** acting at a different point Q and no couple (Fig 4.47b)? The sum of the forces is the same as in system 1. If we can choose the point Q so that $\mathbf{r} \times \mathbf{F} = \mathbf{M}$, where **r** is the vector from P to Q (Fig. 4.47c), the sum of the moments about P is the same as in system 1 and the systems are equivalent. But the vector $\mathbf{r} \times \mathbf{F}$ is perpendicular to **F**, so it can equal **M** only if **M** is perpendicular to **F**. That means that, in general, we can't represent system 1 by the force **F** alone.

However, we *can* represent system 1 by the force **F** acting at a point Q and the *component* of **M** that is parallel to **F**. Figure 4.47(d) shows system 1 with a coordinate system placed so that **F** is along the y axis and **M** is contained in the x-y plane. In terms of this coordinate system, we can express the force and couple as $\mathbf{F} = F\mathbf{j}$ and $\mathbf{M} = M_x\mathbf{i} + M_y\mathbf{j}$. System 2 in Fig. 4.47(e) consists of the force **F** acting at a point on the z axis and the component of **M** parallel to **F**. If we choose the distance D so that $D = M_x/F$, system 2 is equivalent to system 1. The sum of the forces in each system is **F**. The sum of the moments about P in system 1 is **M**, and the sum of the moments about P in system 2 is

$$(\Sigma \mathbf{M}_P)_2 = [(-D\mathbf{k}) \times (F\mathbf{j})] + M_y\mathbf{j} = M_x\mathbf{i} + M_y\mathbf{j} = \mathbf{M}.$$

Figure 4.47

(a) System 1 is a single force and a single couple.

(b) Can system 1 be represented by a single force and no couple?

(c) The moment of **F** about P is $\mathbf{r} \times \mathbf{F}$.

(d) **F** is along the y axis, and **M** is contained in the x-y plane.

(e) System 2 is the force **F** and the component of **M** parallel to **F**.

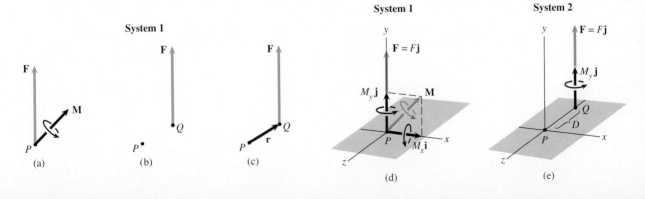

A force **F** and a couple $\mathbf{M_p}$ that is parallel to **F** is called a **wrench**; it is the simplest system that can be equivalent to an arbitrary system of forces and moments.

How can you represent a given system of forces and moments by a wrench? If the system is a single force or a single couple or if it consists of a force **F** and a couple that is parallel to **F**, it is a wrench, and you can't simplify it further. If the system is more complicated than a single force and a single couple, begin by choosing a convenient point P and representing the system by a force **F** acting at P and a couple **M** (Fig. 4.48a). Then representing this system by a wrench requires two steps:

1. Determine the components of **M** parallel and normal to **F** (Fig. 4.48b).

2. The wrench consists of the force **F** acting at a point Q and the parallel component $\mathbf{M}_p$ (Fig. 4.48c). To achieve equivalence, you must choose the point Q so that the moment of **F** about P equals the normal component $\mathbf{M}_n$ (Fig. 4.48d)—that is, so that $\mathbf{r}_{PQ} \times \mathbf{F} = \mathbf{M}_n$.

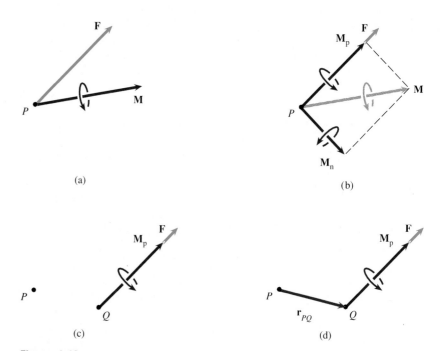

(a)

(b)

(c)

(d)

Figure 4.48

(a) If necessary, first represent the system by a single force and a single couple.
(b) The components of **M** parallel and normal to **F**.
(c) The wrench.
(d) Choose Q so that the moment of **F** about P equals the normal component of **M**.

Example 4.21

The system in Fig. 4.49 consists of the force and couple

$$\mathbf{F} = 3\mathbf{i} + 6\mathbf{j} + 2\mathbf{k} \text{ (N)},$$

$$\mathbf{M} = 12\mathbf{i} + 4\mathbf{j} + 6\mathbf{k} \text{ (N-m)}.$$

Represent it by a wrench, and determine where the line of action of the wrench's force intersects the *x-z* plane.

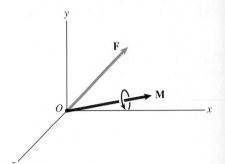

Figure 4.49

STRATEGY

The wrench is the force **F** and the component of **M** parallel to **F** (Figs. a, b). We must choose the point of application *P* so that the moment of **F** about *O* equals the normal component $\mathbf{M}_n$. By letting *P* be an arbitrary point of the *x-z* plane, we can determine where the line of action of **F** intersects that plane.

SOLUTION

Dividing **F** by its magnitude, we obtain a unit vector **e** with the same direction as **F**:

$$\mathbf{e} = \frac{\mathbf{F}}{|\mathbf{F}|} = \frac{3\mathbf{i} + 6\mathbf{j} + 2\mathbf{k}}{\sqrt{(3)^2 + (6)^2 + (2)^2}} = 0.429\mathbf{i} + 0.857\mathbf{j} + 0.286\mathbf{k}.$$

We can use **e** to calculate the component of **M** parallel to **F**:

$$\mathbf{M}_p = (\mathbf{e} \cdot \mathbf{M})\,\mathbf{e} = [(0.429)(12) + (0.857)(4) + (0.286)(6)]\,\mathbf{e}$$

$$= 4.408\mathbf{i} + 8.816\mathbf{j} + 2.939\mathbf{k} \text{ (N-m)}.$$

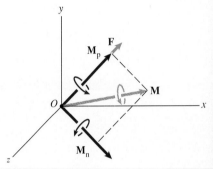

(a) Resolving **M** into components parallel and normal to **F**.

The component of **M** normal to **F** is

$$\mathbf{M}_n = \mathbf{M} - \mathbf{M}_p = 7.592\mathbf{i} - 4.816\mathbf{j} + 3.061\mathbf{k} \text{ (N-m)}.$$

The wrench is shown in Fig. (b). Let the coordinates of *P* be (*x*, 0, *z*). The moment of **F** about *O* is

$$\mathbf{r}_{OP} \times \mathbf{F} = \begin{vmatrix} \mathbf{i} & \mathbf{j} & \mathbf{k} \\ x & 0 & z \\ 3 & 6 & 2 \end{vmatrix} = -6z\mathbf{i} - (2x - 3z)\,\mathbf{j} + 6x\mathbf{k}.$$

By equating this moment to $\mathbf{M}_n$,

$$-6z\mathbf{i} - (2x - 3z)\,\mathbf{j} + 6x\mathbf{k} = 7.592\mathbf{i} - 4.816\mathbf{j} + 3.061\mathbf{k},$$

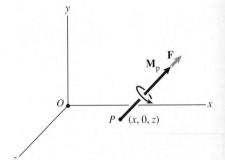

(b) The wrench acting at a point in the *x-z* plane.

we obtain the equations

$$-6z = 7.592,$$

$$-2x + 3z = -4.816,$$

$$6x = 3.061.$$

Solving these equations, we find the coordinates of point *P* are *x* = 0.510 m, *z* = −1.265 m.

Problems

4.128 Two systems of forces act on the beam. Are they equivalent?

Strategy: Check the two conditions for equivalence. The sums of the forces must be equal, and the sums of the moments about an arbitrary point must be equal.

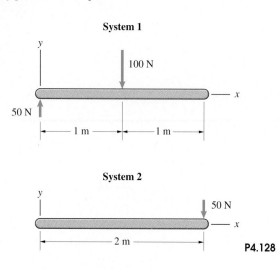

4.129 Two systems of forces and moments act on the beam. Are they equivalent?

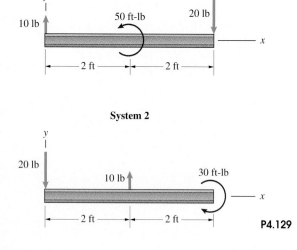

4.130 Four systems of forces and moments act on an 8-ft beam. Which systems are equivalent?

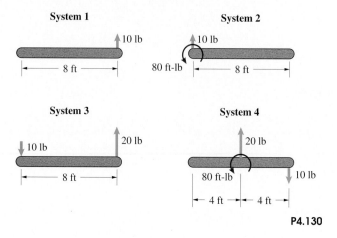

P4.130

4.131 The four systems shown in Problem 4.130 can be made equivalent by adding a single couple to one of the systems. Which system is it, and what couple must be added?

4.132 System 1 is a force **F** acting at a point O. System 2 is the force **F** acting at a different point O' along the same line of action. Explain why these systems are equivalent. (This simple result is called the *principle of transmissibility*.)

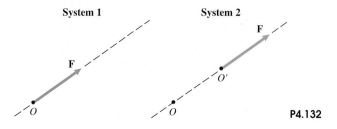

P4.132

4.133 The vector sum of the forces exerted on the log by the cables is the same in the two cases. Show that the systems of forces exerted on the log are equivalent.

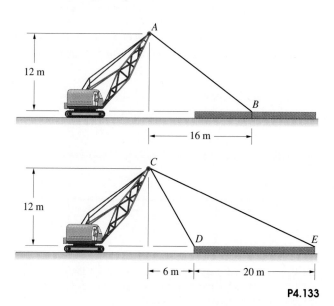

P4.133

4.134 Systems 1 and 2 each consist of a couple. If they are equivalent, what is F?

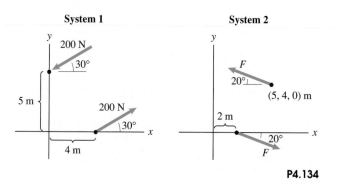

P4.134

4.135 Two equivalent systems of forces and moments act on the L-shaped bar. Determine the forces F_A and F_B and the couple M.

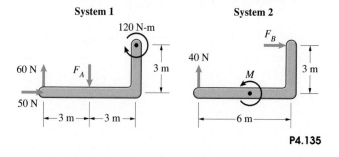

P4.135

4.136 Two equivalent systems of forces and moments act on the plate. Determine the force F and the couple M.

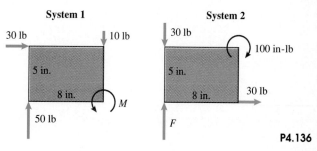

P4.136

4.137 In system 1, four forces act on the rectangular, flat plate. The forces are perpendicular to the plate, and the 400-lb force acts at its midpoint. In system 2, no forces or couples act on the plate. Systems 1 and 2 are equivalent. Determine the forces F_1, F_2, and F_3.

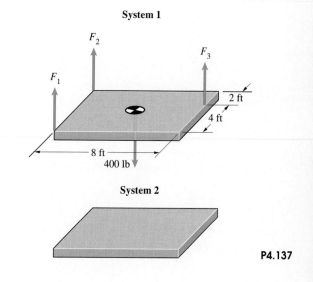

P4.137

4.138 Three forces and a couple are applied to a beam (system 1).
(a) If you represent system 1 by a force applied at A and a couple (system 2), what are $\mathbf{F}$ and M?
(b) If you represent system 1 by the force $\mathbf{F}$ (system 3), what is the distance D?

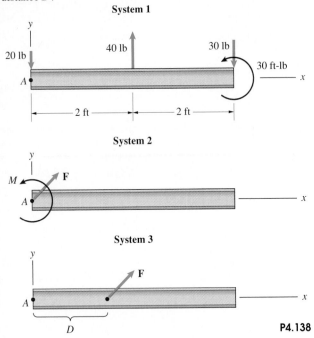

P4.138

4.139 Represent the two forces and couple acting on the beam by a force $\mathbf{F}$. Determine $\mathbf{F}$ and determine where its line of action intersects the x axis.

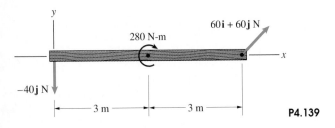

P4.139

4.140 The vector sum of the forces acting on the beam is zero, and the sum of the moments about the left end of the beam is zero.
(a) Determine the forces A_x, A_y, and B.
(b) If you represent the forces A_x, A_y, and B by a force $\mathbf{F}$ acting at the right end of the beam and a couple M, what are $\mathbf{F}$ and M?

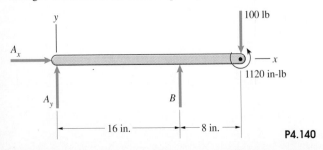

P4.140

4.141 The vector sum of the forces acting on the beam is zero, and the sum of the moments about the left end of the beam is zero.
(a) Determine the forces A_x and A_y, and the couple M_A.
(b) Determine the sum of the moments about the right end of the beam.
(c) If you represent the 600-N force, the 200-N force, and the 30 N-m couple by a force $\mathbf{F}$ acting at the left end of the beam and a couple M, what are $\mathbf{F}$ and M?

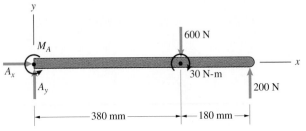

P4.141

4.142 The vector sum of the forces acting on the truss is zero, and the sum of the moments about the origin O is zero.
(a) Determine the forces A_x, A_y, and B.
(b) If you represent the 2-kip, 4-kip, and 6-kip forces by a force $\mathbf{F}$, what is $\mathbf{F}$, and where does its line of action intersect the y axis?
(c) If you replace the 2-kip, 4-kip, and 6-kip forces by the force you determined in (b), what are the vector sum of the forces acting on the truss and the sum of the moments about O?

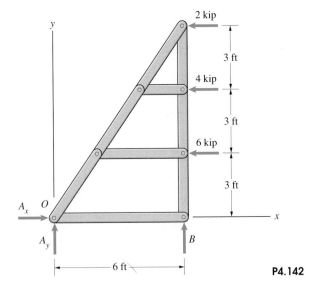

P4.142

4.143 The distributed force exerted on part of a building foundation by the soil is represented by five forces. If you represent them by a force **F**, what is **F**, and where does its line of action intersect the *x* axis?

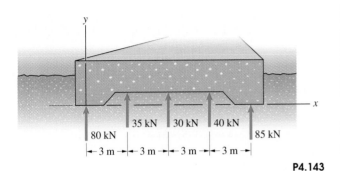

P4.143

4.144 The thrust forces exerted on the airplane by its four engines are parallel to the *x* axis. Their magnitudes are 160 kN, 175 kN, 185 kN, and 160 kN, respectively.
(a) Represent the four thrust forces by a force **F**. What is **F**, and where does its line of action intersect the *y* axis?
(b) The pilot wants to adjust the thrust of engine 1 so that the four thrust forces can be represented by a force acting at the origin. What is the necessary thrust of engine 1?

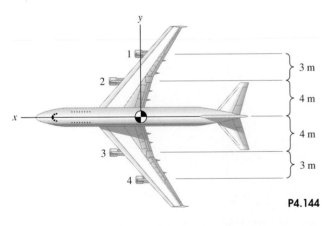

P4.144

4.145 The lift forces on an airplane's wing are represented by eight forces. The magnitude of each force is given in terms of its position *x* on the wing by

$$200\sqrt{1-\left(\frac{x}{17}\right)^2}\ \text{lb.}$$

If you represent the lift forces by a single force as shown, what are the force *F* and the distance *D*?

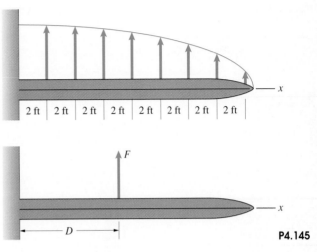

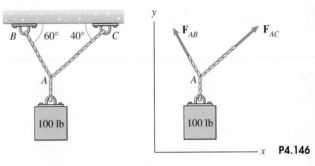

P4.145

4.146 The system is in equilibrium. If you represent the forces $\mathbf{F}_{AB}$ and $\mathbf{F}_{AC}$ by a force **F** acting at *A* and a couple **M**, what are **F** and **M**?

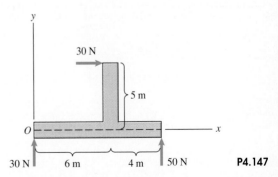

P4.146

4.147 Three forces act on the beam.
(a) Represent the system by a force **F** acting at the origin *O* and a couple *M*.
(b) Represent the system by a single force. Where does the line of action of the force intersect the *x* axis?

4.148 The tension in cable AB is 400 N, and the tension in cable CD is 600 N.
(a) If you represent the forces exerted on the left post by the cables by a force $\mathbf{F}$ acting at the origin O and a couple M, what are $\mathbf{F}$ and M?
(b) If you represent the forces exerted on the left post by the cables by the force $\mathbf{F}$ alone, where does its line of action intersect the y axis?

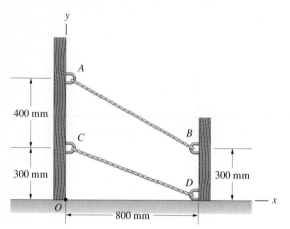

P4.148

4.149 Consider the system shown in Problem 4.148. The tension in each of the cables AB and CD is 400 N. If you represent the forces exerted on the right post by the cables by a force $\mathbf{F}$, what is $\mathbf{F}$, and where does its line of action intersect the y axis?

4.150 If you represent the three forces acting on the beam cross section by a force $\mathbf{F}$, what is $\mathbf{F}$, and where does its line of action intersect the x axis?

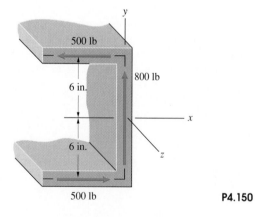

P4.150

4.151 The two systems of forces and moments acting on the beam are equivalent. Determine the force $\mathbf{F}$ and the couple $\mathbf{M}$.

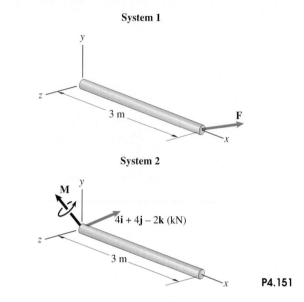

P4.151

4.152 The wall bracket is subjected to the force shown. Determine the moment exerted by the force (a) about the z axis, (b) about the y axis.
(c) If you represent the force by a force $\mathbf{F}$ acting at O and a couple $\mathbf{M}$, what are $\mathbf{F}$ and $\mathbf{M}$?

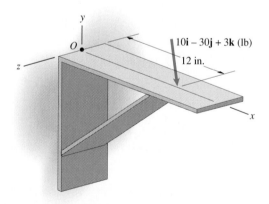

P4.152

4.153 A basketball player executes a "slam dunk" shot, then hangs momentarily on the rim, exerting the two 100-lb forces shown. The dimensions are $h = 14\frac{1}{2}$ in. and $r = 9\frac{1}{2}$ in., and the angle $\alpha = 120°$.
(a) If you represent the forces he exerts by a force $\mathbf{F}$ acting at O and a couple $\mathbf{M}$, what are $\mathbf{F}$ and $\mathbf{M}$?

(b) The glass backboard will shatter if $|\mathbf{M}| > 4000$ in-lb. Does it break?

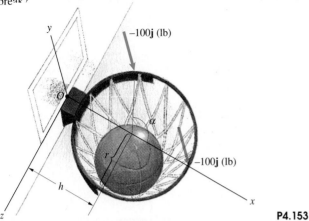

−100**j** (lb)

−100**j** (lb)

P4.153

4.154 The three forces are parallel to the x axis.
(a) If you represent the three forces by a force **F** acting at the origin O and a couple **M**, what are **F** and **M**?
(b) If you represent the forces by a single force, what is the force, and where does its line of action intersect the y-z plane?

Strategy: In (b), assume that the force acts at a point $(0, y, z)$ of the y-z plane, and use the conditions for equivalence to determine the force and the coordinates y and z. (See Example 4.20.)

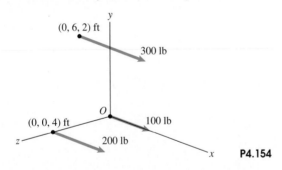

(0, 6, 2) ft

300 lb

(0, 0, 4) ft

O

100 lb

200 lb

P4.154

4.155 The positions and weights of three particles are shown. If you represent the weights by a single force **F**, determine **F** and show that its line of action intersects the x-z plane at

$$x = \frac{\sum_{i=1}^{3} x_i W_i}{\sum_{i=1}^{3} W_i}, \qquad z = \frac{\sum_{i=1}^{3} z_i W_i}{\sum_{i=1}^{3} W_i}.$$

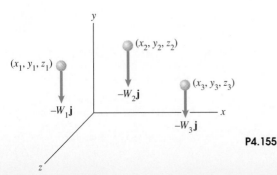

(x_2, y_2, z_2)

(x_1, y_1, z_1)

(x_3, y_3, z_3)

$-W_1\mathbf{j}$

$-W_2\mathbf{j}$

$-W_3\mathbf{j}$

P4.155

4.156 Two forces act on the beam. If you represent them by a force **F** acting at C and a couple **M**, what are **F** and **M**?

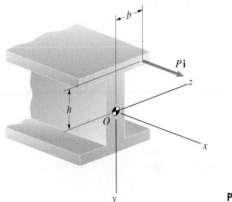

100 N

80 N

3 m

C

P4.156

4.157 An axial force of magnitude P acts on the beam. If you represent it by a force **F** acting at the origin O and a couple **M**, what are **F** and **M**?

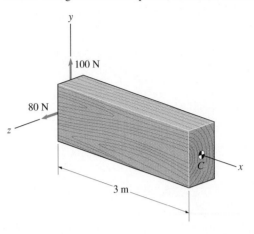

$P\mathbf{i}$

h

O

P4.157

4.158 The brace is being used to remove a screw.
(a) If you represent the forces acting on the brace by a force **F** acting at the origin O and a couple **M**, what are **F** and **M**?
(b) If you represent the forces acting on the brace by a force **F′** acting at a point P with coordinates (x_P, y_P, z_P) and a couple **M′**, what are **F′** and **M′**?

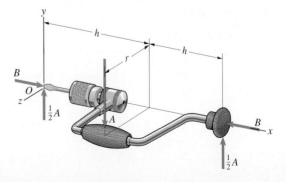

B

O

$\frac{1}{2}A$

h

r

h

A

B

$\frac{1}{2}A$

P4.158

4.159 Two forces and a couple act on the cube. If you represent them by a force **F** acting at point P and a couple **M**, what are **F** and **M**?

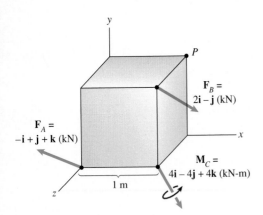

P4.159

4.160 The two shafts are subjected to the torques (couples) shown.
(a) If you represent the two couples by a force **F** acting at the origin O and a couple **M**, what are **F** and **M**?
(b) What is the magnitude of the total moment exerted by the two couples?

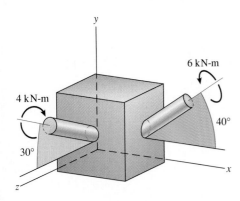

P4.160

4.161 The persons A and B support a bar to which three dogs are tethered. The forces and couples they exert are

$$\mathbf{F}_A = -5\mathbf{i} + 15\mathbf{j} - 10\mathbf{k} \text{ (lb)},$$

$$\mathbf{M}_A = 15\mathbf{j} + 10\mathbf{k} \text{ (ft-lb)},$$

$$\mathbf{F}_B = 5\mathbf{i} + 10\mathbf{j} - 10\mathbf{k} \text{ (lb)},$$

$$\mathbf{M}_B = -10\mathbf{j} - 15\mathbf{k} \text{ (ft-lb)}.$$

If person B let go, person A would have to exert a force **F** and couple **M** equivalent to the system both of them were exerting together. What are **F** and **M**?

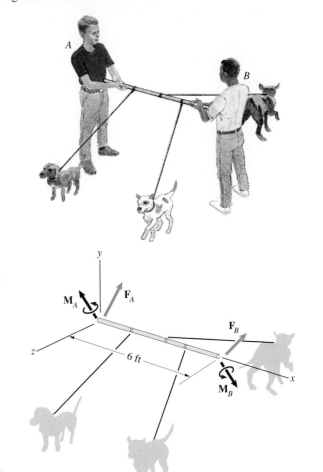

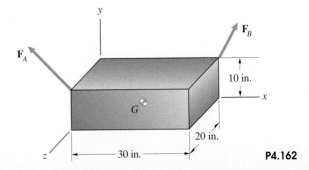

P4.161

4.162 Point G is at the center of the block. The forces are

$$\mathbf{F}_A = -20\mathbf{i} + 10\mathbf{j} + 20\mathbf{k} \text{ (lb)},$$

$$\mathbf{F}_B = 10\mathbf{j} - 10\mathbf{k} \text{ (lb)}.$$

If you represent the two forces by a force **F** acting at G and a couple **M**, what are **F** and **M**?

P4.162

4.163 The engine above the airplane's fuselage exerts a thrust $T_0 = 16$ kip, and each of the engines under the wings exerts a thrust $T_U = 12$ kip. The dimensions are $h = 8$ ft, $c = 12$ ft, and $b = 16$ ft. If you represent the three thrust forces by a force **F** acting at the origin O and a couple **M**, what are **F** and **M**?

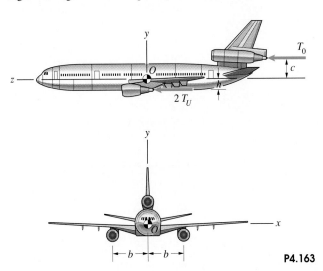

P4.163

4.164 Consider the airplane described in Problem 4.163 and suppose that the engine under the wing to the pilot's right loses thrust. (a) If you represent the two remaining thrust forces by a force **F** acting at the origin O and a couple **M**, what are **F** and **M**? (b) If you represent the two remaining thrust forces by the force **F** alone, where does its line of action intersect the x-y plane?

4.165 The tension in cable AB is 100 lb, and the tension in cable CD is 60 lb. Suppose that you want to replace these two cables by a single cable EF so that the force exerted on the wall at E is equivalent to the two forces exerted by cables AB and CD on the walls at A and C. What is the tension in cable EF, and what are the coordinates of points E and F?

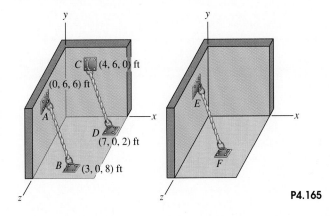

P4.165

4.166 The distance $s = 4$ m. If you represent the force and the 200-N-m couple by a force **F** acting at the origin O and a couple **M**, what are **F** and **M**?

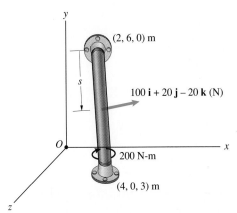

P4.166

4.167 The force **F** and couple **M** in system 1 are

$$F = 12i + 4j - 3k \text{ (lb)},$$

$$M = 4i + 7j + 4k \text{ (ft-lb)}.$$

Suppose you want to represent system 1 by a wrench (system 2). Determine the couple M_p and the coordinates x and z where the line of action of the force intersects the x-z plane.

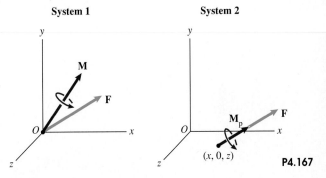

P4.167

4.168 A system consists of a force **F** acting at the origin O and a couple **M**, where

$$F = 10i \text{ (lb)}, \qquad M = 20j \text{ (ft-lb)}.$$

If you represent the system by a wrench consisting of the force **F** and a parallel couple M_p, what is M_p, and where does the line of action of **F** intersect the y-z plane?

4.169 A system consists of a force **F** acting at the origin O and a couple **M**, where

$$F = i + 2j + 5k \text{ (N)}, \qquad M = 10i + 8j - 4k \text{ (N-m)}.$$

If you represent it by a wrench consisting of the force **F** and a parallel couple M_p, (a) determine M_p, and determine where the line of action of **F** intersects (b) the x-z plane, (c) the y-z plane.

4.170   Consider the force **F** acting at the origin O and the couple **M** given in Example 4.21. If you represent this system by a wrench, where does the line of action of the force intersect the x-y plane?

4.171   Consider the force **F** acting at the origin O and the couple **M** given in Example 4.21. If you represent this system by a wrench, where does the line of action of the force intersect the plane $y = 3$ m?

4.172   A wrench consists of a force of magnitude 100 N acting at the origin O and a couple of magnitude 60 N-m. The force and couple point in the direction from O to the point $(1, 1, 2)$ m. If you represent the wrench by a force **F** acting at the point $(5, 3, 1)$ m and a couple **M**, what are **F** and **M**?

4.173   System 1 consists of two forces and a couple. Suppose that you want to represent it by a wrench (system 2). Determine the force **F**, the couple **M**$_\text{p}$, and the coordinates x and z where the line of action of **F** intersects the x-z plane.

4.174   A plumber exerts the two forces shown to loosen a pipe.
(a) What total moment does he exert about the axis of the pipe?
(b) If you represent the two forces by a force **F** acting at O and a couple **M**, what are **F** and **M**?
(c) If you represent the two forces by a wrench consisting of the force **F** and a parallel couple **M**$_\text{p}$, what is **M**$_\text{p}$, and where does the line of action of **F** intersect the x-y plane?

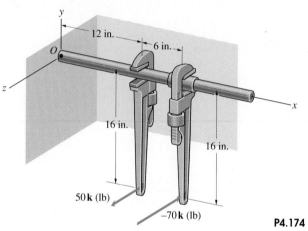

P4.174

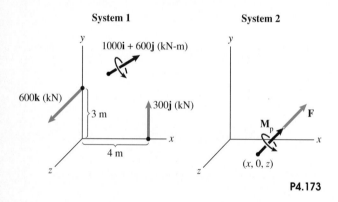

P4.173

Computational Mechanics

The following example and problems are designed for the use of a programmable calculator or computer.

Example 4.22

The radius R of the steering wheel in Fig. 4.50 is 200 mm. The distance from O to C is 1 m. The center C of the steering wheel lies in the x-y plane. The force $\mathbf{F} = \sin \alpha \, (10\mathbf{i} + 10\mathbf{j} - 5\mathbf{k})$ N. Determine the value of α at which the magnitude of the moment of **F** about the shaft OC of the steering wheel is a maximum. What is the maximum magnitude?

STRATEGY

We will determine the moment of **F** about OC in terms of the angle α and obtain a graph of the moment as a function of α.

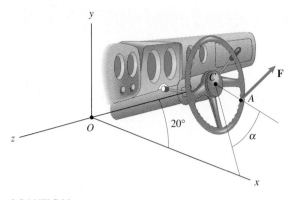

Figure 4.50

SOLUTION

In terms of the vector $\mathbf{r}_{CA}$ from point C on the shaft to the point of application of the force, and the unit vector $\mathbf{e}_{OC}$ that points along the shaft from point O toward point C (Fig. a), the moment of $\mathbf{F}$ about the shaft is

$$\mathbf{M}_{OC} = [\mathbf{e}_{OC} \cdot (\mathbf{r}_{CA} \times \mathbf{F})]\, \mathbf{e}_{OC}.$$

From Fig. (a), the unit vector $\mathbf{e}_{OC}$ is

$$\mathbf{e}_{OC} = \cos 20° \,\mathbf{i} + \sin 20° \,\mathbf{j},$$

and the z component of $\mathbf{r}_{CA}$ is $-R \sin \alpha$. By viewing the steering wheel with the z axis perpendicular to the page (Fig. b), we can see that the x component of $\mathbf{r}_{CA}$ is $R \cos \alpha \sin 20°$ and the y component is $-R \cos \alpha \cos 20°$, so

$$\mathbf{r}_{CA} = R(\cos \alpha \sin 20° \,\mathbf{i} - \cos \alpha \cos 20° \,\mathbf{j} - \sin \alpha \,\mathbf{k}).$$

The magnitude of $\mathbf{M}_{OC}$ is the absolute value of the scalar

$$\mathbf{e}_{OC} \cdot (\mathbf{r}_{CA} \times \mathbf{F}) = \begin{vmatrix} \cos 20° & \sin 20° & 0 \\ R \cos \alpha \sin 20° & -R \cos \alpha \cos 20° & -R \sin \alpha \\ 10 \sin \alpha & 10 \sin \alpha & -5 \sin \alpha \end{vmatrix}$$

$$= R[5 \sin \alpha \cos \alpha + 10(\cos 20° - \sin 20°) \sin^2 \alpha].$$

Computing the absolute value of this expression as a function of α, we obtain the graph shown in Fig. 4.51. The magnitude of the moment is an extremum at values of α of approximately 70° and 250°. By examining the computed results near 70°,

α	$|\mathbf{M}_{OC}|$ (N-m)
67°	1.3725
68°	1.3749
69°	1.3764
70°	1.3769
71°	1.3765
72°	1.3751
73°	1.3728

we can see that the maximum value is approximately 1.38 N-m. The value of the moment at $\alpha = 250°$ is also 1.38 N-m.

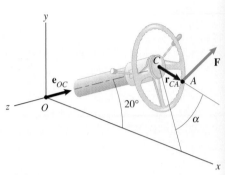

(a) The position vector $\mathbf{r}_{CA}$ and the unit vector $\mathbf{e}_{OC}$.

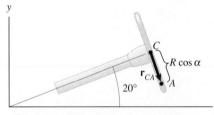

(b) Determining the x and y components of $\mathbf{r}_{CA}$.

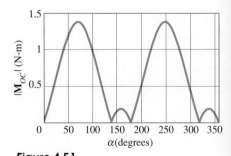

Figure 4.51

Magnitude of the moment as a function of α.

Problems

4.175 Consider the system described in Problem 4.43.
(a) Obtain a graph of the moment about A due to the force exerted by the spring on the circular bar at B for values of the angle α from zero to 90°.
(b) Use the result of (a) to estimate the angle at which the maximum moment occurs and the value of the maximum moment.

4.176 The exercise equipment shown is used by resting the elbow on the fixed pad and rotating the forearm to stretch the elastic cord AB. The cord behaves like a linear spring, and its unstretched length is 1 ft. Suppose you want to design the equipment so that the maximum moment that will be exerted about the elbow joint E as the forearm is rotated will be 60 ft-lb. What should the spring constant k of the elastic cord be?

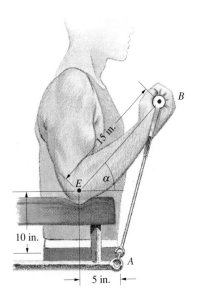

P4.176

4.177 The hydraulic cylinder BC exerts a 2200-lb force on the boom of the crane at C. The force is parallel to the cylinder. Draw a graph of the moment exerted by the force about A as a function of the angle α for $0 \le \alpha \le 90°$, and use it to estimate the values of α for which the moment equals 12,000 ft-lb.

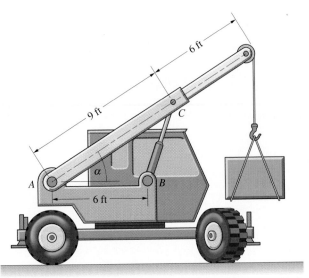

P4.177

4.178 In Problem 4.177, the moment about A exerted by the 2200-lb force exerted by the hydraulic cylinder BC depends on the angle α. Estimate the maximum value of the moment and the angle α at which it occurs.

4.179 The support cable extends from the top of the 3-m column at A to a point B on the line L. The tension in the cable is 2 kN. The line L intersects the ground at the point $(3, 0, 1)$ m and is parallel to the unit vector $\mathbf{e} = \frac{2}{7}\mathbf{i} + \frac{6}{7}\mathbf{j} - \frac{3}{7}\mathbf{k}$. The distance along L from the ground to point B is denoted s. What is the range of values of s for which the magnitude of the moment about O due to the force exerted by the cable at A exceeds 5.6 kN-m?

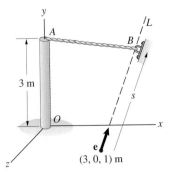

P4.179

4.180 Consider Problem 4.106. Determine the distance d that causes the moment about the z axis due to the force exerted by the cable CD at point C to be a maximum. What is the maximum moment?

4.181 Consider Problem 4.107. The rod AB must exert a moment of magnitude 100 ft-lb about the x axis to support the hood of the car. Draw a graph of the magnitude of the force the rod must exert on the hood at B as a function of d for $1 \leq d \leq 4$ ft. If you were designing the support AB, what value of d would you choose, and what is the magnitude of the force AB must exert on the hood?

4.182 Consider the system shown in Problem 4.148. The forces exerted on the left post by cables AB and CD can be represented by a single force **F**. Determine the tensions in the cables so that $|\mathbf{F}| = 600$ N and the line of action of **F** intersects the y axis at $y = 400$ mm.

4.183 Suppose you want to represent the force and the 200-N-m couple in Problem 4.166 by a force **F** and a couple **M**, and choose the distance s so that the magnitude of **M** is a minimum. Determine s, **F**, and **M**.

Chapter Summary

In this chapter we have defined the moment of a force about a point and about a line and explained how to evaluate them. We introduced the concept of a couple and defined equivalent systems of forces and moments. We can now apply two consequences of equilibrium—the sum of the forces equals zero and the sum of the moments about any point equals zero. We will consider individual objects in Chapter 5 and structures in Chapter 6.

Moment of a Force About a Point

The moment of a force about a point is the measure of the tendency of the force to cause rotation about the point. The **moment** of a force **F** about a point P is the vector

$$\mathbf{M}_P = \mathbf{r} \times \mathbf{F}, \qquad \text{Eq. (4.2)}$$

where **r** is a position vector from P to *any* point on the line of action of **F**. The magnitude of $\mathbf{M}_P$ is equal to the product of the perpendicular distance D from P to the line of action of **F** and the magnitude of **F**:

$$|\mathbf{M}_P| = D|\mathbf{F}|. \qquad \text{Eq. (4.3)}$$

The vector $\mathbf{M}_P$ is perpendicular to the plane containing P and **F**. When the thumb of the right hand points in the direction of $\mathbf{M}_P$, the arc of the fingers indicates the sense of the rotation that **F** tends to cause about P. The dimensions of the moment are (distance) $\times$ (force).

If you resolve a force into components, the moment of the force about a point P is equal to the sum of the moments of its components about P. If the line of action of a force passes through a point P, the moment of the force about P is zero.

When the view is perpendicular to the plane containing the force and the point (Fig. a), the two-dimensional description of the moment is

$$M_P = DF. \qquad \text{Eq. (4.1)}$$

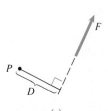

(a)

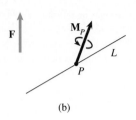

(b)

Moment of a Force About a Line

The moment of a force about a line is the measure of the tendency of the force to cause rotation about the line. Let P be any point on a line L and let $\mathbf{M}_P$ be the moment about P of a force $\mathbf{F}$ (Fig. b). The moment $\mathbf{M}_L$ of $\mathbf{F}$ about L is the vector component of $\mathbf{M}_P$ parallel to L. If $\mathbf{e}$ is a unit vector along L,

$$\mathbf{M}_L = (\mathbf{e} \cdot \mathbf{M}_P)\,\mathbf{e} = [\mathbf{e} \cdot (\mathbf{r} \times \mathbf{F})]\,\mathbf{e}. \qquad \textbf{Eqs. (4.4), (4.5)}$$

When the line of action of $\mathbf{F}$ is perpendicular to a plane containing L, $|\mathbf{M}_L|$ is equal to the product of the magnitude of $\mathbf{F}$ and the perpendicular distance D from L to the point where the line of action intersects the plane. When the line of action of $\mathbf{F}$ is parallel to L or intersects L, $\mathbf{M}_L = 0$.

Couples

Two forces that have equal magnitudes, opposite directions, and do not have the same line of action are called a **couple**. The moment $\mathbf{M}$ of a couple is the same about any point. The magnitude of $\mathbf{M}$ is equal to the product of the magnitude of one of the forces and the perpendicular distance between the lines of action, and its direction is perpendicular to the plane containing the lines of action.

(c)   (d)

Because a couple exerts a moment but no net force, it can be represented by showing the moment vector (Fig. c), or it can be represented in two dimensions by showing the magnitude of the moment and a circular arrow to indicate the sense (Fig. d). The moment represented in this way is called the **moment of a couple**, or simply a **couple**.

Equivalent Systems

Two systems of forces and moments are defined to be **equivalent** if the sums of the forces are equal,

$$(\Sigma\,\mathbf{F})_1 = (\Sigma\,\mathbf{F})_2, \qquad \textbf{Eq. (4.7)}$$

and the sums of the moments about a point P are equal,

$$(\Sigma\,\mathbf{M}_P)_1 = (\Sigma\,\mathbf{M}_P)_2. \qquad \textbf{Eq. (4.8)}$$

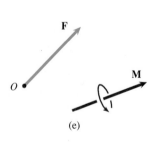

(e)

If the sums of the forces are equal and the sums of the moments about one point are equal, the sums of the moments about any point are equal.

Representing Systems by Equivalent Systems

If the system of forces and moments acting on an object is represented by an equivalent system, the equivalent system exerts the same total force and total moment on the object.

Any system can be represented by an equivalent system consisting of a force $\mathbf{F}$ acting at a given point P and a couple $\mathbf{M}$ (Fig. e). The simplest system that can be equivalent to any system of forces and moments is the **wrench**, which is a force $\mathbf{F}$ and a couple $\mathbf{M}_p$ that is parallel to $\mathbf{F}$ (Fig. f).

A system of concurrent forces can be represented by a single force. A system of parallel forces whose sum is not zero can be represented by a single force.

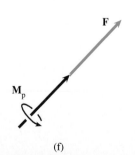

(f)

Review Problems

4.184 Determine the moment of the 200-N force about A.
(a) What is the two-dimensional description of the moment?
(b) Express the moment as a vector.

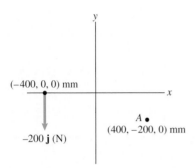

P4.184

4.185 The Leaning Tower of Pisa is approximately 180 ft tall and 23 ft in diameter. The horizontal displacement of the top of the tower from the vertical is approximately 15 ft. If you model the tower as a cylinder and assume that its weight is 7×10^6 lb and acts at the center, what is the magnitude of the moment exerted by the weight about the point at the center of its base?

P4.185

4.186 The device shown has been suggested as a design for a perpetual motion machine. Determine the moment about the axis of rotation due to the four masses as a function of the angle as the device rotates 90° clockwise from the position shown, and indicate whether gravity could cause rotation in that direction.

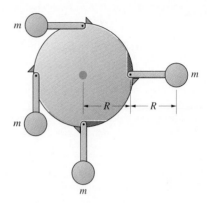

P4.186

4.187 In Problem 4.186, determine whether gravity could cause rotation in the counterclockwise direction.

4.188 Determine the moment of the 400-N force (a) about A, (b) about B.

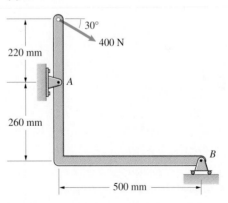

P4.188

4.189   Determine the sum of the moments exerted about A by the three forces and the couple.

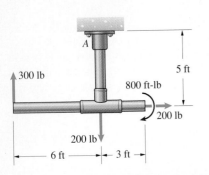

P4.189

4.190   In Problem 4.189, if you represent the three forces and the couple by an equivalent system consisting of a force **F** acting at A and a couple **M**, what are the magnitudes of **F** and **M**?

4.191   The vector sum of the forces acting on the beam is zero, and the sum of the moments about A is zero.
(a) What are the forces A_x, A_y, and B?
(b) What is the sum of the moments about B?

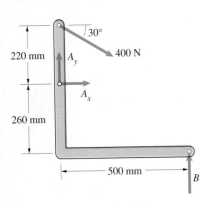

P4.191

4.192   To support the ladder, the force exerted at B by the hydraulic piston AB must exert a moment about C equal in magni-

tude to the moment about C due to the ladder's 450-lb weight. What is the magnitude of the force exerted at B?

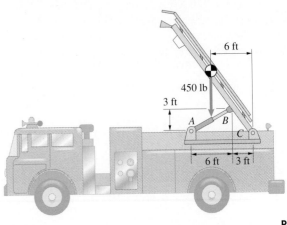

P4.192

4.193   The force $\mathbf{F} = -60\mathbf{i} + 60\mathbf{j}$ (lb).
(a) Determine the moment of **F** about point A.
(b) What is the perpendicular distance from point A to the line of action of **F**?

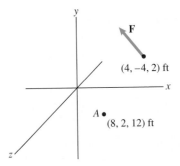

P4.193

4.194   The 20-kg mass is suspended by cables attached to three vertical 2-m posts. Point A is at $(0, 1.2, 0)$ m. Determine the moment about the base E due to the force exerted on the post BE by the cable AB.

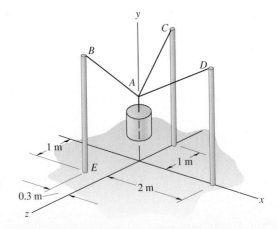

P4.194

4.195 Three forces of equal magnitude are applied parallel to the sides of an equilateral triangle.
(a) Show that the sum of the moments of the forces is the same about any point.
(b) Determine the magnitude of the moment.

Strategy: To do (a), resolve one of the forces into vector components parallel to the other two forces.

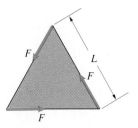

P4.195

4.196 The bar AB supporting the lid of the grand piano exerts a force $\mathbf{F} = -6\mathbf{i} + 35\mathbf{j} - 12\mathbf{k}$ (lb) at B. The coordinates of B are $(3, 4, 3)$ ft. What is the moment of the force about the hinge line of the lid (the x axis)?

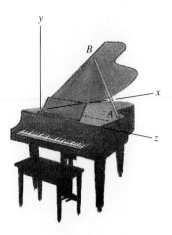

P4.196

4.197 Determine the moment of the vertical 800-lb force about point C.

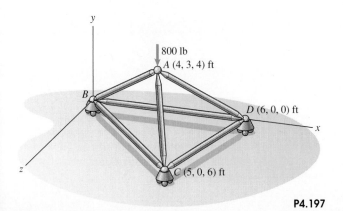

P4.197

4.198 In Problem 4.197, determine the moment of the vertical 800-lb force about the straight line through points C and D.

4.199 The system of cables and pulleys supports the 300-lb weight of the work platform. If you represent the upward force exerted at E by cable EF and the upward force exerted at G by cable GH by a single equivalent force $\mathbf{F}$, what is $\mathbf{F}$, and where does its line of action intersect the x axis?

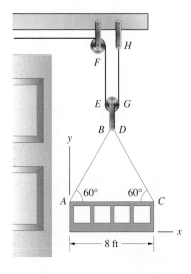

P4.199

4.200 Consider the system in Problem 4.199.
(a) What are the tensions in cables AB and CD?
(b) If you represent the forces exerted by the cables at A and C by a single equivalent force $\mathbf{F}$, what is $\mathbf{F}$, and where does its line of action intersect the x axis?

4.201   The two systems are equivalent. Determine the forces A_x and A_y and the couple M_A.

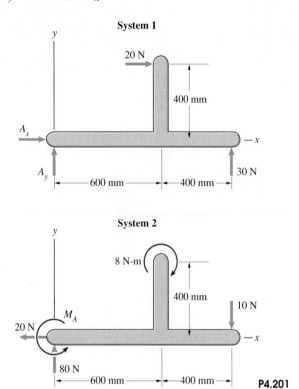

System 1

20 N

400 mm

A_x

A_y

—600 mm—   —400 mm—

30 N

— x

System 2

8 N-m

400 mm

M_A

20 N

80 N

10 N

—600 mm—   —400 mm—

— x

P4.201

4.202   If you represent the equivalent systems in Problem 4.201 by a force **F** acting at the origin and a couple M, what are **F** and M?

4.203   If you represent the equivalent systems in Problem 4.201 by a force **F**, what is **F**, and where does its line of action intersect the x axis?

4.204   The two systems are equivalent. If

$$\mathbf{F} = -100\mathbf{i} + 40\mathbf{j} + 30\mathbf{k} \text{ (lb)},$$

$$\mathbf{M}' = -80\mathbf{i} + 120\mathbf{j} + 40\mathbf{k} \text{ (in-lb)},$$

determine **F′** and **M**.

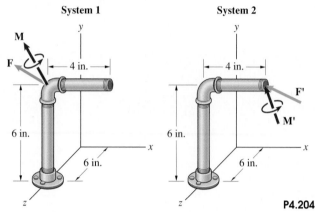

System 1

M

F

4 in.

6 in.

6 in.

System 2

F′

M′

4 in.

6 in.

6 in.

P4.204

4.205   The tugboats A and B exert forces $F_A = 1$ kN and $F_B = 1.2$ kN on the ship. The angle $\theta = 30°$. If you represent the two forces by a force **F** acting at the origin O and a couple M, what are **F** and M?

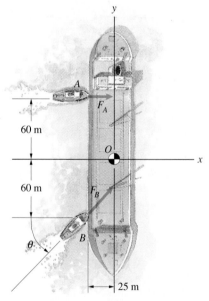

60 m

F_A

O

— x

60 m

F_B

θ B

25 m

P4.205

4.206   The tugboats A and B in Problem 4.205 exert forces $F_A = 600$ N and $F_B = 800$ N on the ship. The angle $\theta = 45°$. If you represent the two forces by a force **F**, what is **F**, and where does its line of action intersect the y axis?

4.207   The tugboats A and B in Problem 4.205 want to exert two forces on the ship that are equivalent to a force **F** acting at the origin O of 2-kN magnitude. If $F_A = 800$ N, determine the necessary values of F_B and θ.

4.208   If you represent the forces exerted by the floor on the table legs by a force **F** acting at the origin O and a couple **M**, what are **F** and **M**?

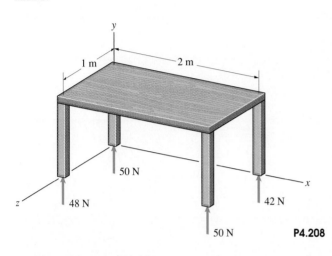

50 N

1 m   2 m

48 N   42 N

50 N   **P4.208**

4.209   If you represent the forces exerted by the floor on the table legs in Problem 4.208 by a force **F**, what is **F**, and where does its line of action intersect the x-z plane?

4.210   Two forces are exerted on the crankshaft by the connecting rods. The direction cosines of $\mathbf{F}_A$ are $\cos \theta_x = -0.182$, $\cos \theta_y = 0.818$, and $\cos \theta_z = 0.545$, and its magnitude is 4 kN. The direction cosines of $\mathbf{F}_B$ are $\cos \theta_x = 0.182$, $\cos \theta_y = 0.818$, and $\cos \theta_z = -0.545$, and its magnitude is 2 kN. If you represent the two forces by a force **F** acting at the origin O and a couple **M**, what are **F** and **M**?

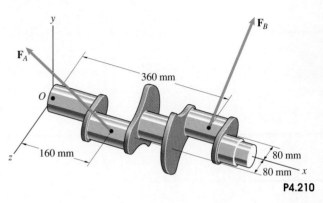

F_B

F_A

O

360 mm

160 mm

80 mm

80 mm

P4.210

4.211   If you represent the two forces exerted on the crankshaft in Problem 4.210 by a wrench consisting of a force **F** and a parallel couple $\mathbf{M_p}$, what are **F** and $\mathbf{M_p}$, and where does the line of action of **F** intersect the x-z plane?

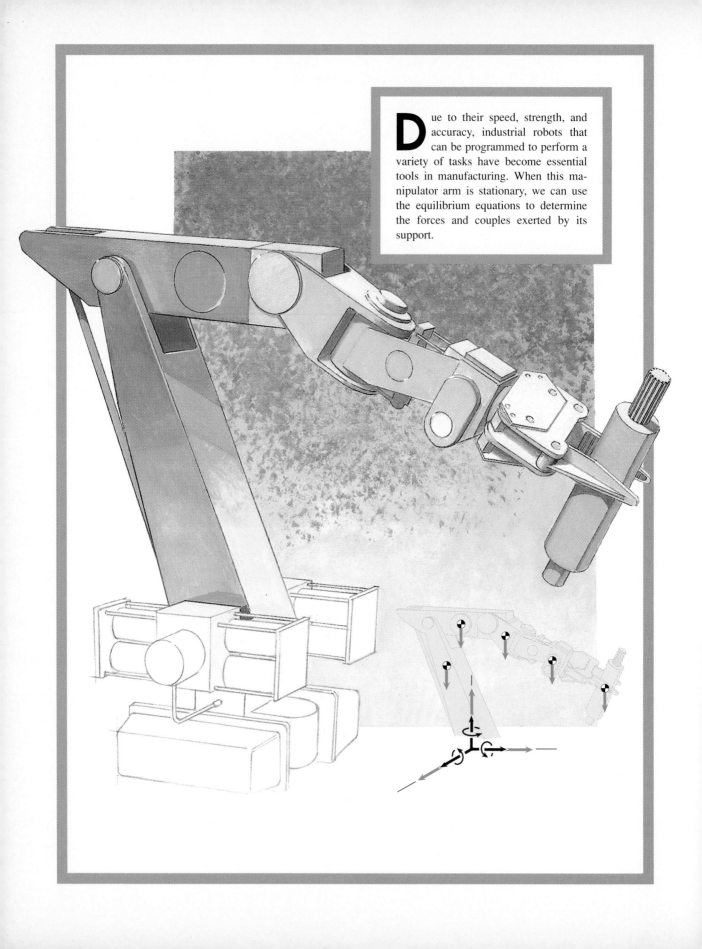

ue to their speed, strength, and accuracy, industrial robots that can be programmed to perform a variety of tasks have become essential tools in manufacturing. When this manipulator arm is stationary, we can use the equilibrium equations to determine the forces and couples exerted by its support.

Chapter 5

Objects in Equilibrium

STATICS is the analysis of objects in equilibrium, from robotic manipulators to bridges, dams, and buildings. Now that we have explained how to calculate moments, we are ready to come to grips with more interesting equilibrium problems.

In this chapter we state the equilibrium equations and describe simple models of the various types of supports used in engineering. We then show how the equilibrium equations can be used to obtain information about the systems of forces and moments acting on objects.

5.1 *The Equilibrium Equations*

In Chapter 3 we defined an object to be in equilibrium when it is stationary or in steady translation relative to an inertial reference frame. When an object acted upon by a system of forces and moments is in equilibrium, the following conditions are satisfied.

1. The sum of the forces is zero:

$$\boxed{\Sigma\, \mathbf{F} = \mathbf{0}.}$$ (5.1)

2. The sum of the moments about any point is zero:

$$\boxed{\Sigma\, \mathbf{M}_{\text{(any point)}} = \mathbf{0}.}$$ (5.2)

Before we consider specific applications, some general observations about these equations are in order. If the sum of the forces acting on an object is zero and the sum of the moments about one point is zero, then the sum of the moments about *every* point is zero. To demonstrate this, let's consider the object shown in Fig. 5.1(a), which is subjected to forces $\mathbf{F}_A$, $\mathbf{F}_B$, and $\mathbf{F}_C$ and couples $\mathbf{M}_D$ and $\mathbf{M}_E$. If the sum of the forces is zero,

$$\Sigma\, \mathbf{F} = \mathbf{F}_A + \mathbf{F}_B + \mathbf{F}_C = \mathbf{0},$$

and the sum of the moments about a point P is zero (Fig. 5.1b),

$$\Sigma\, \mathbf{M}_P = (\mathbf{r}_A \times \mathbf{F}_A) + (\mathbf{r}_B \times \mathbf{F}_B) + (\mathbf{r}_C \times \mathbf{F}_C) + \mathbf{M}_D + \mathbf{M}_E = \mathbf{0},$$

we can choose any other point P' (Fig. 5.1c) and show that the sum of the moments about P' is zero. That is, we will show that

$$\Sigma\, \mathbf{M}_{P'} = (\mathbf{r}'_A \times \mathbf{F}_A) + (\mathbf{r}'_B \times \mathbf{F}_B) + (\mathbf{r}'_C \times \mathbf{F}_C) + \mathbf{M}_D + \mathbf{M}_E$$
$$= \mathbf{0}.$$ (5.3)

In terms of the vector $\mathbf{r}$ from P' to P, notice that

$$\mathbf{r}'_A = \mathbf{r} + \mathbf{r}_A, \qquad \mathbf{r}'_B = \mathbf{r} + \mathbf{r}_B, \qquad \mathbf{r}'_C = \mathbf{r} + \mathbf{r}_C.$$

Substituting these expressions into Eq. (5.3), we obtain

$$\Sigma\, \mathbf{M}_{P'} = [(\mathbf{r} + \mathbf{r}_A) \times \mathbf{F}_A] + [(\mathbf{r} + \mathbf{r}_B) \times \mathbf{F}_B]$$
$$+ [(\mathbf{r} + \mathbf{r}_C) \times \mathbf{F}_C] + \mathbf{M}_D + \mathbf{M}_E$$
$$= [\mathbf{r} \times (\mathbf{F}_A + \mathbf{F}_B + \mathbf{F}_C)]$$
$$+ (\mathbf{r}_A \times \mathbf{F}_A) + (\mathbf{r}_B \times \mathbf{F}_B) + (\mathbf{r}_C \times \mathbf{F}_C) + \mathbf{M}_D + \mathbf{M}_E$$
$$= (\mathbf{r} \times \Sigma\, \mathbf{F}) + \Sigma\, \mathbf{M}_P = \mathbf{0}.$$

Figure 5.2 shows an object subjected to concurrent forces $\mathbf{F}_1$, $\mathbf{F}_2$, ... $\mathbf{F}_N$ and no couples. If the sum of the concurrent forces is zero, Eq. (5.2) is identically satisfied. The sum of the moments of the forces about P is zero, so the

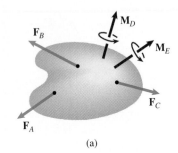

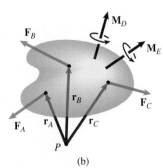

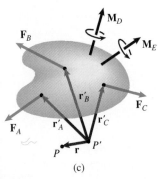

Figure 5.1

(a) An object subjected to a system of forces and moments.
(b) Determining the sum of the moments about P.
(c) Determining the sum of the moments about a different point P'.

sum of the moments about every point is zero. The only condition imposed by equilibrium on a set of concurrent forces is that their sum is zero.

In Chapter 4 we discussed the determination of the moment of a force about a line. To determine the sum of the moments about a line due to a system of forces and moments, you first choose any point P on the line and determine the sum of the moments $\Sigma\,\mathbf{M}_P$ about P. Then the sum of the moments about the line is the component of $\Sigma\,\mathbf{M}_P$ parallel to the line. If an object is in equilibrium, $\Sigma\,\mathbf{M}_P = \mathbf{0}$, so the sum of the moments about any line is also zero. You will find this result useful for certain types of problems.

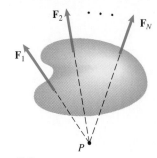

Figure 5.2
An object subjected to concurrent forces.

5.2 Two-Dimensional Applications

Many engineering applications involve two-dimensional systems of forces and moments. These include the forces and moments exerted on many beams and planar structures, pliers, some cranes and other machines, and some types of bridges and dams. In this section we discuss supports, free-body diagrams, and the equilibrium equations for two-dimensional applications.

Supports

When you are standing, the floor supports you. When you sit in a chair, the chair supports you. In this section we are concerned with the ways objects are held in place or are attached to other objects. Forces and couples exerted on an object by its supports are called **reactions**, expressing the fact that the supports "react" to the other forces and couples, or **loads**, acting on the object. For example, a bridge is held up by the reactions exerted by its supports, and the loads are the forces exerted by the weight of the bridge itself, the traffic crossing it, and the wind.

Some very common kinds of supports are represented by stylized models called support conventions. Actual supports often closely resemble the support conventions, but even when they don't, we represent them by these conventions if the actual supports exert the same (or approximately the same) reactions as the models.

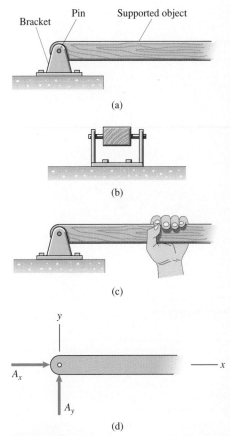

The Pin Support Figure 5.3(a) shows a **pin support**. The diagram represents a bracket to which an object (such as a beam) is attached by a smooth pin that passes through the bracket and the object. The side view is shown in Fig. 5.3(b).

To understand the reactions that a pin support can exert, it's helpful to imagine holding a bar attached to a pin support (Fig. 5.3c). If you try to move the bar without rotating it (that is, translate the bar), the support exerts a reactive force that prevents this movement. However, you can rotate the bar about the axis of the pin. The support cannot exert a couple about the pin axis to prevent rotation. Thus a pin support can't exert a couple about the pin axis, but it can exert a force on an object in any direction, which is usually expressed by representing the force in terms of components (Fig. 5.3d). The arrows indicate the directions of the reactions if A_x and A_y are positive. If you determine A_x or A_y to be negative, the reaction is in the direction opposite to that of the arrow.

Figure 5.3
(a) A pin support.
(b) Side view showing the pin passing through the beam.
(c) Holding a supported bar.
(d) The pin support is capable of exerting two components of force.

The pin support is used to represent any real support capable of exerting a force in any direction but not exerting a couple. You can observe pin supports in many common devices, particularly those designed to allow connected parts to rotate relative to each other (Fig. 5.4).

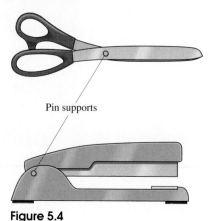

Pin supports

Figure 5.4

Pin supports in a pair of scissors and a stapler.

The Roller Support The convention called a **roller support** (Fig. 5.5a) represents a pin support mounted on wheels. Like the pin support, it cannot exert a couple about the axis of the pin. Since it can move freely in the direction parallel to the surface on which it rolls, it can't exert a force parallel to the surface but can only exert a force normal (perpendicular) to this surface (Fig. 5.5b). Figures 5.5(c)–(e) are other commonly used conventions equivalent to the roller support. The wheels of vehicles and wheels supporting parts of machines are roller supports if the friction forces exerted on them are negligible in comparison to the normal forces. A plane smooth surface can also be modeled by a roller support (Fig. 5.6). Beams and bridges are sometimes supported in this way so that they will be free to undergo thermal expansion and contraction.

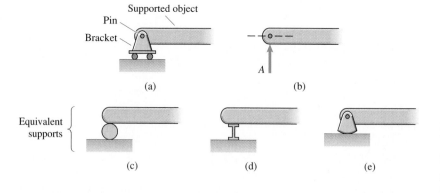

Figure 5.5

(a) A roller support.
(b) The reaction consists of a force normal to the surface.
(c)–(e) Supports equivalent to the roller support.

Figure 5.6

Supporting an object with a plane smooth surface.

The supports shown in Fig. 5.7 are similar to the roller support in that they cannot exert a couple and can only exert a force normal to a particular direction. (Friction is neglected.) In these supports, the supported object is attached to a pin or slider that can move freely in one direction but is constrained in the perpendicular direction. Unlike the roller support, these supports can exert a normal force in either direction.

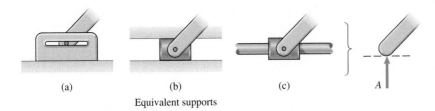

(a) (b) (c) A

Equivalent supports

Figure 5.7

Supports similar to the roller support except that the normal force can be exerted in either direction.
(a) Pin in a slot.
(b) Slider in a slot.
(c) Slider on a shaft.

The Built-In Support   The **built-in support** shows the supported object literally built into a wall (Fig. 5.8a). This convention is also called a **fixed support**. To understand the reactions, imagine holding a bar attached to a built-in support (Fig. 5.8b). If you try to translate the bar, the support exerts a reactive force that prevents translation, and if you try to rotate the bar, the support exerts a reactive couple that prevents rotation. A built-in support can exert two components of force and a couple (Fig. 5.8c). The term M_A is the couple exerted by the support, and the curved arrow indicates its direction. Fence posts and lampposts have built-in supports. The attachments of parts connected so that they cannot move or rotate relative to each other, such as the head of a hammer and its handle, can be modeled as built-in supports.

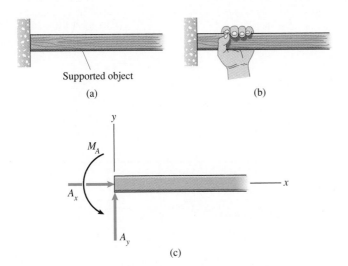

Supported object

(a) (b)

y

M_A

A_x x

A_y

(c)

Figure 5.8

(a) Built-in support.
(b) Holding a supported bar.
(c) The reactions a built-in support is capable of exerting.

Table 5.1 summarizes the support conventions commonly used in two-dimensional applications, including those we discussed in Chapter 3. Although the number of conventions may appear daunting, the examples and problems will help you become familiar with them. You should also observe how various objects you see in your everyday experience are supported and think about whether each support could be represented by one of the conventions.

Table 5.1

Supports		Reactions
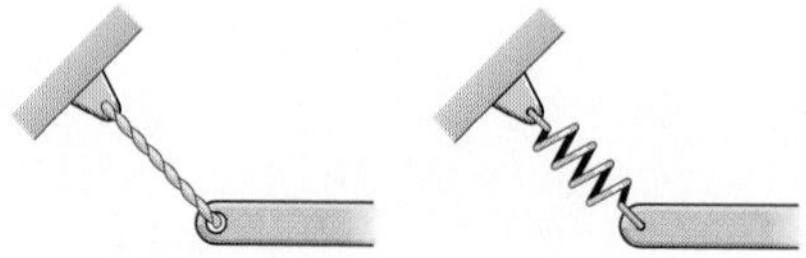 Rope or Cable	Spring	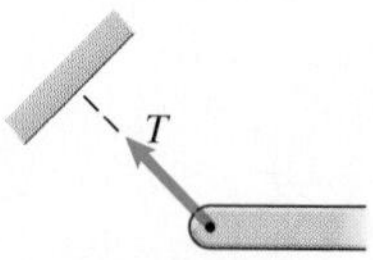 One Collinear Force
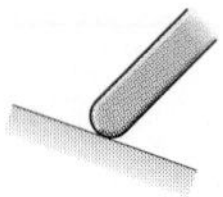 Contact with a Smooth Surface		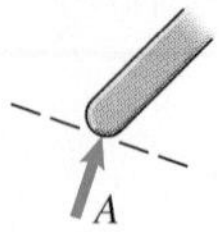 One Force Normal to the Supporting Surface
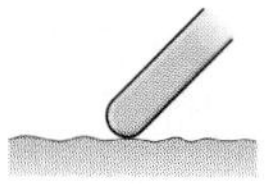 Contact with a Rough Surface		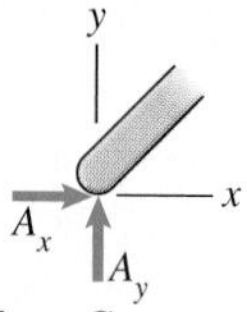 Two Force Components
 Pin Support		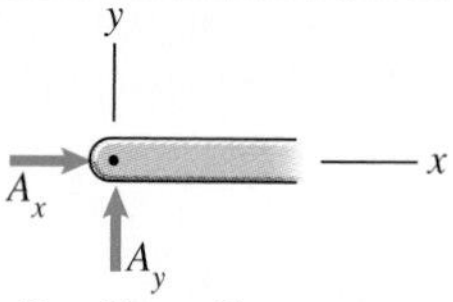 Two Force Components
 Roller Support 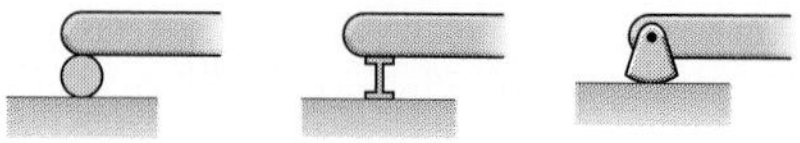 Equivalents		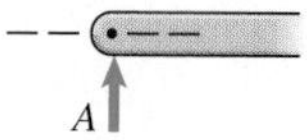 One Force Normal to the Supporting Surface
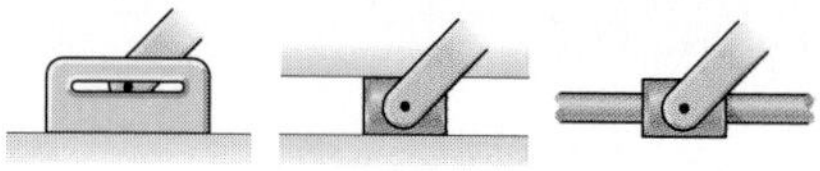 Constrained Pin or Slider		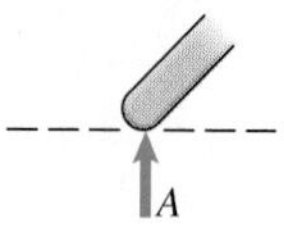 One Normal Force
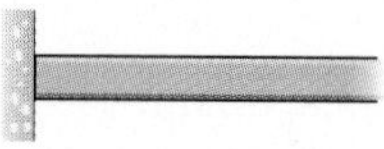 Built-in (Fixed) Support		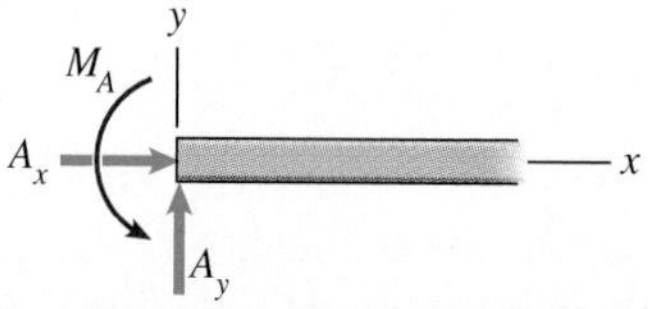 Two Force Components and One Couple

Table 5.1

Supports used in two-dimensional applications.

Free-Body Diagrams

We introduced free-body diagrams in Chapter 3 and used them to determine forces acting on simple objects in equilibrium. By using the support conventions, we can model more elaborate objects and construct their free-body diagrams in a systematic way.

For example, the beam in Fig. 5.9(a) has a pin support at the left end and a roller support at the right end and is loaded by a force F. The roller support rests on a surface inclined at 30° to the horizontal. To obtain the free-body diagram of the beam, we first isolate it from its supports (Fig. 5.9b), since the free-body diagram must contain no object other than the beam. We complete the free-body diagram by showing the reactions that may be exerted on the beam by the supports (Fig. 5.9c). Notice that the reaction B exerted by the roller support is normal to the surface on which the support rests.

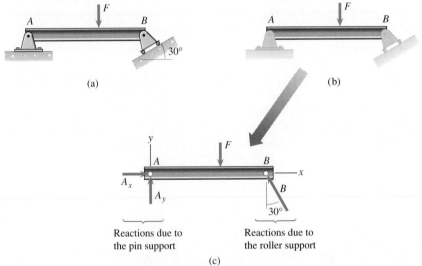

(a)

(b)

Reactions due to
the pin support

Reactions due to
the roller support

(c)

Figure 5.9

(a) A beam with pin and roller supports.
(b) Isolating the beam from its supports.
(c) The completed free-body diagram.

The object in Fig. 5.10(a) has a fixed support at the left end. A cable passing over a pulley is attached to the object at two points. We isolate it from its supports (Fig. 5.10b) and complete the free-body diagram by showing the reactions at the built-in support and the forces exerted by the cable (Fig. 5.10c). *Don't forget the couple at a built-in support.* Since we assume the tension in the cable is the same on both sides of the pulley, the two forces exerted by the cable have the same magnitude T.

Once you have obtained the free-body diagram of an object in equilibrium to identify the loads and reactions acting on it, you can apply the equilibrium equations.

Figure 5.10

(a) An object with a built-in support.
(b) Isolating the object.
(c) The completed free-body diagram.

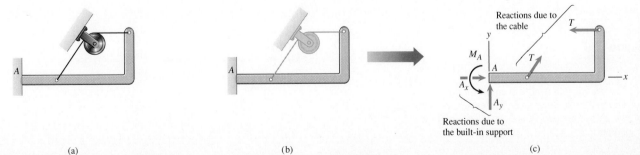

Reactions due to
the cable

Reactions due to
the built-in support

(a)

(b)

(c)

The Scalar Equilibrium Equations

When the loads and reactions on an object in equilibrium form a two-dimensional system of forces and moments, they are related by three scalar equilibrium equations:

$$\Sigma F_x = 0, \tag{5.4}$$

$$\Sigma F_y = 0, \tag{5.5}$$

$$\Sigma M_{(any\ point)} = 0. \tag{5.6}$$

A natural question is whether you can obtain more than one equation from Eq. (5.6) by evaluating the sum of the moments about more than one point. The answer is yes, and in some cases you will find it convenient to do so. But there is a catch—the additional equations will not be independent of Eqs. (5.4)–(5.6). In other words, *you can't obtain more than three independent equilibrium equations from a two-dimensional free-body diagram, which means you can solve for at most three unknown forces or couples.* We discuss this point further in Section 5.3.

The seesaw found on playgrounds, consisting of a board with a pin support at the center that allows it to rotate, is a simple and familiar example that illustrates the role of Eq. (5.6). If two people of unequal weight sit at the seesaw's ends, the heavier person sinks to the ground (Fig. 5.11a). To obtain equilibrium, that person must move closer to the center (Fig. 5.11b).

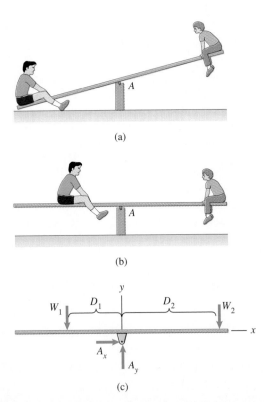

(a)

(b)

(c)

Figure 5.11

(a) If both people sit at the ends of the seesaw, the heavier one sinks.
(b) The seesaw and people in equilibrium.
(c) The free-body diagram of the seesaw, showing the weights of the people and the reactions at the pin support.

We draw the free-body diagram of the seesaw in Fig. 5.11(c), showing the weights of the people W_1 and W_2 and the reactions at the pin support. Evaluating the sum of the moments about A, the equilibrium equations are

$$\Sigma F_x = A_x = 0, \tag{5.7}$$

$$\Sigma F_y = A_y - W_1 - W_2 = 0, \tag{5.8}$$

$$\Sigma M_{(\text{point } A)} = D_1 W_1 - D_2 W_2 = 0. \tag{5.9}$$

Thus $A_x = 0$, $A_y = W_1 + W_2$, and $D_1 W_1 = D_2 W_2$. The last condition indicates the relation between the positions of the two persons necessary for equilibrium.

To demonstrate that an additional independent equation is not obtained by evaluating the sum of the moments about a different point, we can sum the moments about the right end of the seesaw:

$$\Sigma M_{(\text{right end})} = (D_1 + D_2) W_1 - D_2 A_y = 0.$$

This equation is a linear combination of Eqs. (5.8) and (5.9):

$$(D_1 + D_2)W_1 - D_2 A_y = \underbrace{-D_2(A_y - W_1 - W_2)}_{\textbf{Eq. (5.8)}}$$

$$+ \underbrace{(D_1 W_1 - D_2 W_2)}_{\textbf{Eq. (5.9)}} = 0.$$

Until now we have assumed in examples and problems that the tension in a rope or cable is the same on both sides of a pulley. Consider the pulley in Fig. 5.12(a). In its free-body diagram in Fig. 5.12(b), we do not assume that the tensions are equal. Summing the moments about the center of the pulley, we obtain the equilibrium equation

$$\Sigma M_{(\text{point } A)} = RT_1 - RT_2 = 0.$$

The tensions must be equal if the pulley is in equilibrium. However, notice that we have assumed that the pulley's support behaves like a pin support and cannot exert a couple on the pulley. When that is not true—for example, due to friction between the pulley and the support—the tensions are not necessarily equal.

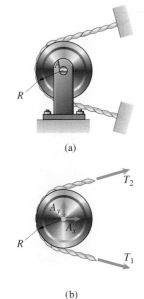

(a)

(b)

Figure 5.12

(a) A pulley of radius R.
(b) Free-body diagram of the pulley and part of the cable.

In the following examples we determine the reactions exerted on objects with various kinds of loads and supports. This typically involves two steps:

1. Draw the free-body diagram—*You must obtain the free-body diagram by isolating the object from its supports and showing the loads and the reactions the supports may exert on the object.*

2. Apply the equilibrium equations—*Use Eqs. (5.4)–(5.6) to determine the reactions.*

Example 5.1

The beam in Fig. 5.13 has pin and roller supports and is subjected to a 2-kN force. What are the reactions at the supports?

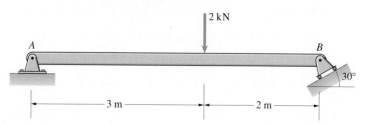

Figure 5.13

SOLUTION

Draw the Free-Body Diagram We isolate the beam from its supports and show the loads and the reactions that may be exerted by the pin and roller supports (Fig. a). There are three unknown reactions: two components of force A_x and A_y at the pin support and a force B at the roller support.

Apply the Equilibrium Equations Summing the moments about point A, the equilibrium equations are

$$\Sigma F_x = A_x - B \sin 30° = 0,$$

$$\Sigma F_y = A_y + B \cos 30° - 2 = 0,$$

$$\Sigma M_{(\text{point } A)} = (5)(B \cos 30°) - (3)(2) = 0.$$

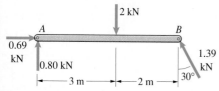

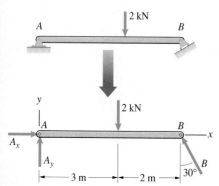

(a) Drawing the free-body diagram of the beam.

Solving these equations, the reactions are $A_x = 0.69$ kN, $A_y = 0.80$ kN, and $B = 1.39$ kN. The load and reactions are shown in Fig. (b). It is good practice to show your answers in this way and confirm that the equilibrium equations are satisfied:

$$\Sigma F_x = 0.69 - 1.39 \sin 30° = 0,$$

$$\Sigma F_y = 0.80 + 1.39 \cos 30° - 2 = 0,$$

$$\Sigma M_{(\text{point } A)} = (5)(1.39 \cos 30°) - (3)(2) = 0.$$

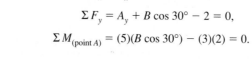

(b) The load and reactions.

DISCUSSION

We drew the arrows indicating the directions of the reactions A_x and A_y in the positive x and y axis directions, but we could have drawn them in either direction. In Fig. (c) we draw the free-body diagram of the beam with the component A_y pointed downward. From this free-body diagram we obtain the equilibrium equations

$$\Sigma F_x = A_x - B \sin 30° = 0,$$

$$\Sigma F_y = -A_y + B \cos 30° - 2 = 0,$$

$$\Sigma M_{(\text{point } A)} = (5)(B \cos 30°) - (2)(3) = 0.$$

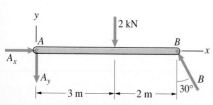

(c) An alternative free-body diagram.

The solutions are $A_x = 0.69$ kN, $A_y = -0.80$ kN, and $B = 1.39$ kN. The negative value of A_y indicates that the vertical force exerted on the beam by the pin support is in the direction opposite to that of the arrow in Fig. (c); that is, the force is 0.80 kN upward. Thus we again obtain the reactions shown in Fig. (b).

Example 5.2

The object in Fig. 5.14 has a built-in support and is subjected to two forces and a couple. What are the reactions at the support?

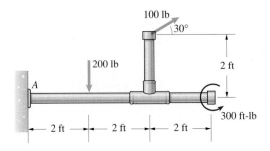

Figure 5.14

SOLUTION

Draw the Free-Body Diagram We isolate the object from its support and show the reactions at the built-in support (Fig. a). There are three unknown reactions: two force components A_x and A_y and a couple M_A. (Remember that we can choose the directions of these arrows arbitrarily.) We also resolve the 100-lb force into its components.

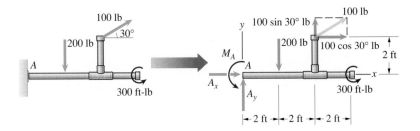

(a) Drawing the free-body diagram.

Apply the Equilibrium Equations Summing the moments about point A, the equilibrium equations are

$$\Sigma F_x = A_x + 100 \cos 30° = 0,$$

$$\Sigma F_y = A_y - 200 + 100 \sin 30° = 0,$$

$$\Sigma M_{(point\ A)} = M_A + 300 - (200)(2) - (100 \cos 30°)(2)$$
$$+ (100 \sin 30°)(4) = 0.$$

Solving these equations, we obtain the reactions $A_x = -86.6$ lb, $A_y = 150.0$ lb, and $M_A = 73.2$ ft-lb.

DISCUSSION

Notice that the 300-ft-lb couple and the couple M_A exerted by the built-in support don't appear in the first two equilibrium equations because a couple exerts no net force. Also, since the moment due to a couple is the same about any point, the moment about point A due to the 300-ft-lb counterclockwise couple is 300 ft-lb counterclockwise.

Example 5.3

The 2800-lb car in Fig. 5.15 is stationary. Determine the normal forces exerted on the front and rear tires by the road.

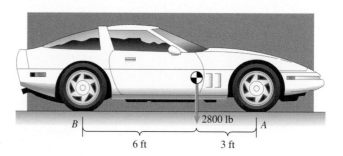

Figure 5.15

SOLUTION

Draw the Free-Body Diagram   In Fig. (a) we isolate the car and show its weight and the reactions exerted by the road. There are two unknown reactions: the forces A and B exerted on the front and rear tires.

Apply the Equilibrium Equations   The forces have no x components. Summing the moments about point B, the equilibrium equations are

$$\Sigma F_y = A + B - 2800 = 0,$$

$$\Sigma M_{(\text{point }B)} = (6)(2800) - 9A = 0.$$

Solving these equations, the reactions are $A = 1867$ lb and $B = 933$ lb.

DISCUSSION

This example doesn't fall strictly within our definition of a two-dimensional system of forces and moments because the forces acting on the car are not coplanar. Let's examine why you can analyze problems of this kind as if they were two-dimensional.

In Fig. (b) we show an oblique view of the free-body diagram of the car. In this view you can see the forces acting on the individual tires. The total normal force on the front tires is $A_L + A_R = A$, and the total normal force on the rear tires is $B_L + B_R = B$. The sum of the forces in the y direction is

$$\Sigma F_y = A_L + A_R + B_L + B_R - 2800 = A + B - 2800 = 0.$$

Since the sum of the moments about any line due to the forces and couples acting on an object in equilibrium is zero, the sum of the moments about the z axis due to the forces acting on the car is zero:

$$\Sigma M_{(z\text{ axis})} = (9)(A_L + A_R) - (6)(2800) = 9A - (6)(2800) = 0.$$

Thus we obtain the same equilibrium equations we did when we solved the problem using a two-dimensional analysis.

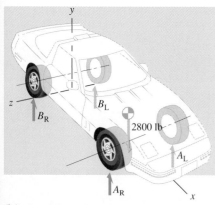

Total normal force exerted on the two rear tires

Total normal force exerted on the two front tires

(a) The free-body diagram.

(b) An oblique view showing the forces on the individual tires.

Example 5.4

The structure AB in Fig. 5.16 supports a suspended 2-Mg (megagram) mass. The structure is attached to a slider in a vertical slot at A and has a pin support at B. What are the reactions at A and B?

SOLUTION

Draw the Free-Body Diagram   We isolate the structure and mass from the supports and show the reactions at the supports and the force exerted by the weight of the 2000-kg mass (Fig. a). The slot at A can exert only a horizontal force on the slider.

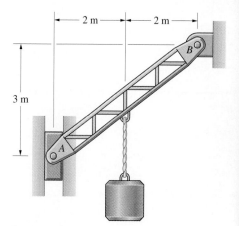

Figure 5.16

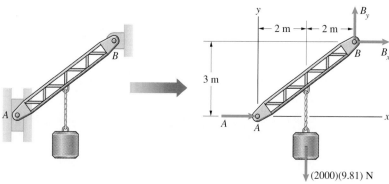

(a) Drawing the free-body diagram.

Apply the Equilibrium Equations   Notice that if we sum the moments about point B, we obtain an equation containing only one unknown reaction, the force A. The equilibrium equations are

$$\Sigma F_x = A + B_x = 0,$$

$$\Sigma F_y = B_y - (2000)(9.81) = 0,$$

$$\Sigma M_{(\text{point } B)} = A(3) + (2000)(9.81)(2) = 0.$$

The reactions are $A = -13.1$ kN, $B_x = 13.1$ kN, and $B_y = 19.6$ kN.

DISCUSSION

You can often simplify equilibrium equations by a careful choice of the point you sum moments about. For example, when you can choose a point where the lines of action of unknown forces intersect, those forces will not appear in your moment equation.

Example 5.5

Application to Engineering

Design for Human Factors

Figure 5.17 shows an airport luggage carrier and its free-body diagram when it is held in equilibrium in the tilted position. If the luggage carrier supports a weight $W = 50$ lb, the angle $\alpha = 30°$, $a = 8$ in., $b = 16$ in., and $d = 48$ in., what force F must the user exert?

STRATEGY

The unknown reactions on the free-body diagram are the force F and the normal force N exerted by the floor. If we sum moments about the center of the wheel C, we obtain an equation in which F is the only unknown reaction.

SOLUTION

Summing moments about C,

$$\Sigma M_{(\text{point } C)} = d(F \cos \alpha) + a(W \sin \alpha) - b(W \cos \alpha) = 0,$$

and solving for F, we obtain

$$F = \frac{(b - a \tan \alpha)\, W}{d}. \tag{5.10}$$

Substituting the values of W, α, a, b, and d, the solution is $F = 11.9$ lb.

DESIGN ISSUES

Design that accounts for human physical dimensions, capabilities, and characteristics is a special challenge. This art is called design for human factors. Here we consider a simple device, the airport luggage carrier in Fig. 5.17, and show how consideration of its potential users *and the constraints imposed by the equilibrium equations* affect its design.

The user moves the carrier by grasping the bar at the top, tilting it, and walking while pulling the carrier. The height of the handle (the dimension h) needs to be comfortable. Since $h = R + d \sin \alpha$, if we choose values of h and the wheel radius R, we obtain a relation between the length of the carrier's handle d and the tilt angle α:

$$d = \frac{h - R}{\sin \alpha}. \tag{5.11}$$

Substituting this expression for d into Eq. (5.10), we obtain

$$F = \frac{\sin \alpha (b - a \tan \alpha) W}{h - R}. \tag{5.12}$$

Suppose that based on statistical data on human dimensions, we decide to design the carrier for convenient use by persons up to 6 ft 2 in. tall, which corresponds to a dimension h of approximately 36 in. Let $R = 3$ in., $a = 6$ in., and $b = 12$ in. The resulting value of F/W as a function of α is shown in Fig. 5.18. At $\alpha = 63°$, the

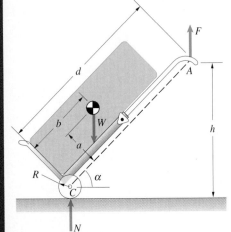

Figure 5.17

force the user must exert is zero, which means the weight of the luggage acts at a point directly above the wheels. This would be the optimum solution if the user could maintain exactly that value of α. However, α inevitably varies, and the resulting changes in F make it difficult to control the carrier. In addition, the relatively steep angle would make the carrier awkward to pull. From this point of view, it is desirable to choose a design within the range of values of α in which F varies slowly, say, $30° \leq \alpha \leq 45°$. (Even though the force the user must exert is large in this range of α in comparison with larger values of α, it is only about 13% of the weight.) Over this range of α, the dimension d varies from 5.5 ft to 3.9 ft. A smaller carrier is desirable for lightness and ease of storage, so we choose $d = 4$ ft for our preliminary design.

We have chosen the dimension d based on particular values of the dimensions R, a, and b. In an actual design study, we would carry out the analysis for expected ranges of values of these parameters. Our final design would also reflect decisions based on safety (for example, there must be adequate means to secure the luggage and no sharp projections), reliability (the frame must be sufficiently strong and the wheels must have adequate and reliable bearings), and the cost of manufacture.

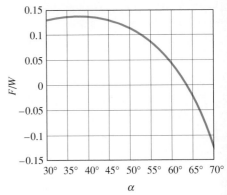

Figure 5.18
Graph of the ratio F/W as a function of α.

Problems

Assume that objects are in equilibrium. In the statements of the answers, *x* components are positive to the right and *y* components are positive upward.

5.1 Draw free-body diagrams of the objects a, b, and c.

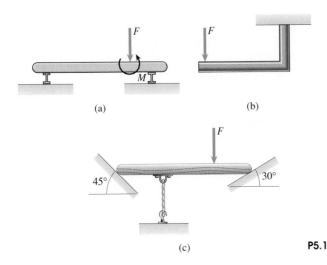

(a)

(b)

(c)

P5.1

5.2 Draw free-body diagrams of the objects a, b, and c.

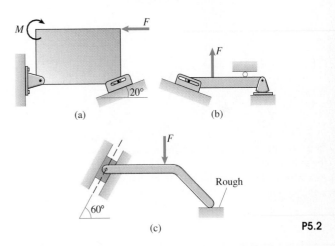

(a)

(b)

(c)

P5.2

5.3 The tension in cable AB is T. Draw a free-body diagram (a) of the pipe assembly and cable AB, treating them as a single object; (b) of the pipe assembly, not including cable AB; (c) of cable AB.

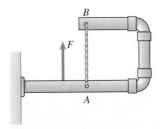

P5.3

5.4 (a) Draw the free-body diagram of the beam.
(b) Determine the reactions at the supports.

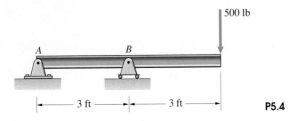

P5.4

5.5 (a) Draw the free-body diagram of the 60-lb drill press, assuming that the surfaces at A and B are smooth.
(b) Determine the reactions at A and B.

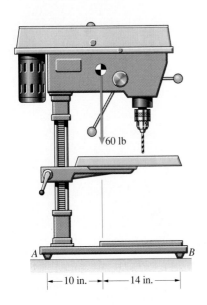

P5.5

5.6 The mass of the person is 80 kg, and the mass of the diving board is 45 kg.
(a) Draw the free-body diagram of the diving board.
(b) Determine the reactions at the supports A and B.

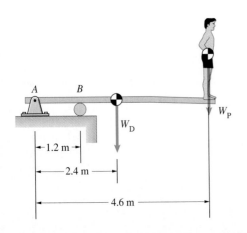

P5.6

5.7 The ironing board has supports at A and B that can be modeled as roller supports.
(a) Draw the free-body diagram of the ironing board.
(b) Determine the reactions at A and B.

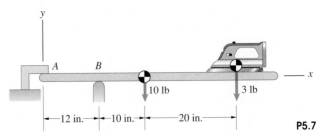

P5.7

5.8 The distance $x = 2$ m.
(a) Draw the free-body diagram of the beam.
(b) Determine the reactions at the supports.

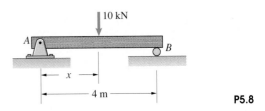

P5.8

5.9 Consider the beam in Problem 5.8. An engineer determines that each support will safely support a force of 7.5 kN. What is the range of values of the distance x at which the 10-kN force can safely be applied?

5.10 (a) Draw the free-body diagram of the beam.
(b) Determine the reactions at the supports.

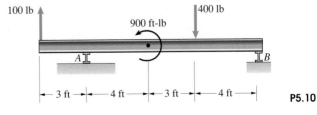

P5.10

5.11 Consider the beam in Problem 5.10. First represent the loads (the 100-lb force, the 400-lb force, and the 900 ft-lb couple) by a single equivalent force; then determine the reactions at the supports.

5.12 (a) Draw the free-body diagram of the beam.
(b) Determine the reactions at the supports.

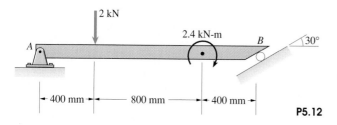

P5.12

5.13 Consider the beam in Problem 5.12. First represent the loads (the 2-kN force and the 24-kN-m couple) by a single equivalent force; then determine the reactions at the supports.

5.14 If the force $F = 40$ kN, what are the reactions at A and B?

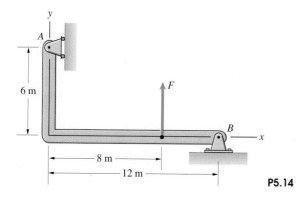

P5.14

5.15 In Problem 5.14, the structural designer determines that the magnitude of the force exerted on the support A by the beam must not exceed 80 kN, and the magnitude of the force exerted on the support B must not exceed 140 kN. Based on these criteria, what is the largest allowable value of the upward load F?

5.16 A man doing push-ups pauses in the position shown. His weight $W = 175$ lb. The distances are $a = 10$ in., $b = 29$ in., and $c = 12$ in. Determine the normal force exerted by the floor (a) on each hand; (b) on each foot.

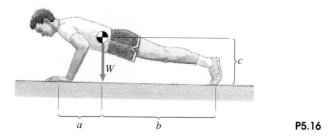

P5.16

5.17 With each of the devices shown you can support a load R by applying a force F. They are called levers of the first, second, and third class.

(a) The ratio R/F is called the *mechanical advantage*. Determine the mechanical advantage of each lever.
(b) Determine the magnitude of the reaction at A for each lever. (Express your answers in terms of F.)

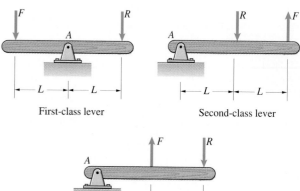

First-class lever   Second-class lever

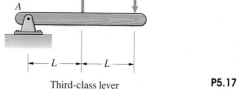

Third-class lever   P5.17

5.18 (a) Draw the free-body diagram of the beam.
(b) Determine the reactions at the support.

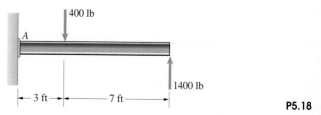

P5.18

5.19 The force $F = 500$ lb. Determine the reactions at the support.

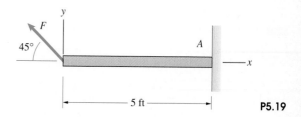

P5.19

5.20 Consider the beam in Problem 5.19. An engineer determines that the built-in support will fail if the magnitude of the total force exerted on the support exceeds 1000 lb or if the magnitude of the couple exerted on the support exceeds 3000 ft-lb. Based on these criteria, what is the maximum force F that can be applied?

5.21   The mobile is in equilibrium. The fish B weighs 27 oz. Determine the weights of the fish A, C, and D. (The weights of the crossbars are negligible.)

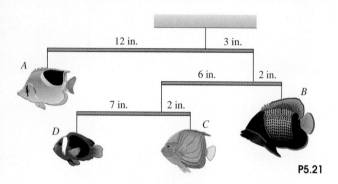

P5.21

5.22   If the total weight of the four fish in Problem 5.21 is 16 oz., what are the weights of the individual fish?

5.23   The lift forces on an airplane's wing are represented by eight forces. The magnitude of each force is given in terms of its position x on the wing by

$$200\sqrt{1 - (x/17)^2} \text{ lb.}$$

The weight of the wing is $W = 400$ lb. Determine the reactions on the wing at the root R.

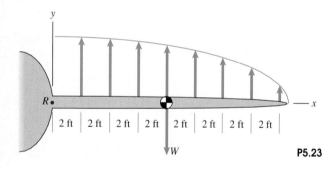

P5.23

5.24   The 14.5-lb chain saw is subjected to the loads at A by the log it cuts. Determine the reactions R, B_x, and B_y that must be applied by the person using the saw to hold it in equilibrium.

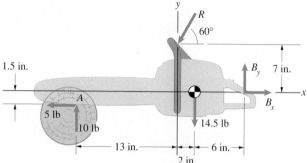

P5.24

5.25   The mass of the trailer is 2.2 Mg (megagrams). The distances $a = 2.5$ m and $b = 5.5$ m. The truck is stationary, and the wheels of the trailer can turn freely, which means the road exerts no horizontal force on them. The hitch at B can be modeled as a pin support.
(a) Draw the free-body diagram of the trailer.
(b) Determine the total normal force exerted on the rear tires at A and the reactions exerted on the trailer at the pin support B.

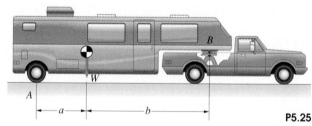

P5.25

5.26 The total weight of the wheelbarrow and its load is $W = 100$ lb.
(a) If $F = 0$, what are the vertical reactions at A and B?
(b) What force F is necessary to lift the support at A off the ground?

5.28 The forklift is stationary. The front wheels are free to turn, and the rear wheels are locked. The distances are $a = 1.25$ m, $b = 0.50$ m, and $c = 1.40$ m. The weight of the load is $W_L = 2$ kN, and the weight of the truck and operator is $W_F = 8$ kN. What are the reactions at A and B?

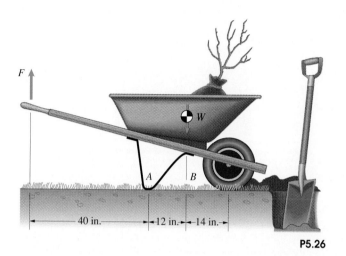

P5.26

P5.28

5.27 The airplane's weight is $W = 2400$ lb. Its brakes keep the rear wheels locked. The front (nose) wheel can turn freely, and so the ground exerts no horizontal force on it. The force T exerted by the airplane's propeller is horizontal.
(a) Draw the free-body diagram of the airplane. Determine the reaction exerted on the nose wheel and the total normal reaction exerted on the rear wheels (b) when $T = 0$; (c) when $T = 250$ lb.

5.29 Consider the stationary forklift shown in Problem 5.28. The front wheels are free to turn, and the rear wheels are locked. The distances are $a = 45$ in., $b = 20$ in., and $c = 50$ in. The weight of the truck and operator is $W_F = 3000$ lb. For safety reasons, a rule is established that the reaction at the rear wheels must be at least 400 lb. If the weight W_L of the load acts at the position shown, what is the maximum safe load?

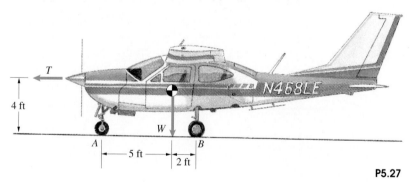

P5.27

5.30 The weight of the fan is $W = 20$ lb. Its base has four equally spaced legs of length $b = 12$ in., and $h = 36$ in. What is the largest thrust T exerted by the fan's propeller for which the fan will remain in equilibrium?

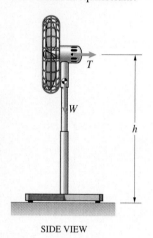
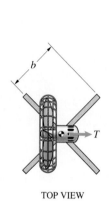

SIDE VIEW   TOP VIEW

P5.30

5.31 Consider the fan described in Problem 5.30. As a safety criterion, an engineer decides that the vertical reaction on any of the fan's legs should not be less than 20% of the fan's weight. If the thrust T is 1 lb when the fan is set on its highest speed, what is the maximum safe value of h?

5.32 To decrease costs, an engineer considers supporting a fan with three equally spaced legs instead of the four-leg configuration shown in Problem 5.30. For the same values of b, h, and W, show that the largest thrust T for which the fan will remain in equilibrium with three legs is related to the value with four legs by

$$T_{\text{(three legs)}} = \left(1/\sqrt{2}\right)T_{\text{(four legs)}}.$$

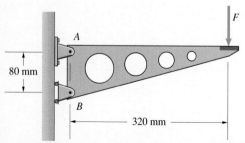

P5.32

5.33 A force $F = 400$ N acts on the bracket. What are the reactions at A and B?

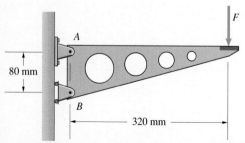

80 mm

320 mm

P5.33

5.34 The hanging sign exerts vertical 25-lb forces at A and B. Determine the tension in the cable and the reactions at the support at C.

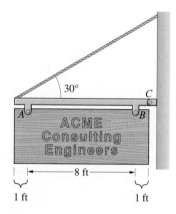

30°

C

A B

ACME
Consulting
Engineers

8 ft

1 ft   1 ft   **P5.34**

5.35 This device, called a *swape* or *shadoof*, is used to help a person lift a heavy load. (It was used in Egypt at least as early as 1550 B.C. and is still in use in various parts of the world today.) The distances are $a = 12$ ft and $b = 4$ ft. If the load being lifted weighs 100 lb and $W = 200$ lb, determine the vertical force the person must exert to support the stationary load (a) when the load is just above the ground (the position shown); (b) when the load is 3 ft above the ground. (Assume that the rope remains vertical.)

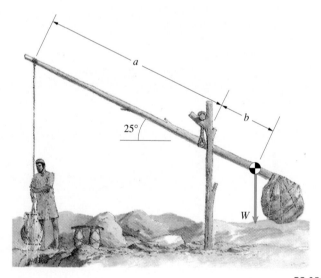

a

b

25°

W

P5.35

5.36 This structure, called a *truss*, has a pin support at A and a roller support at B and is loaded by two forces. Determine the reactions at the supports.

Strategy: Draw a free-body diagram, treating the entire truss as a single object.

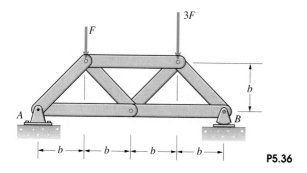

P5.36

5.37 An Olympic gymnast is stationary in the "iron cross" position. The weight of his left arm and the weight of his body *not including his arms* are shown. The distances are $a = b = 9$ in. and $c = 13$ in. Treat his shoulder S as a built-in support, and determine the magnitudes of the reactions at his shoulder. That is, determine the force and couple his shoulder must support.

P5.37

5.38 Determine the reactions at A.

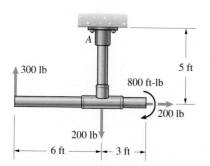

P5.38

5.39 The car's brakes keep the rear wheels locked, and the front wheels are free to turn. Determine the forces exerted on the front and rear wheels by the road when the car is parked (a) on an upslope with $\alpha = 15°$; (b) on a downslope with $\alpha = -15°$.

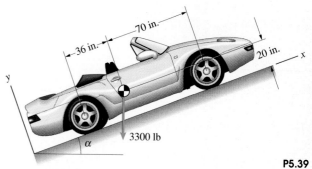

P5.39

5.40 The weight W of the bar acts at its center. The surfaces are smooth. What is the tension in the horizontal string?

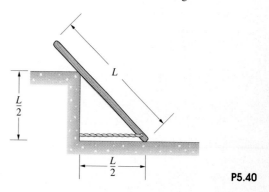

P5.40

5.41 The bar weighs 15 lb, and its weight acts at its midpoint. The spring is unstretched when $\alpha = 0$. The bar is in equilibrium when $\alpha = 30°$. Determine the spring constant k.

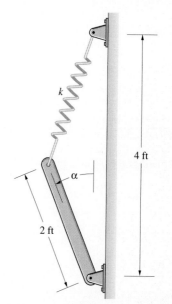

P5.41

5.42   The plate is supported by a pin in a smooth slot at *B*. What are the reactions at the supports?

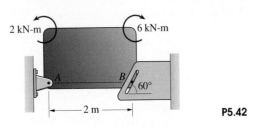

P5.42

5.43   The force $F = 800$ N, and the couple $M = 200$ N-m. The distance $L = 2$ m. What are the reactions at *A* and *B*?

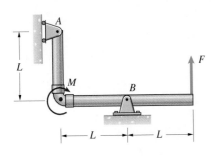

P5.43

5.44   The weight $W = 1.2$ kN. Determine the tension in the cable and the reactions at *A*.

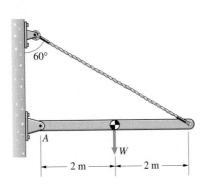

P5.44

5.45   Consider the beam in Problem 5.44. An engineer determines that the cable will safely support a tension of 3 kN and that the support at *A* will safely support a force of magnitude 2.4 kN. Based on these criteria, what is the largest beam weight that can safely be supported?

5.46   The mass of each of the suspended boxes is 80 kg. Determine the reactions at the supports at *A* and *E*.

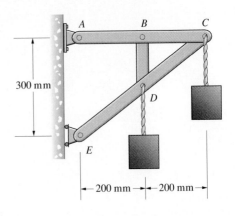

P5.46

5.47   The suspended boxes in Problem 5.46 are each of mass *m*. The supports at *A* and *E* will each safely support a force of 6 kN magnitude. Based on this criterion, what is the largest safe value of *m*?

5.48   The tension in cable *BC* is 100 lb. Determine the reactions at the built-in support.

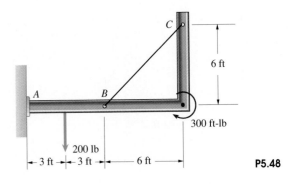

P5.48

5.49   The tension in cable *AB* is 2 kN. What are the reactions at *C* in the two cases?

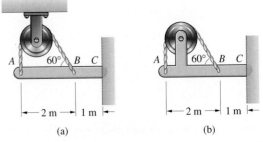

(a)   (b)

P5.49

5.50 Determine the reactions at the supports.

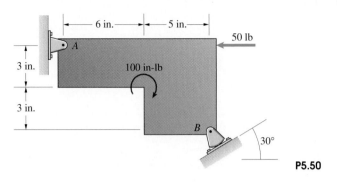

P5.50

5.51 The weight $W = 1000$ lb. Determine the tension in the cable and the reactions at A.

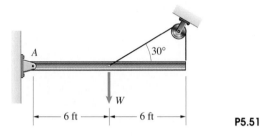

P5.51

5.52 Consider the beam in Problem 5.51. An engineer determines that the cable will safely support a tension of 1000 lb. What is the largest beam weight W that can safely be supported?

5.53 The distance $x = 3$ ft. Determine the tension in the cable and the reactions at A and B.

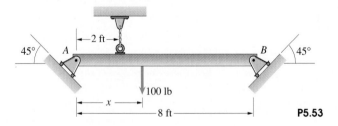

P5.53

5.54 Consider the beam in Problem 5.53. Determine the range of values of the distance x for which the beam can be in equilibrium in the position shown.

5.55 Suppose that you want to design the safety valve to open when the difference between the pressure p in the circular pipe (diameter = 150 mm) and atmospheric pressure is 10 MPa (megapascals; a pascal is 1 N/m²). The spring is compressed 20 mm

when the valve is closed. What should the value of the spring constant be?

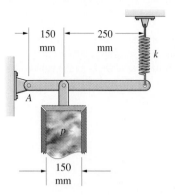

P5.55

5.56 The bar AB is of length L and weight W, and the weight acts at its midpoint. The angle $\alpha = 30°$. What is the tension in the string?

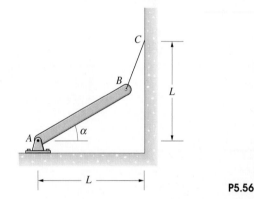

P5.56

5.57 The crane's arm has a pin support at A. The hydraulic cylinder BC exerts a force on the arm at C in the direction parallel to BC. The crane's arm has a mass of 200 kg, and its weight can be assumed to act at a point 2 m to the right of A. If the mass of the suspended box is 800 kg and the system is in equilibrium, what is the magnitude of the force exerted by the hydraulic cylinder?

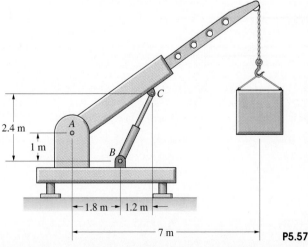

P5.57

5.58 In Problem 5.57, what is the magnitude of the force exerted on the crane's arm by the pin support at A?

5.59 A speaker system is suspended by the cables attached at D and E. The mass of the speaker system is 130 kg, and its weight acts at G. Determine the tensions in the cables and the reactions at A and C.

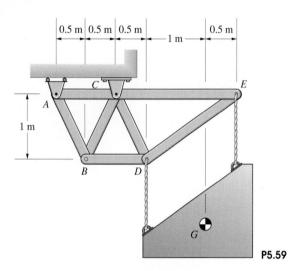

P5.59

5.60 The weight $W_1 = 1000$ lb. Neglect the weight of the bar AB. The cable goes over a pulley at C. Determine the weight W_2 and the reactions at the pin support A.

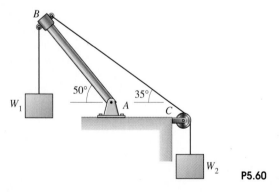

P5.60

5.61 The distances are $a = 6$ ft and $b = 3$ ft. The couple $M = 200$ ft-lb. The spring constant is $k = 1000$ lb/ft. The spring would be unstretched if $h = 0$. The system is in equilibrium when $h = 4$ ft and the beam is horizontal. Determine the force F and the reactions at A.

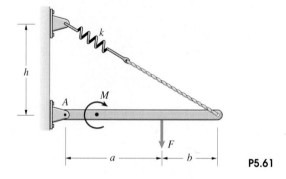

P5.61

5.62 The bar is 1 m long, and its weight W acts at its midpoint. The distance $b = 0.75$ m, and the angle $\alpha = 30°$. The spring constant is $k = 100$ N/m, and the spring is unstretched when the bar is vertical. Determine W and the reactions at A.

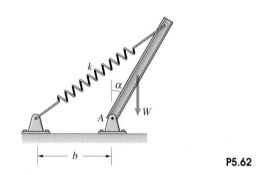

P5.62

5.63 The boom derrick supports a suspended 15-kip load. The booms BC and DE are each 20 ft long. The distances are $a = 15$ ft and $b = 2$ ft, and the angle $\theta = 30°$. Determine the tension in cable AB and the reactions at the pin supports C and D.

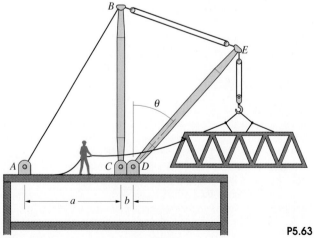

P5.63

5.64 The arrangement shown controls the elevators of an airplane. (The elevators are the horizontal control surfaces in the airplane's tail.) The elevators are attached to member *EDG*. Aerodynamic pressures on the elevators exert a clockwise couple of 120 in-lb. Cable *BG* is slack, and its tension can be neglected. Determine the force *F* and the reactions at the pin support *A*.

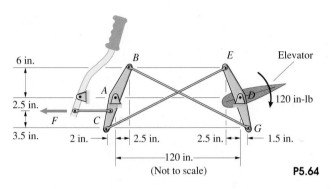

6 in.

2.5 in.

F

3.5 in.   2 in.   2.5 in.   2.5 in.   1.5 in.

120 in.

(Not to scale)

B   *E*   Elevator

A

C   *D*   120 in-lb

G

P5.64

Problems 5.65–5.68 are related to Example 5.5.

5.65 In Fig. 5.17, suppose that $\alpha = 40°$, $d = 1$ m, $a = 200$ mm, $b = 500$ mm, $R = 75$ mm, and the mass of the luggage is 40 kg. Determine *F* and *N*.

5.66 In Fig. 5.17, suppose that $\alpha = 35°$, $d = 46$ in., $a = 10$ in., $b = 14$ in., $R = 3$ in., and you don't want the user to have to exert a force *F* larger than 20 lb. What is the largest luggage weight that can be placed on the carrier?

5.67 One of the difficulties in making design decisions is that you don't know how the user will place the luggage on the carrier in Example 5.5. Suppose you assume that the point where the weight acts may be anywhere within the "envelope" $R \le a \le 0.75c$ and $0 \le b \le 0.75d$. If $\alpha = 30°$, $c = 14$ in., $d = 48$ in., $R = 3$ in., and $W = 80$ lb, what is the largest force *F* the user will have to exert for any luggage placement?

5.68 In our design of the luggage carrier in Example 5.5, we assumed a user that would hold the carrier's handle at $h = 36$ in. above the floor. We assumed that $R = 3$ in., $a = 6$ in., and $b = 12$ in., and we chose the dimension $d = 4$ ft. The resulting ratio of the force the user must exert to the weight of the luggage is $F/W = 0.132$. Suppose that people with a range of heights use this carrier. Obtain a graph of F/W as a function of h for $24 \le h \le 36$ in.

5.3 *Statically Indeterminate Objects*

In Section 5.2 we discussed examples in which we were able to use the equilibrium equations to determine unknown forces and couples acting on objects in equilibrium. You need to be aware of two common situations in which this procedure doesn't lead to a solution.

First, the free-body diagram of an object can have more unknown forces or couples than the number of independent equilibrium equations you can obtain. Since you can write no more than three such equations for a given free-body diagram in a two-dimensional problem, when there are more than three unknowns you can't determine them from the equilibrium equations alone. This occurs, for example, when an object has more supports than the minimum number necessary to maintain it in equilibrium. Such an object is said to have **redundant supports**. The second situation is when the supports of an object are improperly designed such that they cannot maintain equilibrium under the loads acting on it. The object is said to have **improper supports**. In either situation, the object is said to be **statically indeterminate**.

Engineers use redundant supports whenever possible for strength and safety. Some designs, however, require that the object be incompletely supported so that it is free to undergo certain motions. These two situations—more supports than necessary for equilibrium or not enough—are so common that we consider them in detail.

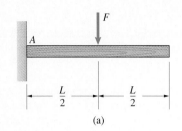

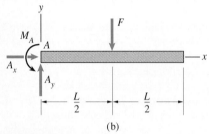

Figure 5.19

(a) A beam with a built-in support.
(b) The free-body diagram has three unknown reactions.

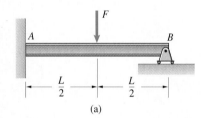

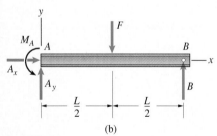

Figure 5.20

(a) A beam with built-in and roller supports.
(b) The free-body diagram has four unknown reactions.

Redundant Supports

Let's consider a beam with a built-in support (Fig. 5.19a). From its free-body diagram (Fig. 5.19b), we obtain the equilibrium equations

$$\Sigma F_x = A_x = 0,$$
$$\Sigma F_y = A_y - F = 0,$$
$$\Sigma M_{(\text{point } A)} = M_A - \left(\frac{L}{2}\right) F = 0.$$

Assuming we know the load F, we have three equations and three unknown reactions, for which we obtain the solutions $A_x = 0$, $A_y = F$, and $M_A = FL/2$.

Now suppose we add a roller support at the right end of the beam (Fig. 5.20a). From the new free-body diagram (Fig. 5.20b), we obtain the equilibrium equations

$$\Sigma F_x = A_x = 0, \tag{5.13}$$
$$\Sigma F_y = A_y - F + B = 0, \tag{5.14}$$
$$\Sigma M_{(\text{point } A)} = M_A - \left(\frac{L}{2}\right) F + LB = 0. \tag{5.15}$$

Now we have three equations and four unknown reactions. Although the first equation tells us that $A_x = 0$, we can't solve the two equations (5.14) and (5.15) for the three reactions A_y, B, and M_A.

When faced with this situation, students often attempt to sum the moments about another point, such as point B, to obtain an additional equation:

$$\Sigma M_{(\text{point } B)} = M_A + \left(\frac{L}{2}\right) F - LA_y = 0.$$

Unfortunately, this doesn't help. This is not an independent equation but is a linear combination of Eqs. (5.14) and (5.15):

$$\Sigma M_{(\text{point } B)} = M_A + \left(\frac{L}{2}\right) F - LA_y$$
$$= \underbrace{M_A - \left(\frac{L}{2}\right) F + LB}_{\text{Eq. (5.15)}} - \underbrace{L(A_y - F + B)}_{\text{Eq. (5.14)}}.$$

As this example demonstrates, each support added to an object results in additional reactions. The difference between the number of reactions and the number of independent equilibrium equations is called the **degree of redundancy**.

Even if an object is statically indeterminate due to redundant supports, it may be possible to determine some of the reactions from the equilibrium equations. Notice that in our previous example we were able to determine the reaction A_x even though we could not determine the other reactions.

Since redundant supports are so ubiquitous, you may wonder why we devote so much effort to teaching you how to analyze objects whose reactions can be determined with the equilibrium equations. We want to develop your understanding of equilibrium and give you practice writing equilibrium equations. The reactions on an object with redundant supports *can* be determined by supplementing the equilibrium equations with additional equations that relate the forces and couples acting on the object to its deformation, or change in shape. Thus obtaining the equilibrium equations is the first step of the solution.

Example 5.6

The beam in Fig. 5.21 has two pin supports and is loaded by a 2-kN force.
(a) Show that the beam is statically indeterminate.
(b) Determine as many reactions as possible.

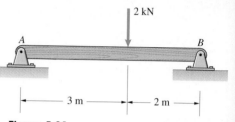

Figure 5.21

STRATEGY

The beam is statically indeterminate if its free-body diagram has more unknown reactions than the number of independent equilibrium equations we can obtain. But even if this is the case, we may be able to solve the equilibrium equations for some of the reactions.

SOLUTION

Draw the Free-Body Diagram   We draw the free-body diagram of the beam in Fig. (a). There are four unknown reactions—A_x, A_y, B_x, and B_y— and we can write only three independent equilibrium equations. Therefore the beam is statically indeterminate.

(a) The free-body diagram of the beam.

Apply the Equilibrium Equations   Summing the moments about point A, the equilibrium equations are

$$\Sigma F_x = A_x + B_x = 0,$$

$$\Sigma F_y = A_y + B_y - 2 = 0,$$

$$\Sigma M_{(\text{point } A)} = 5B_y - (2)(3) = 0.$$

We can solve the third equation for B_y and then solve the second equation for A_y. The results are $A_y = 0.8$ kN and $B_y = 1.2$ kN. The first equation tells us that $B_x = -A_x$, but we can't solve for their values.

DISCUSSION

This example can give you insight into why the reactions on objects with redundant constraints can't be determined from the equilibrium equations alone. The two pin supports can exert horizontal reactions on the beam even in the absence of loads (Fig. b), and these reactions satisfy the equilibrium equations for any value of T ($\Sigma F_x = -T + T = 0$).

(b) The supports can exert reactions on the beam.

Improper Supports

We say that an object has improper supports if it will not remain in equilibrium under the action of the loads exerted on it. Thus an object with improper supports will move when the loads are applied. In two-dimensional problems, this can occur in two ways:

1. *The supports can exert only parallel forces.* This leaves the object free to move in the direction perpendicular to the support forces. If the loads exert a component of force in that direction, the object is not in equilibrium. Figure 5.22(a) shows an example of this situation. The two roller supports can exert only vertical forces, while the force F has a horizontal component. The beam will move horizontally when F is applied. This is particularly apparent from the free-body diagram (Fig. 5.22b). The sum of the forces in the horizontal direction cannot be zero because the roller supports can exert only vertical reactions.

Figure 5.22

(a) A beam with two roller supports is not in equilibrium when subjected to the load shown.
(b) The sum of the forces in the horizontal direction is not zero.

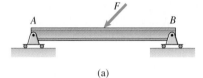

(a)

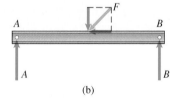

(b)

2. *The supports can exert only concurrent forces.* If the loads exert a moment about the point where the lines of action of the support forces intersect, the object is not in equilibrium. For example, consider the beam in Fig. 5.23(a). From its free-body diagram (Fig. 5.23b) we see that the reactions A and B exert no moment about the point P, where their lines of action intersect, but the load F does. The sum of the moments about point P is not zero, and the beam will rotate when the load is applied.

Figure 5.23

(a) A beam with roller supports on sloped surfaces.
(b) The sum of the moments about point P is not zero.

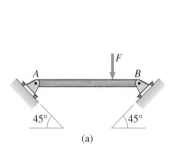

(a)

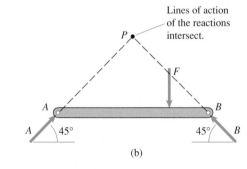

(b)

Except for problems that deal explicitly with improper supports, objects in our examples and problems have proper supports. You should develop the habit of examining objects in equilibrium and thinking about why they are properly supported for the loads acting on them.

Example 5.7

State whether each L-shaped bar in Fig. 5.24 is properly or improperly supported. If a bar is properly supported, determine the reactions at its supports.

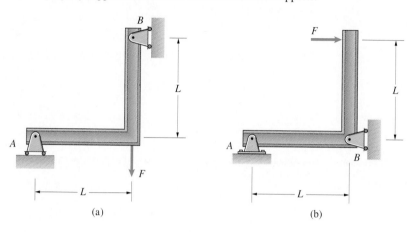
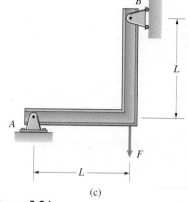

(a) (b) (c)

Figure 5.24

SOLUTION

We draw the free-body diagrams of the bars in Fig. 5.25.

Bar (a)   The lines of action of the reactions due to the two roller supports intersect at P, and the load F exerts a moment about P. This bar is improperly supported.

Bar (b)   The lines of action of the reactions intersect at A, and the load F exerts a moment about A. This bar is also improperly supported.

Bar (c)   The three support forces are neither parallel nor concurrent. This bar is properly supported. The equilibrium equations are

$$\Sigma F_x = A_x - B = 0,$$
$$\Sigma F_y = A_y - F = 0,$$
$$\Sigma M_{(\text{point } A)} = BL - FL = 0.$$

Solving these equations, the reactions are $A_x = F$, $A_y = F$, and $B = F$.

Figure 5.25

Free-body diagrams of the three bars.

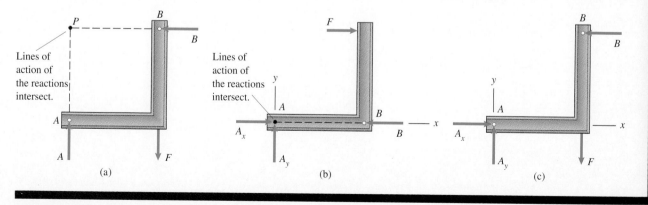

(a) (b) (c)

Problems

5.69 (a) Draw the free-body diagram of the beam, and show that it is statically indeterminate.
(b) Determine as many of the reactions as possible.

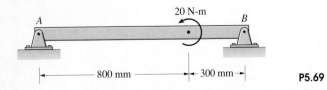

P5.69

5.70 Consider the beam in Problem 5.69. Choose supports at A and B so that it is not statically indeterminate. Determine the reactions at the supports.

5.71 (a) Draw the free-body diagram of the plate, and show that it is statically indeterminate.
(b) Determine as many of the reactions as possible.

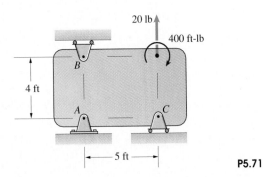

P5.71

5.72 Consider the plate in Problem 5.71. Choose supports at A, B, and C so that it is not statically indeterminate. Determine the reactions at the supports.

5.73 Draw the free-body diagram of the L-shaped pipe assembly, and show that it is statically indeterminate. Determine as many of the reactions as possible.

Strategy: Place the coordinate system so that the x axis passes through points A and B.

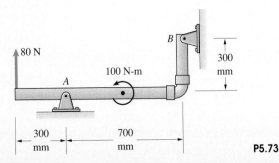

P5.73

5.74 Consider the pipe assembly in Problem 5.73. Choose supports at A and B so that it is not statically indeterminate. Determine the reactions at the supports.

5.75 State whether each of the L-shaped bars shown is properly or improperly supported. If a bar is properly supported, determine the reactions at its supports.

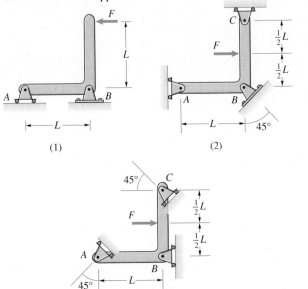

P5.75

5.76 State whether each of the L-shaped bars shown is properly or improperly supported. If a bar is properly supported, determine the reactions at its supports.

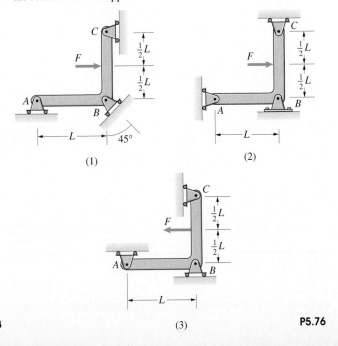

P5.76

5.4 *Three-Dimensional Applications*

You have seen that when an object in equilibrium is subjected to a two-dimensional system of forces and moments, you can obtain no more than three independent equilibrium equations. In the case of a three-dimensional system of forces and moments, you can obtain up to six independent equilibrium equations: The three components of the sum of the forces must equal zero, and the three components of the sum of the moments about any point must equal zero. Your procedure for determining the reactions on objects subjected to three-dimensional systems of forces and moments—drawing the free-body diagram and applying the equilibrium equations—is the same as in two dimensions. You just need to become familiar with the support conventions used in three-dimensional applications.

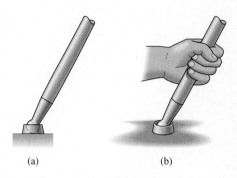

(a)   (b)

Supports

We present five conventions frequently used in three-dimensional problems. Again, even when actual supports do not physically resemble these models, we represent them by the models if they exert the same (or approximately the same) reactions.

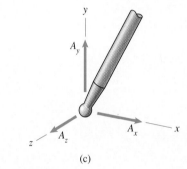

(c)

Figure 5.26

(a) A ball and socket support.
(b) Holding a supported bar.
(c) The ball and socket support can exert three components of force.

The Ball and Socket Support   In the **ball and socket support**, the supported object is attached to a ball enclosed within a spherical socket (Fig. 5.26a). The socket permits the ball to rotate freely (friction is neglected) but prevents it from translating in any direction.

Imagine holding a bar attached to a ball and socket support (Fig. 5.26b). If you try to translate the bar (move it without rotating it) in any direction, the support exerts a reactive force to prevent the motion. However, you can rotate the bar about the support. The support cannot exert a couple to prevent rotation. Thus a ball and socket support can't exert a couple but can exert three components of force (Fig. 5.26c). It is the three-dimensional analog of the two-dimensional pin support.

The human hip joint is an example of a ball and socket support (Fig. 5.27). The support of the gear shift lever of a car can be modeled as a ball and socket support within the lever's range of motion.

The Roller Support   The **roller support** (Fig. 5.28a) is a ball and socket support that can roll freely on a supporting surface. A roller support can exert only a force normal to the supporting surface (Fig. 5.28b). The rolling "casters" sometimes used to support furniture legs are supports of this type.

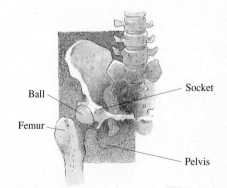

Ball   Socket

Femur

Pelvis

Figure 5.27

The human femur is attached to the pelvis by a ball and socket support.

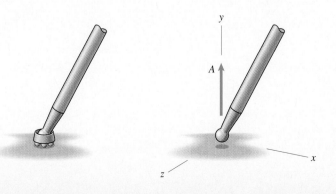

Figure 5.28

(a) A roller support.
(b) The reaction is normal to the supporting surface.

The Hinge   The hinge support is the familiar device used to support doors. It permits the supported object to rotate freely about a line, the **hinge axis**. An object is attached to a hinge in Fig. 5.29(a). The z axis of the coordinate system is aligned with the hinge axis.

If you imagine holding a bar attached to a hinge (Fig. 5.29b), notice that you can rotate the bar *about the hinge axis*. The hinge cannot exert a couple about the hinge axis (the z axis) to prevent rotation. However, you can't rotate the bar about the x or y axis because the hinge can exert couples about those axes to resist the motion. In addition, you can't translate the bar in any direction. The reactions a hinge can exert on an object are shown in Fig. 5.29(c). There are three components of force, A_x, A_y, and A_z, and couples about the x and y axes, M_{Ax} and M_{Ay}.

In some situations, either a hinge exerts no couples on the object it supports, or they are sufficiently small to neglect. An example of the latter case is when the axes of the hinges supporting a door are properly aligned. In these situations the hinge exerts only forces on an object (Fig. 5.29d). Situations also arise in which a hinge exerts no couples on an object and exerts no force in the direction of the hinge axis. (The hinge may actually be designed so that it cannot support a force parallel to the hinge axis.) Then the hinge exerts forces only in the directions perpendicular to the hinge axis (Fig. 5.29e). In examples and problems, we indicate when a hinge does not exert all five of the reactions in Fig. 5.29(c).

Figure 5.29

(a) A hinge. The z axis is aligned with the hinge axis.
(b) Holding a supported bar.
(c) In general, a hinge can exert five reactions: three force components and two couple components.
(d) The reactions when the hinge exerts no couples.
(e) The reactions when the hinge exerts neither couples nor a force parallel to the hinge axis.

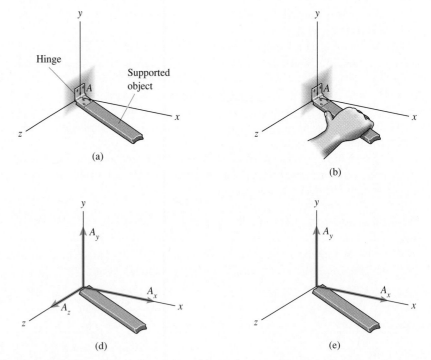

(a)

(b)

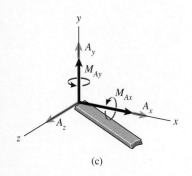

(c)

(d)

(e)

The Bearing   The type of bearing shown in Fig. 5.30(a) supports a circular shaft while permitting it to rotate about its axis. The reactions are identical to those exerted by a hinge. In the most general case (Fig. 5.30b), the bearing can exert a force on the supported shaft in each coordinate direction and can exert couples about axes perpendicular to the shaft but cannot exert a couple about the axis of the shaft.

As in the case of the hinge, situations can occur in which the bearing exerts no couples (Fig. 5.30c) or exerts no couples and no force parallel to the shaft axis (Fig. 5.30d). Some bearings are designed in this way for specific applications. In examples and problems, we indicate when a bearing does not exert all of the reactions in Fig. 5.30(b).

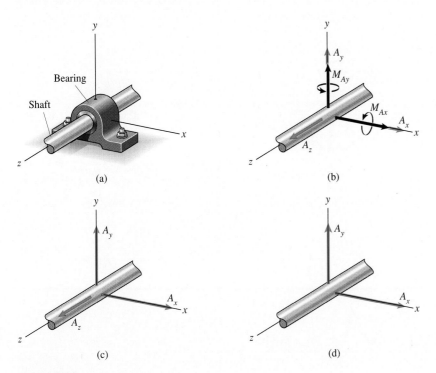

Figure 5.30

(a) A bearing. The z axis is aligned with the axis of the shaft.
(b) In general, a bearing can exert five reactions: three force components and two couple components.
(c) The reactions when the bearing exerts no couples.
(d) The reactions when the bearing exerts neither couples nor a force parallel to the axis of the shaft.

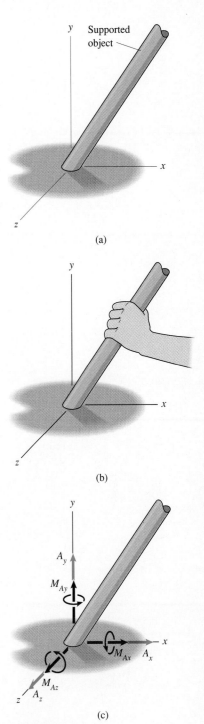

(a)

(b)

(c)

Figure 5.31

(a) A built-in support.
(b) Holding a supported bar.
(c) A built-in support can exert six reactions: three force components and three couple components.

The Built-In Support You are already familiar with the built-in, or fixed, support (Fig. 5.31a). Imagine holding a bar with a built-in support (Fig. 5.31b). You cannot translate it in any direction, and you cannot rotate it about any axis. The support is capable of exerting forces A_x, A_y, and A_z in each coordinate direction and couples M_{Ax}, M_{Ay}, and M_{Az} about each coordinate axis (Fig. 5.31c).

Table 5.2 summarizes the support conventions commonly used in three-dimensional applications.

Table 5.2

Supports used in three-dimensional applications.

Supports	Reactions
 Rope or Cable	 One Collinear Force
 Contact with a Smooth Surface	 One Normal Force
 Contact with a Rough Surface	 Three Force Components
 Ball and Socket Support	 Three Force Components
 Roller Support	 One Normal Force

Table 5.2   *Continued*

Supports	Reactions
  Hinge (The z axis is parallel to the hinge axis.)	 Three Force Components, Two Couple Components (When no couples are exerted)
  Bearing (The z axis is parallel to the axis of the supported shaft.)	 (When no couples and no axial force are exerted)
  Built-in (Fixed) Support	 Three Force Components, Three Couple Components

The Scalar Equilibrium Equations

The loads and reactions on an object in equilibrium satisfy the six scalar equilibrium equations

$$\Sigma F_x = 0, \tag{5.16}$$

$$\Sigma F_y = 0, \tag{5.17}$$

$$\Sigma F_z = 0, \tag{5.18}$$

$$\Sigma M_x = 0, \tag{5.19}$$

$$\Sigma M_y = 0, \tag{5.20}$$

$$\Sigma M_z = 0. \tag{5.21}$$

You can evaluate the sums of the moments about any point. Although you can obtain other equations by summing the moments about additional points, they will not be independent of these equations. *You can't obtain more than six independent equilibrium equations from a given free-body diagram, so you can solve for at most six unknown forces or couples.*

In the following examples we determine the reactions on objects with various kinds of loads and supports. The steps required are familiar from your experience with two-dimensional applications:

1. Draw the free-body diagram—*You must obtain the free-body diagram by isolating the object from its supports and showing the loads and the reactions the supports may exert on the object.*

2. Apply the equilibrium equations—*Use Eqs. (5.16)–(5.21) to determine the reactions.*

Example 5.8

The bar AB in Fig. 5.32 is supported by the cables BC and BD and a ball and socket support at A. Cable BC is parallel to the z axis, and cable BD is parallel to the x axis. The 200-N weight of the bar acts at its midpoint. What are the tensions in the cables and the reactions at A?

STRATEGY

We must obtain the free-body diagram of the bar AB by isolating it from the support at A and the two cables. Then we can use the equilibrium equations to determine the reactions at A and the tensions in the cables.

SOLUTION

Draw the Free-Body Diagram In Fig. (a) we isolate the bar and show the reactions that may be exerted on it. The ball and socket support can exert three components of force, A_x, A_y, and A_z. The terms T_{BC} and T_{BD} represent the tensions in the cables.

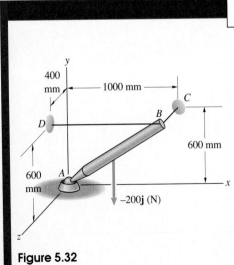

Figure 5.32

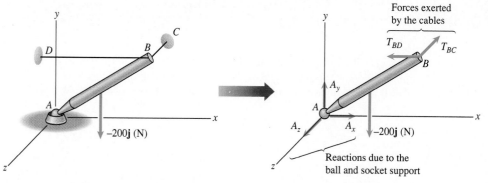

(a) Obtaining the free-body diagram of the bar.

Apply the Equilibrium Equations The sums of the forces in each coordinate direction equal zero:

$$\Sigma F_x = A_x - T_{BD} = 0,$$

$$\Sigma F_y = A_y - 200 = 0, \qquad\qquad (5.22)$$

$$\Sigma F_z = A_z - T_{BC} = 0.$$

Let $\mathbf{r}_{AB}$ be the position vector from A to B. The sum of the moments about A is

$$\Sigma \mathbf{M}_{(\text{point } A)} = [\mathbf{r}_{AB} \times (-T_{BC}\,\mathbf{k})] + [\mathbf{r}_{AB} \times (-T_{BD}\mathbf{i})]$$

$$+ \left[\tfrac{1}{2}\mathbf{r}_{AB} \times (-200\mathbf{j})\right]$$

$$= \begin{vmatrix} \mathbf{i} & \mathbf{j} & \mathbf{k} \\ 1 & 0.6 & 0.4 \\ 0 & 0 & -T_{BC} \end{vmatrix} + \begin{vmatrix} \mathbf{i} & \mathbf{j} & \mathbf{k} \\ 1 & 0.6 & 0.4 \\ -T_{BD} & 0 & 0 \end{vmatrix}$$

$$+ \begin{vmatrix} \mathbf{i} & \mathbf{j} & \mathbf{k} \\ 0.5 & 0.3 & 0.2 \\ 0 & -200 & 0 \end{vmatrix}$$

$$= (-0.6T_{BC} + 40)\,\mathbf{i} + (T_{BC} - 0.4T_{BD})\,\mathbf{j} + (0.6T_{BD} - 100)\,\mathbf{k}.$$

The components of this vector (the sums of the moments about the three coordinate axes) each equal zero:

$$\Sigma M_x = -0.6T_{BC} + 40 = 0,$$

$$\Sigma M_y = T_{BC} - 0.4T_{BD} = 0,$$

$$\Sigma M_z = 0.6T_{BD} - 100 = 0.$$

Solving these equations, we obtain the tensions in the cables:

$$T_{BC} = 66.7 \text{ N}, \qquad T_{BD} = 166.7 \text{ N}.$$

(Notice that we needed only two of the three equations to obtain the two tensions. The third equation is redundant.)

Then from Eqs. (5.22) we obtain the reactions at the ball and socket support:

$$A_x = 166.7 \text{ N}, \qquad A_y = 200 \text{ N}, \qquad A_z = 66.7 \text{ N}.$$

DISCUSSION

Notice that by summing moments about A we obtained equations in which the unknown reactions A_x, A_y, and A_z did not appear. You can often simplify your solutions in this way.

Example 5.9

The bar AC in Fig. 5.33 is 4 ft long and is supported by a hinge at A and the cable BD. The hinge axis is along the z axis. The centerline of the bar lies in the x-y plane, and the cable attachment point B is the midpoint of the bar. Determine the tension in the cable and the reactions exerted on the bar by the hinge.

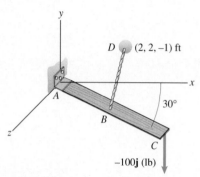

Figure 5.33

SOLUTION

Draw the Free-Body Diagram   We isolate the bar from the hinge support and the cable and show the reactions they exert (Fig. a). The terms A_x, A_y, and A_z are the components of force exerted by the hinge, and the terms M_{Ax} and M_{Ay} are the couples exerted by the hinge about the x and y axes. (Remember that the hinge cannot exert a couple on the bar about the hinge axis.) The term T is the tension in the cable.

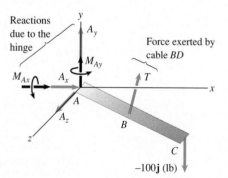

(a) The free-body diagram of the bar.

Apply the Equilibrium Equations   To write the equilibrium equations, we must first express the cable force in terms of its components. The coordinates of point B are $(2 \cos 30°, -2 \sin 30°, 0)$ ft, so the position vector from B to D is

$$\mathbf{r}_{BD} = (2 - 2 \cos 30°)\,\mathbf{i} + [2 - (-2 \sin 30°)]\,\mathbf{j} + (-1 - 0)\,\mathbf{k}$$

$$= 0.268\mathbf{i} + 3\mathbf{j} - \mathbf{k}.$$

We divide this vector by its magnitude to obtain a unit vector $\mathbf{e}_{BD}$ that points from point B toward point D:

$$\mathbf{e}_{BD} = \frac{\mathbf{r}_{BD}}{|\mathbf{r}_{BD}|} = 0.084\mathbf{i} + 0.945\mathbf{j} - 0.315\mathbf{k}.$$

Now we can write the cable force as the product of its magnitude and $\mathbf{e}_{BD}$:

$$T\mathbf{e}_{BD} = T(0.084\mathbf{i} + 0.945\mathbf{j} - 0.315\mathbf{k}).$$

The sums of the forces in each coordinate direction must equal zero:

$$\Sigma F_x = A_x + 0.084T = 0,$$
$$\Sigma F_y = A_y + 0.945T - 100 = 0, \qquad\qquad (5.23)$$
$$\Sigma F_z = A_z - 0.315T = 0.$$

If we sum moments about A, the resulting equations do not contain the unknown reactions A_x, A_y, and A_z. The position vectors from A to B and from A to C are

$$\mathbf{r}_{AB} = 2\cos 30°\,\mathbf{i} - 2\sin 30°\,\mathbf{j},$$
$$\mathbf{r}_{AC} = 4\cos 30°\,\mathbf{i} - 4\sin 30°\,\mathbf{j}.$$

The sum of the moments about A is

$$\Sigma \mathbf{M}_{(\text{point } A)} = M_{Ax}\mathbf{i} + M_{Ay}\mathbf{j} + [\mathbf{r}_{AB} \times (T\mathbf{e}_{BD})] + [\mathbf{r}_{AC} \times (-100\mathbf{j})]$$

$$= M_{Ax}\mathbf{i} + M_{Ay}\mathbf{j} + \begin{vmatrix} \mathbf{i} & \mathbf{j} & \mathbf{k} \\ 1.732 & -1 & 0 \\ 0.084T & 0.945T & -0.315T \end{vmatrix}$$

$$+ \begin{vmatrix} \mathbf{i} & \mathbf{j} & \mathbf{k} \\ 3.464 & -2 & 0 \\ 0 & -100 & 0 \end{vmatrix}$$

$$= (M_{Ax} + 0.315T)\,\mathbf{i} + (M_{Ay} + 0.546T)\,\mathbf{j}$$
$$+ (1.72T - 346)\,\mathbf{k} = 0.$$

From this vector equation we obtain the scalar equations

$$\Sigma M_x = M_{Ax} + 0.315T = 0,$$
$$\Sigma M_y = M_{Ay} + 0.546T = 0,$$
$$\Sigma M_z = 1.72T - 346 = 0.$$

Solving these equations, we obtain the reactions

$$T = 201 \text{ lb}, \qquad M_{Ax} = -63.4 \text{ ft-lb}, \qquad M_{Ay} = -109.8 \text{ ft-lb}.$$

Then from Eqs. (5.23) we obtain the forces exerted on the bar by the hinge:

$$A_x = -17.0 \text{ lb}, \qquad A_y = -90.2 \text{ lb}, \qquad A_z = 63.4 \text{ lb}.$$

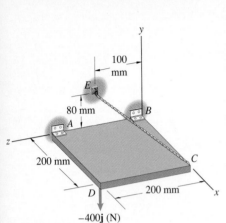

100 mm

E

80 mm

A B

z

200 mm

C

D 200 mm x

$-400\mathbf{j}$ (N)

Figure 5.34

Example 5.10

The plate in Fig. 5.34 is supported by hinges at A and B and the cable CE. The properly aligned hinges do not exert couples on the plate, and the hinge at A does not exert a force on the plate in the direction of the hinge axis. Determine the reactions at the hinges and the tension in the cable.

SOLUTION

Draw the Free-Body Diagram   We isolate the plate and show the reactions at the hinges and the force exerted by the cable (Fig. a). The term T is the force exerted on the plate by cable CE.

Apply the Equilibrium Equations   Since we know the coordinates of points C and E, we can express the cable force as the product of its magnitude T and a unit vector directed from C toward E. The result is

$$T(-0.842\mathbf{i} + 0.337\mathbf{j} + 0.421\mathbf{k}).$$

The sums of the forces in each coordinate direction equal zero:

$$\Sigma F_x = A_x + B_x - 0.842T = 0,$$

$$\Sigma F_y = A_y + B_y + 0.337T - 400 = 0,$$

$$\Sigma F_z = B_z + 0.421T = 0.$$

If we sum the moments about B, the resulting equations will not contain the three unknown reactions at B. The sum of the moments about B is

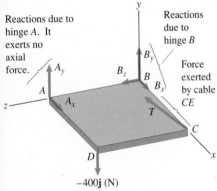

Reactions due to hinge A. It exerts no axial force.

A_y

A_x

A

z

B_z B_y

B B_x

Reactions due to hinge B

Force exerted by cable CE

T

C

D x

$-400\mathbf{j}$ (N)

(a) The free-body diagram of the plate.

$$\Sigma \mathbf{M}_{(\text{point } B)} = \begin{vmatrix} \mathbf{i} & \mathbf{j} & \mathbf{k} \\ 0.2 & 0 & 0 \\ -0.842T & 0.337T & 0.421T \end{vmatrix} + \begin{vmatrix} \mathbf{i} & \mathbf{j} & \mathbf{k} \\ 0 & 0 & 0.2 \\ A_x & A_y & 0 \end{vmatrix}$$

$$+ \begin{vmatrix} \mathbf{i} & \mathbf{j} & \mathbf{k} \\ 0.2 & 0 & 0.2 \\ 0 & -400 & 0 \end{vmatrix}$$

$$= (-0.2A_y + 80)\,\mathbf{i} + (-0.0842T + 0.2A_x)\,\mathbf{j}$$

$$+ (0.0674T - 80)\,\mathbf{k} = 0.$$

The scalar equations are

$$\Sigma M_x = -0.2A_y + 80 = 0,$$

$$\Sigma M_y = -0.0842T + 0.2A_x = 0,$$

$$\Sigma M_z = 0.0674T - 80 = 0.$$

Solving these equations, we obtain the reactions

$$T = 1187 \text{ N}, \qquad A_x = 500 \text{ N}, \qquad A_y = 400 \text{ N}.$$

Then from Eqs. (5.24), the reactions at B are

$$B_x = 500 \text{ N}, \qquad B_y = -400 \text{ N}, \qquad B_z = -500 \text{ N}.$$

DISCUSSION

If our only objective had been to determine the tension T, we could have done so easily by setting the sum of the moments about the line AB (the z axis) equal to zero. Since the reactions at the hinges exert no moment about the z axis, we obtain the equation

$$(0.2)(0.337T) - (0.2)(400) = 0,$$

which yields the result $T = 1187$ N.

Problems

5.77   The bar ABC has a built-in support at A. It is loaded by the forces

$$\mathbf{F}_B = 2\mathbf{i} + 6\mathbf{j} + 3\mathbf{k} \text{ (kN)},$$

$$\mathbf{F}_C = \mathbf{i} - 2\mathbf{j} + 2\mathbf{k} \text{ (kN)},$$

and the couple

$$\mathbf{M}_C = 2\mathbf{i} + \mathbf{j} - 2\mathbf{k} \text{ (kN-m)}.$$

(a) Draw the free-body diagram of the bar.
(b) Determine the reactions at A.

Strategy: Isolate the bar from the built-in support, and show the loads and the reactions at the support. Then use Eqs. (5.16)–(5.21) to determine the reactions. (Remember that you can sum the moments about any point.)

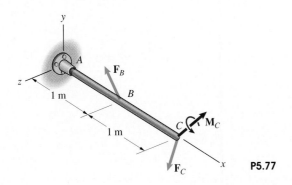

P5.77

5.78   The robotic manipulator is stationary. The weights of the arms AB and BC act at their midpoints. The direction cosines of the centerline of arm AB are $\cos\theta_x = 0.500$, $\cos\theta_y = 0.866$, $\cos\theta_z = 0$, and the direction cosines of the centerline of arm BC are $\cos\theta_x = 0.707$, $\cos\theta_y = 0.619$, $\cos\theta_z = -0.342$. The support at A behaves like a built-in support. What are the reactions at A?

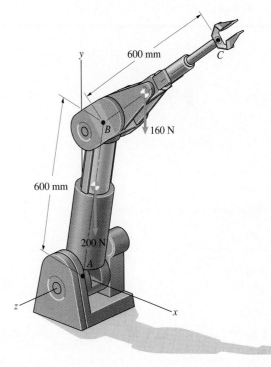

P5.78

5.79 The bar AB has a built-in support at A. The collar at B is fixed to the bar. The tension in the cable BC is 10 kN.
(a) Draw the free-body diagram of the bar.
(b) Determine the reactions at A.

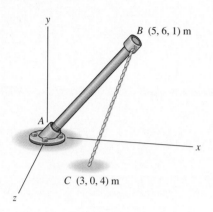

B $(5, 6, 1)$ m

C $(3, 0, 4)$ m

P5.79

5.80 Consider the bar in Problem 5.79. The magnitude of the couple exerted on the bar by the built-in support is 100 kN-m. What is the tension in the cable?

5.81 The force exerted on the highway sign by wind and the sign's weight is $\mathbf{F} = 800\mathbf{i} - 600\mathbf{j}$ (N). Determine the reactions at the built-in support at O.

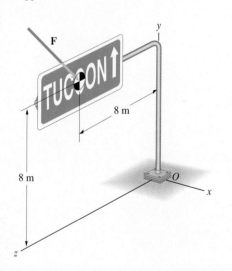

P5.81

5.82 In Problem 5.81, the force exerted on the sign by wind and the sign's weight is $\mathbf{F} = \pm 4.4v^2\mathbf{i} - 600\mathbf{j}$ (N), where v is the component of the wind's velocity perpendicular to the sign in meters per second (m/s). If you want to design the sign to remain standing in hurricane winds with velocities v as high as 70 m/s, what reactions must the built-in support at O be designed to withstand?

5.83 The tension in cable AB is 24 kN. Determine the reactions at the built-in support D.

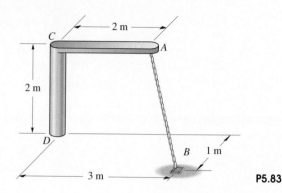

P5.83

5.84 In Problem 5.83, the structural designer determines that the built-in support D can safely be subjected to a couple of magnitude 160 kN-m. Based on this criterion, what is the largest safe value of the tension in cable AB?

5.85 The force exerted on the grip of the exercise machine is $\mathbf{F} = 60\mathbf{i} - 30\mathbf{j}$ (lb). What are the reactions at the built-in support at O?

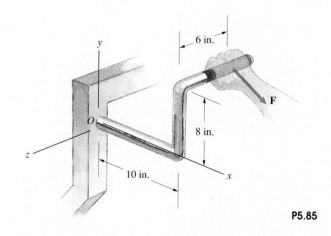

P5.85

5.86 In Problem 5.85, the designer assumes that the force $\mathbf{F}$ exerted on the grip of the exercise machine will be parallel to the x-y plane and that its magnitude will not exceed 200 lb. Based on these criteria, what reactions must the built-in support at O be designed to withstand?

5.87 The boom *AB* is subjected to a force **F** = −1000**j** (lb) at *B* and is supported by a ball and socket support at *A* and the cables *CD* and *CE*. The collar *C* is fixed to the bar.
(a) Draw the free-body diagram of the boom.
(b) Determine the tensions in the cables and the reactions at *A*.

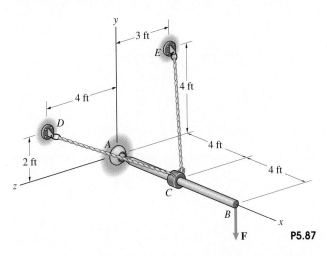

P5.87

5.88 Consider the crane shown in Problem 5.87. The force **F** points downward (parallel to the *y* axis). The cables *CD* and *CE* will each safely support a tension of 5000 lb. Based on this criterion, what is the largest safe magnitude of **F**?

5.89 The suspended load exerts a force *F* = 600 lb at *A*, and the weight of the bar *OA* is negligible. Determine the tensions in the cables and the reactions at the ball and socket support *O*.

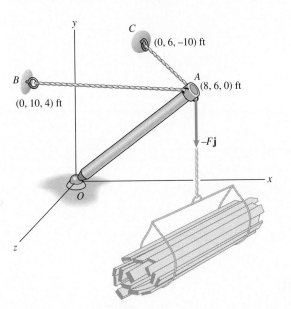

P5.89

5.90 In Problem 5.89, suppose that the suspended load exerts a force *F* = 600 lb at *A* and bar *OA* weighs 200 lb. Assume that the bar's weight acts at its midpoint. Determine the tensions in the cables and the reactions at the ball and socket support *O*.

5.91 The airplane is at rest on the ground (*z* = 0 is ground level). Its landing gear are at points *A*, *B*, and *C*. The coordinates of the point *G* where the 350-kip weight of the plane acts are (10, 1.5, −15) ft. What are the magnitudes of the normal reactions exerted by the ground on the plane's landing gear?

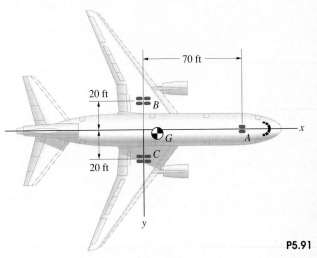

P5.91

5.92 The 800-kg horizontal wall section is supported by the three vertical cables *A*, *B*, and *C*. What are the tensions in the cables?

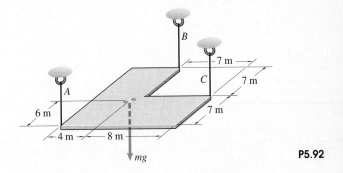

P5.92

5.93 The cables in Problem 5.92 will each safely support a tension of 10 kN. Based on this criterion, what is the largest safe mass of the horizontal wall section?

5.94 An engineer designs a system of pulleys to pull his model trains up and out of the way when they aren't in use. What are the tensions in the three ropes when the system is in equilibrium?

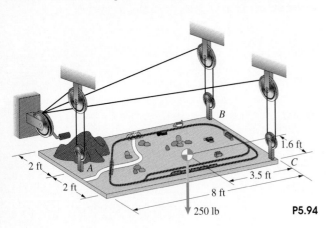

P5.94

5.95 The L-shaped bar is supported by a bearing at A and rests on a smooth horizontal surface at B. The vertical force $F = 800$ lb, and the distance $b = 1.5$ ft. Determine the reactions at A and B.

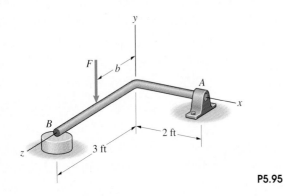

P5.95

5.96 Consider the system in Problem 5.95. The vertical force $F = 800$ lb, and the distance $b = 1.5$ ft. If you represent the reactions exerted at A and B by a single equivalent force, what is the force and where does its line of action intersect the centerline of the bar?

5.97 Consider the system in Problem 5.95. The vertical force $F = 800$ lb. The bearing at A will safely support a force of 500 lb and a couple of 1000 ft-lb. The support at B will safely support a force of 500 lb. Based on these criteria, what is the allowable range of the distance b?

5.98 The 1.1-m bar is supported by a ball and socket support at A and the two smooth walls. The tension in the vertical cable CD is 1 kN.
(a) Draw the free-body diagram of the bar.
(b) Determine the reactions at A and B.

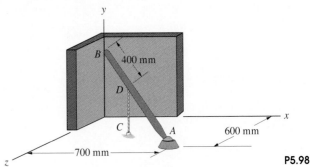

P5.98

5.99 The 8-ft bar is supported by a ball and socket support at A, the cable BD, and a roller support at C. The collar at B is fixed to the bar at its midpoint. The force $\mathbf{F} = -50\mathbf{k}$ (lb). Determine the tension in cable BD and the reactions at A and C.

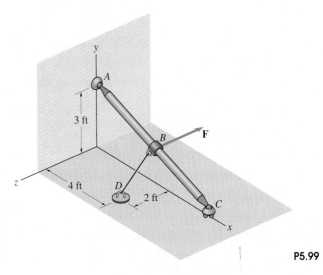

P5.99

5.100 Consider the 8-ft bar in Problem 5.99. The force $\mathbf{F} = F_y\mathbf{j} - 50\mathbf{k}$ (lb). What is the largest value of F_y for which the roller support at C will remain on the floor?

5.101 The tower is 70 m tall. The tension in each cable is 2 kN. Treat the base of the tower *A* as a built-in support. What are the reactions at *A*?

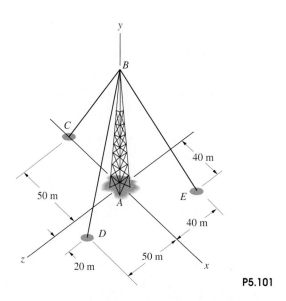

P5.101

5.102 Consider the tower in Problem 5.101. If the tension in cable *BC* is 2 kN, what must the tensions in cables *BD* and *BE* be if you want the couple exerted on the tower by the built-in support at *A* to be zero? What are the resulting reactions at *A*?

5.103 The space truss has roller supports *B*, *C*, and *D* and supports a vertical 800-lb load at *A*. What are the reactions at the three roller supports?

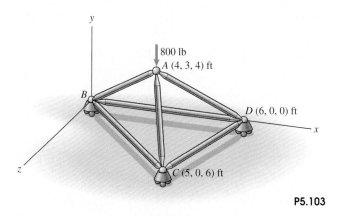

P5.103

5.104 In Problem 5.103, suppose that you don't want the reaction at any of the three roller supports to exceed 1000 lb. Based on this criterion, what is the largest vertical force the space truss can support at *A*?

5.105 The 40-lb door is supported by hinges at *A* and *B*. The *y* axis is vertical. The hinges do not exert couples on the door, and the hinge at *B* does not exert a force parallel to the hinge axis. The weight of the door acts at its midpoint. What are the reactions at *A* and *B*?

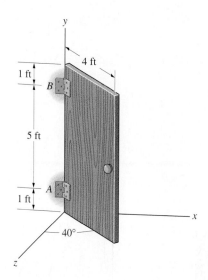

P5.105

5.106 The vertical cable is attached at *A*. Determine the tension in the cable and the reactions at the bearing *B* due to the force $\mathbf{F} = 10\mathbf{i} - 30\mathbf{j} - 10\mathbf{k}$ (N).

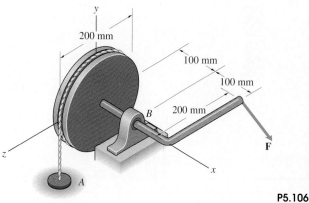

P5.106

5.107 In Problem 5.106, suppose that the *z* component of the force $\mathbf{F}$ is zero, but otherwise $\mathbf{F}$ is unknown. If the couple exerted on the shaft by the bearing at *B* is $\mathbf{M}_B = 6\mathbf{j} - 6\mathbf{k}$ N-m, what are the force $\mathbf{F}$ and the tension in the cable?

5.108 The device in Problem 5.106 is badly designed because of the couples that must be supported by the bearing at B, which would cause the bearing to "bind." (Imagine trying to open a door supported by only one hinge.) In this improved design, the bearings at B and C support no couples, and the bearing at C does not exert a force in the x direction. If the force $\mathbf{F} = 10\mathbf{i} - 30\mathbf{j} - 10\mathbf{k}$ (N), what are the tension in the vertical cable and the reactions at the bearings B and C?

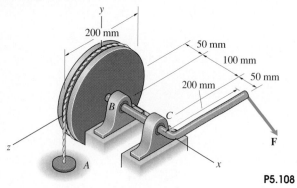

P5.108

5.109 The rocket launcher is supported by the hydraulic jack DE and the bearings A and B. The bearings lie on the x axis and support shafts parallel to the x axis. The hydraulic cylinder DE exerts a force on the launcher that points along the line from D to E. The coordinates of D are $(7, 0, 7)$ ft, and the coordinates of E are $(9, 6, 4)$ ft. The weight $W = 30$ kip acts at $(4.5, 5, 2)$ ft. What is the magnitude of the reaction on the launcher at E?

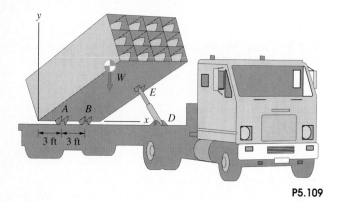

P5.109

5.110 Consider the rocket launcher described in Problem 5.109. The bearings at A and B do not exert couples, and the bearing B does not exert a force in the x direction. Determine the reactions at A and B.

5.111 The crane's cable CD is attached to a stationary object at D. The crane is supported by the bearings E and F and the horizontal cable AB. The tension in cable AB is 8 kN. Determine the tension in the cable CD.

Strategy: Since the reactions exerted on the crane by the bearings do not exert moments about the z axis, the sum of the moments about the z axis due to the forces exerted on the crane by

the cables AB and CD equals zero. (See the discussion at the end of Example 5.10.)

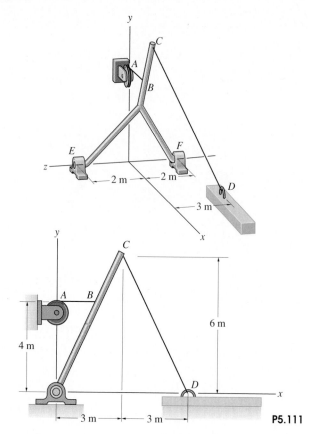

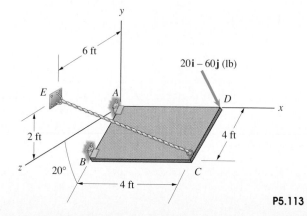

P5.111

5.112 The crane in Problem 5.111 is supported by the horizontal cable AB and the bearings at E and F. The bearings do not exert couples, and the bearing at F does not exert a force in the z direction. The tension in cable AB is 8 kN. Determine the reactions at E and F.

5.113 The plate is supported by hinges at A and B and the cable CE, and is loaded by the force at D. The edge of the plate supported by the hinges lies in the y-z plane. The hinges do not exert couples on the plate. What is the tension in the cable CE?

P5.113

5.114 Consider the plate in Problem 5.113. The hinges do not exert couples on the plate, and the hinge at B does not exert a force parallel to the hinge axis. Determine the magnitudes of the forces exerted on the plate at A and B.

5.115 The bar ABC is supported by ball and socket supports at A and C and the cable BD, and is loaded by the 200-lb suspended weight. What is the tension in cable BD?

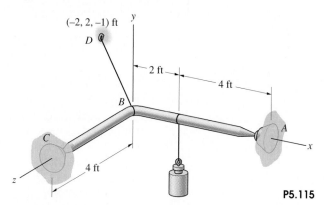

P5.115

5.116 In Problem 5.115, determine the y components of the reactions exerted on the bar ABC by the ball and socket supports at A and C.

5.117 The bearings at A, B, and C do not exert couples on the bar and do not exert forces in the direction of the axis of the bar. Determine the reactions at the bearings due to the two forces on the bar.

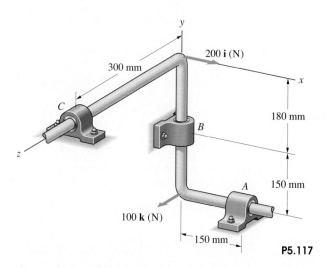

P5.117

5.118 The support that attaches the sailboat's mast to the deck behaves like a ball and socket support. The line that attaches the spinnaker (the sail) to the top of the mast exerts a 200-lb force on the mast. The force is in the horizontal plane at 15° from the centerline of the boat. (See the top view.) The spinnaker pole exerts a 50-lb force on the mast at P. The force is in the horizontal plane at 45° from the centerline. (See the top view.) The mast is supported by two cables, the back stay AB and the port shroud ACD. (The fore stay AE and the starboard shroud AFG are slack, and their ten-

sions can be neglected.) Determine the tensions in the cables AB and CD and the reactions at the bottom of the mast.

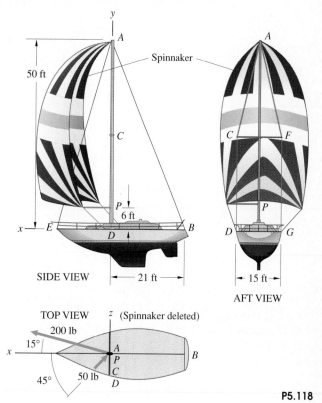

P5.118

5.119 The door is supported by the cable DE and hinges at A and B, and is subjected to a 2-kN force at C. The door's weight is negligible. The hinges do not exert couples on the door, and their axes are aligned with the line from A to B. Determine the tension in the cable.

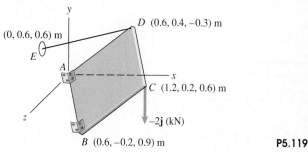

P5.119

5.120 Determine the reactions at the hinges supporting the door in Problem 5.119. Assume that the hinge at B exerts no force parallel to the hinge axis.

Strategy: Express the reactions at the hinges as $\mathbf{A} = A_x\mathbf{i} + A_y\mathbf{j} + A_z\mathbf{k}$ and $\mathbf{B} = B_x\mathbf{i} + B_y\mathbf{j} + B_z\mathbf{k}$. Let $\mathbf{e}_{AB}$ be a unit vector parallel to the hinge axes. Since the hinge at B exerts no force parallel to the hinge axis, you know that $\mathbf{e}_{AB} \cdot \mathbf{B} = 0$.

5.5 Two-Force and Three-Force Members

You have seen how the equilibrium equations are used to analyze objects supported and loaded in different ways. Here we discuss two particular cases that occur so frequently you need to be familiar with them. The first one is especially important and plays a central role in our analysis of structures in the next chapter.

Two-Force Members

If the system of forces and moments acting on an object is equivalent to two forces acting at different points, we refer to the object as a **two-force member**. For example, the object in Fig. 5.35(a) is subjected to two sets of concurrent forces whose lines of action intersect at A and B. Since we can represent them by single forces acting at A and B (Fig. 5.35b), where $\mathbf{F} = \mathbf{F}_1 + \mathbf{F}_2 + \cdots + \mathbf{F}_N$ and $\mathbf{F}' = \mathbf{F}'_1 + \mathbf{F}'_2 + \cdots + \mathbf{F}'_M$, this object is a two-force member.

If the object is in equilibrium, what can we infer about the forces $\mathbf{F}$ and $\mathbf{F}'$? The sum of the forces equals zero only if $\mathbf{F}' = -\mathbf{F}$ (Fig. 5.35c). Furthermore, the forces $\mathbf{F}$ and $-\mathbf{F}$ form a couple, so the sum of the moments is not zero unless the lines of action of the forces lie along the line through the points A and B (Fig. 5.35d). Thus *equilibrium tells us that the two forces are equal in magnitude, are opposite in direction, and have the same line of action.* However, without additional information, we cannot determine their magnitude.

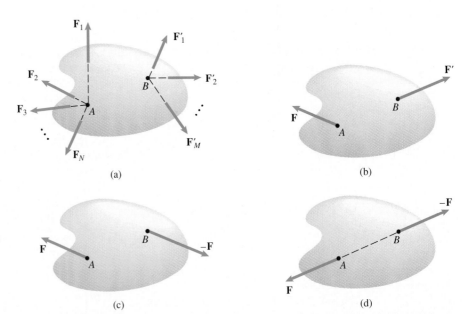

(a)

(b)

(c)

(d)

Figure 5.35

(a) An object subjected to two sets of concurrent forces.
(b) Representing the concurrent forces by two forces $\mathbf{F}$ and $\mathbf{F}'$.
(c) If the object is in equilibrium, the forces must be equal and opposite.
(d) The forces form a couple unless they have the same line of action.

A cable attached at two points (Fig. 5.36a) is a familiar example of a two-force member (Fig. 5.36b). The cable exerts forces on the attachment points that are directed along the line between them (Fig. 5.36c).

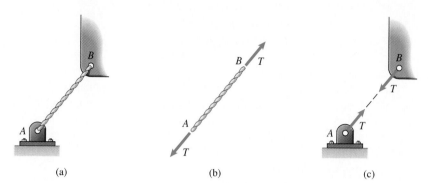

Figure 5.36

(a) A cable attached at A and B.
(b) The cable is a two-force member.
(c) The forces exerted by the cable.

A bar that has two supports that exert only forces on it (no couples) and is not subjected to any loads is a two-force member (Fig. 5.37a). Such bars are often used as supports for other objects. Because the bar is a two-force member, the lines of action of the forces exerted on the bar must lie along the line between the supports (Fig. 5.37b). Notice that, unlike the cable, the bar can exert forces at A and B either in the directions shown in Fig. 5.37(c) or in the opposite directions. (In other words, the cable can only pull on its supports, while the bar can either pull or push.)

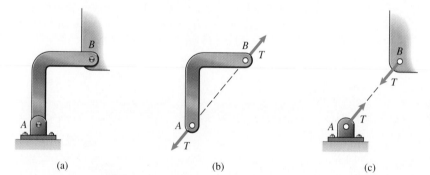

Figure 5.37

(a) The bar AB attaches the object to the pin support.
(b) The bar AB is a two-force member.
(c) The force exerted on the supported object by the bar AB.

In these examples we assumed that the weights of the cable and the bar could be neglected in comparison with the forces exerted on them by their supports. When that is not the case, they are clearly not two-force members.

Three-Force Members

If the system of forces and moments acting on an object is equivalent to three forces acting at different points, we call it a **three-force member**. We can show that if a three-force member is in equilibrium, the three forces are coplanar and are either parallel or concurrent.

We first prove that the forces are coplanar. Let them be called $\mathbf{F}_1$, $\mathbf{F}_2$, and $\mathbf{F}_3$, and let P be the plane containing the three points of application (Fig. 5.38a). Let L be the line through the points of application of $\mathbf{F}_1$ and $\mathbf{F}_2$. Since the moments due to $\mathbf{F}_1$ and $\mathbf{F}_2$ about L are zero, the moment due to $\mathbf{F}_3$ about L must equal zero (Fig. 5.38b):

$$[\mathbf{e} \cdot (\mathbf{r} \times \mathbf{F}_3)] \mathbf{e} = [\mathbf{F}_3 \cdot (\mathbf{e} \times \mathbf{r})] \mathbf{e} = \mathbf{0}.$$

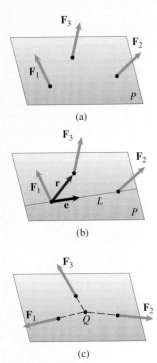

(a)

(b)

(c)

Figure 5.38
(a) The three forces and the plane P.
(b) Determining the moment due to force $\mathbf{F}_3$ about L.
(c) If the forces are not parallel, they must be concurrent.

This equation requires that $\mathbf{F}_3$ be perpendicular to $\mathbf{e} \times \mathbf{r}$, which means that $\mathbf{F}_3$ is contained in P. The same procedure can be used to show that $\mathbf{F}_1$ and $\mathbf{F}_2$ are contained in P, so the forces are coplanar. (A different proof is required if the points of application lie on a straight line, but the result is the same.)

If the three coplanar forces are not parallel, there will be points where their lines of action intersect. Suppose that the lines of action of two of the forces intersect at a point Q. Then the moments of those two forces about Q are zero, and the sum of the moments about Q is zero only if the line of action of the third force also passes through Q. Therefore either the forces are parallel or they are concurrent (Fig. 5.38c).

You can often simplify the analysis of an object in equilibrium by recognizing that it is a two-force or three-force member. However, you are not getting something for nothing. Once you have drawn the free-body diagram of a two-force member as shown in Figs. 5.36(b) and 5.37(b), you cannot obtain any further information about the forces from the equilibrium equations. When you require that the lines of action of nonparallel forces acting on a three-force member be coincident, you have used the fact that the sum of the moments about a point must be zero and cannot obtain any further information from that condition.

Recognizing that an object in equilibrium is a two-force or three-force member provides clues about its free-body diagram that can substantially simplify your analysis, as we demonstrate in the following examples.

Example 5.11

The L-shaped bar in Fig. 5.39 has a pin support at A and is loaded by a 6-kN force at B. Neglect the weight of the bar. Determine the angle α and the reactions at A.

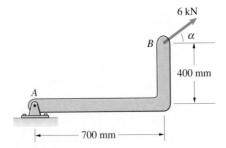

Figure 5.39

STRATEGY

The bar is a two-force member because it is subjected only to the 6-kN force at B and the force exerted by the pin support. (If we could not neglect the weight of the bar, it would not be a two-force member.) We will determine the angle α and the reactions at A in two ways, first by applying the equilibrium equations in the usual way and then by using the fact that the bar is a two-force member.

SOLUTION

Applying the Equilibrium Equations We draw the free-body diagram of the bar in Fig. (a), showing the reactions at the pin support. Summing moments about point A, the equilibrium equations are

$$\Sigma F_x = A_x + 6 \cos \alpha = 0,$$
$$\Sigma F_y = A_y + 6 \sin \alpha = 0,$$
$$\Sigma M_{(\text{point } A)} = (6 \sin \alpha)(0.7) - (6 \cos \alpha)(0.4) = 0.$$

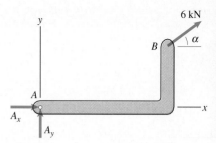

(a) The free-body diagram of the bar.

From the third equation we see that $\alpha = \arctan (0.4/0.7)$. In the range $0 \leq \alpha \leq 360°$, this equation has the two solutions $\alpha = 29.7°$ and $\alpha = 209.7°$. Knowing α, we can determine A_x and A_y from the first two equilibrium equations. The solutions for the two values of α are

$$\alpha = 29.7°, \qquad A_x = -5.21 \text{ kN}, \qquad A_y = -2.98 \text{ kN},$$

and

$$\alpha = 209.7°, \qquad A_x = 5.21 \text{ kN}, \qquad A_y = 2.98 \text{ kN}.$$

Treating the Bar as a Two-Force Member We know that the 6-kN force at B and the force exerted by the pin support must be equal in magnitude, opposite in direction, and directed along the line between points A and B. The two possibilities are shown in Figs. (b) and (c). Thus by recognizing that the bar is a two-force member, we immediately know the possible directions of the forces and the magnitude of the reaction at A.

In Fig. (b) we can see that $\tan \alpha = 0.4/0.7$, so $\alpha = 29.7°$ and the components of the reaction at A are

$$A_x = -6 \cos 29.7° = -5.21 \text{ kN},$$
$$A_y = -6 \sin 29.7° = -2.98 \text{ kN}.$$

In Fig. (c), $\alpha = 180° + 29.7° = 209.7°$, and the components of the reaction at A are

$$A_x = 6 \cos 29.7° = 5.21 \text{ kN},$$
$$A_y = 6 \sin 29.7° = 2.98 \text{ kN}.$$

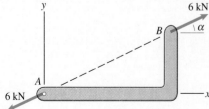
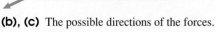

(b), (c) The possible directions of the forces.

Example 5.12

The 100-lb weight of the rectangular plate in Fig. 5.40 acts at its midpoint. Determine the reactions exerted on the plate at B and C.

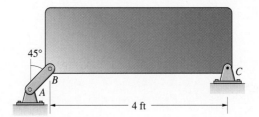

Figure 5.40

STRATEGY

The plate is subjected to its weight and the reactions exerted by the pin supports at B and C, so it is a three-force member. Furthermore, the bar AB is a two-force member, so we know that the line of action of the reaction it exerts on the plate at B is directed along the line between A and B. We can use this information to simplify the free-body diagram of the plate.

SOLUTION

The reaction exerted on the plate by the two-force member AB must be directed along the line between A and B, and the line of action of the weight is vertical. Since the three forces on the plate must be either parallel or concurrent, their lines of action must intersect at the point P shown in Fig. (a). From the equilibrium equations

$$\Sigma F_x = B \sin 45° - C \sin 45° = 0,$$

$$\Sigma F_y = B \cos 45° + C \cos 45° - 100 = 0,$$

we obtain the reactions $B = C = 70.7$ lb.

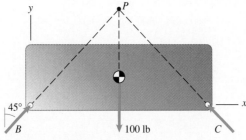

(a) The free-body diagram of the plate. The three forces must be concurrent.

Problems

5.121 The horizontal bar has a mass of 10 kg. Its weight acts at the midpoint of the bar, and it is supported by a roller support at A and the cable BC. Use the fact that the bar is a three-force member to determine the angle α, the tension in the cable BC, and the magnitude of the reaction at A.

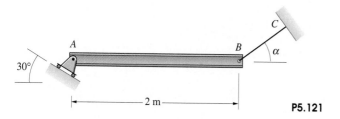

P5.121

5.122 The horizontal bar is of negligible weight. Use the fact that the bar is a three-force member to determine the angle α necessary for equilibrium.

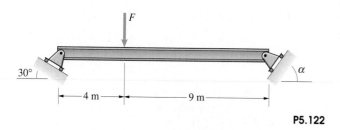

P5.122

5.123 The suspended load weighs 1000 lb. If you neglect its weight, the structure is a three-force member. Use this fact to determine the magnitudes of the reactions at A and B.

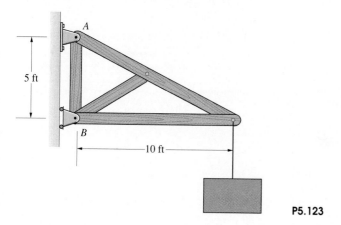

P5.123

5.124 The weight $W = 50$ lb acts at the center of the disk. Use the fact that the disk is a three-force member to determine the tension in the cable and the magnitude of the reaction at the pin support.

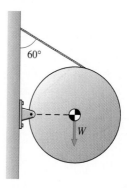

P5.124

5.125 The weight $W = 40$ N acts at the center of the disk. The surfaces are rough. What force F is necessary to lift the disk off the floor?

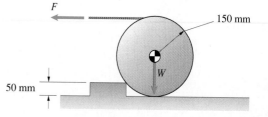

P5.125

5.126 Use the fact that the horizontal bar is a three-force member to determine the angle α and the magnitudes of the reactions at A and B.

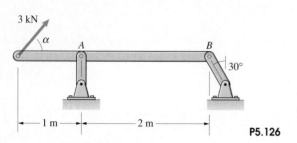

P5.126

5.127 The suspended load weighs 600 lb. Use the fact that ABC is a three-force member to determine the magnitudes of the reactions at A and B.

5.128 (a) Is the L-shaped bar a three-force member?
(b) Determine the magnitudes of the reactions at A and B.
(c) Are the three forces acting on the L-shaped bar concurrent?

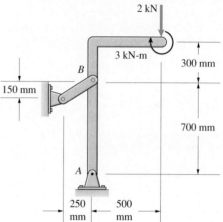

P5.128

5.129 The bucket of the excavator is supported by the two-force member AB and the pin support at C. Its weight is $W = 1500$ lb. What are the reactions at C?

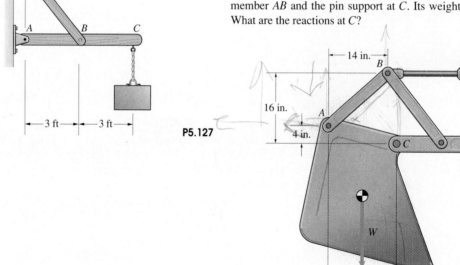

P5.127

P5.129

5.130 The member *ACG* of the front-end loader is subjected to a load $W = 2$ kN and is supported by a pin support at *A* and the hydraulic cylinder *BC*. Treat the hydraulic cylinder as a two-force member.

(a) Draw the free-body diagrams of the hydraulic cylinder and the member *ACG*.

(b) Determine the reactions on the member *ACG*.

5.132 A rectangular plate is subjected to two forces *A* and *B* (Fig. a). In Fig. (b), the two forces are resolved into components. By writing equilibrium equations in terms of the components A_x, A_y, B_x, and B_y, show that the two forces *A* and *B* are equal in magnitude, opposite in direction, and directed along the line between their points of application.

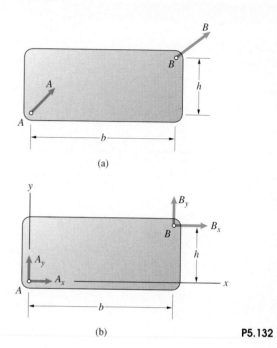

(a)

(b)

P5.132

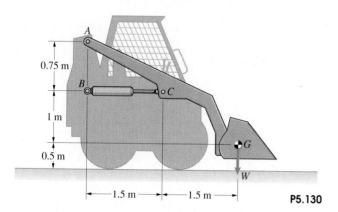

P5.130

5.131 In Problem 5.130, determine the reactions on the member *ACG* by using the fact that it is a three-force member.

5.133 An object in equilibrium is subjected to three forces whose points of application lie on a straight line. Prove that the forces are coplanar.

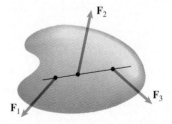

P5.133

Computational Mechanics

The following example and problems are designed for the use of a programmable calculator or computer.

Example 5.13

The beam in Fig. 5.41 weighs 200 lb and is supported by a pin support at A and the wire BC. The wire behaves like a linear spring with spring constant $k = 60$ lb/ft and is unstretched when the beam is in the position shown. Determine the reactions at A and the tension in the wire when the beam is in equilibrium.

Figure 5.41

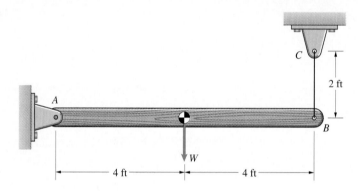

STRATEGY

When the beam is in equilibrium, the sum of the moments about A due to the beam's weight and the force exerted by the wire equals zero. We will obtain a graph of the sum of the moments as a function of the angle of rotation of the beam relative to the horizontal to determine the position of the beam when it is in equilibrium. Once we know the position, we can determine the tension in the wire and the reactions at A.

SOLUTION

Let α be the angle from the horizontal to the centerline of the beam (Fig. a). The distances b and h are

$$b = 8(1 - \cos \alpha),$$

$$h = 2 + 8 \sin \alpha,$$

and the length of the stretched wire is

$$L = \sqrt{b^2 + h^2}.$$

The tension in the wire is

$$T = k(L - 2).$$

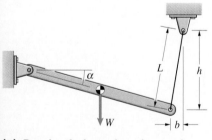

(a) Rotating the beam through an angle α.

We draw the free-body diagram of the beam in Fig. (b). In terms of the components of the force exerted by the wire,

$$T_x = \frac{b}{L}T, \qquad T_y = \frac{h}{L}T,$$

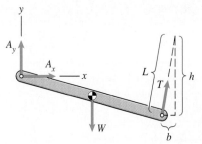

the sum of the moments about A is

$$\Sigma M_A = (8 \sin \alpha)T_x + (8 \cos \alpha)T_y - (4 \cos \alpha)W.$$

(b) The free-body diagram of the beam.

If we choose a value of α, we can sequentially evaluate these quantities. Computing ΣM_A as a function of α, we obtain the graph shown in Fig. 5.42. From the graph we estimate that $\Sigma M_A = 0$ when $\alpha = 12°$. By examining computed results near 12°,

α	ΣM_A (ft-lb)
11.87°	−1.2600
11.88°	−0.5925
11.89°	0.0750
11.90°	0.7424
11.91°	1.4099

we estimate that the beam is in equilibrium when $\alpha = 11.89°$. The corresponding value of the tension in the wire is $T = 99.1$ lb.

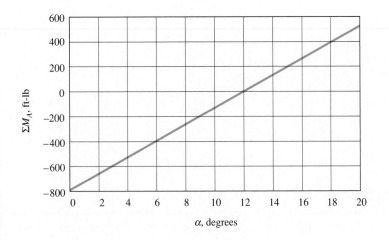

Figure 5.42

The sum of the moments as a function of α.

To determine the reactions at A, we use the equilibrium equations

$$\Sigma F_x = A_x + T_x = 0,$$
$$\Sigma F_y = A_y + T_y - W = 0,$$

obtaining $A_x = -4.7$ lb and $A_y = 101.0$ lb.

Problems

5.134 The rectangular plate is held in equilibrium by the horizontal force F. The weight W acts at the midpoint of the plate. The ratio $b/h = 4$. Determine the angle α at which the plate is in equilibrium for five values of the ratio F/W: 0, 0.5, 1.0, 1.5, and 2. (Assume that $0 \leq \alpha \leq 90°$.)

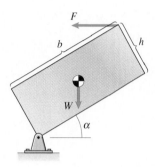

P5.134

5.135 Consider the system shown in Problem 5.41. The bar weighs 15 lb, and its weight acts at its midpoint. The spring is unstretched when $\alpha = 0$, and the spring constant is $k = 6$ lb/ft. Determine the values of α in the range $0 \leq \alpha \leq 90°$ at which the bar is in equilibrium.

5.136 Consider the system shown in Problem 5.61. The distances are $a = 2$ m and $b = 1$ m. The couple $M = 1$ kN-m, and the force $F = 2$ kN. The spring constant is $k = 3$ kN/m. The spring would be unstretched if $h = 0$. Determine the distance h for equilibrium of the horizontal bar and the reactions at A.

5.137 Consider the system shown in Problem 5.62. The bar is 1 m long, and its weight $W = 35$ N acts at its midpoint. The distance $b = 0.75$ m. The spring constant is $k = 100$ N/m, and the spring is unstretched when the bar is vertical. Determine the angle α and the reactions at A.

5.138 Consider the system shown in Problem 5.152. The hydraulic actuator BC exerts a force at C that points along the line from B to C. Treat point A as a pin support. The mass of the suspended load is 4000 kg. If the actuator BC can exert a maximum force of 80 kN, what is the smallest permissible value of α?

5.139 The beam is in equilibrium in the position shown. Each spring has an unstretched length of 1 m. Determine the distance b and the reactions at A.

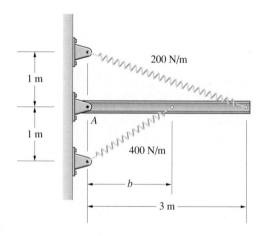

P5.139

Chapter Summary

Building on our discussions of forces in Chapter 3 and moments in Chapter 4, in this chapter we have used the equilibrium equations to analyze the forces and couples acting on many types of objects. We defined the support conventions commonly used in engineering and presented examples of their use. We discussed situations that can result in an object's being statically indeterminate. Finally, we defined two-force and three-force members. In Chapter 6 we will use the concepts and methods developed in this chapter to analyze the individual members of structures, beginning with structures consisting entirely of two-force members.

When an object is in equilibrium, the following conditions are satisfied:

1. The sum of the forces is zero,

$$\Sigma \mathbf{F} = 0. \qquad \textbf{Eq. (5.1)}$$

2. The sum of the moments about any point is zero,

$$\Sigma \mathbf{M}_{(\text{any point})} = 0. \qquad \textbf{Eq. (5.2)}$$

Forces and couples exerted on an object by its supports are called **reactions**. The other forces and couples on the object are the **loads**. Common supports are represented by models called **support conventions**.

Two-Dimensional Applications

When the loads and reactions on an object in equilibrium form a two-dimensional system of forces and moments, they are related by three scalar equilibrium equations:

$$\Sigma F_x = 0,$$
$$\Sigma F_y = 0, \qquad \textbf{Eqs. (5.4)–(5.6)}$$
$$\Sigma M_{(\text{any point})} = 0.$$

No more than three independent equilibrium equations can be obtained from a given two-dimensional free-body diagram.

Support conventions commonly used in two-dimensional applications are summarized in Table 5.1.

Three-Dimensional Applications

The loads and reactions on an object in equilibrium satisfy the six scalar equilibrium equations

$$\Sigma F_x = 0, \qquad \Sigma F_y = 0, \qquad \Sigma F_z = 0,$$
$$\Sigma M_x = 0, \qquad \Sigma M_y = 0, \qquad \Sigma M_z = 0. \qquad \textbf{Eqs. (5.16)–(5.21)}$$

No more than six independent equilibrium equations can be obtained from a given free-body diagram.

Support conventions commonly used in three-dimensional applications are summarized in Table 5.2.

Statically Indeterminate Objects

An object has **redundant supports** when it has more supports than the minimum number necessary to maintain it in equilibrium and **improper supports** when its supports are improperly designed to maintain equilibrium under the applied loads. In either situation, the object is **statically indeterminate**. The difference between the number of reactions and the number of independent equilibrium equations is called the **degree of redundancy**. Even if an object is statically indeterminate due to redundant supports, it may be possible to determine some of the reactions from the equilibrium equations.

Two-Force and Three-Force Members

If the system of forces and moments acting on an object is equivalent to two forces acting at different points, the object is a **two-force member**. If the object is in equilibrium, the two forces are equal in magnitude, opposite in direction, and directed along the line through their points of application. If the system of forces and moments acting on an object is equivalent to three forces acting at different points, it is a **three-force member**. If the object is in equilibrium, the three forces are coplanar and either parallel or concurrent.

Review Problems

5.140 Determine the reactions at *A* and *B*.

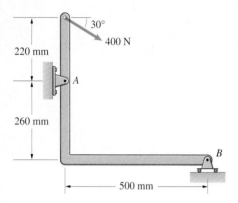

P5.140

5.142 (a) Draw the free-body diagram of the 50-lb plate, and explain why it is statically indeterminate.
(b) Determine as many of the reactions at *A* and *B* as possible.

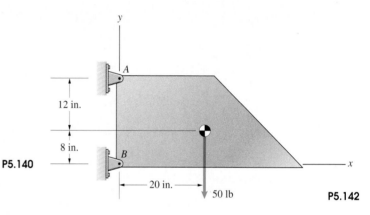

P5.142

5.141 Paleontologists speculate that the stegosaur could stand on its hind limbs for short periods to feed. Based on the free-body diagram shown and assuming that *m* = 2000 kg, determine the magnitudes of the forces *B* and *C* exerted by the ligament-muscle brace and vertebral column, and determine the angle α.

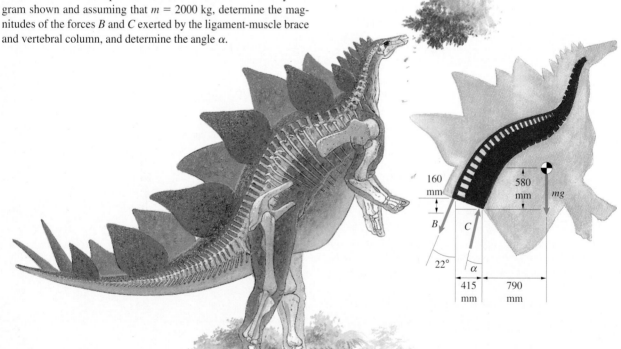

P5.141

5.143 The mass of the truck is 4 Mg. Its wheels are locked, and the tension in its cable is $T = 10$ kN.
(a) Draw the free-body diagram of the truck.
(b) Determine the normal forces exerted on the truck's wheels by the road.

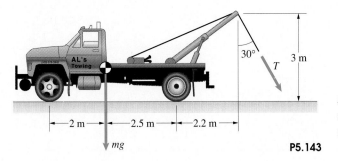

P5.143

5.144 Assume that the force exerted on the head of the nail by the hammer is vertical, and neglect the hammer's weight.
(a) Draw the free-body diagram of the hammer.
(b) If $F = 10$ lb, what are the magnitudes of the force exerted on the nail by the hammer and the normal and friction forces exerted on the floor by the hammer?

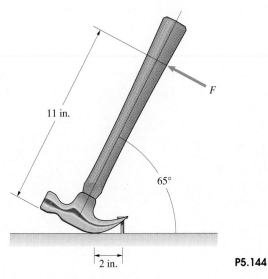

P5.144

5.145 (a) Draw the free-body diagram of the beam.
(b) Determine the reactions at the supports.

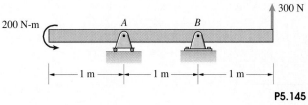

P5.145

5.146 Consider the beam shown in Problem 5.145. First represent the loads (the 300-N force and the 200-N-m couple) by a single equivalent force; then determine the reactions at the supports.

5.147 The truss supports a 90-kg suspended object. What are the reactions at the supports A and B?

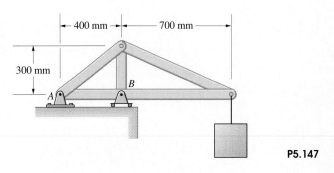

P5.147

5.148 The trailer is parked on a 15° slope. Its wheels are free to turn. The hitch H behaves like a pin support. Determine the reactions at A and H.

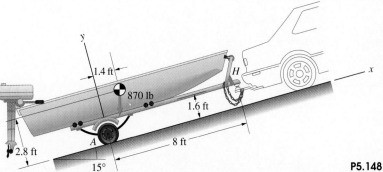

P5.148

5.149 To determine the location of the point where the weight of a car acts (the *center of mass*), an engineer places the car on scales and measures the normal reactions at the wheels for two values of α, obtaining the following results.

α	A_y (kN)	B (kN)
10°	10.134	4.357
20°	10.150	3.677

What are the distances b and h?

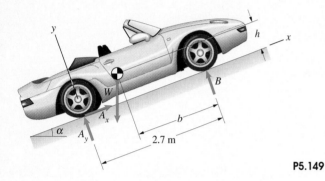

P5.149

5.150 The bar is attached by pin supports to collars that slide on the two fixed bars. Its mass is 10 kg, it is 1 m in length, and its weight acts at its midpoint. Neglect friction and the masses of the collars. The spring is unstretched when the bar is vertical ($\alpha = 0$), and the spring constant is $k = 100$ N/m. Determine the values of α in the range $0 \leq \alpha \leq 60°$ at which the bar is in equilibrium.

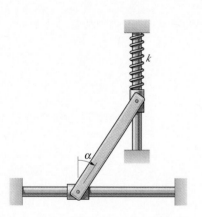

P5.150

5.151 The 450-lb ladder is supported by the hydraulic cylinder AB and the pin support at C. The reaction at B is parallel to the hydraulic cylinder. Determine the reactions on the ladder.

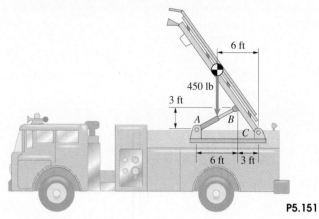

P5.151

5.152 The hydraulic actuator BC exerts a force at C that points along the line from B to C. Treat point A as a pin support. The angle $\alpha = 40°$. The mass of the suspended load is 4000 kg. Determine the force exerted at C and the reactions at A.

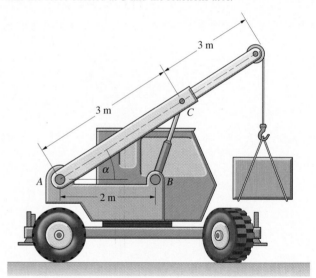

P5.152

5.153 The horizontal rectangular plate weighs 800 N and is suspended by three vertical cables. The weight of the plate acts at its midpoint. What are the tensions in the cables?

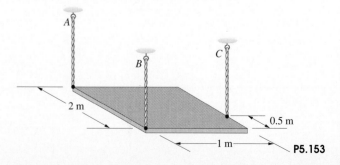

P5.153

5.154 Consider the suspended 800-N plate in Problem 5.153. The weight of the plate acts at its midpoint. If you represent the reactions exerted on the plate by the three cables by a single equivalent force, what is the force, and where does its line of action intersect the plate?

5.155 The 20-kg mass is suspended by cables attached to three vertical 2-m posts. Point A is at $(0, 1.2, 0)$ m. Determine the reactions at the built-in support at E.

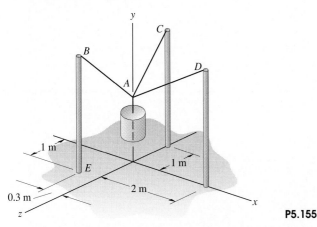

P5.155

5.156 In Problem 5.155, the built-in support of each vertical post will safely support a couple of 800 N-m magnitude. Based on this criterion, what is the maximum safe value of the suspended mass?

5.157 The 80-lb bar is supported by a ball and socket support at A, the smooth wall it leans against, and the cable BC. The weight of the bar acts at its midpoint.
(a) Draw the free-body diagram of the bar.
(b) Determine the tension in cable BC and the reactions at A.

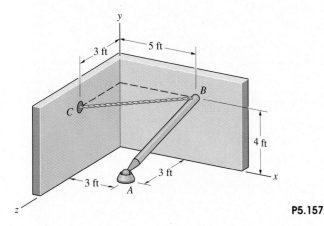

P5.157

5.158 The horizontal bar of weight W is supported by a roller support at A and the cable BC. Use the fact that the bar is a three-force member to determine the angle α, the tension in the cable, and the magnitude of the reaction at A.

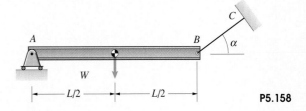

P5.158

5.159 The bicycle brake on the right is pinned to the bicycle's frame at A. Determine the force exerted by the brake pad on the wheel rim at B in terms of the cable tension T.

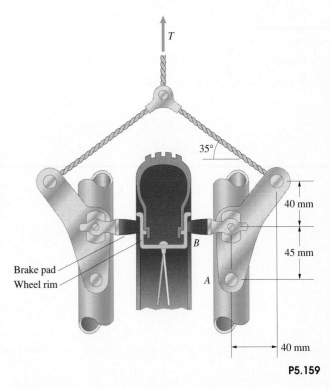

P5.159

Project 5.1 Find five objects in equilibrium whose supports can be modeled by the conventions in Table 5.1. Include at least one example each of a pin support, a roller support, and a built-in support. Draw a sketch of each object and its supports, and draw the free-body diagram of the object.

Project 5.2 Find five objects in equilibrium whose supports can be modeled by the conventions in Table 5.2. Include at least one example of a ball and socket support and one example of a hinge. Draw a sketch of each object and its supports, and draw the free-body diagram of the object.

Large structures, such as buildings, bridges, and dams, must withstand enormous forces and moments due to their weights and the external loads exerted on them. To support these forces and moments, engineers design interior structures, such as the framework of steel bars that stabilizes the bronze sheath of the Statue of Liberty. Designed by Gustave Eiffel in 1879, this framework foreshadowed the structure of the Eiffel Tower, built in 1887–1889. In this chapter we will use free-body diagrams and the equilibrium equations to determine the reactions on individual members of such structures, an essential step in their design.

Chapter 6

Structures in Equilibrium

IN engineering, the term **structure** can refer to any object that has the capacity to support and exert loads. In this chapter we consider structures composed of interconnected parts, or **members**. To design such a structure, or to determine whether an existing one is adequate, you must determine the forces and couples acting on the structure as a whole as well as on its individual members. We first demonstrate how this is done for the structures called trusses, which are composed entirely of two-force members. The familiar frameworks of steel members that support some highway bridges are trusses. We then consider other structures, called frames if they are designed to remain stationary and support loads and machines if they are designed to move and exert loads. For example, the framework of steel members that supports the Statue of Liberty is a frame, and a pair of pliers is a machine. The material we discuss here will prepare you for more advanced structural applications.

6.1 *Trusses*

Figure 6.1

A typical house is supported by trusses made of wood beams.

We can explain the nature of truss structures such as the beams supporting a house (Fig. 6.1) by starting with very simple examples. Suppose we pin three bars together at their ends to form a triangle. If we add supports as shown in Fig. 6.2(a), we obtain a structure that will support a load *F*. We can construct more elaborate structures by adding more triangles (Figs. 6.2b and c). The bars are the members of these structures, and the places where the bars are pinned together are called the **joints**. Even though these examples are quite simple, you can see that Fig. 6.2(c), which is called a *Warren truss*, begins to resemble the structures used to support bridges and the roofs of houses (Fig. 6.3). If these structures are supported and loaded at their joints and we neglect the weights of the bars, each bar is a two-force member. We call such a structure a **truss**.

We draw the free-body diagram of a member of a truss in Fig. 6.4(a). Because it is a two-force member, the forces at the ends, which are the sums of the forces exerted on the member at its joints, must be equal in magnitude, opposite in direction, and directed along the line between the joints. We call the force *T* the **axial force** in the member. When *T* is positive in the direction shown (that is, when the forces are directed away from each other), the member is in **tension**. When the forces are directed toward each other, the member is in **compression**.

In Fig. 6.4(b), we "cut" the member by a plane and draw the free-body diagram of the part of the member on one side of the plane. We represent the sys-

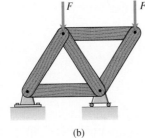

(a)   (b)   (c)

Figure 6.2

Making structures by pinning bars together to form triangles.

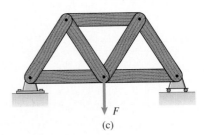
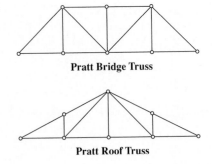

Howe Bridge Truss   **Pratt Bridge Truss**

Howe Roof Truss   **Pratt Roof Truss**

Figure 6.3

Simple examples of bridge and roof structures. (The lines represent members, and the circles represent joints.)

tem of internal forces and moments exerted by the part not included in the free-body diagram by a force **F** acting at the point P where the plane intersects the axis of the member and a couple **M**. The sum of the moments about P must equal zero, so $\mathbf{M} = \mathbf{0}$. Therefore we have a two-force member, which means that **F** must be equal in magnitude and opposite in direction to the force T acting at the joint (Fig. 6.4c). The internal force is a tension or compression equal to the tension or compression exerted at the joint. Notice the similarity to a rope or cable, in which the internal force is a tension equal to the tension applied at the ends.

Although many actual structures, including "roof trusses" and "bridge trusses," consist of bars connected at the ends, very few have pinned joints. For example, if you examine a joint of a bridge truss, you will see that the members are bolted or riveted together so that they are not free to rotate at the joint (Fig. 6.5). It is obvious that such a joint can exert couples on the members. Why are these structures called trusses?

The reason is that they are designed to function as trusses, meaning that they support loads primarily by subjecting their members to axial forces. They can usually be *modeled* as trusses, treating the joints as pinned connections under the assumption that couples they exert on the members are small in comparison to axial forces. When we refer to structures with riveted joints as trusses in problems, we mean that you can model them as trusses.

In the following sections we describe two methods for determining the axial forces in the members of trusses. The method of joints is usually the preferred approach when you need to determine the axial forces in all members of a truss. When you only need to determine the axial forces in a few members, the method of sections often results in a faster solution than the method of joints.

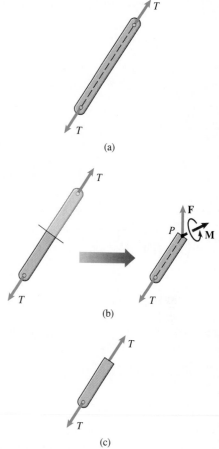

Figure 6.4

(a) Each member of a truss is a two-force member.

(b) Obtaining the free-body diagram of part of the member.

(c) The internal force is equal and opposite to the force acting at the joint, and the internal couple is zero.

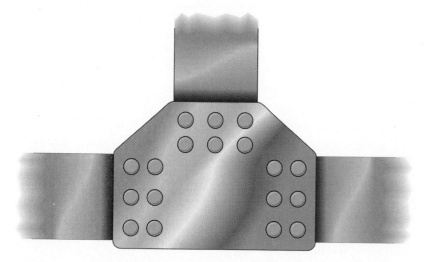

Figure 6.5

A joint of a bridge truss.

6.2 The Method of Joints

The method of joints involves drawing free-body diagrams of the joints of a truss one by one and using the equilibrium equations to determine the axial forces in the members. Before you begin, you usually need to draw a free-body diagram of the entire truss (that is, treat the truss as a single object) and

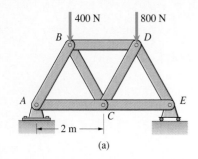

(a)

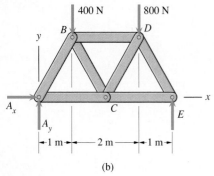

(b)

Figure 6.6

(a) A Warren truss supporting two loads.
(b) Free-body diagram of the truss.

determine the reactions at its supports. For example, let's consider the Warren truss in Fig. 6.6(a), which has members 2 m in length and supports loads at B and D. We draw its free-body diagram in Fig. 6.6(b). From the equilibrium equations,

$$\Sigma F_x = A_x = 0,$$

$$\Sigma F_y = A_y + E - 400 - 800 = 0,$$

$$\Sigma M_{(\text{point } A)} = -(1)(400) - (3)(800) + 4E = 0,$$

we obtain the reactions $A_x = 0$, $A_y = 500$ N, and $E = 700$ N.

Our next step is to choose a joint and draw its free-body diagram. In Fig. 6.7(a), we isolate joint A by cutting members AB and AC. The terms T_{AB} and T_{AC} are the axial forces in members AB and AC, respectively. Although the directions of the arrows representing the unknown axial forces can be chosen arbitrarily, notice that we have chosen them so that a member is in tension if we obtain a positive value for the axial force. We feel that consistently choosing the directions in this way helps avoid errors.

The equilibrium equations for joint A are

$$\Sigma F_x = T_{AC} + T_{AB} \cos 60° = 0,$$

$$\Sigma F_y = T_{AB} \sin 60° + 500 = 0.$$

Solving these equations, we obtain the axial forces $T_{AB} = -577$ N and $T_{AC} = 289$ N. Member AB is in compression, and member AC is in tension (Fig. 6.7b).

Figure 6.7

(a) Obtaining the free-body diagram of joint A.
(b) The axial forces on members AB and AC.
(c) Realistic and simple free-body diagrams of joint A.

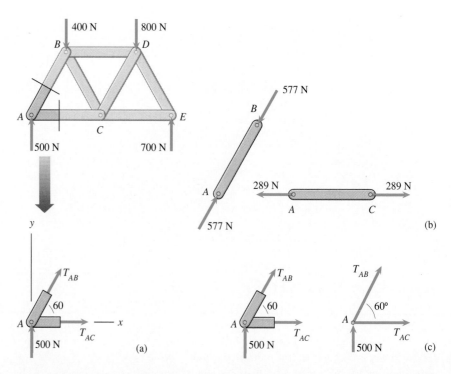

Although we use a realistic figure for the joint in Fig. 6.7(a) to help you understand the free-body diagram, in your own work you can use a simple figure showing only the forces acting on the joint (Fig. 6.7c).

We next obtain a free-body diagram of joint B by cutting members AB, BC, and BD (Fig. 6.8a). From the equilibrium equations for joint B,

$$\Sigma F_x = T_{BD} + T_{BC} \cos 60° + 577 \cos 60° = 0,$$

$$\Sigma F_y = -400 + 577 \sin 60° - T_{BC} \sin 60° = 0,$$

we obtain $T_{BC} = 115$ N and $T_{BD} = -346$ N. Member BC is in tension, and member BD is in compression (Fig. 6.8b). By continuing to draw free-body diagrams of the joints, we can determine the axial forces in all of the members.

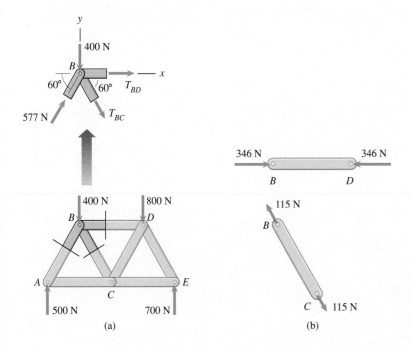

Figure 6.8

(a) Obtaining the free-body diagram of joint B.

(b) Axial forces in members BD and BC.

In two dimensions, you can obtain only two independent equilibrium equations from the free-body diagram of a joint. Summing the moments about a point does not result in an additional independent equation because the forces are concurrent. Therefore when applying the method of joints, you should choose joints to analyze that are subjected to known forces and no more than two unknown forces. In our example, we analyzed joint A first because it was subjected to the known reaction exerted by the pin support and two unknown forces, the axial forces T_{AB} and T_{AC} (Fig. 6.7a). We could then analyze joint B because it was subjected to two known forces and two unknown forces, T_{BC} and T_{BD} (Fig. 6.8a). If we had attempted to analyze joint B first, there would have been three unknown forces.

When you determine the axial forces in the members of a truss, your task will often be simpler if you are familiar with three particular types of joints.

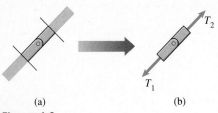

Figure 6.9

(a) A joint with two collinear members and no load.
(b) Free-body diagram of the joint.

- **Truss joints with two collinear members and no load** (Fig. 6.9). The sum of the forces must equal zero, $T_1 = T_2$. *The axial forces are equal.*

- **Truss joints with two noncollinear members and no load** (Fig. 6.10). Because the sum of the forces in the x direction must equal zero, $T_2 = 0$. Therefore T_1 must also equal zero. *The axial forces are zero.*

- **Truss joints with three members, two of which are collinear, and no load** (Fig. 6.11). Because the sum of the forces in the x direction must equal zero, $T_3 = 0$. The sum of the forces in the y direction must equal zero, so $T_1 = T_2$. *The axial forces in the collinear members are equal, and the axial force in the third member is zero.*

Figure 6.10

(a) A joint with two noncollinear members and no load.
(b) Free-body diagram of the joint.

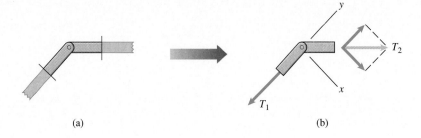

Figure 6.11

(a) A joint with three members, two of which are collinear, and no load.
(b) Free-body diagram of the joint.

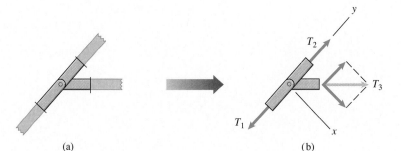

In the following example we use the method of joints to determine the axial forces in the members of a truss. This typically involves three steps:

1. Determine the reactions at the supports—*You usually need to draw the free-body diagram of the entire truss and determine the reactions at its supports.*

2. Identify special joints—*Examine the truss to see if it has any of the types of joints discussed in the preceding list. Although it is not essential, this step can simplify your solution.*

3. Analyze the joints—*Draw free-body diagrams of the joints, and apply the equilibrium equations to determine the axial forces in the members. Choose joints that require solving for no more than two unknown forces.*

Example 6.1

Determine the axial forces in the members of the truss in Fig. 6.12.

SOLUTION

Determine the Reactions at the Supports We first draw the free-body diagram of the entire truss (Fig. a). From the equilibrium equations,

$$\Sigma F_x = A_x + B = 0,$$

$$\Sigma F_y = A_y - 2 = 0,$$

$$\Sigma M_{(\text{point } B)} = -6A_x - (10)(2) = 0,$$

we obtain the reactions $A_x = -3.33$ kip, $A_y = 2$ kip, and $B = 3.33$ kip.

Identify Special Joints Because joint C has three members, two of which are collinear, and no load, the axial force in member BC is zero, $T_{BC} = 0$, and the axial forces in the collinear members AC and CD are equal, $T_{AC} = T_{CD}$.

Draw Free-Body Diagrams of the Joints We know the reaction exerted on joint A by the support, and joint A is subjected to only two unknown forces, the axial forces in members AB and AC. We draw its free-body diagram in Fig. (b). The angle $\alpha = \arctan(5/3) = 59.0°$. The equilibrium equations for joint A are

$$\Sigma F_x = T_{AC} \sin \alpha - 3.33 = 0,$$

$$\Sigma F_y = 2 - T_{AB} - T_{AC} \cos \alpha = 0.$$

Solving these equations, we obtain $T_{AB} = 0$ and $T_{AC} = 3.89$ kip. Because the axial forces in members AC and CD are equal, $T_{CD} = 3.89$ kip.

Now we draw the free-body diagram of joint B in Fig. (c). (We already know that the axial forces in members AB and BC are zero.) From the equilibrium equation

$$\Sigma F_x = T_{BD} + 3.33 = 0,$$

we obtain $T_{BD} = -3.33$ kip. The negative sign indicates that member BD is in compression.

The axial forces in the members are

AB: 0,

AC: 3.89 kip in tension (T),

BC: 0,

BD: 3.33 kip in compression (C),

CD: 3.89 kip in tension (T).

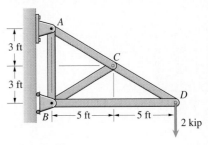

Figure 6.12

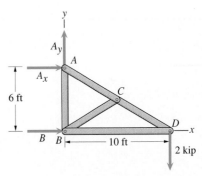

(a) Free-body diagram of the entire truss.

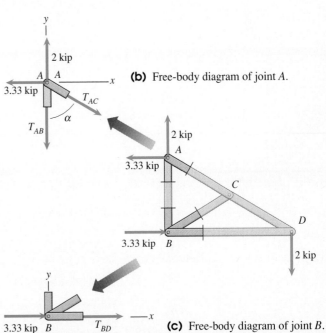

(b) Free-body diagram of joint A.

(c) Free-body diagram of joint B.

Example 6.2

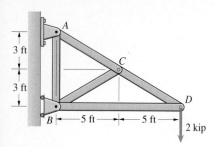

Figure 6.13

Each member of the truss in Fig. 6.13 will safely support a tensile force of 10 kip and a compressive force of 2 kip. What is the largest downward load F that the truss will safely support?

STRATEGY

This truss is identical to the one we analyzed in Example 6.1. By applying the method of joints in the same way, the axial forces in the members can be determined in terms of the load F. The smallest value of F that will cause a tensile force of 10 kip or a compressive force of 2 kip in any of the members is the largest value of F that the truss will support.

SOLUTION

By using the method of joints in the same way as in Example 6.1, we obtain the axial forces

$$AB: 0,$$
$$AC: 1.94F \text{ (T)},$$
$$BC: 0,$$
$$BD: 1.67F \text{ (C)},$$
$$CD: 1.94F \text{ (T)}.$$

For a given load F, the largest tensile force is $1.94F$ (in members AC and CD) and the largest compressive force is $1.67F$ (in member BD). The largest safe tensile force would occur when $1.94F = 10$ kip or when $F = 5.14$ kip. The largest safe compressive force would occur when $1.67F = 2$ kip or when $F = 1.20$ kip. Therefore the largest load F that the truss will safely support is 1.20 kip.

Example 6.3

Application to Engineering

Bridge Design

The loads a bridge structure must support and pin supports where the structure is to be attached are shown in Fig. 6.14(1). Assigned to design the structure, a civil engineering student proposes the structure shown in Fig. 6.14(2). What are the axial forces in the members?

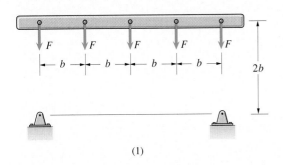

(1)

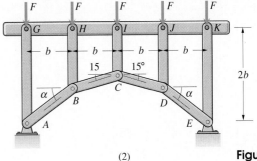

(2)

Figure 6.14

SOLUTION

The vertical members AG, BH, CI, DJ, and EK are subjected to compressive forces of magnitude F. From the free-body diagram of joint C, we obtain $T_{BC} = T_{CD} = -1.93F$. We draw the free-body diagram of joint B in Fig. (a). From the equilibrium equations

$$\Sigma F_x = -T_{AB} \cos \alpha + T_{BC} \cos 15° = 0,$$

$$\Sigma F_y = -T_{AB} \sin \alpha + T_{BC} \sin 15° - F = 0,$$

we obtain $T_{AB} = -2.39F$ and $\alpha = 38.8°$. By symmetry, $T_{DE} = T_{AB}$. The axial forces in the members are shown in Table 6.1.

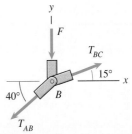

(a) Free-body diagram of joint B.

Table 6.1 Axial forces in the members of the bridge structure.

Members	Axial Force	
AG, BH, CI, DJ, EK	F	(C)
AB, DE	$2.39F$	(C)
BC, CD	$1.93F$	(C)

Figure 6.15

This bridge along Highway 1 in California is supported by a concrete arch anchored in rocky cliffs.

DESIGN ISSUES

The bridge was an early application of engineering. A practical need—crossing a chasm or body of water—was met with a technological solution. Although initially the solution was as primitive as laying a log between the banks, engineers constructed surprisingly elaborate bridges in the remote past. For example, archaeologists have identified foundations of the seven piers of a 120-m (400-ft) highway bridge over the Euphrates that existed in Babylon at the time of Nebuchadnezzar II (reigned 605–562 B.C.).

A bridge must support people and vehicles continuously as they travel between banks. The difficulty is that a single beam extended between the banks will fail if the distance between banks, or *span*, is too large. To meet the need for bridges of increasing strength and span, civil engineers created ingenious and aesthetic designs in antiquity and continue to do so today.

The bridge structure proposed by the student in Example 6.3, called an *arch*, is an ancient design. Notice in Table 6.1 that all the members of the structure are in compression. Because masonry (stone, brick, or concrete) is weak in tension but very strong in compression, many bridges made of these materials were designed with arched spans in the past. For the same reason, modern concrete bridges are often built with arched spans (Fig. 6.15).

Unlike masonry, wood and steel can support substantial forces in both compression and tension. Beginning with the wooden truss bridges designed by the architect Andrea Palladio (1518–1580), both of these materials have been used to construct a large variety of trusses to support bridges. For example, the forces in Fig. 6.14(1) can be supported by the Pratt truss shown in Fig. 6.16. Its members are subjected to both tension and compression (Table 6.2). The Forth Bridge (Fig. 6.17) has a truss structure.

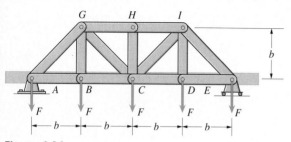

Figure 6.16

A Pratt truss supporting a bridge.

Table 6.2 Axial forces in the members of the Pratt truss.

Members	Axial Force	
AB, BC, CD, DE	1.5F	(T)
AG, EI	2.12F	(C)
CG, CI	0.71F	(T)
GH, HI	2F	(C)
BG, DI	F	(T)
CH	0	

Figure 6.17

The Forth Bridge (Scotland, 1890) is an example of a large truss bridge. Each main span is 520 m long.

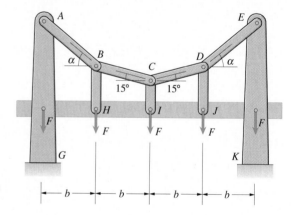

Figure 6.18

A suspension structure supporting a bridge.

Table 6.3 Axial forces in the members of the suspension structure.

Members	Axial Force	
BH, CI, DJ	F	(T)
AB, DE	2.39F	(T)
BC, CD	1.93F	(T)

Figure 6.19

The Golden Gate Bridge (California) has a central suspended span 1280 m (4200 ft) in length.

Truss structures are too heavy for the largest bridges. (The Forth Bridge contains 58,000 tons of steel.) By taking advantage of the ability of relatively light cables to support large tensile forces, civil engineers use *suspension* structures to bridge very large spans. The system of five forces we are using as an example can be supported by the simple suspension structure in Fig. 6.18. In effect, the compression arch used since antiquity is inverted. (Compare Figs. 6.14(2) and 6.18.) The loads in Fig. 6.18 are "suspended" from members *AB*, *BC*, *CD*, and *DE*. Every member of this structure *except the towers AG and EK* is in tension (Table 6.3). The largest existing bridges, such as the Golden Gate Bridge (Fig. 6.19), consist of cable-suspended spans supported by towers.

Problems

6.1 The weight $W = 600$ lb.
(a) Draw the free-body diagram of joint A.
(b) Use the free-body diagram you obtained in (a) to determine the axial forces in members AB and AC. Indicate whether they are in tension (T) or compression (C).

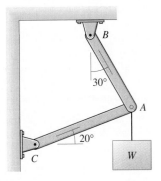

P6.1

6.2 The truss supports a 10-kN load at C.
(a) Draw the free-body diagram of the entire truss, and determine the reactions at its supports.
(b) Determine the axial forces in the members. Indicate whether they are in tension (T) or compression (C).

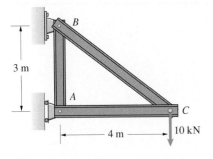

P6.2

6.3 Consider the truss in Problem 6.2. An engineer determines that the members will safely support 30-kN axial forces in either tension or compression. What is the largest downward load the truss will safely support at C?

6.4 Determine the axial forces in the members of the truss.

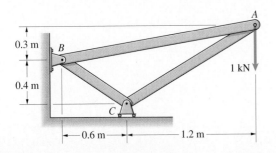

P6.4

6.5 (a) Let the dimension $h = 0.1$ m. Determine the axial forces in the members, and show that in this case this truss is equivalent to the one in Problem 6.4.
(b) Let the dimension $h = 0.5$ m. Determine the axial forces in the members. Compare the results to (a), and observe the dramatic effect of this simple change in design on the maximum tensile and compressive forces to which the members are subjected.

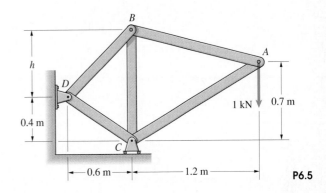

P6.5

6.6 The load $F = 10$ kN. Determine the axial forces in the members.

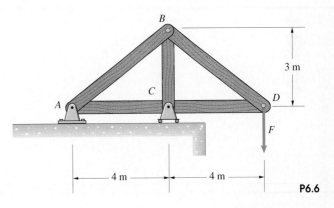

P6.6

6.7 Consider the truss in Problem 6.6. Each member will safely support a tensile force of 150 kN and a compressive force of 30 kN. What is the largest downward load F that the truss will safely support at D?

6.8   The Howe and Pratt bridge trusses are subjected to identical loads.
(a) In which truss does the largest tensile force occur? In what member(s) does it occur, and what is its value?
(b) In which truss does the largest compressive force occur? In what member(s) does it occur, and what is its value?

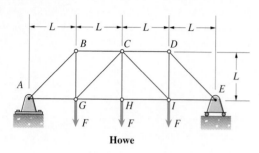

Howe

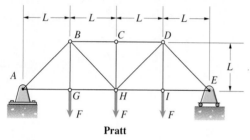

Pratt   **P6.8**

6.9   The truss shown is part of an airplane's internal structure. Determine the axial forces in members BC, BD, and BE.

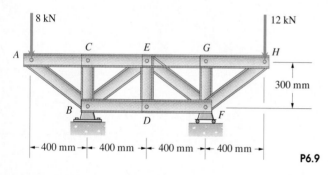

P6.9

6.10   For the truss in Problem 6.9, determine the axial forces in members DF, EF, and FG.

6.11   The loads $F_1 = F_2 = 10$ kip. Determine the axial forces in members BD, BE, and BG.

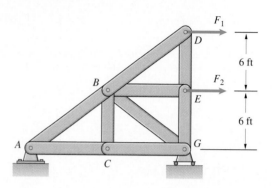

P6.11

6.12   Consider the truss in Problem 6.11. The loads $F_1 = 60$ lb and $F_2 = 120$ lb. Determine the axial forces in members AB, BC, and BD.

6.13   The truss supports loads at C and E. If $F = 3$ kN, what are the axial forces in members BC and BE?

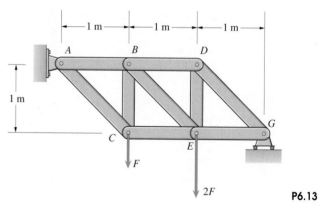

P6.13

6.14   Consider the truss in Problem 6.13. Each member will safely support a tensile force of 28 kN and a compressive force of 12 kN. Taking this criterion into account, what is the largest safe (positive) value of F?

6.15 The truss attaches one end of a stretcher to a rescue helicopter. If the stretcher exerts 60-lb vertical loads at A and B, what are the axial forces in members CF, DF, and FG?

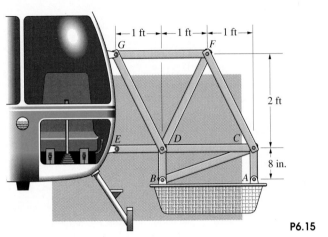

P6.15

6.16 Consider the truss in Problem 6.15. If the stretcher exerts 60-lb vertical loads at A and B, what are the axial forces in members AC, CD, and DG?

6.17 Determine the axial forces in the members in terms of the weight W.

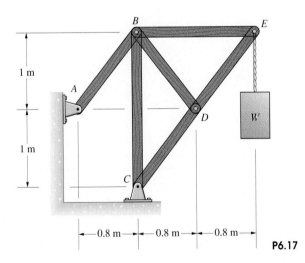

P6.17

6.18 Consider the truss in Problem 6.17. Each member will safely support a tensile force of 6 kN and a compressive force of 2 kN. Use this criterion to determine the largest weight W the truss will safely support.

6.19 The loads $F_1 = 600$ lb and $F_2 = 300$ lb. Determine the axial forces in members AE, BD, and CD.

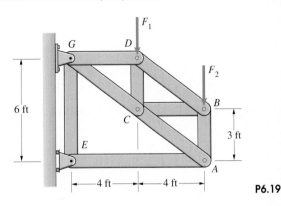

P6.19

6.20 Consider the truss in Problem 6.19. The loads $F_1 = 450$ lb and $F_2 = 150$ lb. Determine the axial forces in members AB, AC, and BC.

6.21 Each member of the truss will safely support a tensile force of 4 kN and a compressive force of 1 kN. Determine the largest mass m that can safely be suspended.

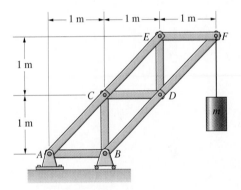

P6.21

6.22 The Warren truss supports a walkway between two buildings. The walkway exerts vertical 12-kip loads at B, D, F, and H. Model the supports at A and I as roller supports. Determine the axial forces in members BC, CD, and CE.

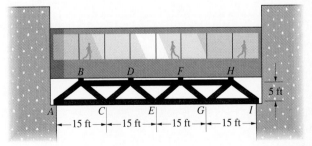

P6.22

6.23 For the Warren truss in Problem 6.22, determine the axial forces in members DF, EF, and FG.

6.24 The Pratt bridge truss supports five forces ($F = 300$ kN). The dimension $L = 8$ m. Determine the axial forces in members BC, BI, and BJ.

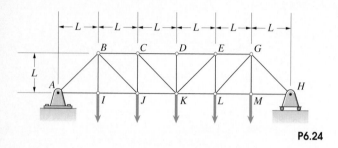

P6.24

6.25 For the Pratt bridge truss in Problem 6.24, determine the axial forces in members CD, CJ, and CK.

6.26 The Howe truss helps support a roof. Model the supports at A and G as roller supports. Determine the axial forces in members AB, BC, and CD.

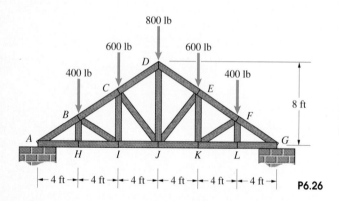

P6.26

6.27 The plane truss forms part of the supports of a crane on an offshore oil platform. The crane exerts vertical 75-kN forces on the truss at B, C, and D. You can model the support at A as a pin support and model the support at E as a roller support that can exert a force normal to the dashed line but cannot exert a force parallel to it. The angle $\alpha = 45°$. Determine the axial forces in the members of the truss.

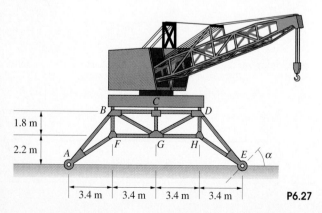

P6.27

6.28 (a) Design a truss attached to the supports A and B that supports the loads applied at points C and D.
(b) Determine the axial forces in the members of the truss you designed in (a).

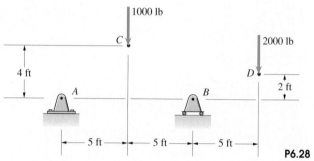

P6.28

6.29 (a) Design a truss attached to the supports A and B that supports the loads applied at points C and D.
(b) Determine the axial forces in the members of the truss you designed in (a).

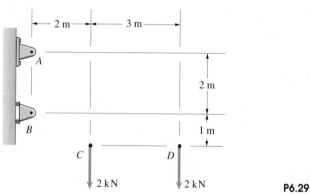

P6.29

6.30 Suppose that you want to design a truss supported at A and B (Fig. a) that provides support for a chinning bar at C that will subject the truss to a maximum downward load of 300 lb. If you use the simplest design (Fig. b), you obtain a 500-lb tensile force in member AC. Improve on this design by finding a truss for which the largest tensile force is less than 500 lb.

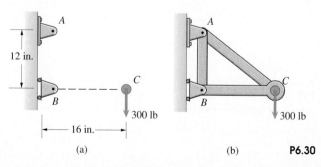

(a)   (b)   **P6.30**

6.31 The bars are of equal length. The spring is unstretched when $\alpha = 90°$ and the horizontal surface is smooth. For what value of α in the range $0 < \alpha < 90°$ will the system remain in equilibrium?

Problems 6.32 and 6.33 are related to Example 6.3.

6.32 Determine the axial forces in the Pratt truss in Fig. 6.16, and confirm the values in Table 6.2.

6.33 Determine the axial forces in the suspension bridge structure in Fig. 6.18, including the reactions exerted on the towers, and confirm the values in Table 6.3.

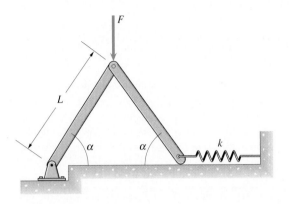

P6.31

6.3 *The Method of Sections*

When you need to know the axial forces only in certain members of a truss, you often can determine them more quickly using the method of sections than using the method of joints. For example, let's reconsider the Warren truss we used to introduce the method of joints (Fig. 6.20a). It supports loads at B and D, and each member is 2 m in length. Suppose that we need to determine only the axial force in member BC.

Just as in the method of joints, we begin by drawing a free-body diagram of the entire truss and determining the reactions at the supports. The results of this step are shown in Fig. 6.20(b). Our next step is to cut the members AC,

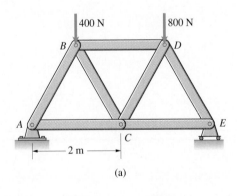

(a)

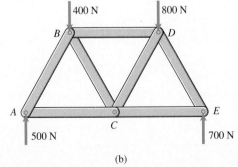

(b)

Figure 6.20

(a) A Warren truss supporting two loads.
(b) Free-body diagram of the truss, showing the reactions at the supports.

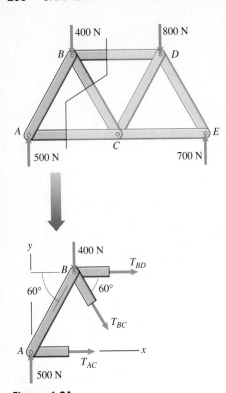

Figure 6.21

Obtaining a free-body diagram of a section of the truss.

BC, and *BD* to obtain a free-body diagram of a part, or **section**, of the truss (Fig. 6.21). Summing moments about point *B*, the equilibrium equations for the section are

$$\Sigma F_x = T_{AC} + T_{BD} + T_{BC} \cos 60° = 0,$$

$$\Sigma F_y = 500 - 400 - T_{BC} \sin 60° = 0,$$

$$\Sigma M_{(\text{point } B)} = T_{AC} (2 \sin 60°) - (500)(2 \cos 60°) = 0.$$

Solving them, we obtain $T_{AC} = 289$ N, $T_{BC} = 115$ N, and $T_{BD} = -346$ N.

You can see how similar this method is to the method of joints. Both methods involve cutting members to obtain free-body diagrams of parts of a truss. In the method of joints, you move from joint to joint, drawing free-body diagrams of the joints and determining the axial forces in the members as you go. In the method of sections, you try to obtain a single free-body diagram that allows you to determine the axial forces in specific members. In our example, we obtained a free-body diagram by cutting three members, including the one (member *BC*) whose axial force we wanted to determine.

In contrast to the free-body diagrams of joints, the forces on the free-body diagrams used in the method of sections are not usually concurrent, and as in our example, you can obtain three independent equilibrium equations. Although there are exceptions, you must usually choose a section that requires cutting no more than three members, or you will have more unknown axial forces than equilibrium equations.

In the following examples we use the method of sections to determine the axial forces in particular members of trusses. This typically involves two steps:

1. Determine the reactions at the supports—*You usually need to draw the free-body diagram of the entire truss and determine the reactions at its supports. (This step is not always necessary; see Example 6.4.)*

2. Choose a section—*Obtain a free-body diagram of part of the truss by cutting members, including those whose axial forces you want to determine. Some trial and error and cleverness may be needed to find a free-body diagram that leads to the results you need.*

Example 6.4

The truss in Fig. 6.22 supports a 100-kN load. The horizontal members are each 1 m in length. Determine the axial force in member CJ, and state whether it is in tension or compression.

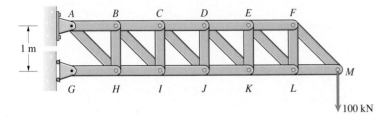

Figure 6.22

STRATEGY

We need to obtain a section by cutting members that include member CJ. By cutting members CD, CJ, and IJ, we will obtain a free-body diagram with three unknown axial forces.

SOLUTION

To obtain a section (Fig. a), we cut members CD, CJ, and IJ and draw the free-body diagram of the part of the truss on the right side of the cuts. From the equilibrium equation

$$\Sigma F_y = T_{CJ} \sin 45° - 100 = 0,$$

we obtain $T_{CJ} = 141.4$ kN. The axial force in member CJ is 141.4 kN (T).

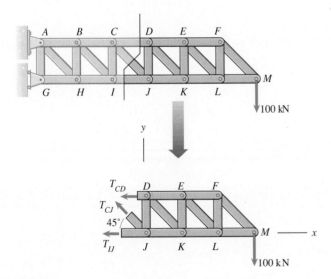

(a) Obtaining the section.

DISCUSSION

Notice that by using the section on the right side of the cuts, we did not need to determine the reactions at the supports A and G.

Example 6.5

Determine the axial forces in members DG and BE of the truss in Fig. 6.23.

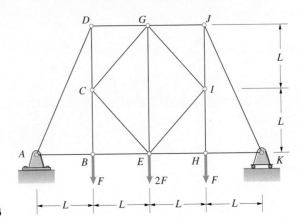

Figure 6.23

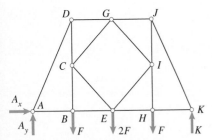

(a) Free-body diagram of the entire truss.

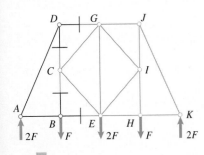

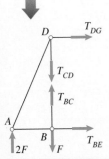

(b) A section of the truss obtained by passing planes through members DG, CD, BC, and BE.

STRATEGY

An appropriate choice of section is not obvious, and it isn't clear beforehand that we can determine the requested information by the method of sections. We can't obtain a section that involves cutting members DG and BE without cutting more than three members. However, cutting members DG, BE, CD, and BC results in a section with which we can determine the axial forces in members DG and BE even though the resulting free-body diagram is statically indeterminate.

SOLUTION

Determine the Reactions at the Supports   We draw the free-body diagram of the entire truss in Fig. (a). From the equilibrium equations,

$$\Sigma F_x = A_x = 0,$$
$$\Sigma F_y = A_y + K - F - 2F - F = 0,$$
$$\Sigma M_{\text{(point A)}} = -FL - 2F(2L) - F(3L) + K(4L) = 0,$$

we obtain the reactions $A_x = 0$, $A_y = 2F$, and $K = 2F$.

Choose a Section   In Fig. (b), we obtain a section by cutting members DG, CD, BC, and BE. Because the lines of action of T_{BE}, T_{BC}, and T_{CD} pass through point B, we can determine T_{DG} by summing moments about B:

$$\Sigma M_{\text{(point B)}} = -2FL - T_{DG}(2L) = 0.$$

The axial force $T_{DG} = -F$. Then from the equilibrium equation

$$\Sigma F_x = T_{DG} + T_{BE} = 0,$$

we see that $T_{BE} = -T_{DG} = F$. Member DG is in compression, and member BE is in tension.

Problems

6.34 In Example 6.4, what are the axial forces in members *CD* and *IJ*?

6.35 The truss supports loads at *C* and *E*. Use the method of sections to determine the axial forces in members *BE* and *CE*.

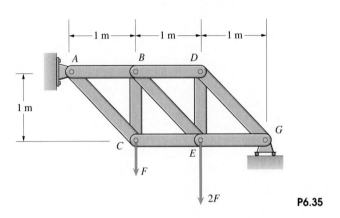

P6.35

6.36 For the truss in Problem 6.35, use the method of sections to determine the axial forces in members *BD* and *DE*.

6.37 The truss supports loads at *A* and *H*.
(a) Draw the free-body diagram of the entire truss, and determine the reactions at the supports *B* and *F*.
(b) Obtain a section by cutting members *CE*, *BE*, and *BD*, and draw its free-body diagram.
(c) Use the section you obtained in (b) to determine the axial forces in members *CE*, *BE*, and *BD*.

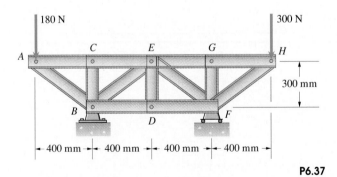

P6.37

6.38 Consider the truss in Problem 6.37. Use the method of sections to determine the axial forces in members *EG*, *EF*, and *DF*.

6.39 For the Howe and Pratt trusses, use the method of sections to determine the axial force in member *BC*.

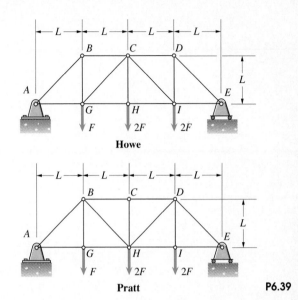

P6.39

6.40 For the Howe and Pratt trusses in Problem 6.39, determine the axial force in member *HI*.

6.41 The Pratt bridge truss supports five forces (*F* = 75 kip). The dimension *L* = 25 ft. Use the method of sections to determine the axial force in member *JK*.

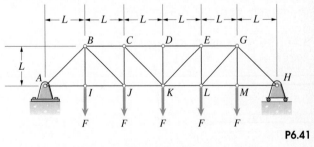

P6.41

6.42 Consider the bridge truss in Problem 6.41. The load *F* = 40 kip, and the length *L* = 30 ft. Use the method of sections to determine the axial force in member *EK*.

6.43 The Warren truss supports a walkway between two buildings. The walkway exerts vertical 10-kip loads at *B*, *D*, *F*, and *H*. Model the supports at *A* and *I* as roller supports.
(a) Use the method of joints to determine the axial force in member *BC*.

(b) Use the method of sections to determine the axial force in member *BC*.

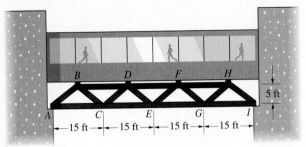

P6.43

6.44 Consider the truss described in Problem 6.43. Use the method of sections to determine the axial forces in members *CE*, *DE*, and *DF*.

6.45 The mass *m* = 120 kg. Use the method of sections to determine the axial forces in members *BD*, *CD*, and *CE*.

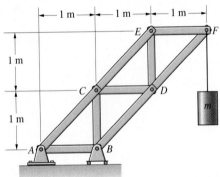

P6.45

6.46 For the truss in Problem 6.45, use the method of sections to determine the axial forces in members *AC*, *BC*, and *BD*.

6.47 The Howe truss helps support a roof. Model the supports at *A* and *G* as roller supports.
(a) Use the method of joints to determine the axial force in member *BI*.
(b) Use the method of sections to determine the axial force in member *BI*.

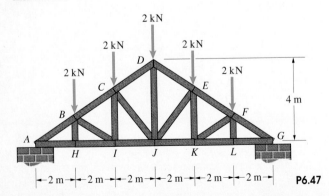

P6.47

6.48 Consider the truss in Problem 6.47. Use the method of sections to determine the axial force in member *EJ*.

6.49 Use the method of sections to determine the axial force in member *EF*.

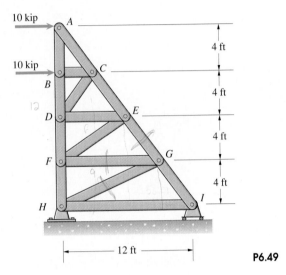

P6.49

6.50 Consider the truss in Problem 6.49. Use the method of sections to determine the axial force in member *FG*.

6.51 The load *F* = 20 kN and the dimension *L* = 2 m. Use the method of sections to determine the axial force in member *HK*.
   Strategy: Obtain a section by cutting members *HK*, *HI*, *IJ*, and *JM*. You can determine the axial forces in members *HK* and *JM* even though the resulting free-body diagram is statically indeterminate.

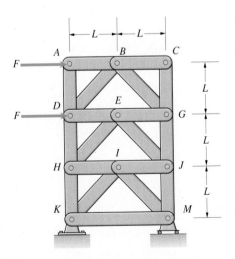

P6.51

6.52 The weight of the bucket is $W = 1000$ lb. The cable passes over pulleys at A and D.
(a) Determine the axial forces in members FG and HI.
(b) By drawing free-body diagrams of sections, explain why the axial forces in members FG and HI are equal.

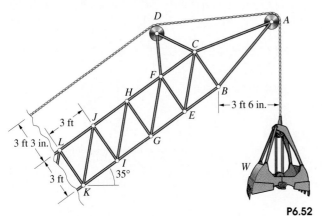

P6.52

6.53 Consider the truss in Problem 6.52. The weight of the bucket is $W = 1000$ lb. The cable passes over pulleys at A and D. Determine the axial forces in members IK and JL.

6.54 The truss supports loads at N, P, and R. Determine the axial forces in members IL and KM.

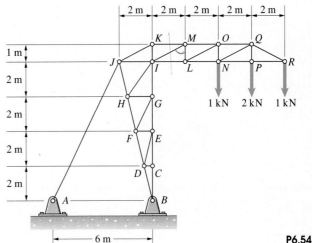

P6.54

6.55 Consider the truss in Problem 6.54. Determine the axial forces in members HJ and GI.

6.56 Consider the truss in Problem 6.54. By drawing free-body diagrams of sections, explain why the axial forces in members DE, FG, and HI are zero.

6.4 *Space Trusses*

We can form a simple three-dimensional structure by connecting six bars at their ends to obtain a *tetrahedron*, as shown in Fig. 6.24(a). By adding members, we can obtain more elaborate structures (Figs. 6.24b and c). Three-dimensional structures such as these are called **space trusses** if they have joints that do not exert couples on the members (that is, the joints behave like ball and socket supports) and they are loaded and supported at their joints. Space trusses are analyzed by the same methods we described for two-dimensional trusses. The only difference is the need to cope with the more complicated geometry.

(a)

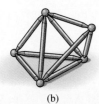
(b)

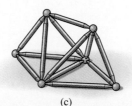
(c)

Figure 6.24
Space trusses with 6, 9, and 12 members.

Let's consider the space truss in Fig. 6.25(a). Suppose that the load $\mathbf{F} = -2\mathbf{i} - 6\mathbf{j} - \mathbf{k}$ (kN). The joints A, B, and C rest on the smooth floor. Joint A is supported by the corner where the smooth walls meet, and joint C rests against the back wall. We can apply the method of joints to this truss.

First we must determine the reactions exerted by the supports (the floor and walls). We draw the free-body diagram of the entire truss in Fig. 6.25(b). The corner can exert three components of force at A, the floor and wall can exert two components of force at C, and the floor can exert a normal force at B. Summing moments about A, the equilibrium equations are

$$\Sigma F_x = A_x - 2 = 0,$$

$$\Sigma F_y = A_y + B_y + C_y - 6 = 0,$$

$$\Sigma F_z = A_z + C_z - 1 = 0,$$

$$\Sigma M_{(\text{point } A)} = (\mathbf{r}_{AB} \times B_y\mathbf{j}) + [\mathbf{r}_{AC} \times (C_y\mathbf{j} + C_z\mathbf{k})] + (\mathbf{r}_{AD} \times \mathbf{F})$$

$$= \begin{vmatrix} \mathbf{i} & \mathbf{j} & \mathbf{k} \\ 2 & 0 & 3 \\ 0 & B_y & 0 \end{vmatrix} + \begin{vmatrix} \mathbf{i} & \mathbf{j} & \mathbf{k} \\ 4 & 0 & 0 \\ 0 & C_y & C_z \end{vmatrix}$$

$$+ \begin{vmatrix} \mathbf{i} & \mathbf{j} & \mathbf{k} \\ 2 & 3 & 1 \\ -2 & -6 & -1 \end{vmatrix}$$

$$= (-3B_y + 3)\,\mathbf{i} + (-4C_z)\,\mathbf{j}$$

$$+ (2B_y + 4C_y - 6)\,\mathbf{k} = 0.$$

Solving these equations, we obtain the reactions $A_x = 2$ kN, $A_y = 4$ kN, $A_z = 1$ kN, $B_y = 1$ kN, $C_y = 1$ kN, and $C_z = 0$.

In this example, we can determine the axial forces in members AC, BC, and CD from the free-body diagram of joint C (Fig. 6.25c). To write the equilibrium equations for the joint, we must express the three axial forces in terms of their components. Because member AC lies along the x axis, we express the force exerted on joint C by the axial force T_{AC} as the vector $-T_{AC}\mathbf{i}$. Let $\mathbf{r}_{CB}$ be the position vector from C to B:

$$\mathbf{r}_{CB} = (2 - 4)\,\mathbf{i} + (0 - 0)\,\mathbf{j} + (3 - 0)\,\mathbf{k} = -2\mathbf{i} + 3\mathbf{k} \text{ (m)}.$$

Dividing this vector by its magnitude to obtain a unit vector that points from C toward B,

$$\mathbf{e}_{CB} = \frac{\mathbf{r}_{CB}}{|\mathbf{r}_{CB}|} = -0.555\mathbf{i} + 0.832\mathbf{k},$$

we express the force exerted on joint C by the axial force T_{BC} as the vector

$$T_{BC}\mathbf{e}_{CB} = T_{BC}(-0.555\mathbf{i} + 0.832\mathbf{k}).$$

In the same way, we express the force exerted on joint C by the axial force T_{CD} as the vector

$$T_{CD}(-0.535\mathbf{i} + 0.802\mathbf{j} + 0.267\mathbf{k}).$$

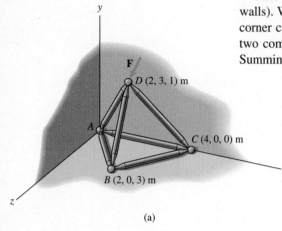

(a)

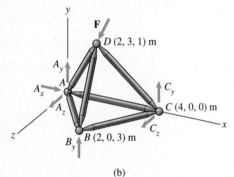

(b)

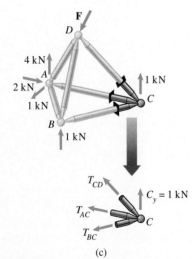

(c)

Figure 6.25

(a) A space truss supporting a load $\mathbf{F}$.
(b) Free-body diagram of the entire truss.
(c) Obtaining the free-body diagram of joint C.

Setting the sum of the forces on the joint equal to zero,

$$-T_{AC}\mathbf{i} + T_{BC}(-0.555\mathbf{i} + 0.832\mathbf{k})$$
$$+ T_{CD}(-0.535\mathbf{i} + 0.802\mathbf{j} + 0.267\mathbf{k}) + (1)\mathbf{j} = 0,$$

we obtain the three equilibrium equations

$$\Sigma F_x = -T_{AC} - 0.555T_{BC} - 0.535T_{CD} = 0,$$
$$\Sigma F_y = 0.802T_{CD} + 1 = 0,$$
$$\Sigma F_z = 0.832T_{BC} + 0.267T_{CD} = 0.$$

Solving these equations, the axial forces are $T_{AC} = 0.444$ kN, $T_{BC} = 0.401$ kN, and $T_{CD} = -1.247$ kN. Members AC and BC are in tension, and member CD is in compression. By continuing to draw free-body diagrams of the joints, we can determine the axial forces in all the members.

As our example demonstrates, you can obtain three equilibrium equations from the free-body diagram of a joint in three dimensions, so you should

Problems

choose joints to analyze that are subjected to known forces and no more than three unknown forces.

6.57 The suspended weight $W = 600$ lb. The bars AB and AC have ball and socket joints at each end. Determine the tension in the cable AD and the axial forces in members AB and AC.

Strategy: Draw the free-body diagram of joint A.

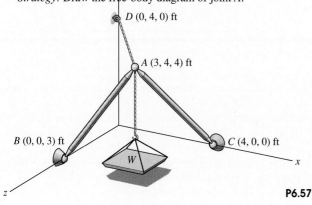

P6.57

6.58 The space truss supports a vertical 10-kN load at D. The reactions at the supports at joints A, B, and C are shown. What are the axial forces in members AD, BD, and CD?

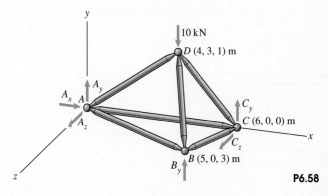

P6.58

6.59 Consider the space truss in Problem 6.58. The reactions at the supports at joints A, B, and C are shown. What are the axial forces in members AB, AC, and AD?

6.60 Consider the space truss in Problem 6.58. The reactions at the supports at joints A, B, and C are shown. What are the axial forces in members AB, BC, and BD?

6.61 The space truss supports a vertical load F at A. Each member is of length L, and the truss has roller supports at B, C, and D. What are the axial forces in the members?

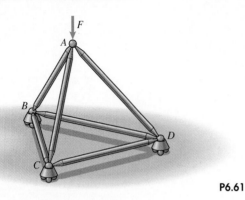

P6.61

6.62 The space truss has roller supports at B, C, and D and supports a vertical 800-lb load at A. What are the axial forces in members AB, AC, and AD?

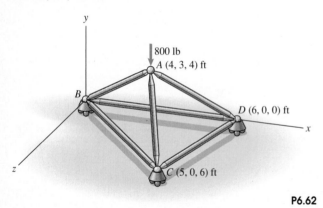

P6.62

6.63 The space truss shown models an airplane's landing gear. It has ball and socket supports at C, D, and E. If the force exerted at A by the wheel is $\mathbf{F} = 40\mathbf{j}$ (kN), what are the axial forces in members AB, AC, and AD?

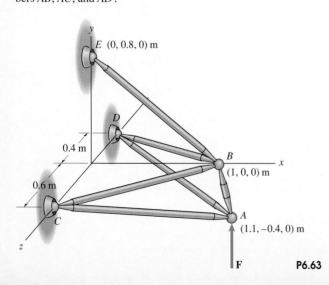

P6.63

6.64 If the force exerted at point A of the truss in Problem 6.63 is $\mathbf{F} = 10\mathbf{i} + 60\mathbf{j} + 20\mathbf{k}$ (kN), what are the axial forces in members BC, BD, and BE?

6.65 The space truss is supported by roller supports on the horizontal surface at C and D and a ball and socket support at E. The y axis points upward. The mass of the suspended object is 120 kg. The coordinates of the joints of the truss are A: (1.6, 0.4, 0) m, B: (1.0, 1.0, −0.2) m, C: (0.9, 0, 0.9) m, D: (0.9, 0, −0.6) m, and E: (0, 0.8, 0) m. Determine the axial forces in members AB, AC, and AD.

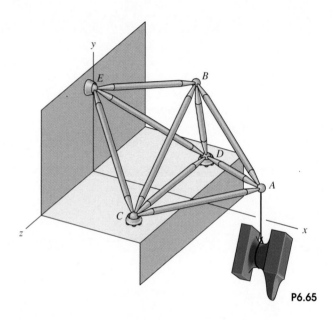

P6.65

6.66 (a) Draw the free-body diagram of the entire space truss in Problem 6.65, and determine the reactions at C, D, and E.
(b) Determine the axial forces in members BE, CE, and DE.

6.67 For the space truss in Problem 6.65, determine the axial forces in members BC, BD, and CD.

6.68 The mirror housing of the telescope is supported by a 6-bar space truss. The mass of the housing is 3 Mg (megagrams), and its weight acts at G. The distance from the axis of the telescope to points A, B, and C is 1 m, and the distance from the axis to points D, E, and F is 2.5 m. If the telescope axis is vertical ($\alpha = 90°$), what are the axial forces in the members of the truss?

6.69 Consider the telescope described in Problem 6.68. Determine the axial forces in the members of the truss if the angle α between the horizontal and the telescope axis is 20°.

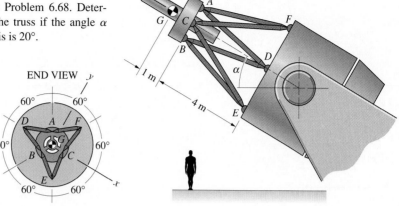

END VIEW

P6.68

6.5 Frames and Machines

Many structures, such as the frame of a car and the human structure of bones, tendons, and muscles (Fig. 6.26), are not composed entirely of two-force members and thus cannot be modeled as trusses. In this section we consider structures of interconnected members that do not satisfy the definition of a truss. Such structures are called **frames** if they are designed to remain stationary and support loads and **machines** if they are designed to move and apply loads.

When trusses are analyzed by cutting members to obtain free-body diagrams of joints or sections, the internal forces acting at the "cuts" are simple axial forces (see Fig. 6.4). This is not generally true for frames or machines, and a different method of analysis is necessary. Instead of cutting members, you isolate entire members, or in some cases combinations of members, from the structure.

To begin analyzing a frame or machine, you draw a free-body diagram of the entire structure (that is, treat the structure as a single object) and determine the reactions at its supports. In some cases you will find that the entire structure is statically indeterminate, but you should determine as many of the reactions as possible. You then draw free-body diagrams of individual members,

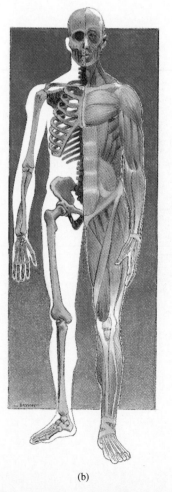

Figure 6.26

A car's frame (a) and the internal structure of a person (b) are not trusses.

(a)

(b)

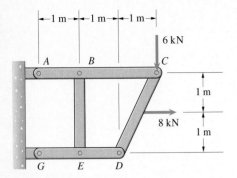

Figure 6.27

A frame supporting two loads.

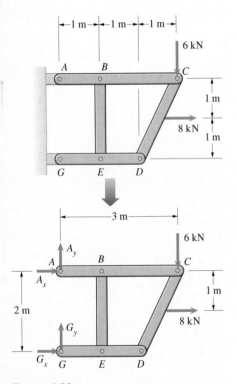

Figure 6.28

Obtaining the free-body diagram of the entire frame.

or selected combinations of members, and apply the equilibrium equations to determine the forces and couples acting on them. For example, let's consider the stationary structure in Fig. 6.27. Member *BE* is a two-force member, but the other three members—*ABC*, *CD*, and *DEG*—are not. This structure is a frame. Our objective is to determine the forces on its members.

Analyzing the Entire Structure

We draw the free-body diagram of the entire frame in Fig. 6.28. It is statically indeterminate: There are four unknown reactions, A_x, A_y, G_x, and G_y, whereas we can write only three independent equilibrium equations. However, notice that the lines of action of three of the unknown reactions intersect at *A*. By summing moments about *A*,

$$\Sigma M_{(\text{point } A)} = 2G_x + (1)(8) - (3)(6) = 0,$$

we obtain the reaction $G_x = 5$ kN. Then from the equilibrium equation

$$\Sigma F_x = A_x + G_x + 8 = 0,$$

we obtain the reaction $A_x = -13$ kN. Although we cannot determine A_y or G_y from the free-body diagram of the entire structure, we can do so by analyzing the individual members.

Analyzing the Members

Our next step is to draw free-body diagrams of the members. To do so, we treat the attachment of a member to another member just as if it were a support. Looked at in this way, you can think of each member as a supported object of the kind you analyzed in Chapter 5. Furthermore, the forces and couples the members exert on one another are *equal in magnitude and opposite in direction*. A simple demonstration is instructive. If you clasp your hands as shown in Fig. 6.29(a) and exert a force on your left hand with your right hand, your left hand exerts an equal and opposite force on your right hand (Fig. 6.29b). Similarly, if you exert a couple on your left hand, your left hand exerts an equal and opposite couple on your right hand.

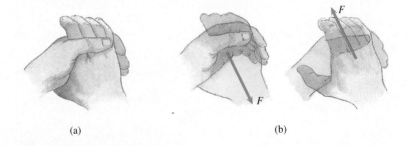

(a)   (b)

Figure 6.29

Demonstrating Newton's third law:
(a) Clasp your hands and pull on your left hand.
(b) Your hands exert equal and opposite forces.

In Fig. 6.30 we "disassemble" the frame and draw free-body diagrams of its members. Observe that the forces exerted on one another by the members are equal and opposite. For example, at point C on the free-body diagram of member ABC, the force exerted by member CD is denoted by the components C_x and C_y. We can choose the directions of these unknown forces arbitrarily, but once we have done so, the forces exerted by member ABC on member CD at point C must be equal and opposite, as shown.

We need to discuss two important aspects of these free-body diagrams before completing the analysis.

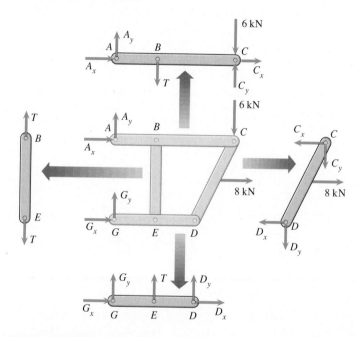

Figure 6.30

Obtaining the free-body diagrams of the members.

Two-Force Members   Member BE is a two-force member, and we have taken this into account in drawing its free-body diagram in Fig. 6.30. The force T is the axial force in member BE, and an equal and opposite force is subjected on member ABC at B and on member GED at E.

Recognizing two-force members in frames and machines and drawing their free-body diagrams as we have done will reduce the number of unknowns you must determine and will greatly simplify your analysis. In our example, if we did not treat member BE as a two-force member, its free-body diagram would have four unknown forces (Fig. 6.31a). By treating it as a two-force member (Fig. 6.31b), we reduce the number of unknown forces by three.

Loads Applied at Joints   A question arises when a load is applied at a joint: Where do you put the load when you draw the free-body diagrams of the individual members? The answer is that you can place the load on *any one* of the members attached at the joint. For example, in Fig. 6.27, the 6-kN load acts at the joint where members ABC and CD are connected. In drawing the free-body diagrams of the individual members (Fig. 6.30), we assumed that the 6-kN load acted on member ABC. The force components C_x and C_y on the free-body diagram of member ABC are the forces exerted by the member CD.

To explain why we can draw the free-body diagrams in this way, let us assume that the 6-kN force acts on the pin connecting members ABC and CD, and draw separate free-body diagrams of the pin and the two members

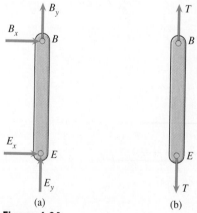

Figure 6.31

Free-body diagram of member BE:
(a) Not treating it as a two-force member.
(b) Treating it as a two-force member.

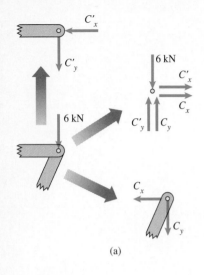

(a)

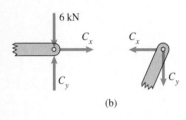

(b)

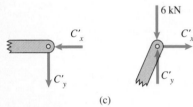

(c)

Figure 6.32

(a) Drawing free-body diagrams of the pin and the two members.
(b) Superimposing the pin on member ABC.
(c) Superimposing the pin on member CD.

(Fig. 6.32a). The force components C'_x and C'_y are the forces exerted by the pin on member ABC, and C_x and C_y are the forces exerted by the pin on member CD. If we superimpose the free-body diagrams of the pin and member ABC, we obtain the two free-body diagrams in Fig. 6.32(b), which is the way we drew them in Fig. 6.30. Alternatively, by superimposing the free-body diagrams of the pin and member CD, we obtain the two free-body diagrams in Fig. 6.32(c).

Thus if a load acts at a joint, you can place it on any one of the members attached at the joint when you draw the free-body diagrams of the individual members. Just make sure you don't place it on more than one member.

To detect errors in your free-body diagrams of the members, you should "reassemble" them (Fig. 6.33a). The forces at the connections between the members cancel (they are internal forces once the members are reassembled), and you should recover the free-body diagram of the entire structure (Fig. 6.33b).

Our final step is to apply the equilibrium equations to the free-body diagrams of the members (Fig. 6.34). In two dimensions, we can obtain three independent equilibrium equations from the free-body diagram of each mem-

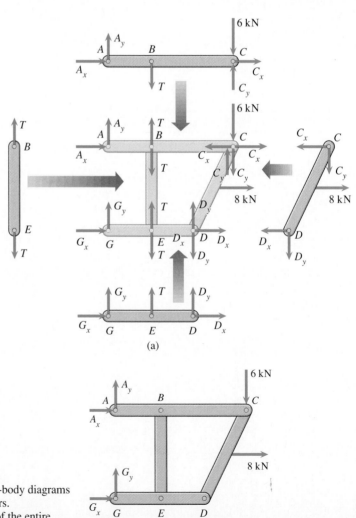

(a)

Figure 6.33

(a) "Reassembling" the free-body diagrams of the individual members.
(b) The free-body diagram of the entire frame is recovered.

(b)

ber of a structure that we do not treat as a two-force member. (By assuming that the forces on a two-force member are equal and opposite axial forces, we have already used the three equilibrium equations for that member.) In this example, there are three members in addition to the two-force member, so we can write $(3)(3) = 9$ independent equilibrium equations, and there are 9 unknown forces: A_x, A_y, C_x, C_y, D_x, D_y, G_x, G_y, and T.

Recall that we determined that $A_x = -13$ kN and $G_x = 5$ kN from our analysis of the entire structure. The equilibrium equations we obtained from the free-body diagram of the entire structure are not independent of the equilibrium equations obtained from the free-body diagrams of the members, but by using them to determine A_x and G_x, we get a head start on solving the equations for the members. Consider the free-body diagram of member ABC (Fig. 6.34a). Because we know A_x, we can determine C_x from the equation

$$\Sigma F_x = A_x + C_x = 0,$$

obtaining $C_x = -A_x = 13$ kN. Now consider the free-body diagram of GED (Fig. 6.34b). We can determine D_x from the equation

$$\Sigma F_x = G_x + D_x = 0,$$

obtaining $D_x = -G_x = -5$ kN. Now consider the free-body diagram of member CD (Fig. 6.34c). Because we know C_x, we can determine C_y by summing moments about D:

$$\Sigma M_{(\text{point } D)} = (2)C_x - (1)C_y - (1)(8) = 0.$$

We obtain $C_y = 18$ kN. Then from the equation

$$\Sigma F_y = -C_y - D_y = 0,$$

we find that $D_y = -C_y = -18$ kN. Now we can return to the free-body diagrams of members ABC and GED to determine A_y and G_y. Summing moments about point B of member ABC,

$$\Sigma M_{(\text{point } B)} = -(1)A_y + (2)C_y - (2)(6) = 0,$$

we obtain $A_y = 2C_y - 12 = 24$ kN. Then by summing moments about point E of member GED,

$$\Sigma M_{(\text{point } E)} = (1)D_y - (1)G_y = 0,$$

we obtain $G_y = D_y = -18$ kN. Finally, from the free-body diagram of member GED, we use the equilibrium equation

$$\Sigma F_y = D_y + G_y + T = 0,$$

which gives us the result $T = -D_y - G_y = 36$ kN. The forces on the members are shown in Fig. 6.35. You can often identify errors by presenting the forces in this way and confirming that each member is in equilibrium.

As this example demonstrates, you can often simplify your determination of the forces on the members by carefully choosing the order in which you solve the equilibrium equations.

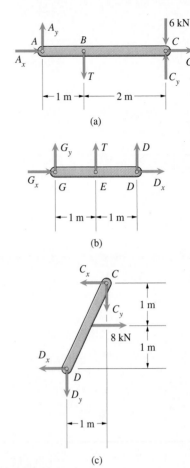

(a)

(b)

(c)

Figure 6.34

Free-body diagrams of the members.

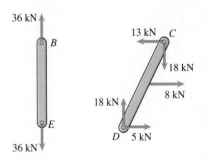

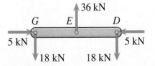

Figure 6.35

Forces on the members of the frame.

In the following examples we determine forces and couples on the members of frames and machines. This typically requires two steps:

1. Determine the reactions at the supports—*Draw the free-body diagram of the entire structure, and determine the reactions at its supports. This step can greatly simplify your analysis of the members. If the free-body diagram is statically indeterminate, determine as many of the reactions as possible.*

2. Analyze the members—*Draw free-body diagrams of the members, and apply the equilibrium equations to determine the forces acting on them. You can simplify this step by identifying two-force members. If a load acts at a joint of the structure, you can place the load on the free-body diagram of any one of the members attached at that joint.*

Example 6.6

The frame in Fig. 6.36 is subjected to a 200-N-m couple. Determine the forces and couples on its members.

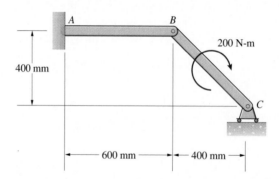

Figure 6.36

SOLUTION

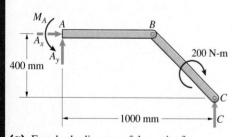

(a) Free-body diagram of the entire frame.

Determine the Reactions at the Supports We draw the free-body diagram of the entire frame in Fig. (a). The term M_A is the couple exerted by the built-in support. From the equilibrium equations

$$\Sigma F_x = A_x = 0,$$

$$\Sigma F_y = A_y + C = 0,$$

$$\Sigma M_{(\text{point } A)} = M_A - 200 + (1)C = 0,$$

we obtain the reaction $A_x = 0$. We can't determine A_y, M_A, or C from this free-body diagram.

Analyze the Members We "disassemble" the frame to obtain the free-body diagrams of the members in Fig. (b). The equilibrium equations for member BC are

$$\Sigma F_x = -B_x = 0,$$

$$\Sigma F_y = -B_y + C = 0,$$

$$\Sigma M_{(\text{point } B)} = -200 + (0.4)C = 0.$$

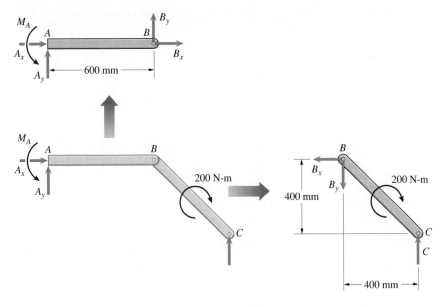

(b) Obtaining the free-body diagrams of the members.

Solving these equations, we obtain $B_x = 0$, $B_y = 500$ N, and $C = 500$ N. The equilibrium equations for member AB are

$$\Sigma F_x = A_x + B_x = 0,$$

$$\Sigma F_y = A_y + B_y = 0,$$

$$\Sigma M_{(\text{point } A)} = M_A + (0.6)B_y = 0.$$

Because we already know A_x, B_x, and B_y, we can solve these equations for A_y and M_A. The results are $A_y = -500$ N and $M_A = -300$ N-m. This completes the solution (Fig. c).

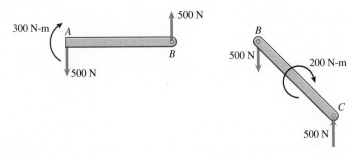

(c) Forces and couples on the members.

DISCUSSION

We were able to solve the equilibrium equations for member BC without having to consider the free-body diagram of member AB. We were then able to solve the equilibrium equations for member AB. By choosing the members with the fewest unknowns to analyze first, you will often be able to solve them sequentially, but in some cases you will have to solve the equilibrium equations for the members simultaneously.

Even though we were unable to determine the four reactions A_x, A_y, M_A, and C with the three equilibrium equations obtained from the free-body diagram of the entire frame, we were able to determine them from the free-body diagrams of the individual members. By drawing free-body diagrams of the members, we gained three equations because we obtained three equilibrium equations from each member but only two new unknowns, B_x and B_y.

Example 6.7

The frame in Fig. 6.37 supports a suspended weight $W = 40$ lb. Determine the forces on members $ABCD$ and CEG.

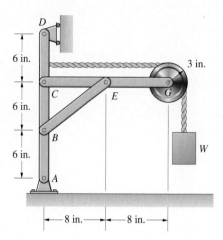

Figure 6.37

SOLUTION

Determine the Reactions at the Supports   We draw the free-body diagram of the entire frame in Fig. (a). From the equilibrium equations

$$\Sigma F_x = A_x - D = 0,$$

$$\Sigma F_y = A_y - 40 = 0,$$

$$\Sigma M_{(\text{point } A)} = (18)D - (19)(40) = 0,$$

we obtain the reactions $A_x = 42.2$ lb, $A_y = 40$ lb, and $D = 42.2$ lb.

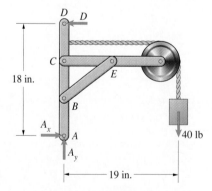

(a) Free-body diagram of the entire frame.

Analyze the Members   We obtain the free-body diagrams of the members in Fig. (b). Notice that BE is a two-force member. The angle $\alpha = \arctan(6/8) = 36.9°$.

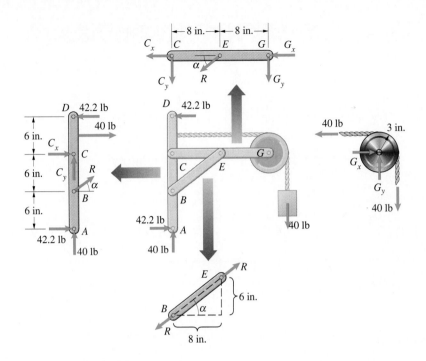

(b) Obtaining the free-body diagrams of the members.

The free-body diagram of the pulley has only two unknown forces. From the equilibrium equations

$$\Sigma F_x = G_x - 40 = 0,$$
$$\Sigma F_y = G_y - 40 = 0,$$

we obtain $G_x = 40$ lb and $G_y = 40$ lb. There are now only three unknown forces on the free-body diagram of member CEG. From the equilibrium equations

$$\Sigma F_x = -C_x - R \cos \alpha - 40 = 0,$$
$$\Sigma F_y = -C_y - R \sin \alpha - 40 = 0,$$
$$\Sigma M_{(\text{point } C)} = -(8)R \sin \alpha - (16)(40) = 0,$$

we obtain $C_x = 66.7$ lb, $C_y = 40$ lb, and $R = -133.3$ lb, completing the solution (Fig. c).

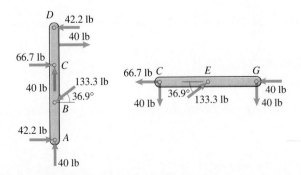

(c) Forces on members $ABCD$ and CEG.

Example 6.8

Determine the forces on the members of the frame in Fig. 6.38.

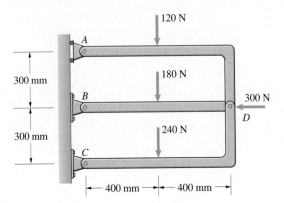

Figure 6.38

STRATEGY

You can confirm that no information can be obtained from the free-body diagram of the entire frame. To analyze the members, we must deal with an interesting challenge at joint D, where a load acts and three members are connected. We will obtain the free-body diagrams of the members by first isolating member AD, then separating members BD and CD.

SOLUTION

We first isolate member AD from the rest of the structure, introducing the reactions D_x and D_y (Fig. a). We then separate members BD and CD, introducing equal and opposite forces E_x and E_y (Fig. b). In this step we could have placed the 300-N load and the forces D_x and D_y on either free-body diagram.

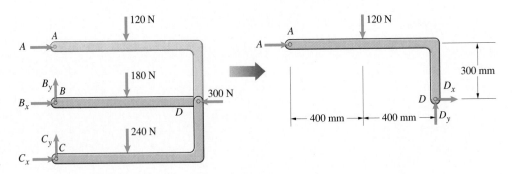

(a) Isolating member AD.

Only three unknown forces act on member AD. From the equilibrium equations

$$\Sigma F_x = A + D_x = 0,$$

$$\Sigma F_y = D_y - 120 = 0,$$

$$\Sigma M_{(\text{point } D)} = -(0.3)A + (0.4)(120) = 0,$$

we obtain $A = 160$ N, $D_x = -160$ N, and $D_y = 120$ N. Now we consider the free-body diagram of member BD. From the equation

$$\Sigma M_{(\text{point } D)} = -(0.8)B_y + (0.4)(180) = 0,$$

we obtain $B_y = 90$ N. Now we use the equation

$$\Sigma F_y = B_y - D_y + E_y - 180 = 90 - 120 + E_y - 180 = 0,$$

obtaining $E_y = 210$ N. Now that we know E_y, there are only three unknown forces on the free-body diagram of member CD. From the equilibrium equations

$$\Sigma F_x = C_x - E_x = 0,$$
$$\Sigma F_y = C_y - E_y - 240 = C_y - 210 - 240 = 0,$$
$$\Sigma M_{(\text{point } C)} = (0.3)E_x - (0.8)E_y - (0.4)(240)$$
$$= (0.3)E_x - (0.8)(210) - (0.4)(240) = 0,$$

we obtain $C_x = 880$ N, $C_y = 450$ N, and $E_x = 880$ N. Finally, we return to the free-body diagram of member BD and use the equation

$$\Sigma F_x = B_x + E_x - D_x - 300 = B_x + 880 + 160 - 300 = 0$$

to obtain $B_x = -740$ N, completing the solution (Fig. c).

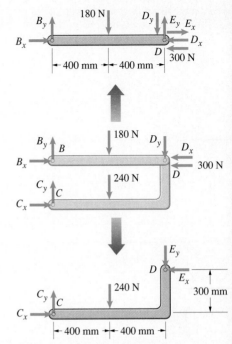

(b) Separating members BD and CD.

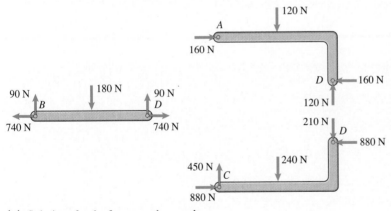

(c) Solutions for the forces on the members.

Example 6.9

The truck in Fig. 6.39 is parked on a $10°$ slope. Its brakes prevent the wheels at B from turning, but the wheels at C and the wheels of the trailer at A can turn freely. The trailer hitch at D behaves like a pin support. Determine the forces exerted on the truck at B, C, and D.

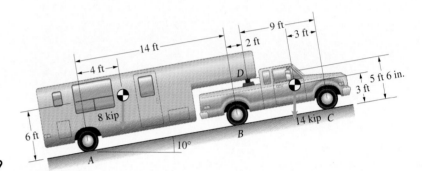

Figure 6.39

STRATEGY

We can treat this example as a structure whose "members" are the truck and trailer. We must isolate the truck and trailer and draw their individual free-body diagrams to determine the forces acting on the truck.

SOLUTION

Determine the Reactions at the Supports The reactions in this example are the forces exerted on the truck and trailer by the road. We draw the free-body diagram of the connected truck and trailer in Fig. (a). Because the tires at B are locked, the road can exert both a normal force and a friction force, but only normal forces are exerted at A and C. The equilibrium equations are

$$\Sigma F_x = B_x - 8 \sin 10° - 14 \sin 10° = 0,$$

$$\Sigma F_y = A + B_y + C - 8 \cos 10° - 14 \cos 10° = 0,$$

$$\Sigma M_{(point\ A)} = 14B_y + 25C + (6)(8 \sin 10°)$$
$$- (4)(8 \cos 10°) + (3)(14 \sin 10°)$$
$$- (22)(14 \cos 10°) = 0.$$

From the first equation we obtain the reaction $B_x = 3.82$ kip, but we can't solve the other two equations for the three reactions A, B_y, and C.

Analyze the Members We draw the free-body diagrams of the trailer and truck in Figs. (b) and (c), showing the forces D_x and D_y exerted at the hitch. Only three unknown forces appear on the free-body diagram of the trailer. From the equilibrium equations for the trailer,

$$\Sigma F_x = D_x - 8 \sin 10° = 0,$$

$$\Sigma F_y = A + D_y - 8 \cos 10° = 0,$$

$$\Sigma M_{(point\ D)} = (0.5)(8 \sin 10°) + (12)(8 \cos 10°) - 16A = 0,$$

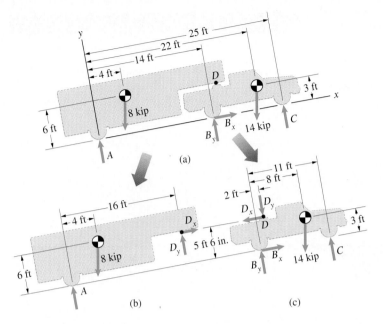

(a)

(a) Free-body diagram of the combined truck and trailer.

(b)

(c)

(b), (c) The individual free-body diagrams.

we obtain $A = 5.95$ kip, $D_x = 1.39$ kip, and $D_y = 1.93$ kip. (Notice that by summing moments about D, we obtained an equation containing only one unknown force.)

The equilibrium equations for the truck are

$$\Sigma F_x = B_x - D_x - 14 \sin 10° = 0,$$

$$\Sigma F_y = B_y + C - D_y - 14 \cos 10° = 0,$$

$$\Sigma M_{(point\ B)} = 11C + 5.5D_x - 2D_y + (3)(14 \sin 10°)$$
$$- (8)(14 \cos 10°) = 0.$$

Using the known values of D_x and D_y, we can solve these equations, obtaining $B_x = 3.82$ kip, $B_y = 6.69$ kip, and $C = 9.02$ kip.

DISCUSSION

We were unable to solve two of the equilibrium equations for the connected truck and trailer. When that happens, you can use the equilibrium equations for the entire structure to check your results:

$$\Sigma F_x = B_x - 8 \sin 10° - 14 \sin 10°$$
$$= 3.82 - 8 \sin 10° - 14 \sin 10° = 0,$$

$$\Sigma F_y = A + B_y + C - 8 \cos 10° - 14 \cos 10°$$
$$= 5.95 + 6.69 + 9.02 - 8 \cos 10° - 14 \cos 10° = 0,$$

$$\Sigma M_{(point\ A)} = 14B_y + 25C + (6)(8 \sin 10°)$$
$$- (4)(8 \cos 10°) + (3)(14 \sin 10°) - (22)(14 \cos 10°)$$
$$= (14)(6.69) + (25)(9.02) + (6)(8 \sin 10°)$$
$$- (4)(8 \cos 10°) + (3)(14 \sin 10°)$$
$$- (22)(14 \cos 10°) = 0.$$

Example 6.10

What forces are exerted on the bolt at E in Fig. 6.40 as a result of the 150-N forces on the pliers?

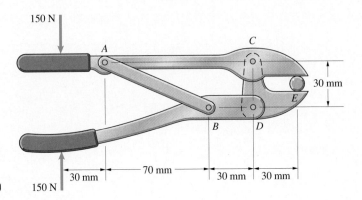

Figure 6.40

STRATEGY

A pair of pliers is a simple example of a machine, a structure designed to move and exert forces. The interconnections of the members are designed to create a mechanical advantage, subjecting an object to forces greater than the forces exerted by the user.

In this case there is no information to be gained from the free-body diagram of the entire structure. We must determine the forces exerted on the bolt by drawing free-body diagrams of the members.

SOLUTION

We "disassemble" the pliers in Fig. (a) to obtain the free-body diagrams of the members, labeled (1), (2), and (3). The force R on free-body diagrams (1) and (3) is exerted by the two-force member AB. The angle $\alpha = \arctan(30/70) = 23.2°$. Our objective is to determine the force E exerted by the bolt.

The free-body diagram of member (3) has only three unknown forces and the 150-N load, so we can determine R, D_x, and D_y from this free-body diagram alone. The equilibrium equations are

$$\Sigma F_x = D_x + R \cos \alpha = 0,$$

$$\Sigma F_y = D_y - R \sin \alpha + 150 = 0,$$

$$\Sigma M_{(\text{point } B)} = 30D_y - (100)(150) = 0.$$

Solving these equations, we obtain $D_x = -1517$ N, $D_y = 500$ N, and $R = 1650$ N. Knowing D_x, we can determine E from the free-body diagram of member (2) by summing moments about C,

$$\Sigma M_{(\text{point } C)} = -30E - 30D_x = 0.$$

The force exerted on the bolt by the pliers is $E = -D_x = 1517$ N. The mechanical advantage of the pliers is $(1517 \text{ N})/(150 \text{ N}) = 10.1$.

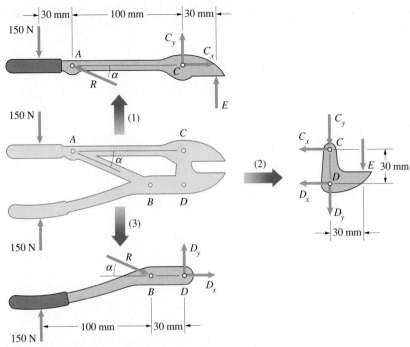

(a) Obtaining the free-body diagrams of the members.

DISCUSSION

Notice that we did not need to use the free-body diagram of member (1) to determine E. When this happens, you can use the "leftover" free-body diagram to check your work. Using our results for R and E, we can confirm that the sum of the moments about point C of member (1) is zero:

$$\Sigma M_{(\text{point } C)} = (130)(150) - 100R \sin \alpha + 30E$$
$$= (130)(150) - (100)(1650) \sin 23.2° + (30)(1517) = 0.$$

Problems

6.70 Draw the free-body diagram of (a) the entire frame; (b) the individual members AB and BC. (c) Determine the forces and couples on member AB.

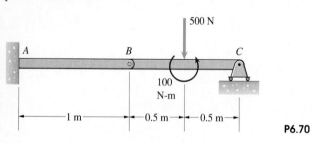

P6.70

6.71 (a) Determine the forces and couples on member AB for cases (1) and (2).
(b) You know that the moment of a couple is the same about any point. Explain why the answers are not the same in cases (1) and (2).

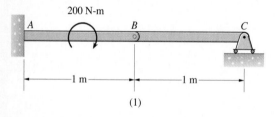

(1)

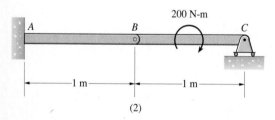

(2)   P6.71

6.72 For the frame shown, determine the reactions at the built-in support A and the forces exerted on member AB at B.

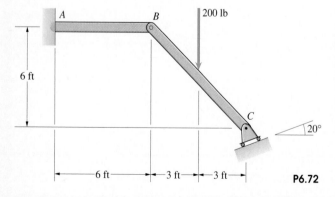

P6.72

6.73 The force $F = 10$ kN. Determine the forces on member ABC, presenting your answers as shown in Fig. 6.35.

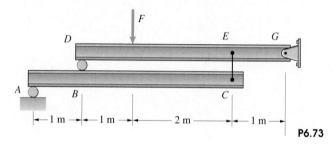

P6.73

6.74 Consider the frame in Problem 6.73. The cable CE will safely support a tension of 10 kN. Based on this criterion, what is the largest downward force F that can be applied to the frame?

6.75 The hydraulic actuator BD exerts a 6-kN force on member ABC. The force is parallel to BD, and the actuator is in compression. Determine the forces on member ABC, presenting your answers as shown in Fig. 6.35.

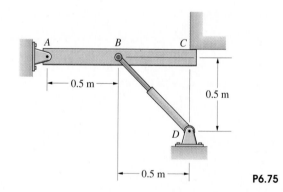

P6.75

6.76 The simple hydraulic jack shown in Problem 6.75 is designed to exert a vertical force at point C. The hydraulic actuator BD exerts a force on the beam ABC that is parallel to BD. The largest lifting force the jack can exert is limited by the pin support A, which will safely support a force of magnitude 20 kN. What is the largest lifting force the jack can exert at C, and what is the resulting axial force in the hydraulic actuator?

6.77 Determine the forces on member BC and the axial force in member AC.

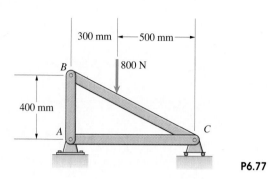

300 mm — 500 mm

800 N

B

400 mm

A

C

P6.77

6.78 A football player works out with a squat thrust machine. To rotate the bar ABD, he must exert a vertical force at A such that the axial force in the two-force member BC is 400 lb. When bar ABD is on the verge of rotating, what are the reactions on member CDE?

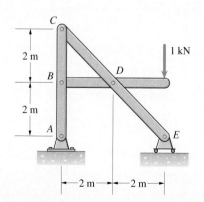

2 ft — 2 ft

C

1 ft 5 in.

A B

D

5 ft 5 in.

E

P6.78

6.79 The frame supports a 1-kN load. Determine the forces on member CDE.

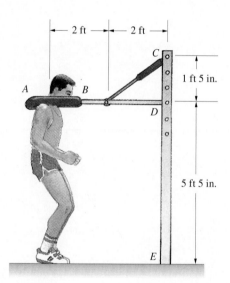

C

2 m

B D

1 kN

2 m

A E

2 m — 2 m

P6.79

6.80 The mass $m = 120$ kg. Determine the forces on member ABC, presenting your answers as shown in Fig. 6.35.

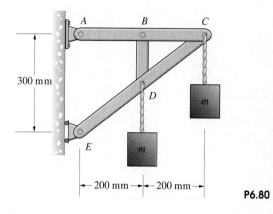

A B C

300 mm

D m

E

m

200 mm — 200 mm

P6.80

6.81 The tension in cable BD is 500 lb. Determine the reactions at A for cases (1) and (2).

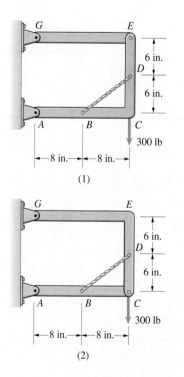

G E

6 in.

D

6 in.

A B C

300 lb

8 in. — 8 in.

(1)

G E

6 in.

D

6 in.

A B C

300 lb

8 in. — 8 in.

(2)   **P6.81**

6.82 Determine the forces on member *ABCD*, presenting your answers as shown in Fig. 6.35.

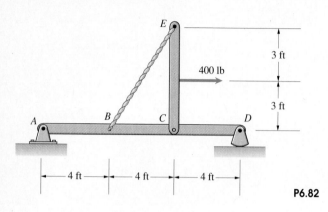

P6.82

6.83 The mass *m* = 50 kg. Determine the forces on member *ABCD*, presenting your answers as shown in Fig. 6.35.

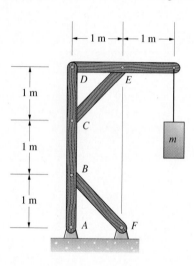

P6.83

6.84 Determine the forces on member *BCD*.

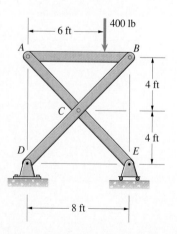

P6.84

6.85 Determine the forces on member *ABC*.

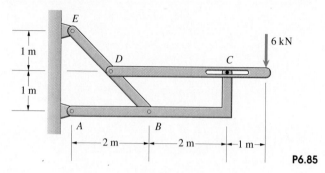

P6.85

6.86 Determine the forces on member *ABD*.

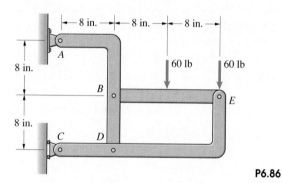

P6.86

6.87 The mass *m* = 12 kg. Determine the forces on member *CDE*.

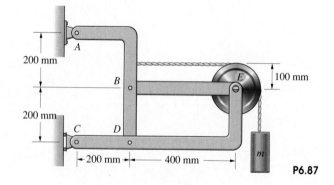

P6.87

6.88 The weight $W = 80$ lb. Determine the forces on member *ABCD*.

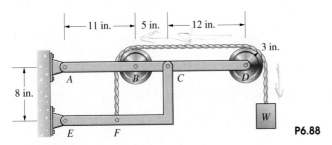

P6.88

6.89 The man using the exercise machine is holding the 80-lb weight stationary in the position shown. What are the reactions at the built-in support *E* and the pin support *F*? (*A* and *C* are pinned connections.)

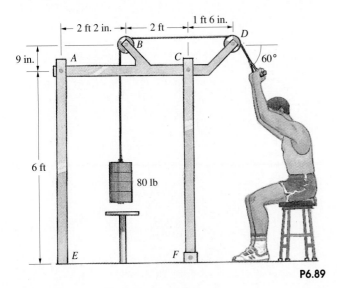

P6.89

6.90 The force $F = 800$ N. Determine the forces on member *ABC*.

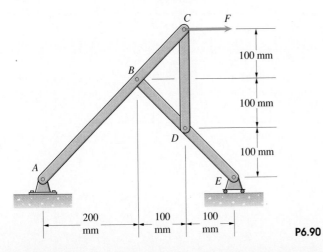

P6.90

6.91 Consider the frame in Problem 6.90. Member *CD* will safely support a compressive axial force of 2 kN. Based on this criterion, what is the maximum force *F* the frame will support?

6.92 The unstretched length of the spring is L_0. Show that when the system is in equilibrium the angle α satisfies the relation $\sin \alpha = 2(L_0 - 2F/k)/L$.

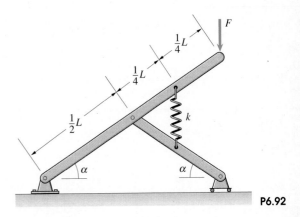

P6.92

6.93 The pin support *B* will safely support a force of 24-kN magnitude. Based on this criterion, what is the largest mass *m* that the frame will safely support?

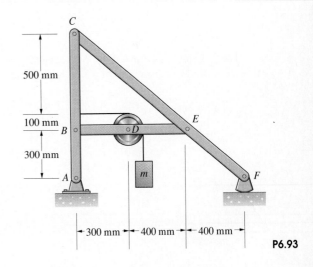

P6.93

MR ≅ 3

6.94 Determine the reactions at A and C.

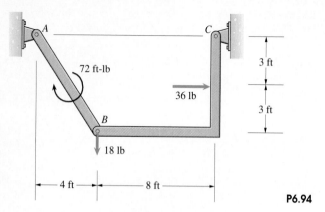

P6.94

6.95 Determine the forces on member AD.

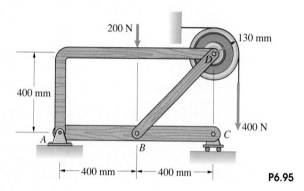

P6.95

6.96 The frame shown is used to support high-tension wires. If b = 3 ft, α = 30°, and W = 200 lb, what is the axial force in member HJ?

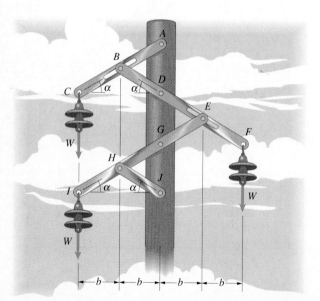

P6.96

6.97 What are the magnitudes of the forces exerted by the pliers on the bolt at A when 30-lb forces are applied as shown? (B is a pinned connection.)

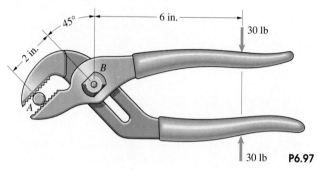

P6.97

6.98 The weight W = 60 kip. What is the magnitude of the force the members exert on each other at D?

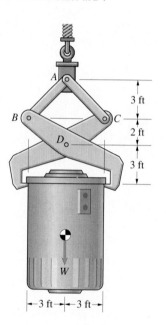

P6.98

6.99 Figure (a) is a diagram of the bones and biceps muscle of a person's arm supporting a mass. Tension in the biceps muscle holds the forearm in the horizontal position, as illustrated in the simple mechanical model in Fig. (b). The weight of the forearm is 9 N, and the mass m = 2 kg.

(a) Determine the tension in the biceps muscle *AB*.
(b) Determine the magnitude of the force exerted on the upper arm by the forearm at the elbow joint *C*.

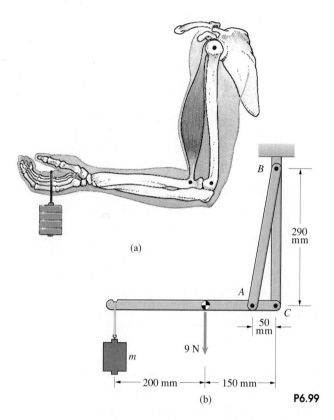

(a)

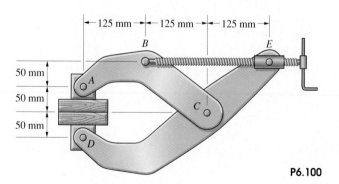

(b)

P6.99

6.100 The clamp presses two blocks of wood together. Determine the magnitude of the force the members exert on each other at *C* if the blocks are pressed together with a force of 200 N.

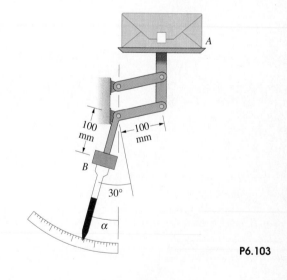

P6.100

6.101 The pressure force exerted on the piston is 2 kN toward the left. Determine the couple *M* necessary to keep the system in equilibrium.

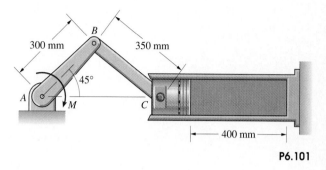

P6.101

6.102 In Problem 6.101, determine the forces on member *AB* at *A* and *B*.

6.103 This mechanism is used to weigh mail. A package placed at *A* causes the weighted pointer to rotate through an angle α. Neglect the weights of the members except for the counterweight at *B*, which has a mass of 4 kg. If $\alpha = 20°$, what is the mass of the package at *A*?

P6.103

6.104 The scoop C of the front-end loader is supported by two identical arms, one on each side of the loader. One of the two arms (ABC) is visible in the figure. It is supported by a pin support at A and the hydraulic actuator BD. The sum of the other loads exerted on the arm, including its own weight, is $F = 1.6$ kN. Determine the axial force in the actuator BD and the magnitude of the reaction at A.

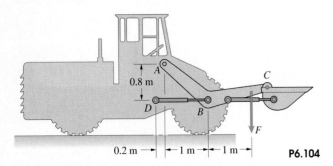

P6.104

6.105 The mass of the scoop is 220 kg, and its weight acts at G. Both the scoop and the hydraulic actuator BC are pinned to the horizontal member at B. The hydraulic actuator can be treated as a two-force member. Determine the forces exerted on the scoop at B and D.

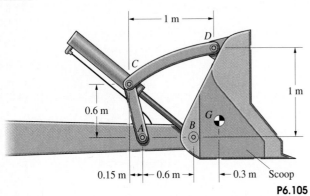

P6.105

6.106 In Problem 6.105, determine the axial force in the hydraulic actuator BC.

6.107 Determine the force exerted on the bolt by the bolt cutters.

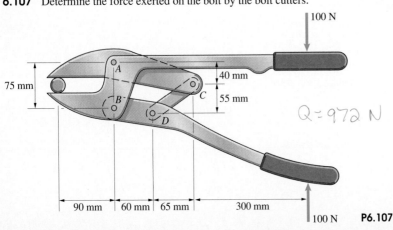

Q = 972 N

P6.107

6.108 For the bolt cutters in Problem 6.107, determine the magnitude of the force the members exert on each other at the pin connection B and the axial force in the two-force member CD.

6.109 This device is designed to exert a large force on the horizontal bar at A for a stamping operation. If the hydraulic cylinder DE exerts an axial force of 800 N and $\alpha = 80°$, what horizontal force is exerted on the horizontal bar at A?

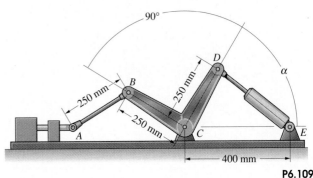

P6.109

6.110 This device raises a load W by extending the hydraulic actuator DE. The bars AD and BC are 4 ft long, and the distances $b = 2.5$ ft and $h = 1.5$ ft. If $W = 300$ lb, what force must the actuator exert to hold the load in equilibrium?

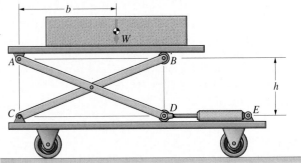

P6.110

6.111 The linkage is in equilibrium under the action of the couples M_A and M_B. If $\alpha_A = 60°$ and $\alpha_B = 70°$, what is the ratio M_A/M_B?

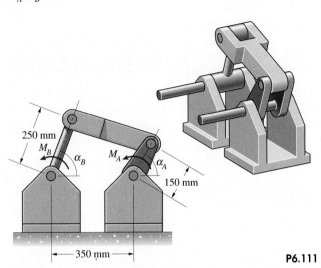

P6.111

6.112 A load $W = 2$ kN is supported by the member ACG and the hydraulic actuator BC. Determine the reactions at A and the compressive axial force in the actuator BC.

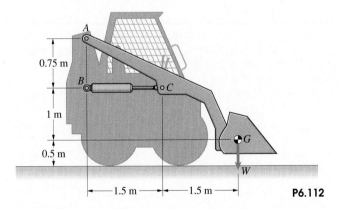

P6.112

6.113 The weights $W_1 = 1500$ lb and $W_2 = 2000$ lb. Determine the forces on member BCD at points B and D.

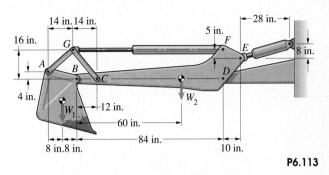

P6.113

6.114 The mechanical typewriter mechanism is in equilibrium under the action of the 2-N force on the key at A and the force F on the typebar at N. Determine the force F and the reactions at D, I, and M.

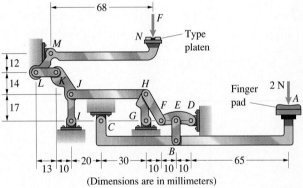

(Dimensions are in millimeters)

P6.114

Computational Mechanics

The following example and problems are designed for the use of a programmable calculator or computer.

Example 6.11

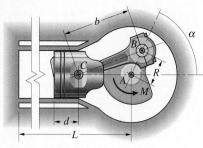

Figure 6.41

The device in Fig. 6.41 is used to compress air in a cylinder by applying a couple M to the arm AB. The pressure p in the cylinder and the net force F exerted on the piston by pressure are

$$p = p_{atm}\left(\frac{V_0}{V}\right),$$

$$F = Ap_{atm}\left(\frac{V_0}{V} - 1\right),$$

where $A = 0.02$ m^2 is the cross-sectional area of the piston, $p_{atm} = 10^5$ Pa (Pascals, or N/m^2) is atmospheric pressure, V is the volume of air in the cylinder, and V_0 is the value of V when $\alpha = 0$. The dimensions $R = 150$ mm, $b = 350$ mm, $d = 150$ mm, and $L = 1050$ mm. If M and α are initially zero and M is slowly increased until its value is 40 N-m, what are the resulting values of α and p?

STRATEGY

By expressing the volume of air in the cylinder in terms of α, we will determine the force exerted on the cylinder by pressure in terms of α. From the free-body diagram of the piston we will determine the axial force in the two-force member BC in terms of the pressure force on the cylinder. Then from the free-body diagram of the arm AB we will obtain a relation between M and α.

SOLUTION

From the geometry of the arms AB and BC (Fig. a), the volume of air in the cylinder is

$$V = A\left(L - d - \sqrt{b^2 - R^2 \sin^2 \alpha} + R \cos \alpha\right).$$

When $\alpha = 0$, the volume is

$$V_0 = A(L - d - b + R).$$

Therefore the force exerted on the piston by pressure is

$$F = Ap_{atm}\left(\frac{V_0}{V} - 1\right)$$

$$= Ap_{atm}\left(\frac{L - d - b + R}{L - d - \sqrt{b^2 - R^2 \sin^2 \alpha} + R \cos \alpha} - 1\right).$$

We draw the free-body diagrams of the piston and the arm AB in Figs. (b) and (c), where N is the force exerted on the piston by the cylinder (friction is neglected), Q is the axial force in the two-force member BC, and A_x and A_y are the reactions due to the pin support A. From Fig. (b), we obtain the equilibrium equation

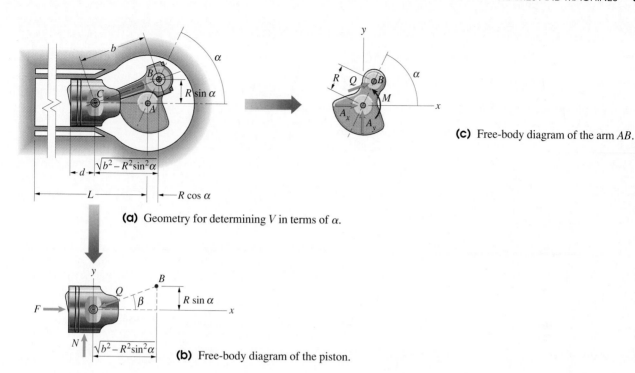

(c) Free-body diagram of the arm AB.

(a) Geometry for determining V in terms of α.

(b) Free-body diagram of the piston.

$$\Sigma F_x = F - Q \cos \beta = 0,$$

where

$$\beta = \arctan \left(\frac{R \sin \alpha}{\sqrt{b^2 - R^2 \sin^2 \alpha}} \right).$$

The force exerted on the arm AB at B is

$$Q \cos \beta \, \mathbf{i} + Q \sin \beta \, \mathbf{j}.$$

The moment of this force about A is

$$\mathbf{r}_{AB} \times (Q \cos \beta \, \mathbf{i} + Q \sin \beta \, \mathbf{j}) = \begin{vmatrix} \mathbf{i} & \mathbf{j} & \mathbf{k} \\ R \cos \alpha & R \sin \alpha & 0 \\ Q \cos \beta & Q \sin \beta & 0 \end{vmatrix}$$

$$= QR(\cos \alpha \sin \beta - \sin \alpha \cos \beta) \, \mathbf{k}.$$

Using this result, the sum of the moments about A is

$$\Sigma M_{(\text{point } A)} = M + QR(\cos \alpha \sin \beta - \sin \alpha \cos \beta) = 0.$$

If we choose a value of α, we can sequentially calculate $V, F, \beta, Q,$ and M. Computing M as a function of α, we obtain the graph shown in Fig. 6.42. The moment $M = 40$ N-m at approximately $\alpha = 80°$. By examining computed results near $80°$ (see table), we estimate that $\alpha = 79.61°$ when $M = 40$ N-m. Once we know α, we can calculate V and then p, obtaining

$$p = 1.148 p_{\text{atm}} = 1.148 \times 10^5 \text{ Pa}.$$

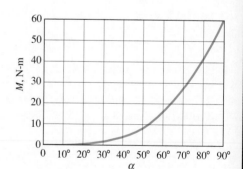

Figure 6.42

The moment M as a function of α.

α	M (N-m)
79.59°	39.9601
79.60°	39.9769
79.61°	39.9937
79.62°	40.0105
79.63°	40.0272

Problems

6.115 (a) For each member of the truss, obtain a graph of (axial force)/F as a function of x for $0 \leq x \leq 2$ m.
(b) If you were designing this truss, what value of x would you choose based on your results in (a)?

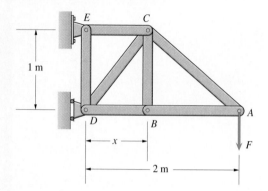

P6.115

6.116 Consider the mechanism for weighing mail described in Problem 6.103.
(a) Obtain a graph of the angle α as a function of the mass of the mail for values of the mass from 0 to 2 kg.
(b) Use the results of (a) to estimate the value of α when the mass is 1 kg.

6.117 Consider the system described in Example 6.11.
(a) Obtain a graph of M as a function of α for values of α from 0° to 180°.
(b) Use the results of (a) to estimate the maximum value of M and the angle α at which it occurs.

6.118 Consider the system described in Problem 6.109. The hydraulic cylinder DE exerts an axial force of 800 N.
(a) Obtain a graph of the horizontal component of force exerted on the horizontal bar at A by the rod AB for values of α from 45° to 85°.
(b) Use the results of (a) to estimate the value of α for which the horizontal force is 2 kN.

6.119 The weight of the suspended object is 10 kN. The two members have equal cross-sectional areas A, and each will safely support an axial force of $40A$ MN, where A is in square meters. Determine the value of h that minimizes the total volume of material in the two members.

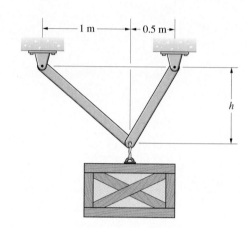

P6.119

6.120 Consider the device shown in Problem 6.110. The bars AD and BC are 4 ft long, the distance $b = 2.5$ ft, and $W = 300$ lb. If the largest force the hydraulic actuator DE can exert is 1000 lb, what is the smallest height h at which the load can be supported?

6.121 The linkage in Problem 6.111 is in equilibrium under the action of the couples M_A and M_B. When $\alpha_A = 60°$, $\alpha_B = 70°$. For the range $0 \leq \alpha_A \leq 180°$, estimate the maximum positive and negative values of M_A/M_B and the values of α_A at which they occur.

6.122 Consider the front-end loader in Problem 6.112. A load $W = 2$ kN is supported by the member ACG and the hydraulic actuator BC. If the actuator BC can exert a maximum axial force of 12 kN, what is the largest height above the ground at which the center of mass G can be supported?

6.123 Consider the truss in Problem 6.27. The crane exerts vertical 75-kN forces on the truss at B, C, and D. You can model the support at A as a pin support and model the support at E as a roller support that can exert a force normal to the dashed line but cannot exert a force parallel to it. Determine the value of the angle α for which the largest compressive force in any of the members is as small as possible. What are the resulting axial forces in the members?

6.124 Draw graphs of the magnitudes of the axial forces in the members BC and BD as functions of the dimension h for $0.5 \leq h \leq 1.5$ m.

6.125 For the truss in Problem 6.124, determine the value of the dimension h in the range $0.5 \leq h \leq 1.5$ m so that the magnitude of the largest axial force in any of the members, tensile or compressive, is a minimum. What are the resulting axial forces in the members?

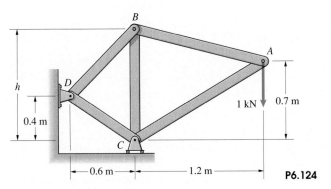

P6.124

Chapter Summary

A structure of **members** interconnected at **joints** is a **truss** if it is composed entirely of two-force members. Otherwise, it is a **frame** if it is designed to remain stationary and support loads and a **machine** if it is designed to move and exert loads.

Trusses

A member of a truss is in **tension** if the **axial forces** at the ends are directed away from each other and is in **compression** if the axial forces are directed toward each other. Before beginning to determine the axial forces in the members of a truss, it is usually necessary to draw a free-body diagram of the entire truss and determine the reactions at its supports. The axial forces in the members can be determined by two methods. The **method of joints** involves drawing free-body diagrams of the joints of a truss one by one and using the equilibrium equations to determine the axial forces in the members. In two dimensions, choose joints to analyze that are subjected to known forces and no more than two unknown forces. The **method of sections** involves drawing free-body diagrams of parts, or **sections**, of a truss and using the equilibrium equations to determine the axial forces in selected members.

A **space truss** is a three-dimensional truss. Space trusses are analyzed by the same methods used for two-dimensional trusses. Choose joints to analyze that are subjected to known forces and no more than three unknown forces.

Frames and Machines

Begin analyzing a frame or machine by drawing a free-body diagram of the entire structure and determining the reactions at its supports. If the entire structure is statically indeterminate, determine as many reactions as possible. Then draw free-body diagrams of individual members, or selected combinations of members, and apply the equilibrium equations to determine the forces and couples acting on them. Recognizing two-force members will reduce the number of unknown forces that must be determined. If a load is applied at a joint, it can be placed on the free-body diagram of *any one* of the members attached at the joint.

Review Problems

6.126 The loads $F_1 = 60$ N and $F_2 = 40$ N.
(a) Draw the free-body diagram of the entire truss, and determine the reactions at its supports.
(b) Determine the axial forces in the members. Indicate whether they are in tension (T) or compression (C).

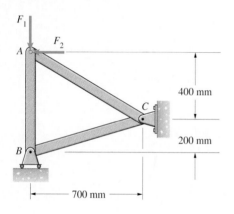

P6.126

6.127 Consider the truss in Problem 6.126. The loads $F_1 = 440$ N and $F_2 = 160$ N. Determine the axial forces in the members. Indicate whether they are in tension (T) or compression (C).

6.128 The load $F = 600$ lb. Determine the axial forces in the members.

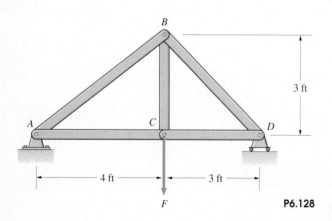

P6.128

6.129 Consider the truss in Problem 6.128. Each member will safely support a tensile force of 2000 lb and a compressive force of 1000 lb. Based on this criterion, determine the largest downward load F that can safely be applied at C.

6.130 The Pratt bridge truss supports loads at F, G, and H. Determine the axial forces in members BC, BG, and FG.

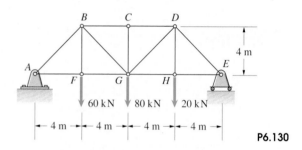

P6.130

6.131 Consider the truss in Problem 6.130. Determine the axial forces in members CD, GD, and GH.

6.132 The truss supports loads at F and H. Determine the axial forces in members AB, AC, BC, BD, CD, and CE.

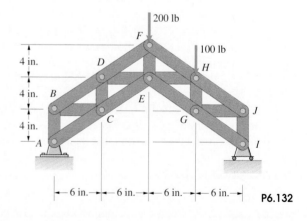

P6.132

6.133 Consider the truss in Problem 6.132. Determine the axial forces in members EH and FH.

6.134 Determine the axial forces in members *BD*, *CD*, and *DE*.

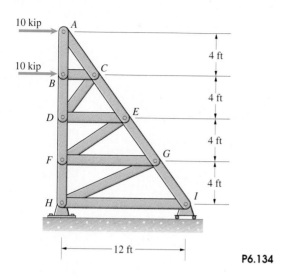

P6.134

6.135 For the truss in Problem 6.134, determine the axial forces in members *DF*, *EF*, and *FG*.

6.136 The truss supports a 400-N load at *G*. Determine the axial forces in members *AC*, *CD*, and *CF*.

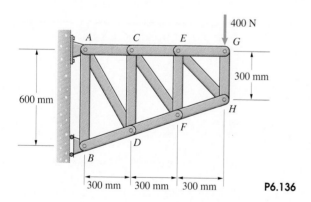

P6.136

6.137 Consider the truss in Problem 6.136. Determine the axial forces in members *CE*, *EF*, and *EH*.

6.138 Consider the truss in Problem 6.136. Which members have the largest tensile and compressive forces, and what are their values?

6.139 The Howe truss helps support a roof. Model the supports at *A* and *G* as roller supports. Use the method of joints to determine the axial forces in members *BC*, *CD*, *CI*, and *CJ*.

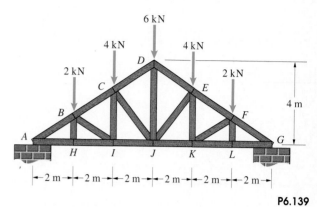

P6.139

6.140 For the roof truss in Problem 6.139, use the method of sections to determine the axial forces in members *CD*, *CJ*, and *IJ*.

6.141 A speaker system is suspended from the truss by cables attached at *D* and *E*. The mass of the speaker system is 130 kg, and its weight acts at *G*. Determine the axial forces in members *BC* and *CD*.

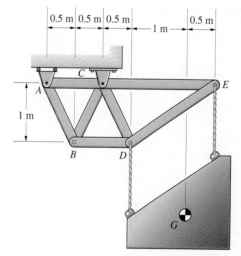

P6.141

6.142 Consider the system described in Problem 6.141. If each member of the truss will safely support a tensile force of 5 kN and a compressive force of 3 kN, what is the maximum safe value of the mass of the speaker system?

6.143 Determine the forces on member *ABC*, presenting your answers as shown in Fig. 6.35. Obtain the answers in two ways:
(a) When you draw the free-body diagrams of the individual members, place the 400-lb load on the free-body diagram of member *ABC*.
(b) When you draw the free-body diagrams of the individual members, place the 400-lb load on the free-body diagram of member *CD*.

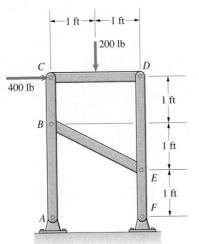

P6.143

6.144 The mass $m = 120$ kg. Determine the forces on member *ABC*.

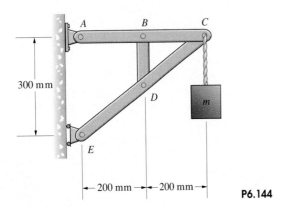

P6.144

6.145 Determine the forces on member *ABC*, presenting your answers as shown in Fig. 6.35.

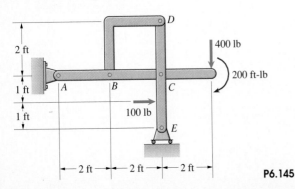

P6.145

6.146 Determine the force exerted on the bolt by the bolt cutters and the magnitude of the force the members exert on each other at the pin connection *A*.

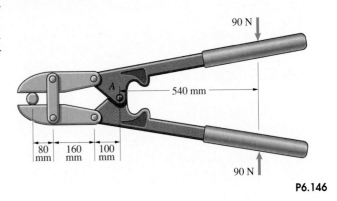

P6.146

6.147 The 600-lb weight of the scoop acts at a point 1 ft 6 in. to the right of the vertical line *CE*. The line *ADE* is horizontal. The hydraulic actuator *AB* can be treated as a two-force member. Determine the axial force in the hydraulic actuator *AB* and the forces exerted on the scoop at *C* and *E*.

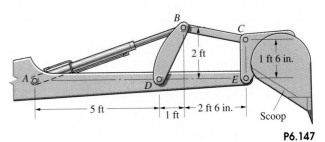

P6.147

6.148 This structure supports a conveyer belt used in a lignite mining operation. The cables connected to the belt exert the force *F* at *J*. As a result of the counterweight *W* = 8 kip, the reaction at *E* and the vertical reaction at *D* are equal. Determine *F* and the axial forces in members *BG* and *EF*.

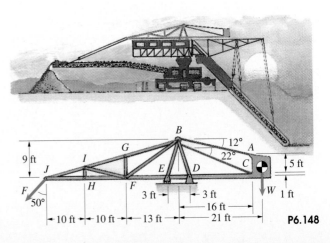

P6.148

6.149 Consider the structure described in Problem 6.148. The counterweight $W = 8$ kip is pinned at D and is supported by the cable ABC, which passes over a pulley at A. What is the tension in the cable, and what forces are exerted on the counterweight at D?

6.150 The weights $W_1 = 4$ kN and $W_2 = 10$ kN. Determine the forces on member $ACDE$ at points A and E.

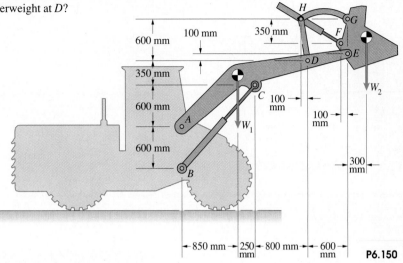

P6.150

The design of a car—especially the placement of the engine—determines the position of its center of mass, the point where the car's weight effectively acts. The position of the center of mass determines the forces exerted on the car's tires when it is in equilibrium and also affects the car's dynamic behavior. In this chapter we show how to locate centers of mass and use them in determining the forces and moments acting on objects in equilibrium.

Chapter 7

Centroids and Centers of Mass

An object's weight does not act at a single point—it is distributed over the entire volume of the object. But we can *represent* the weight by a single equivalent force acting at a point called the center of mass. For example, each part of a car has weight, but we can represent its total weight by a single force acting at its center of mass.

In this chapter we define the center of mass and show how it is determined for various kinds of objects. Along the way, we also introduce definitions that can be interpreted as the average positions of areas, volumes, and lines. These average positions are called *centroids*. Centroids coincide with the centers of mass of particular classes of objects, but they also arise in many other applications. For example, in Chapter 10 we express the hydrostatic force exerted on a submerged plate in terms of the centroid of the plate's area.

Centroids

Because centroids have such varied applications, we first define them using the general concept of a "weighted average." Let's begin with the familiar idea of an average position. Suppose we want to determine the average position of a group of students sitting in a room. First, we introduce a coordinate system so that we can specify the position of each student. For example, we can align the axes with the walls of the room (Fig. 7.1a). We number the students from 1 to N and denote the position of student 1 by x_1, y_1, the position of student 2 by x_2, y_2, and so on. The average x coordinate $\bar{x}$ is the sum of their x coordinates divided by N,

$$\bar{x} = \frac{x_1 + x_2 + \cdots + x_N}{N} = \frac{\sum_i x_i}{N}, \tag{7.1}$$

where the summation sign stands for "the sum from $i = 1$ to N." The average y coordinate is

$$\bar{y} = \frac{\sum_i y_i}{N}. \tag{7.2}$$

We indicate the average position by the symbol shown in Fig. 7.1(b).

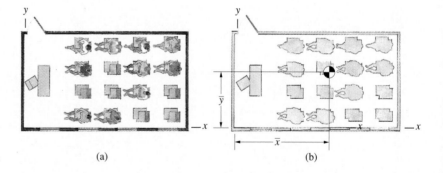

Figure 7.1

(a) A group of students in a classroom.
(b) Their average position.

(a)

(b)

Now suppose that we pass out some pennies to the students. Let the number of coins given to student 1 be c_1, the number given to student 2 be c_2, and so on. What is the average position of the coins in the room? Clearly, the average position of the coins may not be the same as the average position of the students. For example, if the students in the front of the room have more coins, the average position of the coins will be closer to the front of the room than the average position of the students.

To determine the x coordinate of the average position of the coins, we need to sum the x coordinates of the coins and divide by the number of coins. We can obtain the sum of the x coordinates of the coins by multiplying the number of coins each student has by his or her x coordinate and summing. We can obtain the number of coins by summing the numbers c_1, c_2, $\ldots$. Thus the average x coordinate of the coins is

$$\bar{x} = \frac{\sum_i x_i c_i}{\sum_i c_i}. \tag{7.3}$$

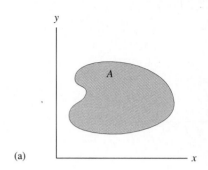

(a)

We can determine the average y coordinate of the coins in the same way:

$$\bar{y} = \frac{\sum_i y_i c_i}{\sum_i c_i}. \tag{7.4}$$

By assigning other meanings to c_1, c_2, $\ldots$, we can determine the average positions of other measures associated with the students. For example, we could determine the average position of their age or the average position of their height.

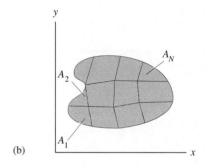

(b)

More generally, we can use Eqs. (7.3) and (7.4) to determine the average position of *any* set of quantities with which we can associate positions. An average position obtained from these equations is called a *weighted average position*, or **centroid**. The "weight" associated with position x_1, y_1 is c_1, the weight associated with position x_2, y_2 is c_2, and so on. In Eqs. (7.1) and (7.2), the weight associated with the position of each student is 1. When the census is taken, the centroid of the population of the United States—the average position of the population—is determined in this way. In the next section we use Eqs. (7.3) and (7.4) to determine centroids of areas.

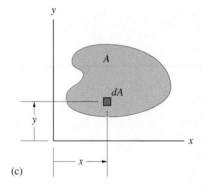

(c)

7.1 *Centroids of Areas*

Consider an arbitrary area A in the x-y plane (Fig. 7.2a). Let us divide the area into parts A_1, A_2, $\ldots$, A_N (Fig. 7.2b) and denote the positions of the parts by (x_1, y_1), (x_2, y_2), $\ldots$, (x_N, y_N). We can obtain the centroid, or average position of the area, by using Eqs. (7.3) and (7.4) with the areas of the parts as the weights:

$$\bar{x} = \frac{\sum_i x_i A_i}{\sum_i A_i}, \qquad \bar{y} = \frac{\sum_i y_i A_i}{\sum_i A_i}. \tag{7.5}$$

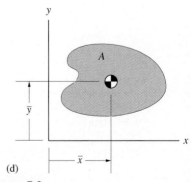

(d)

Figure 7.2

(a) The area A.
(b) Dividing A into N parts.
(c) A differential element of area dA with coordinates x, y.
(d) The centroid of the area.

A question arises if we try to carry out this procedure: What are the exact positions of the areas $A_1, A_2, \ldots, A_N$? We could reduce the uncertainty in their positions by dividing A into smaller parts, but we would still obtain only approximate values for $\bar{x}$ and $\bar{y}$. To determine the exact location of the centroid, we must take the limit as the sizes of the parts approach zero. We obtain this limit by replacing Eqs. (7.5) by the integrals

$$\bar{x} = \frac{\displaystyle\int_A x \, dA}{\displaystyle\int_A dA}, \tag{7.6}$$

$$\bar{y} = \frac{\displaystyle\int_A y \, dA}{\displaystyle\int_A dA}, \tag{7.7}$$

where x and y are the coordinates of the differential element of area dA (Fig. 7.2c). The subscript A on the integral signs means the integration is carried out over the entire area. The centroid of the area is shown in Fig. 7.2(d).

Keeping in mind that the centroid of an area is its average position will often help you locate it. For example, the centroid of a circular area or a rectangular area obviously lies at the center of the area. If an area has "mirror image" symmetry about an axis, the centroid lies on the axis (Fig. 7.3a), and if an area is symmetric about two axes, the centroid lies at their intersection (Fig. 7.3b).

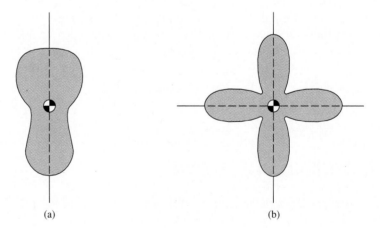

Figure 7.3

(a) An area that is symmetric about an axis.
(b) An area with two axes of symmetry.

(a)

(b)

The following examples demonstrate the use of Eqs. (7.6) and (7.7) to determine centroids of areas. You must choose a differential element of area dA and choose limits so that you integrate over the desired area.

Example 7.1

Determine the centroid of the triangular area in Fig. 7.4.

STRATEGY

We will determine the coordinates of the centroid by using an element of area dA in the form of a "strip" of width dx.

SOLUTION

Let dA be the vertical strip in Fig. (a). The height of the strip is $(h/b)x$, so $dA = (h/b)x\,dx$. To integrate over the entire area, we must integrate with respect to x from $x = 0$ to $x = b$. The x coordinate of the centroid is

$$\bar{x} = \frac{\displaystyle\int_A x\,dA}{\displaystyle\int_A dA} = \frac{\displaystyle\int_0^b x\left(\frac{h}{b}x\,dx\right)}{\displaystyle\int_0^b \frac{h}{b}x\,dx} = \frac{\dfrac{h}{b}\left[\dfrac{x^3}{3}\right]_0^b}{\dfrac{h}{b}\left[\dfrac{x^2}{2}\right]_0^b} = \frac{2}{3}b.$$

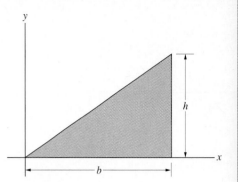

Figure 7.4

To determine $\bar{y}$, we let y in Eq. (7.7) be the y coordinate of the midpoint of the strip (Fig. b):

$$\bar{y} = \frac{\displaystyle\int_A y\,dA}{\displaystyle\int_A dA} = \frac{\displaystyle\int_0^b \frac{1}{2}\left(\frac{h}{b}x\right)\left(\frac{h}{b}x\,dx\right)}{\displaystyle\int_0^b \frac{h}{b}x\,dx} = \frac{\dfrac{1}{2}\left(\dfrac{h}{b}\right)^2\left[\dfrac{x^3}{3}\right]_0^b}{\dfrac{h}{b}\left[\dfrac{x^2}{2}\right]_0^b} = \frac{1}{3}h.$$

The centroid is shown in Fig. (c).

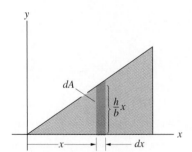

(a) An element dA in the form of a strip.

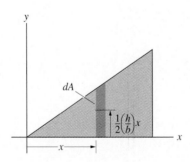

(b) The y coordinate of the midpoint of the strip is $\frac{1}{2}(h/b)x$.

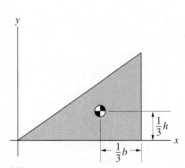

(c) Centroid of the area.

DISCUSSION

You should always be alert for opportunities to check your results. In this example we should make sure that our integration procedure gives the correct result for the area of the triangle:

$$\int_A dA = \int_0^b \frac{h}{b}x\,dx = \frac{h}{b}\left[\frac{x^2}{2}\right]_0^b = \frac{1}{2}bh.$$

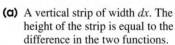

Example 7.2

Determine the centroid of the area in Fig. 7.5.

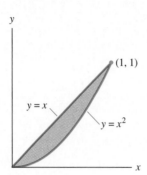

Figure 7.5

SOLUTION

Let dA be the vertical strip in Fig. (a). The height of the strip is $x - x^2$, so $dA = (x - x^2)\, dx$. The x coordinate of the centroid is

$$\bar{x} = \frac{\displaystyle\int_A x\, dA}{\displaystyle\int_A dA} = \frac{\displaystyle\int_0^1 x(x - x^2)\, dx}{\displaystyle\int_0^1 (x - x^2)\, dx} = \frac{\left[\dfrac{x^3}{3} - \dfrac{x^4}{4}\right]_0^1}{\left[\dfrac{x^2}{2} - \dfrac{x^3}{3}\right]_0^1} = \frac{1}{2}.$$

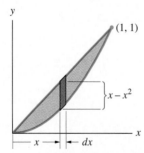

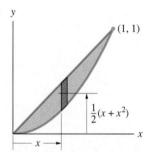

(a) A vertical strip of width dx. The height of the strip is equal to the difference in the two functions.

(b) The y coordinate of the midpoint of the strip.

The y coordinate of the midpoint of the strip is $x^2 + \frac{1}{2}(x - x^2) = \frac{1}{2}(x + x^2)$ (Fig. b). Substituting this expression for y in Eq. (7.7), we obtain the y coordinate of the centroid:

$$\bar{y} = \frac{\displaystyle\int_A y\, dA}{\displaystyle\int_A dA} = \frac{\displaystyle\int_0^1 \left[\frac{1}{2}(x + x^2)\right](x - x^2)\, dx}{\displaystyle\int_0^1 (x - x^2)\, dx} = \frac{\dfrac{1}{2}\left[\dfrac{x^3}{3} - \dfrac{x^5}{5}\right]_0^1}{\left[\dfrac{x^2}{2} - \dfrac{x^3}{3}\right]_0^1} = \frac{2}{5}.$$

Problems

7.1 Determine the x coordinate of the centroid of the area.

   Strategy: The x coordinate of the centroid is given by Eq. (7.6). You can use a differential element of area dA in the form of a vertical strip of width dx. See Example 7.1.

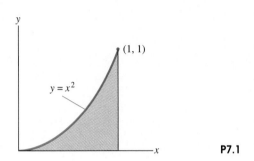

P7.1

7.2 Determine the y coordinate of the centroid of the area in Problem 7.1.

7.3 If $a = 2$, what are the coordinates of the centroid of the area?

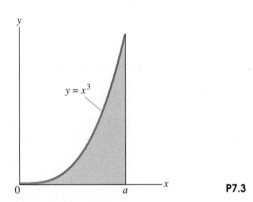

P7.3

7.4 Consider the area in Problem 7.3. If the y coordinate of the centroid is $\bar{y} = 5$, what are a and $\bar{x}$?

7.5 Consider the area in Problem 7.3. The "center of area" is defined to be the point for which there is as much area to the right of the point as to the left of it and as much area above the point as below it. If $a = 4$, what are the x coordinate of the center of area and the x coordinate of the centroid?

7.6 Determine the x coordinate of the centroid of the area.

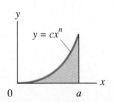

P7.6

7.7 Determine the y coordinate of the centroid of the area in Problem 7.6.

7.8 Suppose that an art student wants to paint a panel of wood as shown, with the horizontal and vertical lines passing through the centroid of the painted area, and asks you to determine the coordinates of the centroid. What are they?

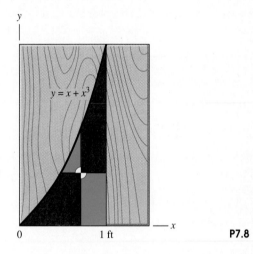

P7.8

7.9 Determine the coordinates of the centroid of the area.

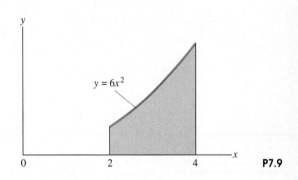

P7.9

7.10 Determine the coordinates of the centroid of the metal plate's cross-sectional area.

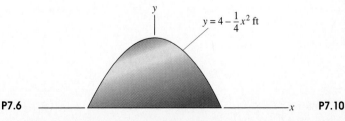

P7.10

7.11 An architect wants to build a wall with the profile shown. To estimate the effects of wind loads on the wall, he must determine the area of the wall and the coordinates of its centroid. What are they?

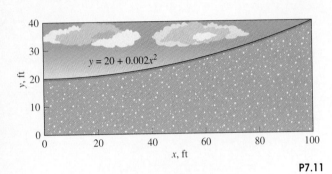

P7.11

7.12 Determine the x coordinate of the centroid of the area.

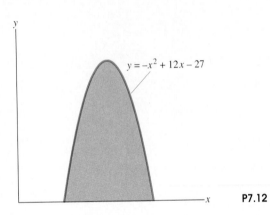

P7.12

7.13 Determine the y coordinate of the centroid of the area in Problem 7.12.

7.14 Determine the x coordinate of the centroid of the area.

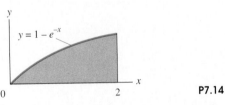

P7.14

7.15 Determine the y coordinate of the centroid of the area in Problem 7.14.

7.16 Determine the coordinates of the centroid of the area.

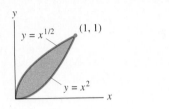

P7.16

7.17 Determine the x coordinate of the centroid of the area.

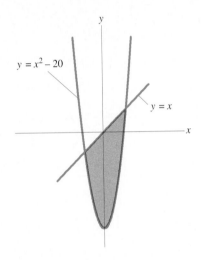

P7.17

7.18 Determine the y coordinate of the centroid of the area in Problem 7.17.

7.19 Determine the y coordinate of the centroid of the area.

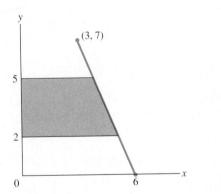

P7.19

7.20 Determine the x coordinate of the centroid of the area in Problem 7.19.

7.21. An agronomist wants to measure the rainfall at the centroid of a plowed field between two roads. What are the coordinates of the point where the rain gauge should be placed?

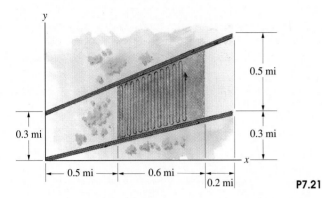

P7.21

7.22 The cross section of an earth-fill dam is shown. Determine the coefficients a and b so that the y coordinate of the centroid of the cross section is 10 m.

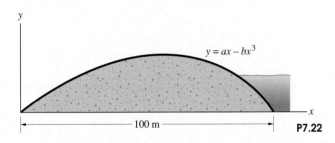

P7.22

7.23 The airplane wing has an elliptical profile. Determine its centroid.

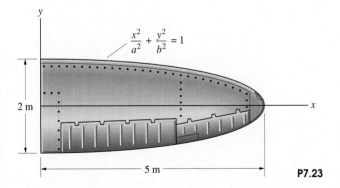

P7.23

7.24 Determine the coordinates of the centroid of the area.

Strategy: Write the equation for the circular boundary in the form $y = (R^2 - x^2)^{1/2}$ and use a vertical "strip" of width dx as the element of area dA.

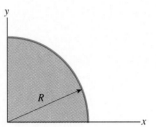

P7.24

7.25 Determine the x coordinate of the centroid of the area. By setting $h = 0$, confirm the answer to Problem 7.24.

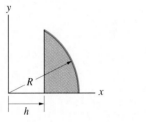

P7.25

7.26 Determine the y coordinate of the centroid of the area in Problem 7.25.

7.2 Centroids of Composite Areas

Although you can determine the centroids of areas by integration, the process becomes difficult and tedious for complicated areas. In this section we describe a much easier approach you can use if the area consists of a combination of simple areas, which we refer to as a **composite area**. You can determine the centroid of a composite area without using integration if you know the centroids of its parts.

The area in Fig. 7.6(a) consists of a triangle, a rectangle, and a semicircle, which we call parts 1, 2, and 3. The x coordinate of the centroid of the composite area is

$$\bar{x} = \frac{\int_A x \, dA}{\int_A dA} = \frac{\int_{A_1} x \, dA + \int_{A_2} x \, dA + \int_{A_3} x \, dA}{\int_{A_1} dA + \int_{A_2} dA + \int_{A_3} dA}. \tag{7.8}$$

The x coordinates of the centroids of the parts are shown in Fig. 7.6(b). From the equation for the x coordinate of the centroid of part 1,

$$\bar{x}_1 = \frac{\int_{A_1} x \, dA}{\int_{A_1} dA},$$

we obtain

$$\int_{A_1} x \, dA = \bar{x}_1 A_1.$$

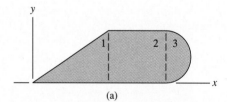

(a)

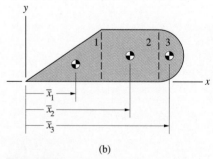

(b)

Figure 7.6

(a) A composite area composed of three simple areas.
(b) The centroids of the parts.

Using this equation and equivalent equations for parts 2 and 3, we can write Eq. (7.8) as

$$\bar{x} = \frac{\bar{x}_1 A_1 + \bar{x}_2 A_2 + \bar{x}_3 A_3}{A_1 + A_2 + A_3}.$$

We have obtained an equation for the x coordinate of the composite area in terms of those of its parts. The coordinates of the centroid of a composite area with an arbitrary number of parts are

$$\bar{x} = \frac{\sum_i \bar{x}_i A_i}{\sum_i A_i}, \qquad \bar{y} = \frac{\sum_i \bar{y}_i A_i}{\sum_i A_i}. \tag{7.9}$$

When you can divide an area into parts whose centroids are known, you can use these expressions to determine its centroid. The centroids of some simple areas are tabulated in Appendix B.

We began our discussion of the centroid of an area by dividing an area into finite parts and writing equations for its weighted average position. The results, Eqs. (7.5), are approximate because of the uncertainty in the positions of the parts of the area. The exact Eqs. (7.9) are identical except that the positions of the parts are their centroids.

The area in Fig. 7.7(a) consists of a triangular area with a circular hole, or cutout. Designating the triangular area (without the cutout) as part 1 of the composite area (Fig. 7.7b) and the area of the cutout as part 2 (Fig. 7.7c), we obtain the x coordinate of the centroid of the composite area:

$$\bar{x} = \frac{\displaystyle\int_{A_1} x\, dA - \int_{A_2} x\, dA}{\displaystyle\int_{A_1} dA - \int_{A_2} dA} = \frac{\bar{x}_1 A_1 - \bar{x}_2 A_2}{A_1 - A_2}.$$

This equation is identical in form to the first of Eqs. (7.9) except that the terms corresponding to the cutout are negative. As this example demonstrates, you can use Eqs. (7.9) to determine the centroids of composite areas containing cutouts by treating the cutouts as negative areas.

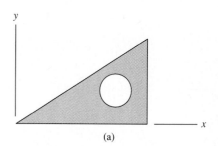

(a)

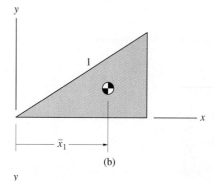

(b)

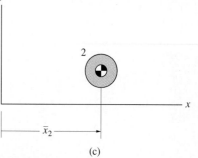

(c)

Figure 7.7

(a) An area with a cutout.
(b) The triangular area.
(c) The area of the cutout.

In the following examples we determine the centroids of composite areas. Three steps are typically required.

1. Choose the parts—*Try to divide the composite area into parts whose centroids you know or can easily determine.*

2. Determine the values for the parts—*Determine the centroid and the area of each part. Watch for instances of symmetry that can simplify your task.*

3. Calculate the centroid—*Use Eqs. (7.9) to determine the centroid of the composite area.*

Example 7.3

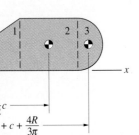

Figure 7.8

Determine the centroid of the area in Fig. 7.8.

SOLUTION

Choose the Parts We can divide the area into a triangle, a rectangle, and a semicircle, which we call parts 1, 2, and 3, respectively.

Determine the Values for the Parts The x coordinates of the centroids of the parts are shown in Fig. (a). The x coordinates, the areas of the parts, and their products are summarized in Table 7.1.

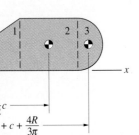

(a) The x coordinates of the centroids of the parts.

Table 7.1 Information for determining the x coordinate of the centroid.

	$\bar{x}_i$	A_i	$\bar{x}_i A_i$
Part 1 (triangle)	$\frac{2}{3}b$	$\frac{1}{2}b(2R)$	$\left(\frac{2}{3}b\right)\left[\frac{1}{2}b(2R)\right]$
Part 2 (rectangle)	$b + \frac{1}{2}c$	$c(2R)$	$\left(b + \frac{1}{2}c\right)[c(2R)]$
Part 3 (semicircle)	$b + c + \dfrac{4R}{3\pi}$	$\frac{1}{2}\pi R^2$	$\left(b + c + \dfrac{4R}{3\pi}\right)\left(\frac{1}{2}\pi R^2\right)$

Calculate the Centroid The x coordinate of the centroid of the composite area is

$$\bar{x} = \frac{\bar{x}_1 A_1 + \bar{x}_2 A_2 + \bar{x}_3 A_3}{A_1 + A_2 + A_3}$$

$$= \frac{\left(\frac{2}{3}b\right)\left[\frac{1}{2}b(2R)\right] + \left(b + \frac{1}{2}c\right)[c(2R)] + \left(b + c + \dfrac{4R}{3\pi}\right)\left(\frac{1}{2}\pi R^2\right)}{\frac{1}{2}b(2R) + c(2R) + \frac{1}{2}\pi R^2}.$$

We repeat the last two steps to determine the y coordinate of the centroid. The y coordinates of the centroids of the parts are shown in Fig. (b). Using the information summarized in Table 7.2, we obtain

$$\bar{y} = \frac{\bar{y}_1 A_1 + \bar{y}_2 A_2 + \bar{y}_3 A_3}{A_1 + A_2 + A_3}$$

$$= \frac{\left[\frac{1}{3}(2R)\right]\left[\frac{1}{2}b(2R)\right] + R[c(2R)] + R\left(\frac{1}{2}\pi R^2\right)}{\frac{1}{2}b(2R) + c(2R) + \frac{1}{2}\pi R^2}.$$

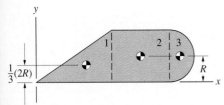

(b) The y coordinates of the centroids of the parts.

Table 7.2 Information for determining the y coordinate of the centroid.

	$\bar{y}_i$	A_i	$\bar{y}_i A_i$
Part 1 (triangle)	$\frac{1}{3}(2R)$	$\frac{1}{2}b(2R)$	$\left[\frac{1}{3}(2R)\right]\left[\frac{1}{2}b(2R)\right]$
Part 2 (rectangle)	R	$c(2R)$	$R[c(2R)]$
Part 3 (semicircle)	R	$\frac{1}{2}\pi R^2$	$R\left(\frac{1}{2}\pi R^2\right)$

Example 7.4

Determine the centroid of the area in Fig. 7.9.

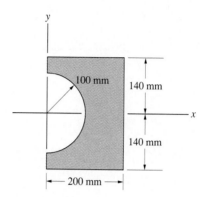

Figure 7.9

SOLUTION

Choose the Parts We will treat the area as a composite area consisting of the rectangle without the semicircular cutout and the area of the cutout, which we call parts 1 and 2, respectively (Fig. a).

Determine the Values for the Parts From Appendix B, the x coordinate of the centroid of the cutout is

$$\bar{x}_2 = \frac{4R}{3\pi} = \frac{4(100)}{3\pi} \text{ mm.}$$

The information for determining the x coordinate of the centroid is summarized in Table 7.3. Notice that we treat the cutout as a negative area.

Table 7.3 Information for determining $\bar{x}$.

	$\bar{x}_i$ (mm)	A_i (mm^2)	$\bar{x}_i A_i$ (mm^3)
Part 1 (rectangle)	100	(200)(280)	(100)[(200)(280)]
Part 2 (cutout)	$\frac{4(100)}{3\pi}$	$-\frac{1}{2}\pi(100)^2$	$-\frac{4(100)}{3\pi}\left[\frac{1}{2}\pi(100)^2\right]$

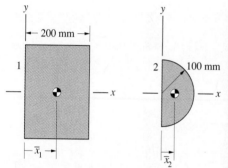

(a) The rectangle and the semicircular cutout.

Calculate the Centroid The x coordinate of the centroid is

$$\bar{x} = \frac{\bar{x}_1 A_1 + \bar{x}_2 A_2}{A_1 + A_2} = \frac{(100)[(200)(280)] - \frac{4(100)}{3\pi}\left[\frac{1}{2}\pi(100)^2\right]}{(200)(280) - \frac{1}{2}\pi(100)^2}$$

$$= 122 \text{ mm.}$$

Because of the symmetry of the area, $\bar{y} = 0$.

Problems

For Problems 7.27–7.36, determine the centroids of the areas.

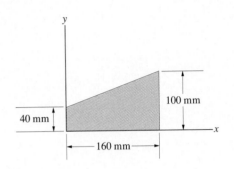

P7.27

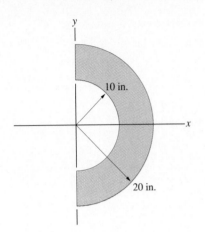

P7.30

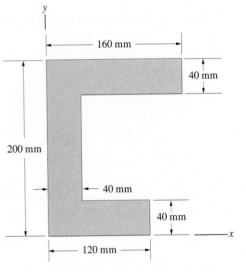

P7.28

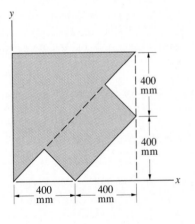

P7.31

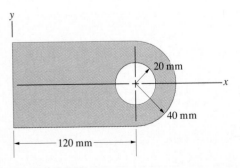

P7.29

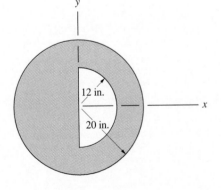

P7.32

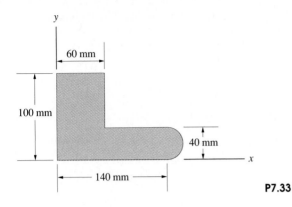

60 mm

100 mm

40 mm

140 mm

x

P7.33

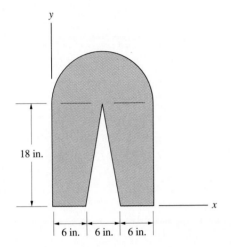

18 in.

6 in. | 6 in. | 6 in.

x

P7.34

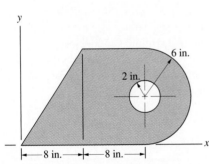

6 in.

2 in.

8 in. 8 in.

x

P7.35

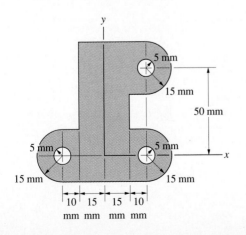

5 mm

15 mm

50 mm

5 mm

5 mm

15 mm

15 mm

10 15 15 10
mm mm mm mm

P7.36

7.37 The dimensions $b = 40$ mm and $h = 20$ mm. Determine the y coordinate of the centroid of the area.

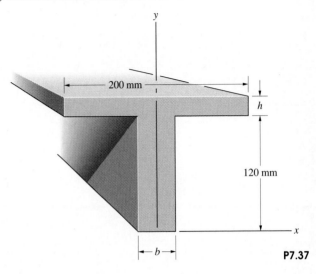

200 mm

h

120 mm

b

x

P7.37

7.38 The area in Problem 7.37 is the cross section of a beam. If the cross-sectional area of the beam is $A = 0.01$ m^2 and the y coordinate of the centroid of the area is $\bar{y} = 80$ mm, what are the dimensions b and h?

7.39 Determine the x coordinate of the centroid of the Boeing 747's vertical stabilizer.

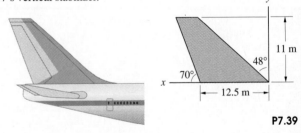

11 m

48°

70°

x

12.5 m

P7.39

7.40 Determine the y coordinate of the centroid of the vertical stabilizer in Problem 7.39.

7.41 Determine the x coordinate of the centroid of the concrete masonry unit.

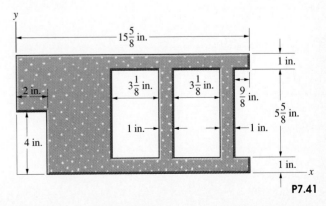

$15\frac{5}{8}$ in.

1 in.

$3\frac{1}{8}$ in. $3\frac{1}{8}$ in.

$\frac{9}{8}$ in.

2 in.

$5\frac{5}{8}$ in.

1 in. 1 in.

4 in.

1 in.

x

P7.41

7.42 Determine the y coordinate of the centroid of the concrete masonry unit in Problem 7.41.

7.43 Three sails of a New York pilot schooner are shown. The coordinates of the points are in feet. Determine the centroid of sail 1.

7.44 Determine the centroid of sail 2 in Problem 7.43.

7.45 Determine the centroid of sail 3 in Problem 7.43.

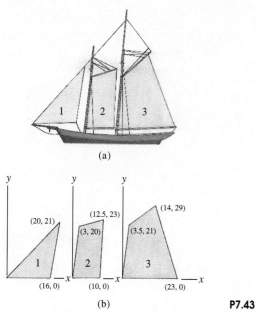

(a)

(b)

P7.43

7.3 Distributed Loads

The load exerted on a beam (stringer) supporting a floor of a building is distributed over the beam's length (Fig. 7.10a). The load exerted by wind on a television transmission tower is distributed along the tower's height (Fig. 7.10b). In many engineering applications, loads are continuously distributed along lines. We will show that the concept of the centroid of an area can be useful in the analysis of objects subjected to such loads.

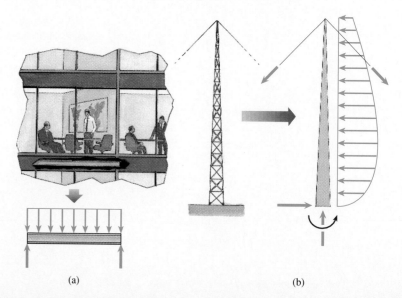

Figure 7.10

Examples of distributed forces: (a) Uniformly distributed load exerted on a beam of a building's frame by the floor. (b) Wind load distributed along the height of a tower.

(a)

(b)

Describing a Distributed Load

We can use a simple example to demonstrate how such loads are expressed analytically. Suppose that we pile bags of sand on a beam, as shown in Fig. 7.11(a). You can see that the load exerted by the bags is distributed over the length of the beam and that its magnitude at a given position x depends on how high the bags are piled at that position. To describe the load, we define a function w such that the *downward* force exerted on an infinitesimal element dx of the beam is $w\, dx$. With this function we can model the varying magnitude of the load exerted by the sand bags (Fig. 7.11b). The arrows in the figure indicate that the load acts in the downward direction. Loads distributed along lines, from simple examples such as a beam's own weight to complicated ones such as the lift distributed along the length of an airplane's wing, are modeled by the function w. Since the product of w and dx is a force, the dimensions of w are (force)/(length). For example, w can be expressed in newtons per meter in SI units or in pounds per foot in U.S. Customary units.

Determining Force and Moment

Let's assume that the function w describing a particular distributed load is known (Fig. 7.12a). The graph of w is called the **loading curve**. Since the force acting on an element dx of the line is $w\, dx$, we can determine the total force F exerted by the distributed load by integrating the loading curve with respect to x:

$$F = \int_L w\, dx. \tag{7.10}$$

We can also integrate to determine the moment about a point exerted by the distributed load. For example, the moment about the origin due to the force exerted on the element dx is $xw\, dx$, so the total moment about the origin due to the distributed load is

$$M = \int_L xw\, dx. \tag{7.11}$$

When you are concerned only with the total force and moment exerted by a distributed load, you can represent it by a single equivalent force F (Fig. 7.12b). For equivalence, the force must act at a position $\bar{x}$ on the x axis such that the moment of F about the origin is equal to the moment of the distributed load about the origin:

$$\bar{x}F = \int_L xw\, dx.$$

Therefore the force F is equivalent to the distributed load if we place it at the position

$$\bar{x} = \frac{\displaystyle\int_L xw\, dx}{\displaystyle\int_L w\, dx}. \tag{7.12}$$

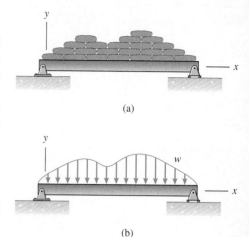

(a)

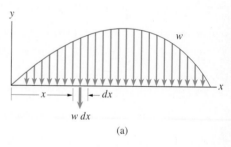

(b)

Figure 7.11

(a) Loading a beam with bags of sand.
(b) The distributed load w models the load exerted by the bags.

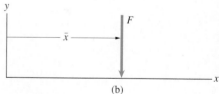

(a)

(b)

Figure 7.12

(a) A distributed load and the force exerted on a differential element dx.
(b) The equivalent force.

The Area Analogy

Notice that the term $w\,dx$ is equal to an element of "area" dA between the loading curve and the x axis (Fig. 7.13a). (We use quotation marks because $w\,dx$ is actually a force and not an area.) Interpreted in this way, Eq. (7.10) states that *the total force exerted by the distributed load is equal to the "area" A between the loading curve and the x axis:*

$$F = \int_L w\,dx = \int_A dA = A. \tag{7.13}$$

Substituting $w\,dx = dA$ into Eq. (7.12), we obtain

$$\bar{x} = \frac{\displaystyle\int_L xw\,dx}{\displaystyle\int_L w\,dx} = \frac{\displaystyle\int_A x\,dA}{\displaystyle\int_A dA}. \tag{7.14}$$

The single force equivalent to the distributed load acts at the centroid, or weighted average position, of the "area" between the loading curve and the x axis (Fig. 7.13b). You will find that this analogy between distributed load and area is useful when you want to represent a simple distributed load by an equivalent force.

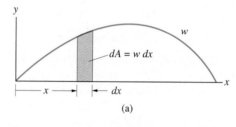

(a)

Figure 7.13

(a) Determining the "area" between the function w and the x axis.

(b) The equivalent force is equal to the "area," and the line of action passes through its centroid.

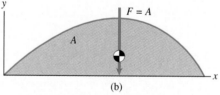

(b)

The following examples demonstrate how to determine the reactions at the supports of beams subjected to distributed loads. You can represent a distributed load by a single equivalent force whose magnitude equals the "area" between the loading curve and the x axis and whose line of action passes through the centroid of the "area." The "area" is given in terms of the loading function w by Eq. (7.13), and the location of the centroid is given by Eq. (7.14).

Example 7.5

The beam in Fig. 7.14 is subjected to a "triangular" distributed load whose value at B is 100 N/m.
(a) Represent the distributed load by a single equivalent force.
(b) Determine the reactions at A and B.

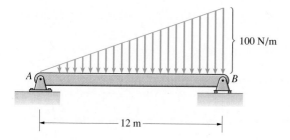

100 N/m

A B

|← 12 m →|

Figure 7.14

STRATEGY

(a) The magnitude of the force is equal to the "area" under the triangular loading curve, and the equivalent force acts at the centroid of the triangular "area." (b) Once the distributed load is represented by a single equivalent force, we can apply the equilibrium equations to determine the reactions.

SOLUTION

(a) The "area" of the triangular distributed load is one-half its base times its height, or $\frac{1}{2}(12 \text{ m}) \times (100 \text{ N/m}) = 600 \text{ N}$. The centroid of the triangular "area" is located at $\bar{x} = \frac{2}{3}(12 \text{ m}) = 8 \text{ m}$. We can therefore represent the distributed load by an equivalent downward force of 600-N magnitude acting at $x = 8$ m (Fig. a).
(b) From the equilibrium equations

$$\Sigma F_x = A_x = 0,$$

$$\Sigma F_y = A_y + B - 600 = 0,$$

$$\Sigma M_{(\text{point } A)} = 12B - (8)(600) = 0,$$

we obtain $A_x = 0$, $A_y = 200$ N, and $B = 400$ N.

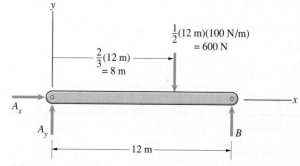

(a) Representing the distributed load by an equivalent force.

DISCUSSION

The loading curve in this example was sufficiently simple that we did not need to integrate to determine its area and centroid. In the following example we must integrate to determine the area and centroid.

Example 7.6

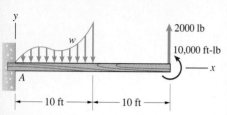

Figure 7.15

The beam in Fig. 7.15 is subjected to a distributed load, a force, and a couple. The distributed load is $w = 300x - 50x^2 + 0.3x^4$ lb/ft.
(a) Represent the distributed load by a single equivalent force.
(b) Determine the reactions at the built-in support A.

STRATEGY

(a) Since we know the function w, we can use Eq. (7.13) to determine the "area" under the loading curve, which is equal to the total force exerted by the distributed load. The x coordinate of the centroid is given by Eq. (7.14).
(b) Once the distributed load is represented by a single equivalent force, we can apply the equilibrium equations to determine the reactions at the built-in support.

SOLUTION

(a) The downward force exerted by the distributed load is

$$F = \int_L w \, dx = \int_0^{10} (300x - 50x^2 + 0.3x^4) \, dx = 4330 \text{ lb.}$$

The x coordinate of the centroid of the distributed load is

$$\bar{x} = \frac{\int_L xw \, dx}{\int_L w \, dx} = \frac{\int_0^{10} x(300x - 50x^2 + 0.3x^4) \, dx}{\int_0^{10} (300x - 50x^2 + 0.3x^4) \, dx}$$

$$= \frac{25{,}000}{4330} = 5.77 \text{ ft.}$$

The distributed load is equivalent to a downward force of 4330-lb magnitude acting at $x = 5.77$ ft.
(b) In Fig. (a), we draw the free-body diagram of the beam with the distributed force represented by the single equivalent force. From the equilibrium equations

$$\Sigma F_x = A_x = 0,$$

$$\Sigma F_y = A_y + 2000 - 4330 = 0,$$

$$\Sigma M_{(\text{point } A)} = (20)(2000) + 10{,}000 - (5.77)(4330) + M_A = 0,$$

we obtain $A_x = 0$, $A_y = 2330$ lb, and $M_A = -25{,}000$ ft-lb.

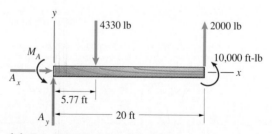

(a) Free-body diagram of the beam.

Example 7.7

The beam in Fig. 7.16 is subjected to two distributed loads. Determine the reactions at A and B.

STRATEGY

We can easily represent the uniform distributed load on the right by an equivalent force. We can treat the distributed load on the left as the sum of uniform and triangular distributed loads and represent each load by an equivalent force.

Figure 7.16

SOLUTION

We draw the free-body diagram of the beam in Fig. (a), expressing the left distributed load as the sum of uniform and triangular loads. In Fig. (b), we represent the three distributed loads by equivalent forces. The "area" of the uniform distributed load on the right is $(6 \text{ m}) \times (400 \text{ N/m}) = 2400 \text{ N}$, and its centroid is 3 m from B. The area of the uniform distributed load on the vertical part of the beam is $(6 \text{ m}) \times (400 \text{ N/m}) = 2400 \text{ N}$, and its centroid is located at $y = 3$ m. The area of the triangular distributed load is $\frac{1}{2}(6 \text{ m}) \times (400 \text{ N/m}) = 1200 \text{ N}$, and its centroid is located at $y = \frac{1}{3}(6 \text{ m}) = 2$ m.

From the equilibrium equations

$$\Sigma F_x = A_x + 1200 + 2400 = 0,$$

$$\Sigma F_y = A_y + B - 2400 = 0,$$

$$\Sigma M_{(\text{point } A)} = 6B - (3)(2400) - (2)(1200) - (3)(2400) = 0,$$

we obtain $A_x = -3600$ N, $A_y = -400$ N, and $B = 2800$ N.

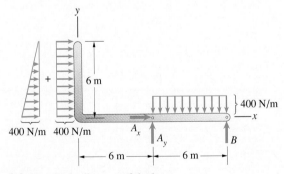

(a) Free-body diagram of the beam.

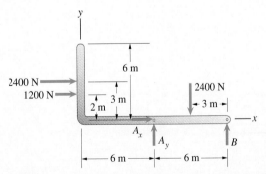

(b) Representing the distributed loads by equivalent forces.

Problems

7.46 The value of the distributed load w at $x = 0$ is 36 lb/ft.
(a) Determine the function w.
(b) Use Eq. (7.10) to determine the total force exerted by the distributed load.
(c) Determine the total force exerted by the distributed load by calculating the "area" between the loading curve and the x axis.

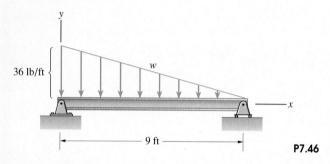

36 lb/ft

w

9 ft

P7.46

7.47 In preliminary design studies for a pedestrian bridge, an engineer models the weight of the bridge and its maximum expected load by the distributed load shown. What reactions are exerted on the supports of the bridge at A and B?

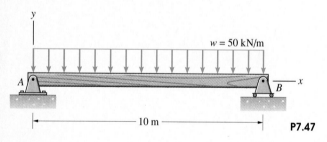

$w = 50$ kN/m

10 m

P7.47

7.48 Determine the reactions at the built-in support A.

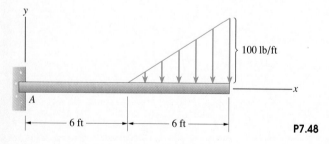

100 lb/ft

6 ft

6 ft

P7.48

7.49 Determine the reactions at A and B.

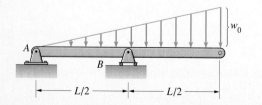

w_0

$L/2$

$L/2$

P7.49

7.50 Determine the reactions at the built-in support A.

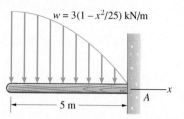

$w = 3(1 - x^2/25)$ kN/m

5 m

P7.50

7.51 An engineer measures the forces exerted by the soil on a 10-m section of a building foundation and finds that they are described by the distributed load $w = -10x - x^2 + 0.2x^3$ kN/m.
(a) Determine the magnitude of the total force exerted on the foundation by the distributed load.
(b) Determine the magnitude of the moment about A due to the distributed load.

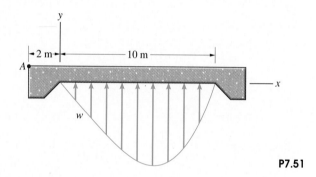

2 m

10 m

A

w

P7.51

7.52 The distributed load is $w = 10x^2$ lb/ft. Determine the reactions at A and B.

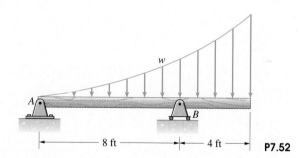

w

8 ft

4 ft

P7.52

7.53 The aerodynamic lift of the wing is described by the distributed load $w = -300\sqrt{1 - 0.04x^2}$ N/m. The mass of the wing is 27 kg, and its center of mass is located 2 m from the wing root R.
(a) Determine the magnitudes of the force and the moment about R exerted by the lift of the wing.

(b) Determine the reactions on the wing at R.

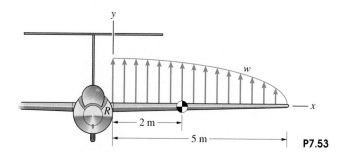

P7.53

7.54 The force $F = 2000$ lb. Determine the reactions at A and B.

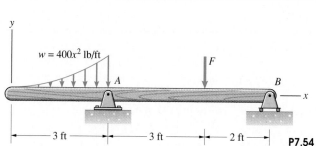

P7.54

7.55 If $w_B = 100$ lb/ft, what are the reactions at A and B?

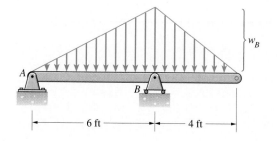

P7.55

7.56 If each beam support in Problem 7.55 will safely support a vertical load of 1000 lb, what is the largest safe value of w_B?

7.57 Determine the reactions at A and B.

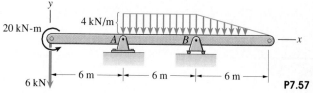

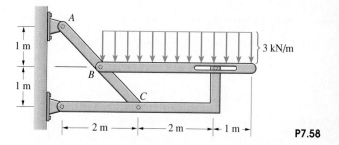

P7.57

7.58 Determine the forces on member ABC of the frame.

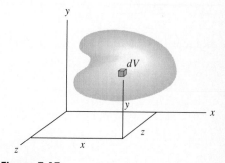

P7.58

7.4 *Centroids of Volumes and Lines*

Here we define the centroids, or average positions, of volumes and lines, and show how to determine the centroids of composite volumes and lines. We will show in Section 7.7 that knowing the centroids of volumes and lines allows you to determine the centers of mass of certain types of objects, which tells you where their weights effectively act.

Figure 7.17

A volume V and differential element dV.

Definitions

Volumes   Consider a volume V, and let dV be a differential element of V with coordinates x, y, and z (Fig. 7.17). By analogy with Eqs. (7.6) and (7.7), the coordinates of the centroid of V are

$$
\bar{x} = \frac{\displaystyle\int_V x\,dV}{\displaystyle\int_V dV}, \qquad \bar{y} = \frac{\displaystyle\int_V y\,dV}{\displaystyle\int_V dV}, \qquad \bar{z} = \frac{\displaystyle\int_V z\,dV}{\displaystyle\int_V dV}. \tag{7.15}
$$

The subscript V on the integral signs means that the integration is carried out over the entire volume.

If a volume has the form of a plate with uniform thickness and cross-sectional area A (Fig. 7.18a), its centroid coincides with the centroid of A and lies at the midpoint between the two faces. To show that this is true, we obtain a volume element dV by projecting an element dA of the cross-sectional area through the thickness T of the volume, so that $dV = T\,dA$ (Fig. 7.18b). Then the x and y coordinates of the centroid of the volume are

$$
\bar{x} = \frac{\displaystyle\int_V x\,dV}{\displaystyle\int_V dV} = \frac{\displaystyle\int_A xT\,dA}{\displaystyle\int_A T\,dA} = \frac{\displaystyle\int_A x\,dA}{\displaystyle\int_A dA},
$$

$$
\bar{y} = \frac{\displaystyle\int_V y\,dV}{\displaystyle\int_V dV} = \frac{\displaystyle\int_A yT\,dA}{\displaystyle\int_A T\,dA} = \frac{\displaystyle\int_A y\,dA}{\displaystyle\int_A dA}.
$$

The coordinate $\bar{z} = 0$ by symmetry. Thus you know the centroid of this type of volume if you know (or can determine) the centroid of its cross-sectional area.

Lines   The coordinates of the centroid of a line L are

$$
\bar{x} = \frac{\displaystyle\int_L x\,dL}{\displaystyle\int_L dL}, \qquad \bar{y} = \frac{\displaystyle\int_L y\,dL}{\displaystyle\int_L dL}, \qquad \bar{z} = \frac{\displaystyle\int_L z\,dL}{\displaystyle\int_L dL}, \tag{7.16}
$$

where dL is a differential length of the line with coordinates x, y, and z (Fig. 7.19).

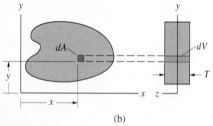

Front View   Side View

(a)

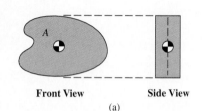

(b)

Figure 7.18

(a) A volume of uniform thickness.
(b) Obtaining dV by projecting dA through the volume.

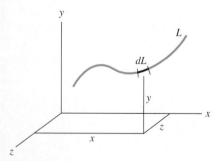

Figure 7.19

A line L and differential element dL.

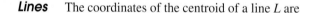

The following examples demonstrate the use of Eqs. (7.15) and (7.16) to determine centroids of volumes and lines. As in determining centroids of areas, you must choose a differential element of volume or length and choose limits so that you integrate over the desired volume or line.

Example 7.8

Determine the centroid of the cone in Fig. 7.20.

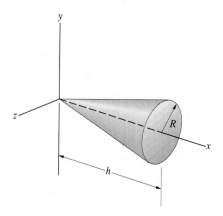

Figure 7.20

STRATEGY

The centroid must lie on the x axis because of symmetry. We will determine its x coordinate by using an element of volume dV in the form of a "disk" of width dx.

SOLUTION

Let dV be the disk in Fig. (a). The radius of the disk is $(R/h)x$ (Fig. b), and its volume equals the product of the area of the disk and its thickness, $dV = \pi[(R/h)x]^2 dx$. To integrate over the entire volume, we must integrate with respect to x from $x = 0$ to $x = h$. The x coordinate of the centroid is

$$\bar{x} = \frac{\int_V x\,dV}{\int_V dV} = \frac{\int_0^h x\pi \frac{R^2}{h^2} x^2\,dx}{\int_0^h \pi \frac{R^2}{h^2} x^2\,dx} = \frac{3}{4}h.$$

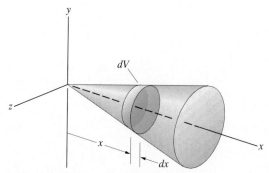

(a) An element dV in the form of a disk.

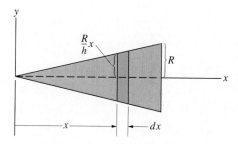

(b) The radius of the element is $(R/h)x$.

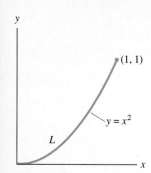

Figure 7.21

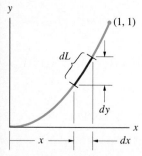

(a) A differential line element dL.

Example 7.9

The line L in Fig. 7.21 is defined by the function $y = x^2$. Determine the x coordinate of its centroid.

SOLUTION

We can express a differential element dL of the line (Fig. a) in terms of dx and dy:

$$dL = \sqrt{dx^2 + dy^2} = \sqrt{1 + \left(\frac{dy}{dx}\right)^2}\, dx.$$

From the equation describing the line, the derivative $dy/dx = 2x$, so we obtain an expression for dL in terms of x:

$$dL = \sqrt{1 + 4x^2}\, dx.$$

To integrate over the entire line, we must integrate from $x = 0$ to $x = 1$. The x coordinate of the centroid is

$$\bar{x} = \frac{\displaystyle\int_L x\, dL}{\displaystyle\int_L dL} = \frac{\displaystyle\int_0^1 x\sqrt{1 + 4x^2}\, dx}{\displaystyle\int_0^1 \sqrt{1 + 4x^2}\, dx} = 0.574.$$

Example 7.10

Determine the centroid of the semicircular line in Fig. 7.22.

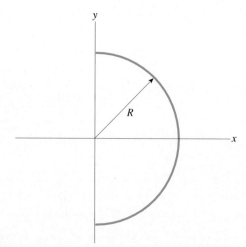

Figure 7.22

STRATEGY

Because of the symmetry of the line, the centroid lies on the x axis. To determine $\bar{x}$, we will integrate in terms of polar coordinates.

SOLUTION

By letting θ change by an amount $d\theta$, we obtain a differential line element of length $dL = R\,d\theta$ (Fig. a). The x coordinate of dL is $x = R\cos\theta$. To integrate over the entire line, we must integrate with respect to θ from $\theta = -\pi/2$ to $\theta = +\pi/2$:

$$\bar{x} = \frac{\int_L x\,dL}{\int_L dL} = \frac{\int_{-\pi/2}^{\pi/2} (R\cos\theta)R\,d\theta}{\int_{-\pi/2}^{\pi/2} R\,d\theta} = \frac{R^2\,[\sin\theta]_{-\pi/2}^{\pi/2}}{R[\theta]_{-\pi/2}^{\pi/2}} = \frac{2R}{\pi}.$$

DISCUSSION

Notice that our integration procedure gives the correct length of the line:

$$\int_L dL = \int_{-\pi/2}^{\pi/2} R\,d\theta = R[\theta]_{-\pi/2}^{\pi/2} = \pi R.$$

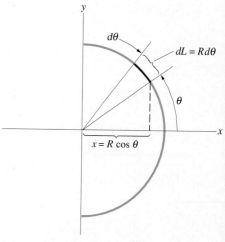

(a) A differential line element $dL = R\,d\theta$.

Problems

7.59 Use the method described in Example 7.8 to confirm that the x component of the centroid of the cylindrical volume is $\bar{x} = \frac{1}{2}L$.

7.60 A volume of revolution is obtained by revolving the curve $y = x^{1/2}$ about the x axis. Determine its centroid.

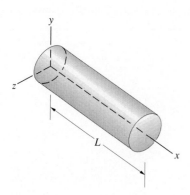

P7.59

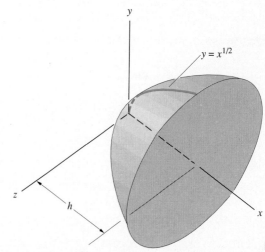

P7.60

7.61 A volume of revolution is obtained by revolving the curve $y = \frac{1}{4}x^2$ about the x axis. Its volume is 10 m^3. Determine its centroid.

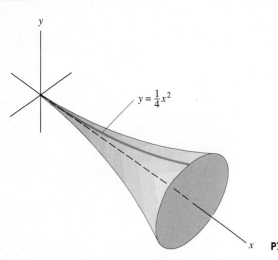

P7.61

7.62 Determine the volume and centroid of the pyramid.

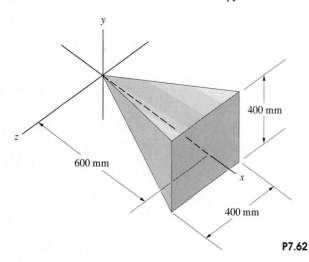

400 mm

600 mm

400 mm

P7.62

7.63 Determine the centroid of the hemispherical volume.

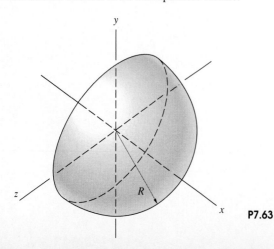

R

P7.63

7.64 The volume consists of a segment of a sphere of radius R. Determine its centroid.

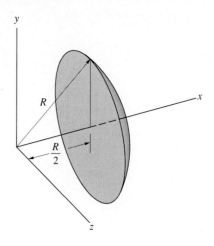

R

$\dfrac{R}{2}$

P7.64

7.65 A volume of revolution is obtained by revolving the curve $x^2/a^2 + y^2/b^2 = 1$ about the x axis. Determine its centroid.

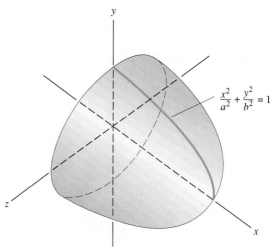

$\dfrac{x^2}{a^2} + \dfrac{y^2}{b^2} = 1$

P7.65

7.66 The volume of revolution has a cylindrical hole of radius R. Determine its centroid.

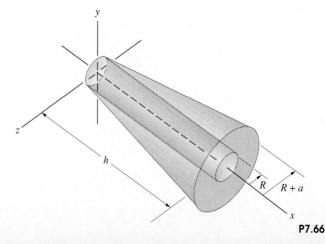

h

R $R+a$

P7.66

7.67 Determine the y coordinate of the centroid of the line (see Example 7.9).

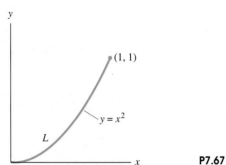

P7.67

7.68 Determine the x coordinate of the centroid of the line.

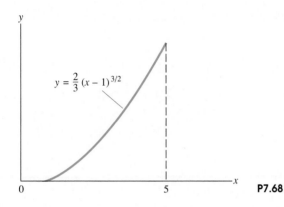

P7.68

7.69 Determine the x coordinate of the centroid of the line.

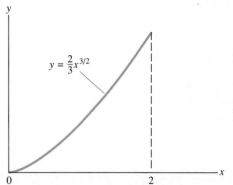

P7.69

7.70 Determine the centroid of the circular arc.

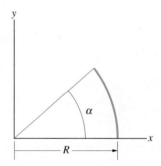

P7.70

Centroids of Composite Volumes and Lines

The centroids of composite volumes and lines can be derived using the same approach we applied to areas. The coordinates of the centroid of a composite volume are

$$\bar{x} = \frac{\sum_i \bar{x}_i V_i}{\sum_i V_i}, \qquad \bar{y} = \frac{\sum_i \bar{y}_i V_i}{\sum_i V_i}, \qquad \bar{z} = \frac{\sum_i \bar{z}_i V_i}{\sum_i V_i}, \qquad (7.17)$$

and the coordinates of the centroid of a composite line are

$$\bar{x} = \frac{\sum_i \bar{x}_i L_i}{\sum_i L_i}, \qquad \bar{y} = \frac{\sum_i \bar{y}_i L_i}{\sum_i L_i}, \qquad \bar{z} = \frac{\sum_i \bar{z}_i L_i}{\sum_i L_i}. \qquad (7.18)$$

The centroids of some simple volumes and lines are tabulated in Appendixes B and C.

In the following examples we determine the centroids of composite volumes and lines. Three steps are typically required.

1. Choose the parts—*Try to divide the composite into parts whose centroids you know or can easily determine.*

2. Determine the values for the parts—*Determine the centroid and the volume or length of each part. Watch for instances of symmetry that can simplify your task.*

3. Calculate the centroid—*Use Eqs. (7.17) or (7.18) to determine the centroid of the composite volume or line.*

Example 7.11

Determine the centroid of the volume in Fig. 7.23.

SOLUTION

Choose the Parts The volume consists of a cone and a cylinder, which we call parts 1 and 2, respectively.

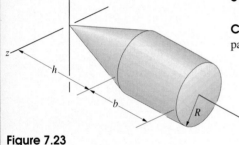

Figure 7.23

Determine the Values for the Parts The centroid and volume of the cone are given in Appendix C. The x coordinates of the centroids of the parts are shown in Fig. (a), and the information for determining the x coordinate of the centroid is summarized in Table 7.4.

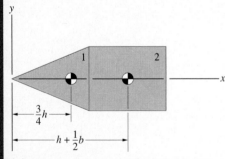

(a) The x coordinates of the centroids of the cone and cylinder.

Table 7.4 Information for determining $\bar{x}$.

	$\bar{x}_i$	V_i	$\bar{x}_i V_i$
Part 1 (cone)	$\frac{3}{4}h$	$\frac{1}{3}\pi R^2 h$	$\left(\frac{3}{4}h\right)\left(\frac{1}{3}\pi R^2 h\right)$
Part 2 (cylinder)	$h + \frac{1}{2}b$	$\pi R^2 b$	$\left(h + \frac{1}{2}b\right)\left(\pi R^2 b\right)$

Calculate the Centroid The x coordinate of the centroid of the composite volume is

$$\bar{x} = \frac{\bar{x}_1 V_1 + \bar{x}_2 V_2}{V_1 + V_2} = \frac{\left(\frac{3}{4}h\right)\left(\frac{1}{3}\pi R^2 h\right) + \left(h + \frac{1}{2}b\right)\left(\pi R^2 b\right)}{\frac{1}{3}\pi R^2 h + \pi R^2 b}.$$

Because of symmetry, $\bar{y} = 0$ and $\bar{z} = 0$.

Example 7.12

Determine the centroid of the volume in Fig. 7.24.

SOLUTION

Choose the Parts We can divide the volume into the five simple parts shown in Fig. (a). Part 5 is the volume of the 20-mm diameter hole.

Determine the Values for the Parts The centroids of parts 1 and 3 are located at the centroids of their semicircular cross sections (Fig. b). The information for determining the x coordinate of the centroid is summarized in Table 7.5. Part 5 is a negative volume.

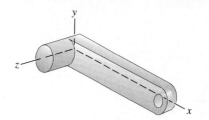

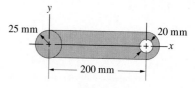

25 mm   20 mm

200 mm

SIDE VIEW

40   20 mm

mm

END VIEW

Figure 7.24

Table 7.5 Information for determining $\bar{x}$.

	$\bar{x}_i$ (mm)	V_i (mm^3)	$\bar{x}_i V_i$ (mm^4)
Part 1	$-\dfrac{4(25)}{3\pi}$	$\dfrac{\pi(25)^2}{2}(20)$	$\left[-\dfrac{4(25)}{3\pi}\right]\left[\dfrac{\pi(25)^2}{2}(20)\right]$
Part 2	100	$(200)(50)(20)$	$(100)[(200)(50)(20)]$
Part 3	$200 + \dfrac{4(25)}{3\pi}$	$\dfrac{\pi(25)^2}{2}(20)$	$\left[200 + \dfrac{4(25)}{3\pi}\right]\left[\dfrac{\pi(25)^2}{2}(20)\right]$
Part 4	0	$\pi(25)^2(40)$	0
Part 5	200	$-\pi(10)^2(20)$	$-(200)[\pi(10)^2(20)]$

Calculate the Centroid The x coordinate of the centroid of the composite volume is

$$\bar{x} = \frac{\bar{x}_1 V_1 + \bar{x}_2 V_2 + \bar{x}_3 V_3 + \bar{x}_4 V_4 + \bar{x}_5 V_5}{V_1 + V_2 + V_3 + V_4 + V_5}$$

$$= \frac{\left[-\dfrac{4(25)}{3\pi}\right]\left[\dfrac{\pi(25)^2}{2}(20)\right] + (100)[(200)(50)(20)] + \left[200 + \dfrac{4(25)}{3\pi}\right]\left[\dfrac{\pi(25)^2}{2}(20)\right] + 0 - (200)[\pi(10)^2(20)]}{\dfrac{\pi(25)^2}{2}(20) + (200)(50)(20) + \dfrac{\pi(25)^2}{2}(20) + \pi(25)^2(40) - \pi(10)^2(20)}$$

$$= 72.77 \text{ mm}.$$

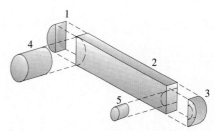

(a) Dividing the volume into five parts.

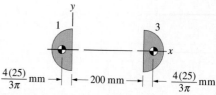

$\dfrac{4(25)}{3\pi}$ mm ⟵ ⟵ 200 mm ⟶ $\dfrac{4(25)}{3\pi}$ mm

(b) Positions of the centroids of parts 1 and 3.

The z coordinates of the centroids of the parts are zero except $\bar{z}_4 = 30$ mm. Therefore the z coordinate of the centroid of the composite volume is

$$\bar{z} = \frac{\bar{z}_4 V_4}{V_1 + V_2 + V_3 + V_4 + V_5}$$

$$= \frac{30[\pi(25)^2(40)]}{\dfrac{\pi(25)^2}{2}(20) + (200)(50)(20) + \dfrac{\pi(25)^2}{2}(20) + \pi(25)^2(40) - \pi(10)^2(20)}$$

$$= 7.56 \text{ mm}.$$

Because of symmetry, $\bar{y} = 0$.

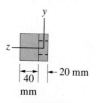

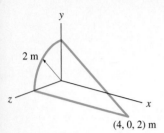

Figure 7.25

Example 7.13

Determine the centroid of the line in Fig. 7.25. The quarter-circular arc lies in the y-z plane.

SOLUTION

Choose the Parts   The line consists of a quarter-circular arc and two straight segments, which we call parts 1, 2, and 3 (Fig. a).

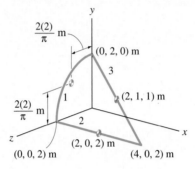

(a) Dividing the line into three parts.

Determine the Values for the Parts   From Appendix B, the coordinates of the centroid of the quarter-circular arc are $\bar{x}_1 = 0$, $\bar{y}_1 = \bar{z}_1 = 2(2)/\pi$ m. The centroids of the straight segments lie at their midpoints. For segment 2, $\bar{x}_2 = 2$ m, $\bar{y}_2 = 0$, and $\bar{z}_2 = 2$ m, and for segment 3, $\bar{x}_3 = 2$ m, $\bar{y}_3 = 1$ m, and $\bar{z}_3 = 1$ m. The length of segment 3 is $L_3 = \sqrt{(4)^2 + (2)^2 + (2)^2} = 4.90$ m. This information is summarized in Table 7.6.

Table 7.6   Information for determining the centroid.

	$\bar{x}_i$	$\bar{y}_i$	$\bar{z}_i$	L_i
Part 1	0	$2(2)/\pi$	$2(2)/\pi$	$\pi(2)/2$
Part 2	2	0	2	4
Part 3	2	1	1	4.90

Calculate the Centroid   The coordinates of the centroid of the composite line are

$$\bar{x} = \frac{\bar{x}_1 L_1 + \bar{x}_2 L_2 + \bar{x}_3 L_3}{L_1 + L_2 + L_3} = \frac{0 + (2)(4) + (2)(4.90)}{\pi + 4 + 4.90} = 1.478 \text{ m},$$

$$\bar{y} = \frac{\bar{y}_1 L_1 + \bar{y}_2 L_2 + \bar{y}_3 L_3}{L_1 + L_2 + L_3} = \frac{[2(2)/\pi][\pi(2)/2] + 0 + (1)(4.90)}{\pi + 4 + 4.90} = 0.739 \text{ m},$$

$$\bar{z} = \frac{\bar{z}_1 L_1 + \bar{z}_2 L_2 + \bar{z}_3 L_3}{L_1 + L_2 + L_3} = \frac{[2(2)/\pi][\pi(2)/2] + (2)(4) + (1)(4.90)}{\pi + 4 + 4.90} = 1.404 \text{ m}.$$

Problems

For Problems 7.71–7.78, determine the centroids of the volumes.

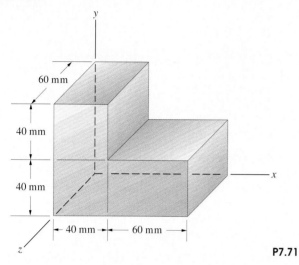

P7.71

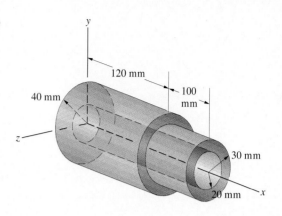

P7.74

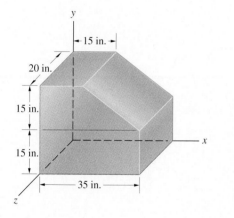

P7.72

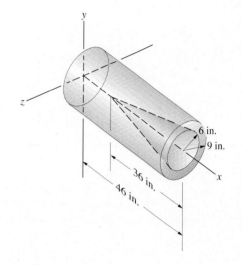

P7.75

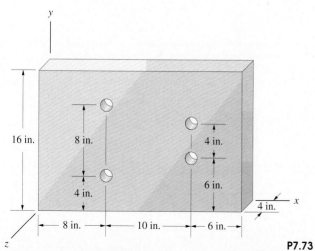

P7.73

The holes are 1 in. in diameter.

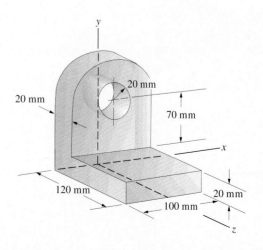

P7.76

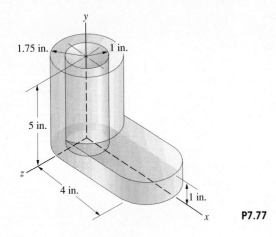

1.75 in.   1 in.

5 in.

z

4 in.   1 in.

x

P7.77

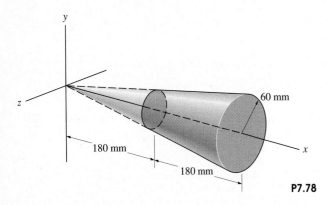

y

z

60 mm

180 mm

180 mm

x

P7.78

7.79 The dimensions of the *Gemini* spacecraft (in meters) are $a = 0.70$, $b = 0.88$, $c = 0.74$, $d = 0.98$, $e = 1.82$, $f = 2.20$, $g = 2.24$, and $h = 2.98$. Determine the centroid of its volume.

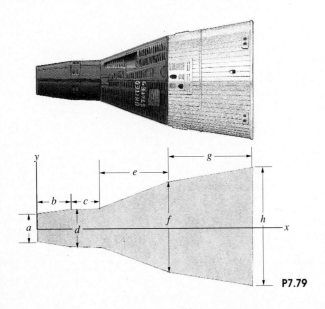

y

b   *c*   *e*   *g*

a   *d*   *f*   *h*   *x*

P7.79

7.80 Two views of a machine element are shown. Determine the centroid of its volume.

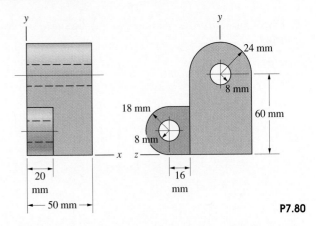

y

24 mm

8 mm

18 mm

8 mm

60 mm

x   *z*

20 mm

16 mm

50 mm

P7.80

For Problems 7.81–7.83, determine the centroids of the lines.

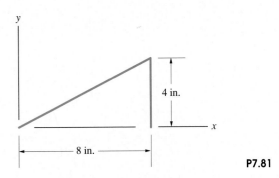

y

4 in.

8 in.

x

P7.81

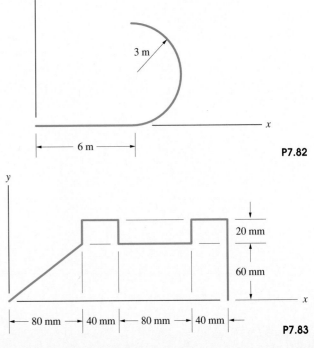

y

3 m

6 m

x

P7.82

y

20 mm

60 mm

x

80 mm   40 mm   80 mm   40 mm

P7.83

7.84 The semicircular part of the line lies in the *x*-*z* plane. Determine the centroid of the line.

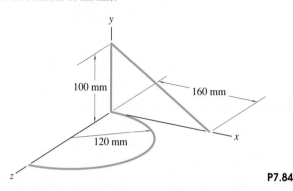

P7.84

7.85 The following theorem is *not true*: "The centroid of any area is coincident with the centroid of the line forming its boundary." Disprove it by finding a counterexample. That is, find an example for which it is not true.

7.5 *The Pappus-Guldinus Theorems*

In this section we discuss two simple and useful theorems relating surfaces and volumes of revolution to the centroids of the lines and areas that generate them.

First Theorem

Consider a line *L* in the *x*-*y* plane that does not intersect the *x* axis (Fig. 7.26a). Let the coordinates of the centroid of the line be $\bar{x}$, $\bar{y}$. We can generate a surface by revolving the line about the *x* axis (Fig. 7.26b). As the line revolves about the *x* axis, the centroid of the line moves in a circular path of radius $\bar{y}$.

The first Pappus-Guldinus theorem states that the area of the surface of revolution is equal to the product of the distance through which the centroid of the line moves and the length of the line:

$$A = 2\pi \bar{y}L. \tag{7.19}$$

To prove this result, we observe that as the line revolves about the *x* axis, the area *dA* generated by an element *dL* of the line is $dA = 2\pi y\, dL$, where *y* is the *y* coordinate of the element *dL* (Fig. 7.26c). Therefore the total area of the surface of revolution is

$$A = 2\pi \int_L y\, dL. \tag{7.20}$$

From the definition of the *y* coordinate of the centroid of the line,

$$\bar{y} = \frac{\int_L y\, dL}{\int_L dL},$$

we obtain

$$\int_L y\, dL = \bar{y}L.$$

Substituting this result into Eq. (7.20), we obtain Eq. (7.19).

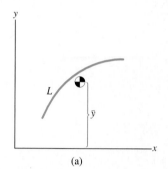

(a)

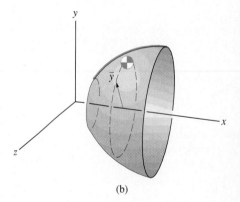

(b)

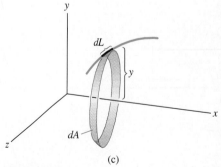

(c)

Figure 7.26
(a) A line *L* and the *y* coordinate of its centroid.
(b) The surface generated by revolving the line *L* about the *x* axis and the path followed by the centroid of the line.
(c) An element *dL* of the line and the element of area *dA* it generates.

Second Theorem

Consider an area A in the x-y plane that does not intersect the x axis (Fig. 7.27a). Let the coordinates of the centroid of the area be $\bar{x}, \bar{y}$. We can generate a volume by revolving the area about the x axis (Fig. 7.27b). As the area revolves about the x axis, the centroid of the area moves in a circular path of length $2\pi\bar{y}$.

The second Pappus-Guldinus theorem states that the volume V of the volume of revolution is equal to the product of the distance through which the centroid of the area moves and the area:

$$V = 2\pi\bar{y}A. \tag{7.21}$$

As the area revolves about the x axis, the volume dV generated by an element dA of the area is $dV = 2\pi y\, dA$, where y is the y coordinate of the element dA (Fig. 7.27c). Therefore the total volume is

$$V = 2\pi \int_A y\, dA. \tag{7.22}$$

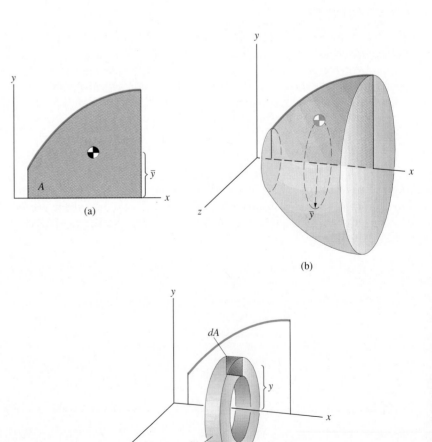

(a)

(b)

(c)

Figure 7.27

(a) An area A and the y coordinate of its centroid.
(b) The volume generated by revolving the area A about the x axis and the path followed by the centroid of the area.
(c) An element dA of the area and the element of volume dV it generates.

From the definition of the y coordinate of the centroid of the area,

$$\bar{y} = \frac{\int_A y \, dA}{A},$$

we obtain

$$\int_A y \, dA = \bar{y}A.$$

Substituting this result into Eq. (7.22), we obtain Eq. (7.21).

Example 7.14

Use the Pappus-Guldinus theorems to determine the surface area A and volume V of the cone in Fig. 7.28.

STRATEGY

We can generate the curved surface of the cone by revolving a straight line about an axis, and we can generate its volume by revolving a right triangular area about the axis. Since we know the centroids of the straight line and the triangular area, we can use the Pappus-Guldinus theorems to determine the area and volume of the cone.

SOLUTION

Revolving the straight line in Fig. (a) about the x axis generates the curved surface of the cone. The y coordinate of the centroid of the line is $\bar{y}_L = \frac{1}{2}R$, and its length is $L = \sqrt{h^2 + R^2}$. The centroid of the line moves a distance $2\pi \bar{y}_L$ as the line revolves about the x axis, so the area of the curved surface is

$$(2\pi \bar{y})L = \pi R \sqrt{h^2 + R^2}.$$

We obtain the total surface area A of the cone by adding the area of the base,

$$A = \pi R \sqrt{h^2 + R^2} + \pi R^2.$$

Revolving the triangular area in Fig. (b) about the x axis generates the volume V. The y coordinate of its centroid is $\bar{y}_T = \frac{1}{3}R$, and its area is $A = \frac{1}{2}hR$, so the volume of the cone is

$$V = (2\pi \bar{y}_T) A = \frac{1}{3}\pi hR^2.$$

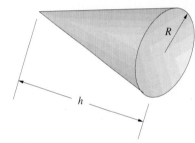

Figure 7.28

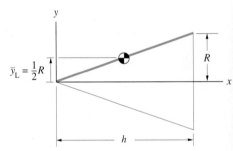

(a) The straight line that generates the curved surface of the cone.

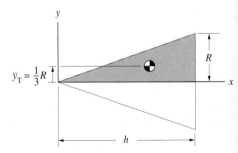

(b) The area that generates the volume of the cone.

Example 7.15

The circumference of a sphere of radius R is $2\pi R$, its surface area is $4\pi R^2$, and its volume is $\frac{4}{3}\pi R^3$. Use this information to determine (a) the centroid of a semicircular line; (b) the centroid of a semicircular area.

STRATEGY

Revolving a semicircular line about an axis generates a spherical area, and revolving a semicircular area around an axis generates a spherical volume. Knowing the area and volume, we can use the Pappus-Guldinus theorems to determine the centroids of the generating line and area.

SOLUTION

(a) Revolving the semicircular line in Fig. (a) about the x axis generates the surface area of a sphere. The length of the line is $L = \pi R$, and $\bar{y}_L$ is the y coordinate of its centroid. The centroid of the line moves a distance $2\pi\,\bar{y}_L$, so the surface area of the sphere is

$$(2\pi\,\bar{y}_L)L = 2\pi^2 R\bar{y}_L.$$

By equating this expression to the surface area $4\pi R^2$, we determine $\bar{y}_L$:

$$\bar{y}_L = \frac{2R}{\pi}.$$

(b) Revolving the semicircular area in Fig. (b) generates the sphere's volume. The area of the semicircle is $A = \frac{1}{2}\pi R^2$, and $\bar{y}_S$ is the y coordinate of its centroid. The centroid moves a distance $2\pi\,\bar{y}_S$, so the volume of the sphere is

$$(2\pi\,\bar{y}_S)A = \pi^2 R^2\,\bar{y}_S.$$

Equating this expression to the volume $\frac{4}{3}\pi R^3$, we obtain

$$\bar{y}_S = \frac{4R}{3\pi}.$$

DISCUSSION

If you can obtain a result by using the Pappus-Guldinus theorems, you will often save time and effort in comparison with other approaches. Compare this example with Example 7.10, in which we use integration to determine the centroid of a semicircular line.

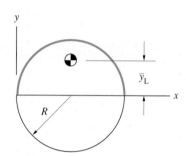

(a) Revolving a semicircular line about the x axis.

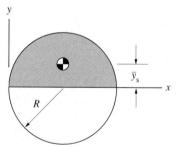

(b) Revolving a semicircular area about the x axis.

Problems

7.86 Use the second Pappus-Guldinus theorem to determine the volume generated by revolving the line about the x axis.

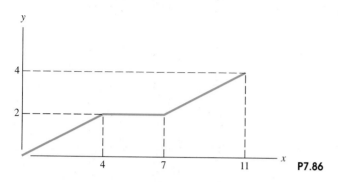

P7.86

7.87 Use the first Pappus-Guldinus theorem to determine the area of the surface generated by revolving the line in Problem 7.86 about the x axis.

Refer to P7.88 for Problems 7.88–7.91.

7.88 Use the second Pappus-Guldinus theorem to determine the volume generated by revolving the curve about the x axis.

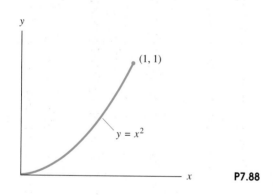

P7.88

7.89 Use the second Pappus-Guldinus theorem to determine the volume generated by revolving the curve about the y axis.

7.90 The length of the curve is $L = 1.479$, and the area generated by rotating it about the x axis is $A = 3.810$. Use the first Pappus-Guldinus theorem to determine the y coordinate of the centroid of the curve.

7.91 Use the first Pappus-Guldinus theorem to determine the area of the surface generated by revolving the curve about the y axis.

7.92 A nozzle for a large rocket engine is designed by revolving the function $y = \frac{2}{3}(x - 1)^{3/2}$ about the y axis. Use the first Pappus-Guldinus theorem to determine the surface area of the nozzle.

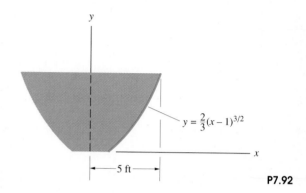

P7.92

7.93 A volume of revolution is obtained by revolving the area between the function $y = \frac{2}{3}x^{3/2}$ about the y axis. Use the second Pappus-Guldinus theorem to determine its volume.

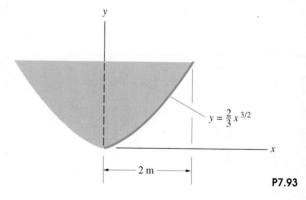

P7.93

7.94 Use the first Pappus-Guldinus theorem to determine the area of the curved surface of the volume of revolution in Problem 7.93.

7.95 The volume of revolution contains a hole of radius R.
(a) Use integration to determine its volume.
(b) Use the second Pappus-Guldinus theorem to determine its volume.

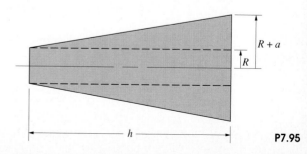

P7.95

7.96 Determine the volume of the volume of revolution.

7.97 Determine the surface area of the volume of revolution in Problem 7.96.

7.98 Determine the area of the surface of revolution obtained by revolving the line in Problem 7.83 about the *x* axis.

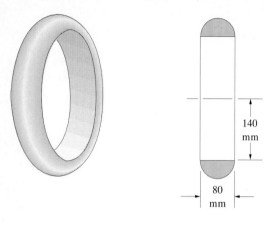

140 mm

80 mm

P7.96

Centers of Mass

The **center of mass** of an object is the centroid, or average position, of its mass. In the following section we give the analytical definition of the center of mass and demonstrate one of its most important properties: *An object's weight can be represented by a single equivalent force acting at its center of mass.* We then discuss how to locate centers of mass and show that for particular classes of objects, the center of mass coincides with the centroid of a volume, area, or line. Finally, we show how to locate centers of mass of composite objects.

7.6 Definition of the Center of Mass

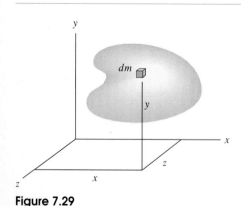

Figure 7.29

An object and differential element of mass *dm*.

The center of mass of an object is defined by

$$\bar{x} = \frac{\displaystyle\int_m x \, dm}{\displaystyle\int_m dm}, \qquad \bar{y} = \frac{\displaystyle\int_m y \, dm}{\displaystyle\int_m dm}, \qquad \bar{z} = \frac{\displaystyle\int_m z \, dm}{\displaystyle\int_m dm}, \qquad (7.23)$$

where *x*, *y*, and *z* are the coordinates of the differential element of mass *dm* (Fig. 7.29). The subscripts *m* indicate that the integration must be carried out over the entire mass of the object.

Before considering how to determine the center of mass of an object, we will demonstrate that the weight of an object can be represented by a single equivalent force acting at its center of mass. Consider an element of mass *dm* of an object (Fig. 7.30a). If the *y* axis of the coordinate system points upward, the weight of *dm* is $-dm \, g\mathbf{j}$. Integrating this expression over the mass *m*, we obtain the total weight of the object,

$$\int_m -g\mathbf{j}\, dm = -mg\mathbf{j} = -W\mathbf{j}.$$

The moment of the weight of the element dm about the origin is

$$(x\mathbf{i} + y\mathbf{j} + z\mathbf{k}) \times (-dm\, g\mathbf{j}) = gz\mathbf{i}\, dm - gx\mathbf{k}\, dm.$$

Integrating this expression over m, we obtain the total moment about the origin due to the weight of the object:

$$\int_m (gz\mathbf{i}\, dm - gx\mathbf{k}\, dm) = mg\bar{z}\mathbf{i} - mg\bar{x}\mathbf{k} = W\bar{z}\mathbf{i} - W\bar{x}\mathbf{k}.$$

If we represent the weight of the object by the force $-W\mathbf{j}$ acting at the center of mass (Fig. 7.30b), the moment of this force about the origin is equal to the total moment due to the weight:

$$(\bar{x}\mathbf{i} + \bar{y}\mathbf{j} + \bar{z}\mathbf{k}) \times (-W\mathbf{j}) = W\bar{z}\mathbf{i} - W\bar{x}\mathbf{k}.$$

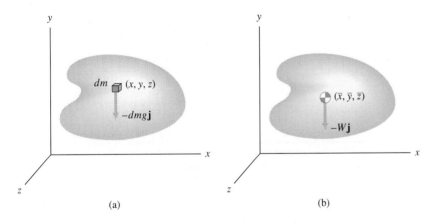

Figure 7.30
(a) Weight of the element dm.
(b) Representing the weight by a single force at the center of mass.

This result shows that when you are concerned only with the total force and total moment exerted by the weight of an object, you can assume that its weight acts at the center of mass.

7.7 Centers of Mass of Objects

To apply Eqs. (7.23) to specific objects, we will change the variable of integration from mass to volume by introducing the mass density.

The **mass density** ρ of an object is defined such that the mass of a differential element of its volume is $dm = \rho\, dV$. The dimensions of ρ are therefore (mass)/(volume). For example, it can be expressed in kg/m^3 in SI units or in slug/ft^3 in U.S. Customary units. The total mass of an object is

$$m = \int_m dm = \int_V \rho\, dV. \qquad (7.24)$$

An object whose mass density is uniform throughout its volume is said to be **homogeneous**. In this case, the total mass equals the product of the mass density and the volume:

$$m = \rho \int_V dV = \rho V. \qquad \text{\textbf{Homogeneous object}} \qquad (7.25)$$

The **weight density** $\gamma = g\rho$. It can be expressed in N/m^3 in SI units or in lb/ft^3 in U.S. Customary units. The weight of an element of volume dV of an object is $dW = \gamma \, dV$, and the total weight of a homogeneous object equals γV.

By substituting $dm = \rho \, dV$ into Eqs. (7.23), we can express the coordinates of the center of mass in terms of volume integrals:

$$\bar{x} = \frac{\int_V \rho x \, dV}{\int_V \rho \, dV}, \qquad \bar{y} = \frac{\int_V \rho y \, dV}{\int_V \rho \, dV}, \qquad \bar{z} = \frac{\int_V \rho z \, dV}{\int_V \rho \, dV}. \qquad (7.26)$$

If ρ is known as a function of position in an object, these integrals determine its center of mass. Furthermore, we can use them to show that the centers of mass of particular classes of objects coincide with centroids of volumes, areas, and lines:

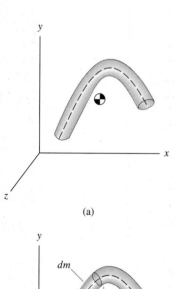

Front View   **Side View**

Figure 7.31

A plate of uniform thickness.

- **The center of mass of a homogeneous object coincides with the centroid of its volume.** If an object is homogeneous, $\rho =$ constant and Eqs. (7.26) become the equations for the centroid of the volume,

$$\bar{x} = \frac{\int_V x \, dV}{\int_V dV}, \qquad \bar{y} = \frac{\int_V y \, dV}{\int_V dV}, \qquad \bar{z} = \frac{\int_V z \, dV}{\int_V dV}.$$

- **The center of mass of a homogeneous plate of uniform thickness coincides with the centroid of its cross-sectional area** (Fig. 7.31). The center of mass of the plate coincides with the centroid of its volume, and we showed in Section 7.4 that the centroid of the volume of a plate of uniform thickness coincides with the centroid of its cross-sectional area.

- **The center of mass of a homogeneous slender bar of uniform cross-sectional area coincides approximately with the centroid of the axis of the bar** (Fig. 7.32a). The axis of the bar is defined to be the line through the centroid of its cross section. Let $dm = \rho A \, dL$, where A is the cross-sectional area of the bar and dL is a differential element of length of its axis (Fig. 7.32b). If we substitute this expression into Eqs. (7.26), they become the equations for the centroid of the axis:

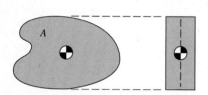

(a)

$$\bar{x} = \frac{\int_L x \, dL}{\int_L dL}, \qquad \bar{y} = \frac{\int_L y \, dL}{\int_L dL}, \qquad \bar{z} = \frac{\int_L z \, dL}{\int_L dL}.$$

This result is approximate because the center of mass of the element dm does not coincide with the centroid of the cross section in regions where the bar is curved.

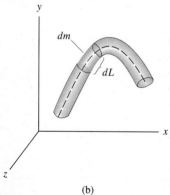

(b)

Figure 7.32

(a) A slender bar and the centroid of its axis.
(b) The element dm.

We have shown that you can locate the centers of mass of homogeneous objects by determining centroids of volumes, areas, and lines. In the following example we determine the center of mass of a homogeneous slender bar by calculating the centroid of its axis. We then demonstrate the use of Eqs. (7.26) to determine the center of mass of a nonhomogeneous object.

Example 7.16

The mass of the homogeneous slender bar in Fig. 7.33 is 80 kg. What are the reactions at A and B?

STRATEGY

We determine the reactions in two ways.

First Method We represent the weight of each straight segment of the bar by a force acting at the center of mass of the segment.

Second Method We determine the center of mass of the bar by determining the centroid of its axis and represent the weight of the bar by a single force acting at the center of mass.

SOLUTION

First Method In the free-body diagram in Fig. (a), we place half of the weight of the bar at the center of mass of each straight segment. From the equilibrium equations

$$\Sigma F_x = A_x - B = 0,$$

$$\Sigma F_y = A_y - (40)(9.81) - (40)(9.81) = 0,$$

$$\Sigma M_{(\text{point } A)} = (1)B - (1)(40)(9.81) - (0.5)(40)(9.81) = 0,$$

we obtain $A_x = 589$ N, $A_y = 785$ N, and $B = 589$ N.

Second Method We can treat the centerline of the bar as a composite line composed of two straight segments (Fig. b). The coordinates of the centroid of the composite line are

$$\bar{x} = \frac{\bar{x}_1 L_1 + \bar{x}_2 L_2}{L_1 + L_2} = \frac{(0.5)(1) + (1)(1)}{1 + 1} = 0.75 \text{ m},$$

$$\bar{y} = \frac{\bar{y}_1 L_1 + \bar{y}_2 L_2}{L_1 + L_2} = \frac{(0)(1) + (0.5)(1)}{1 + 1} = 0.25 \text{ m}.$$

In the free-body diagram in Fig. (c), we place the weight of the bar at its center of mass. From the equilibrium equations

$$\Sigma F_x = A_x - B = 0,$$

$$\Sigma F_y = A_y - (80)(9.81) = 0,$$

$$\Sigma M_{(\text{point } A)} = (1)B - (0.75)(80)(9.81) = 0,$$

we again obtain $A_x = 589$ N, $A_y = 785$ N, and $B = 589$ N.

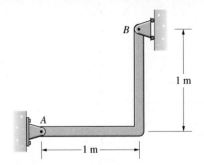

Figure 7.33

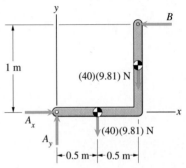

(a) Placing the weights of the straight segments at their centers of mass.

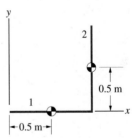

(b) Centroids of the straight segments of the axis.

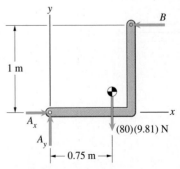

(c) Placing the weight of the bar at its center of mass.

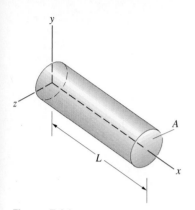

Figure 7.34

Example 7.17

Determine the mass of the cylinder in Fig. 7.34 and the position of its center of mass if (a) it is homogeneous with mass density ρ_0; (b) its density is given by the equation $\rho = \rho_0(1 + x/L)$.

STRATEGY

In (a), the mass of the cylinder is simply the product of its mass density and its volume and the center of mass is located at the centroid of its volume. In (b), the cylinder is nonhomogeneous and we must use Eqs. (7.24) and (7.26) to determine its mass and center of mass.

SOLUTION

(a) The volume of the cylinder is LA, so its mass is $\rho_0 LA$. Since the center of mass is coincident with the centroid of the volume of the cylinder, the coordinates of the center of mass are $\bar{x} = \frac{1}{2}L, \bar{y} = 0, \bar{z} = 0$.

(b) We can determine the mass of the cylinder by using an element of volume dV in the form of a disk of thickness dx (Fig. a). The volume $dV = A\,dx$. The mass of the cylinder is

$$m = \int_V \rho\,dV = \int_0^L \rho_0\left(1 + \frac{x}{L}\right)A\,dx = \frac{3}{2}\rho_0 AL.$$

The x coordinate of the center of mass is

$$\bar{x} = \frac{\displaystyle\int_V x\rho\,dV}{\displaystyle\int_V \rho\,dV} = \frac{\displaystyle\int_0^L \rho_0\left(x + \frac{x^2}{L}\right)A\,dx}{\frac{3}{2}\rho_0 AL} = \frac{5}{9}L.$$

Because the density does not depend on y or z, we know from symmetry that $\bar{y} = 0$ and $\bar{z} = 0$.

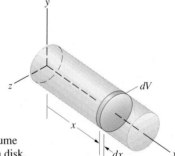

(a) An element of volume dV in the form of a disk.

DISCUSSION

Notice that the center of mass of the nonhomogeneous cylinder is *not* located at the centroid of its volume.

Problems

7.99 The homogeneous flat plate weighs 200 lb. What are the reactions at A and B?

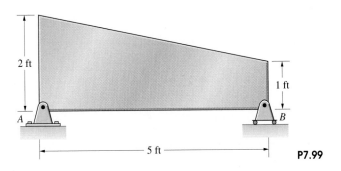

P7.99

7.100 The mass of the homogeneous flat plate is 50 kg. Determine the reactions at the supports A and B.

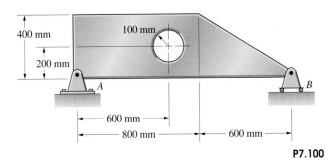

P7.100

7.101 The suspended sign is a homogeneous flat plate that has a mass of 130 kg. Determine the axial forces in members AD and CE. (Notice that the y axis is positive downward.)

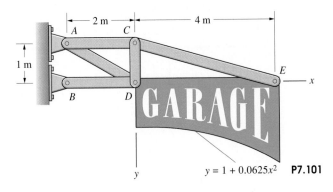

$y = 1 + 0.0625x^2$ **P7.101**

7.102 The circular slender bar has a mass of 20 kg. What are the reactions at A and B?

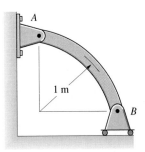

P7.102

7.103 The semicircular part of the homogeneous slender bar lies in the x-z plane. Determine the center of mass of the bar.

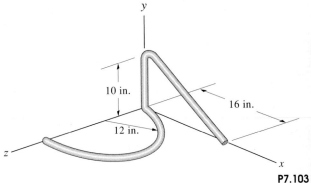

P7.103

7.104 When the truck is unloaded, the total reactions at the front and rear wheels are $A = 12$ kip and $B = 8$ kip. The load of gravel weighs 100 lb/ft³. The dimension of the load in the z direction is 10 ft, and its surface profile, given by the function shown, does not depend on z. What are the total reactions at the front and rear wheels of the loaded truck?

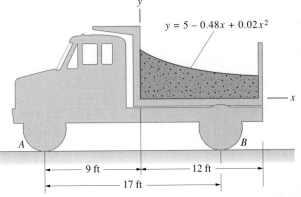

$y = 5 - 0.48x + 0.02x^2$

P7.104

7.105 The 10-ft horizontal cylinder with 1-ft radius is supported at A and B. Its weight density is $\gamma = 100(1 - 0.002x^2)$ lb/ft^3. What are the reactions at A and B?

7.106 A horizontal cone with 800-mm length and 200-mm radius has a built-in support at A. Its mass density is $\rho = 6000(1 + 0.4x^2)$ kg/m^3, where x is in meters. What are the reactions at A?

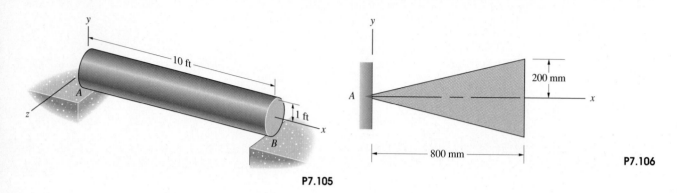

P7.105

P7.106

7.8 *Centers of Mass of Composite Objects*

You can easily determine the center of mass of an object consisting of a combination of parts if you know the centers of mass of its parts. The coordinates of the center of mass of a composite object composed of parts with masses m_1, $m_2, \ldots$, are

$$\bar{x} = \frac{\sum\limits_i \bar{x}_i m_i}{\sum\limits_i m_i}, \qquad \bar{y} = \frac{\sum\limits_i \bar{y}_i m_i}{\sum\limits_i m_i}, \qquad \bar{z} = \frac{\sum\limits_i \bar{z}_i m_i}{\sum\limits_i m_i}, \qquad (7.27)$$

where $\bar{x}_i, \bar{y}_i, \bar{z}_i$ are the coordinates of the centers of mass of the parts. Because the weights of the parts are related to their masses by $W_i = gm_i$, Eqs. (7.27) can also be expressed as

$$\bar{x} = \frac{\sum\limits_i \bar{x}_i W_i}{\sum\limits_i W_i}, \qquad \bar{y} = \frac{\sum\limits_i \bar{y}_i W_i}{\sum\limits_i W_i}, \qquad \bar{z} = \frac{\sum\limits_i \bar{z}_i W_i}{\sum\limits_i W_i}. \qquad (7.28)$$

When you know the masses or weights and the centers of mass of the parts of a composite object, you can use these equations to determine its center of mass.

In the following examples we determine the centers of mass of composite objects. Three steps are typically required.

1. Choose the parts—*Try to divide the object into parts whose centers of mass you know or can easily determine.*

2. Determine the values for the parts—*Determine the center of mass and the mass or weight of each part. Watch for instances of symmetry that can simplify your task.*

3. Calculate the center of mass—*Use Eqs. (7.27) or (7.28) to determine the center of mass of the composite object.*

Example 7.18

The L-shaped machine part in Fig. 7.35 is composed of two homogeneous bars. Bar 1 is tungsten alloy with mass density 14,000 kg/m³, and bar 2 is steel with mass density 7800 kg/m³. Determine the center of mass of the machine part.

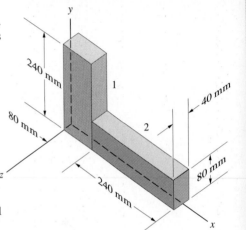

Figure 7.35

SOLUTION

The volume of bar 1 is

$$(80)(240)(40) = 7.68 \times 10^5 \text{ mm}^3 = 7.68 \times 10^{-4} \text{ m}^3,$$

so its mass is $(7.68 \times 10^{-4})(1.4 \times 10^4) = 10.75$ kg. The center of mass of bar 1 coincides with the centroid of its volume: $\bar{x}_1 = 40$ mm, $\bar{y}_1 = 120$ mm, $\bar{z}_1 = 0$.

Bar 2 has the same volume as bar 1, so its mass is $(7.68 \times 10^{-4})(7.8 \times 10^3) = 5.99$ kg. The coordinates of its center of mass are $\bar{x}_2 = 200$ mm, $\bar{y}_2 = 40$ mm, $\bar{z}_2 = 0$. Using the information summarized in Table 7.7, we obtain the x coordinate of the center of mass,

$$\bar{x} = \frac{\bar{x}_1 m_1 + \bar{x}_2 m_2}{m_1 + m_2} = \frac{(40)(10.75) + (200)(5.99)}{10.75 + 5.99} = 97.2 \text{ mm},$$

and the y coordinate,

$$\bar{y} = \frac{\bar{y}_1 m_1 + \bar{y}_2 m_2}{m_1 + m_2} = \frac{(120)(10.75) + (40)(5.99)}{10.75 + 5.99} = 91.4 \text{ mm}.$$

Because of the symmetry of the object, $\bar{z} = 0$.

Table 7.7   Information for determining the center of mass.

	m_i (kg)	$\bar{x}_i$ (mm)	$\bar{x}_i m_i$ (mm-kg)	$\bar{y}_i$ (mm)	$\bar{y}_i m_i$ (mm-kg)
Bar 1	10.75	40	(40)(10.75)	120	(120)(10.75)
Bar 2	5.99	200	(200)(5.99)	40	(40)(5.99)

Example 7.19

The composite object in Fig. 7.36 consists of a bar welded to a cylinder. The homogeneous bar is aluminum (weight density 168 lb/ft^3), and the homogeneous cylinder is bronze (weight density 530 lb/ft^3). Determine the center of mass of the object.

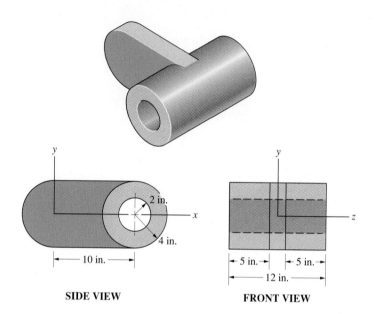

Figure 7.36   **SIDE VIEW**   **FRONT VIEW**

STRATEGY

We can determine the weight of each homogeneous part by multiplying its volume by its weight density. We also know that the center of mass of each part coincides with the centroid of its volume. The centroid of the cylinder is located at its center, but we must determine the location of the centroid of the bar by treating it as a composite volume.

SOLUTION

The volume of the cylinder is $12[\pi(4)^2 - \pi(2)^2] = 452$ in^3 $= 0.262$ ft^3, so its weight is

$$W_{(cylinder)} = (0.262)(530) = 138.8 \text{ lb}.$$

The x coordinate of its center of mass is $\bar{x}_{(cylinder)} = 10$ in.

The volume of the bar is $(10)(8)(2) + \frac{1}{2}\pi(4)^2(2) - \frac{1}{2}\pi(4)^2(2) = 160$ in^3 $= 0.0926$ ft^3, and its weight is

$$W_{(bar)} = (0.0926)(168) = 15.6 \text{ lb}.$$

We can determine the centroid of the volume of the bar by treating it as a composite volume consisting of three parts (Fig. a). Part 3 is a semicircular "cutout." The centroids of part 1 and the semicircular cutout 3 are located at the centroids of their semicircular cross sections (Fig. b). Using the information summarized in Table 7.8, we have

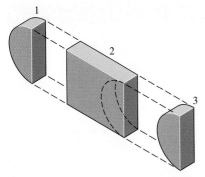

(a) Dividing the bar into three parts.

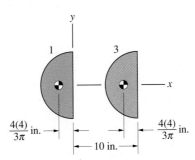

(b) The centroids of the two semicircular parts.

Table 7.8   Information for determining the x coordinate of the centroid of the bar.

	$\bar{x}_i$ (in.)	V_i (in^3)	$\bar{x}_i V_i$ (in^4)
Part 1	$-\dfrac{4(4)}{3\pi}$	$\frac{1}{2}\pi(4)^2(2)$	$-\dfrac{4(4)}{3\pi}\left[\frac{1}{2}\pi(4)^2(2)\right]$
Part 2	5	$(10)(8)(2)$	$5[(10)(8)(2)]$
Part 3	$10 - \dfrac{4(4)}{3\pi}$	$-\frac{1}{2}\pi(4)^2(2)$	$-\left[10 - \dfrac{4(4)}{3\pi}\right]\left[\frac{1}{2}\pi(4)^2(2)\right]$

$$\bar{x}_{(bar)} = \frac{\bar{x}_1 V_1 + \bar{x}_2 V_2 + \bar{x}_3 V_3}{V_1 + V_2 + V_3}$$

$$= \frac{-\dfrac{4(4)}{3\pi}\left[\frac{1}{2}\pi(4)^2(2)\right] + 5[(10)(8)(2)] - \left[10 - \dfrac{4(4)}{3\pi}\right]\left[\frac{1}{2}\pi(4)^2(2)\right]}{\frac{1}{2}\pi(4)^2(2) + (10)(8)(2) - \frac{1}{2}\pi(4)^2(2)}$$

$$= 1.86 \text{ in.}$$

Therefore the x coordinate of the center of mass of the composite object is

$$\bar{x} = \frac{\bar{x}_{(bar)} W_{(bar)} + \bar{x}_{(cylinder)} W_{(cylinder)}}{W_{(bar)} + W_{(cylinder)}}$$

$$= \frac{(1.86)(15.6) + (10)(138.8)}{15.6 + 138.8} = 9.18 \text{ in.}$$

Because of the symmetry of the bar, the y and z coordinates of its center of mass are $\bar{y} = 0$ and $\bar{z} = 0$.

Example 7.20

Application to Engineering

Centers of Mass of Vehicles

A car is placed on a platform that measures the normal force exerted by each tire independently (Fig. 7.37). Measurements made with the platform horizontal and with the platform tilted at $\alpha = 15°$ are shown in Table 7.9. Determine the position of the car's center of mass.

Table 7.9   Measurements of the normal forces exerted by the tires.

Wheelbase = 111 in. Track = 60.9 in.	Measured Loads (lb)	
	$\alpha = 0$	$\alpha = 15°$
Left front wheel, N_{LF}	1147.5	1003.4
Right front wheel, N_{RF}	1130.2	988.3
Left rear wheel, N_{LR}	812.2	889.5
Right rear wheel, N_{RR}	800.1	876.3

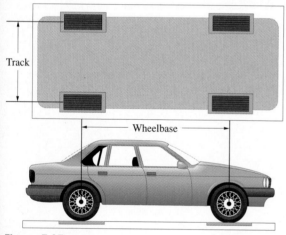

Track

Wheelbase

Figure 7.37

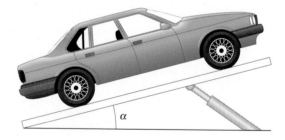

α

SOLUTION

We draw the free-body diagram of the car when the platform is in the horizontal position in Figs. (a) and (b). The car's weight is

$$W = N_{LF} + N_{RF} + N_{LR} + N_{RR} = 1147.5 + 1130.2 + 812.2 + 800.1$$
$$= 3890.0 \text{ lb.}$$

From Fig. (a), we obtain the equilibrium equation

$$\Sigma M_{(z \text{ axis})} = (\text{wheelbase})(N_{LF} + N_{RF}) - \bar{x}W = 0,$$

which we can solve for $\bar{x}$:

$$\bar{x} = \frac{(\text{wheelbase})(N_{LF} + N_{RF})}{W} = \frac{(111)(1147.5 + 1130.2)}{3890.0} = 65.0 \text{ in.}$$

From Fig. (b),

$$\Sigma M_{(x \text{ axis})} = \bar{z}W - (\text{track})(N_{RF} + N_{RR}) = 0,$$

which we can solve for $\bar{z}$:

$$\bar{z} = \frac{(\text{track})(N_{RF} + N_{RR})}{W} = \frac{(60.9)(1130.2 + 800.1)}{3890.0} = 30.2 \text{ in.}$$

Now that we know $\bar{x}$, we can determine $\bar{y}$ from the free-body diagram of the car when the platform is in the tilted position (Fig. c). From the equilibrium equation

$$\Sigma M_{(z \text{ axis})} = (\text{wheelbase})(N_{LF} + N_{RF}) + \bar{y}W \sin 15° - \bar{x}W \cos 15°$$
$$= 0,$$

we obtain

$$\bar{y} = \frac{\bar{x}W \cos 15° - (\text{wheelbase})(N_{LF} + N_{RF})}{W \sin 15°}$$
$$= \frac{(65.0)(3890.0) \cos 15° - (111)(1003.4 + 988.3)}{3890.0 \sin 15°} = 23.0 \text{ in.}$$

Notice that we could not have determined $\bar{y}$ without the measurements made with the car in the tilted position.

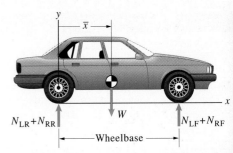

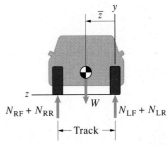

(a) Side view of the free-body diagram with the platform horizontal.

(b) Front view of the free-body diagram with the platform horizontal.

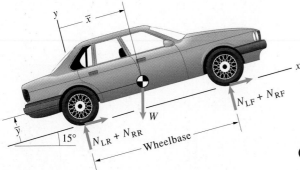

(c) Side view of the free-body diagram with the platform tilted.

DESIGN ISSUES

The center of mass of a vehicle affects its operation and performance. The forces exerted on the suspensions and wheels of cars and train coaches, the tractions their wheels create, and their dynamic behaviors are affected by the locations of their centers of mass. Not only are the performances of airplanes affected by the locations of their centers of mass, but they cannot fly unless their centers of mass lie within prescribed bounds. For engineers who design vehicles, the position of the center of mass is one of the principal parameters governing decisions about the configuration of the vehicle and the layout of its contents. In testing new designs of both land vehicles and airplanes, the position of the center of mass is affected by the configuration of the particular vehicle and the weights and locations of stowage and passengers. It is often necessary to locate the center of mass experimentally by a technique such as the one we have described. Such experimental measurements are also used to confirm center of mass locations predicted by calculations made during design.

Problems

7.107 The circular cylinder is made of aluminum (Al) with mass density 2700 kg/m³ and iron (Fe) with mass density 7860 kg/m³.
(a) Determine the centroid of the volume of the cylinder.
(b) Determine the center of mass of the cylinder.

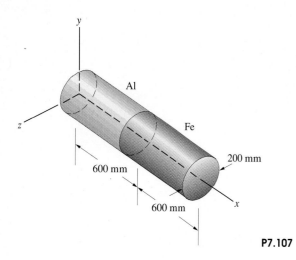

P7.107

7.108 The cylindrical tube is made of aluminum with mass density 2700 kg/m³. The cylindrical plug is made of steel with mass density 7800 kg/m³. Determine the coordinates of the center of mass of the composite object.

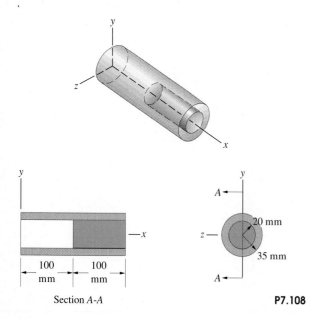

Section A-A

P7.108

7.109 A machine consists of three parts. The masses and the locations of the centers of mass of the parts are

Part	Mass (kg)	$\bar{x}$ (mm)	$\bar{y}$ (mm)	$\bar{z}$ (mm)
1	2.0	100	50	−20
2	4.5	150	70	0
3	2.5	180	30	0

Determine the coordinates of the center of mass of the machine.

7.110 A machine consists of three parts. The masses and the locations of the centers of mass of two of the parts are

Part	Mass (kg)	$\bar{x}$ (mm)	$\bar{y}$ (mm)	$\bar{z}$ (mm)
1	2.0	100	50	−20
2	4.5	150	70	0

The mass of part 3 is 2.5 kg. The design engineer wants to position part 3 so that the center of mass location of the machine is $\bar{x} = 120$ mm, $\bar{y} = 80$ mm, $\bar{z} = 0$. Determine the necessary position of the center of mass of part 3.

7.111 Two views of a machine element are shown. Part 1 is aluminum alloy with mass density 2800 kg/m³, and part 2 is steel with mass density 7800 kg/m³. Determine the x coordinate of its center of mass.

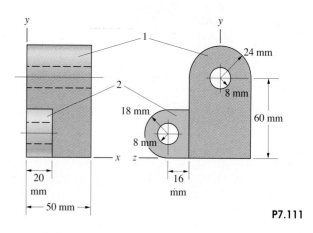

P7.111

7.112 Determine the y and z coordinates of the center of mass of the machine element in Problem 7.111.

7.113 With its engine removed, the mass of the car is 1100 kg and its center of mass is at C. The mass of the engine is 220 kg.
(a) Suppose that you want to place the center of mass E of the engine so that the center of mass of the car is midway between the front wheels A and the rear wheels B. What is the distance b?

(b) If the car is parked on a 15° slope facing up the slope, what total normal force is exerted by the road on the rear wheels B?

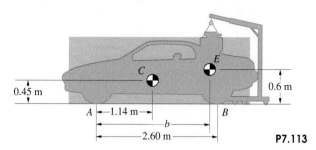

P7.113

7.114 The airplane is parked with its landing gear resting on scales. The weights measured at A, B, and C are 30 kN, 140 kN, and 146 kN, respectively. After a crate is loaded onto the plane, the weights measured at A, B, and C are 31 kN, 142 kN, and 147 kN, respectively. Determine the mass and the x and y coordinates of the center of mass of the crate.

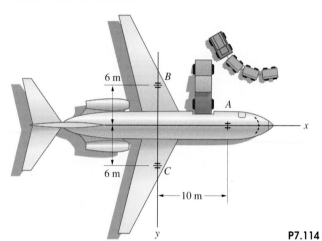

P7.114

Problems 7.115 and 7.116 are related to Example 7.20.

7.115 A suitcase weighing 200 lb is put into the trunk of the car described in Example 7.20. The position of the center of mass of the suitcase is $\bar{x}_s = -21.0$ in., $\bar{y}_s = 30.0$ in., and $\bar{z}_s = -12.0$ in. If you consider the suitcase a part of the car, what is the position of the car's center of mass?

7.116 A group of engineering students constructs a miniature device of the kind described in Example 7.20 and uses it to determine the center of mass of a miniature vehicle. The data they obtain are shown in the following table:

Wheelbase = 36 in. Track = 30 in.	Measured Loads (lb)	
	$\alpha = 0$	$\alpha = 10°$
Left front wheel, N_{LF}	35	32
Right front wheel, N_{RF}	36	33
Left rear wheel, N_{LR}	27	34
Right rear wheel, N_{RR}	29	30

Determine the center of mass of the vehicle. Use the same coordinate system as in Example 7.20.

Chapter Summary

Centroids

A **centroid** is a weighted average position. The coordinates of the centroid of an area A in the x-y plane are

$$\bar{x} = \frac{\int_A x \, dA}{\int_A dA}, \qquad \bar{y} = \frac{\int_A y \, dA}{\int_A dA}. \qquad \textbf{Eqs. (7.6), (7.7)}$$

The coordinates of the centroid of a **composite area** composed of parts A_1, A_2, . . . , are

$$\bar{x} = \frac{\sum_i \bar{x}_i A_i}{\sum_i A_i}, \qquad \bar{y} = \frac{\sum_i \bar{y}_i A_i}{\sum_i A_i}. \qquad \textbf{Eq. (7.9)}$$

Similar equations define the centroids of volumes (Eqs. 7.15 and 7.17) and lines (Eqs. 7.16 and 7.18).

Distributed Forces

A force distributed along a line is described by a function w, defined such that the force on a differential element dx of the line is $w\,dx$. The force exerted by a distributed load is

$$F = \int_L w\,dx, \qquad \textbf{Eq. (7.10)}$$

and the moment about the origin is

$$M = \int_L xw\,dx. \qquad \textbf{Eq. (7.11)}$$

The force F is equal to the "area" between the function w and the x axis and is equivalent to the distributed load if it is placed at the centroid of the "area."

The Pappus-Guldinus Theorems

First Theorem Consider a line of length L in the x-y plane with centroid $\bar{x}$, $\bar{y}$. The area A of the surface generated by revolving the line about the x axis is

$$A = 2\pi\,\bar{y}L. \qquad \textbf{Eq. (7.19)}$$

Second Theorem Let A be an area in the x-y plane with centroid $\bar{x}$, $\bar{y}$. The volume V generated by revolving A about the x axis is

$$V = 2\pi\,\bar{y}A. \qquad \textbf{Eq. (7.21)}$$

Centers of Mass

The **center of mass** of an object is the centroid of its mass. The weight of an object can be represented by a single equivalent force acting at its center of mass.

The **mass density** ρ is defined such that the mass of a differential element of volume is $dm = \rho\,dV$. An object whose mass density is uniform throughout its volume is said to be **homogeneous**. The **weight density** $\gamma = g\rho$.

The coordinates of the center of mass of an object are

$$\bar{x} = \frac{\int_V \rho x\,dV}{\int_V \rho\,dV}, \qquad \bar{y} = \frac{\int_V \rho y\,dV}{\int_V \rho\,dv}, \qquad \bar{z} = \frac{\int_V \rho z\,dV}{\int_V \rho\,dV}. \qquad \textbf{Eq. (7.26)}$$

The center of mass of a homogeneous object coincides with the centroid of its volume. The center of mass of a homogeneous plate of uniform thickness coincides with the centroid of its cross-sectional area. The center of mass of a homogeneous slender bar of uniform cross-sectional area coincides approximately with the centroid of the axis of the bar.

Review Problems

7.117 Determine the centroid of the area by letting dA be a vertical strip of width dx.

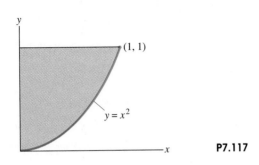

P7.117

7.118 Determine the centroid of the area in Problem 7.117 by letting dA be a horizontal strip of height dy.

7.119 Determine the centroid of the area.

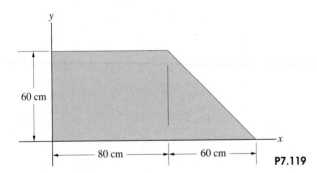

P7.119

7.120 Determine the centroid of the area.

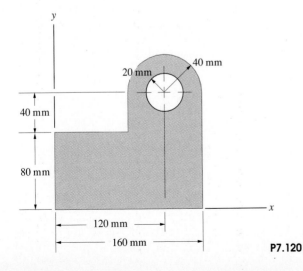

P7.120

7.121 The cantilever beam is subjected to a triangular distributed load. What are the reactions at A?

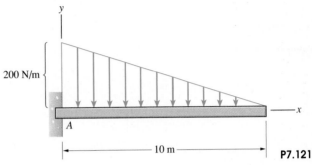

P7.121

7.122 What is the axial load in member BD of the frame?

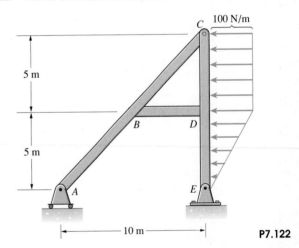

P7.122

7.123 An engineer estimates that the maximum wind load on the 40-m tower in Fig. (a) is described by the distributed load in Fig. (b). The tower is supported by three cables, A, B, and C, from the top of the tower to equally spaced points 15 m from the bottom of the tower (Fig. c). If the wind blows from the west and cables B and C are slack, what is the tension in cable A? (Model the base of the tower as a ball and socket support.)

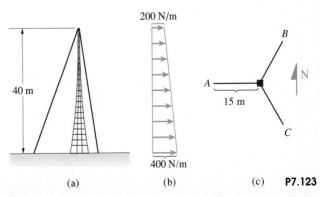

(a)   (b)   (c)   **P7.123**

7.124   If the wind in Problem 7.123 blows from the east and cable *A* is slack, what are the tensions in cables *B* and *C*?

7.125   Estimate the centroid of the volume of the *Apollo* lunar return configuration (not including its rocket nozzle) by treating it as a cone and a cylinder.

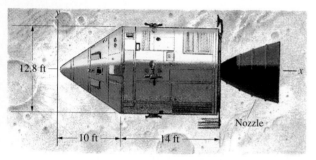

P7.125

7.126   The shape of the rocket nozzle of the *Apollo* lunar return configuration is approximated by revolving the curve shown around the *x* axis. In terms of the coordinate system shown, determine the centroid of the volume of the nozzle.

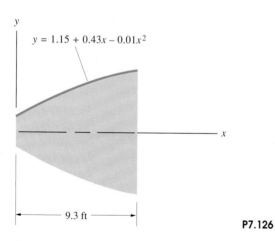

$y = 1.15 + 0.43x - 0.01x^2$

9.3 ft

P7.126

7.127   Determine the volume of the volume of revolution.

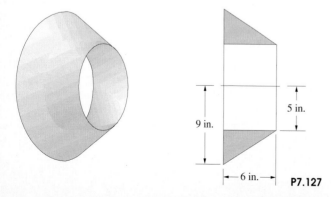

9 in.

5 in.

6 in.

P7.127

7.128   Determine the surface area of the volume of revolution in Problem 7.127.

7.129   Determine the *y* coordinate of the center of mass of the homogeneous steel plate.

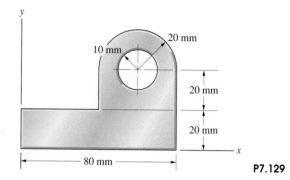

20 mm

10 mm

20 mm

20 mm

80 mm

P7.129

7.130   Determine the *x* coordinate of the center of mass of the homogeneous steel plate.

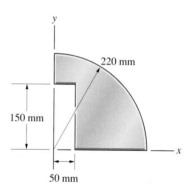

220 mm

150 mm

50 mm

P7.130

7.131   The area of the homogeneous plate is 10 ft². The vertical reactions on the plate at *A* and *B* are 80 lb and 84 lb, respectively. Suppose that you want to equalize the reactions at *A* and *B* by drilling a 1-ft diameter hole in the plate. What horizontal distance from *A* should the center of the hole be? What are the resulting reactions at *A* and *B*?

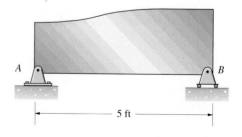

A

B

5 ft

P7.131

7.132 The plate is of uniform thickness and is made of homogeneous material whose mass per unit area of the plate is 2 kg/m². The vertical reactions at A and B are 6 N and 10 N, respectively. What is the x coordinate of the centroid of the hole?

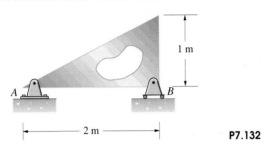

P7.132

7.133 Determine the center of mass of the homogeneous sheet of metal.

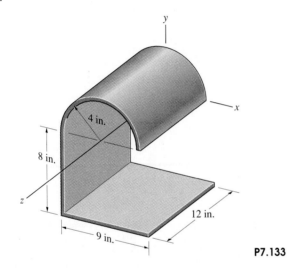

P7.133

7.134 Determine the center of mass of the homogeneous object.

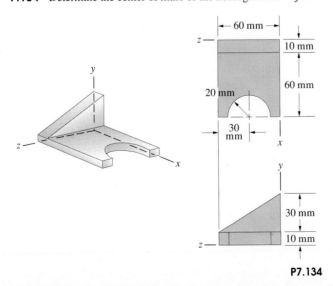

P7.134

7.135 Determine the center of mass of the homogeneous object.

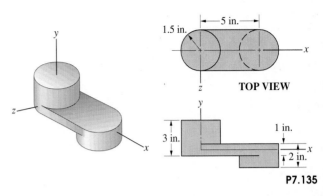

P7.135

7.136 The arrangement shown can be used to determine the position of the center of mass of a person. A horizontal board has a pin support at A and rests on a scale that measures weight at B. The distance from point A to point B is 90 in. When the person is not on the board, the scale measures 20 lb.
(a) When a 135-lb person is in position (1), the scale measures 98 lb. What is the x coordinate of the person's center of mass?
(b) When the person is in position (2), the scale measures 94 lb. What is the x coordinate of the person's center of mass?

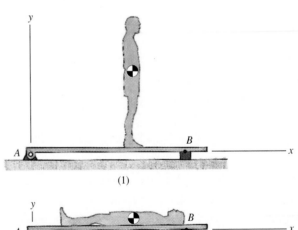

(1)

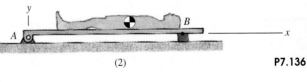

(2)

P7.136

7.137 If a string is tied to the slender bar at A and the bar is allowed to hang freely, what will be the angle between AB and the vertical?

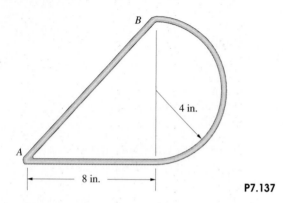

4 in.

8 in.

P7.137

7.138 The positions of the centers of three homogeneous spheres of equal radii are shown. The mass density of sphere 1 is ρ_0, the mass density of sphere 2 is $1.2\rho_0$, and the mass density of sphere 3 is $1.4\rho_0$.
(a) Determine the centroid of the volume of the three spheres.
(b) Determine the center of mass of the three spheres.

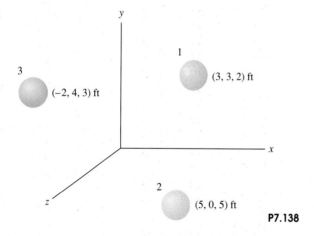

3

$(-2, 4, 3)$ ft

1

$(3, 3, 2)$ ft

2

$(5, 0, 5)$ ft

P7.138

7.139 The mass of the moon is 0.0123 times the mass of the earth. If the moon's center of mass is 238,000 mi from the center of mass of the earth, what is the distance from the center of mass of the earth to the center of mass of the earth-moon system?

7.140 Engineering students participating in a charity benefit take part in an event in which a team of four students must stand on a circular platform with a ball and socket support at the center and additional supports at the edges. Students A and B stand at the points shown. Knowing the coordinates of the points where A and B stand, the team must calculate the coordinates of points where students C and D should stand so that the center of mass of the team is at the center of the platform. The supports at the edges of the platform are then removed, and if the team has miscalculated, the platform will tilt and they will slide into a vat of mud. The students' weights are $W_A = 182$ lb, $W_B = 122$ lb, $W_C = 165$ lb, and $W_D = 124$ lb. Determine the coordinates of points where C and D can safely stand.

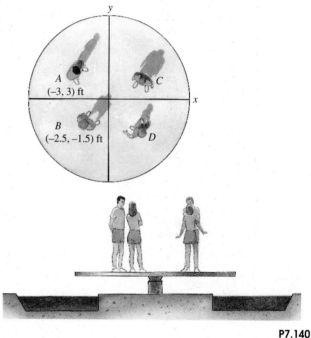

A
$(-3, 3)$ ft

C

B
$(-2.5, -1.5)$ ft

D

P7.140

Project 7.1 Construct a homogeneous thin flat plate with the shape shown. (Use the cardboard back of a pad of paper to construct the plate. Choose your dimensions so that the plate is as large as possible.) Calculate the location of the center of mass of the plate. Measuring as carefully as possible, mark the center of mass clearly on both sides of the plate. Then carry out the following experiments.

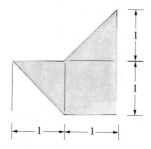

Proj. 7.1

(a) Balance the plate on your finger (Fig. a) and observe that it balances at its center of mass. Explain the result of this experiment by drawing a free-body diagram of the plate.

(b) This experiment requires a needle or slender nail, a length of string, and a small weight. Tie the weight to one end of the string and make a small loop at the other end. Stick the needle through the plate at any point other than its center of mass. Hold the needle horizontal so that the plate hangs freely from it (Fig. b). Use the loop to hang the weight from the needle, and let the weight hang freely so that the string lies along the face of the plate. Observe that the string passes through the center of mass of the plate. Repeat this experiment several times, sticking the needle through various points on the plate. Explain the results of this experiment by drawing a free-body diagram of the plate.

(c) Hold the plate so that the plane of the plate is vertical, and throw the plate upward, spinning it like a Frisbee. Observe that the plate spins about its center of mass.

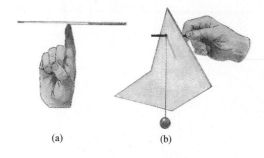

(a)   (b)

uppose that the two beams are made of the same material and have the same length and cross-sectional area. If we subject them to equal vertical loads as shown, the left beam will deflect much less because of the shape of its cross section. The property of the cross section that determines a beam's resistance to bending about a particular axis is called its moment of inertia, which we discuss in this chapter.

Chapter 8

Moments of Inertia

THE quantities called moments of inertia arise repeatedly in analyses of engineering problems. Moments of inertia of areas are used in the study of distributed forces and in calculating deflections of beams. The moment exerted by the pressure on a submerged flat plate can be expressed in terms of the moment of inertia of the plate's area. In dynamics, mass moments of inertia are used in calculating the rotational motions of objects. To be prepared to study these applications, moments of inertia must be part of your technical repertoire.

We introduce both moments of inertia of areas and mass moments of inertia in this chapter. We show how to calculate the moments of inertia of simple areas or objects and then use results called parallel-axis theorems to calculate moments of inertia of more complex areas or objects.

Areas

8.1 Definitions

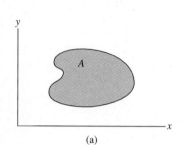

(a)

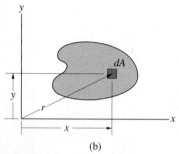

(b)

Figure 8.1

(a) An area A in the x-y plane.
(b) A differential element of A.

The moments of inertia of an area are integrals similar in form to those used to determine the centroid of an area. Consider an area A in the x-y plane (Fig. 8.1a). We define four moments of inertia of A:

1. **Moment of inertia about the x axis:**

$$I_x = \int_A y^2 \, dA, \qquad (8.1)$$

where y is the y coordinate of the differential element of area dA (Fig. 8.1b). This moment of inertia is sometimes expressed in terms of the **radius of gyration** about the x axis, k_x, defined by

$$I_x = k_x^2 A. \qquad (8.2)$$

2. **Moment of inertia about the y axis:**

$$I_y = \int_A x^2 \, dA, \qquad (8.3)$$

where x is the x coordinate of the element dA (Fig. 8.1b). The radius of gyration about the y axis, k_y, is defined by

$$I_y = k_y^2 A. \qquad (8.4)$$

3. **Product of inertia:**

$$I_{xy} = \int_A xy \, dA. \qquad (8.5)$$

4. **Polar moment of inertia:**

$$J_O = \int_A r^2 \, dA, \qquad (8.6)$$

where r is the radial distance from the origin O of the coordinate system to dA (Fig. 8.1b). The radius of gyration about O, k_O, is defined by

$$J_O = k_O^2 A. \qquad (8.7)$$

The polar moment of inertia is equal to the sum of the moments of inertia about the x and y axes:

$$J_O = \int_A r^2 \, dA = \int_A (y^2 + x^2) \, dA = I_x + I_y.$$

Substituting the expressions for the moments of inertia in terms of the radii of gyration into this equation, we obtain

$$k_O^2 = k_x^2 + k_y^2.$$

The dimensions of the moments of inertia of an area are (length)^4, and the radii of gyration have dimensions of length. You can see that the definitions of the moments of inertia I_x, I_y, and J_O and the radii of gyration imply that they have positive values for any area. They cannot be negative or zero. We can also make some qualitative deductions about the values of these moments of inertia based on their definitions. In Fig. 8.2, $A_1 = A_2$. But because the contribution of a given element dA to the integral for I_x is proportional to the *square* of its perpendicular distance from the x axis, the value of I_x is larger for A_2 than for A_1: $(I_x)_2 > (I_x)_1$. For the same reason, $(I_y)_2 < (I_y)_1$ and $(J_O)_2 > (J_O)_1$. The circular areas in Fig. 8.3 are identical, but because of their positions relative to the coordinate system, $(I_x)_2 > (I_x)_1$, $(I_y)_2 > (I_y)_1$, and $(J_O)_2 > (J_O)_1$.

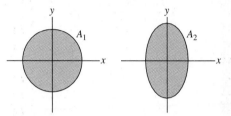

Figure 8.2

The areas are equal, $A_1 = A_2$. Based on their shapes, we can make conclusions about the relative sizes of their moments of inertia.

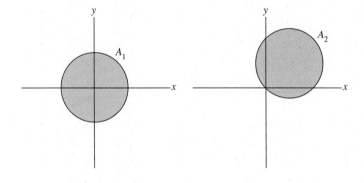

Figure 8.3

Identical areas that have different moments of inertia in terms of the xy coordinate system.

The areas A_2 and A_3 in Fig. 8.4 are obtained by rotating A_1 about the y and x axes, respectively. We can see from the definitions that the moments of inertia I_x, I_y, and J_O of these areas are equal. The products of inertia $(I_{xy})_2 = -(I_{xy})_1$ and $(I_{xy})_3 = -(I_{xy})_1$: For each element dA of A_1 with coordinates (x, y), there is a corresponding element of A_2 with coordinates $(-x, y)$ and a corresponding element of A_3 with coordinates $(x, -y)$. These results also imply that if an area is symmetric about either the x axis or the y axis, its product of inertia is zero.

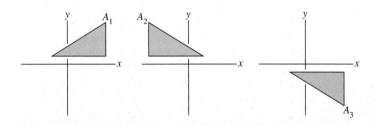

Figure 8.4

Areas obtained by rotating an area about the y and x axes.

The following examples demonstrate the use of Eqs. (8.1)–(8.7) to determine the moments of inertia of areas. Because the forms of the integrals are so similar to those used to determine the centroids of areas (Section 7.1), you can often use the same methods to evaluate them.

Example 8.1

Determine the moments of inertia and radii of gyration of the triangular area in Fig. 8.5.

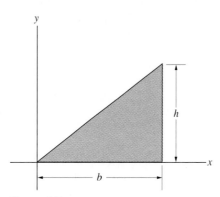

Figure 8.5

STRATEGY

Equation 8.3 for the moment of inertia about the y axis is very similar in form to the equation for the x coordinate of the centroid of an area, and we can evaluate it for this triangular area in exactly the same way: by using a differential element of area dA in the form of a vertical strip of width dx. We can then show that I_x and I_{xy} can be evaluated using the same element of area. The polar moment of inertia J_O is the sum of I_x and I_y.

SOLUTION

Let dA be the vertical strip in Fig. (a). The height of the strip is $(h/b)x$, so $dA = (h/b)x\,dx$. To integrate over the entire area, we must integrate with respect to x from $x = 0$ to $x = b$.

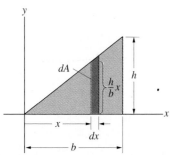

(a) An element dA in the form of a strip.

Moment of Inertia About the *y* Axis

$$I_y = \int_A x^2\,dA = \int_0^b x^2\left(\frac{h}{b}x\right)dx = \frac{h}{b}\left[\frac{x^4}{4}\right]_0^b = \frac{1}{4}hb^3.$$

The radius of gyration k_y is

$$k_y = \sqrt{\frac{I_y}{A}} = \sqrt{\frac{(1/4)hb^3}{(1/2)bh}} = \frac{1}{\sqrt{2}}b.$$

Moment of Inertia About the x Axis   We will first determine the moment of inertia of the strip dA about the x axis while holding x and dx fixed. In terms of the element of area $dA_s = dx\,dy$ shown in Fig. (b),

$$(I_x)_{\text{strip}} = \int_{\text{strip}} y^2\, dA_s = \int_0^{(h/b)x} (y^2\, dx)\, dy$$

$$= \left[\frac{y^3}{3} \right]_0^{(h/b)x} dx = \frac{h^3}{3b^3} x^3\, dx.$$

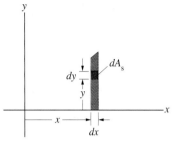

(b) An element of the strip element dA.

Integrating this expression with respect to x from $x = 0$ to $x = b$, we obtain the value of I_x for the entire area:

$$I_x = \int_0^b \frac{h^3}{3b^3} x^3\, dx = \frac{1}{12} bh^3.$$

The radius of gyration k_x is

$$k_x = \sqrt{\frac{I_x}{A}} = \sqrt{\frac{(1/12)bh^3}{(1/2)bh}} = \frac{1}{\sqrt{6}} h.$$

Product of Inertia   We can determine I_{xy} the same way we determined I_x. We first evaluate the product of inertia of the strip dA, holding x and dx fixed (Fig. b):

$$(I_{xy})_{\text{strip}} = \int_{\text{strip}} xy\, dA_s = \int_0^{(h/b)x} (xy\, dx)\, dy$$

$$= \left[\frac{y^2}{2} \right]_0^{(h/b)x} x\, dx = \frac{h^2}{2b^2} x^3\, dx.$$

Integrating this expression with respect to x from $x = 0$ to $x = b$, we obtain the value of I_{xy} for the entire area:

$$I_{xy} = \int_0^b \frac{h^2}{2b^2} x^3\, dx = \frac{1}{8} b^2 h^2.$$

Polar Moment of Inertia

$$J_O = I_x + I_y = \frac{1}{12} bh^3 + \frac{1}{4} hb^3.$$

The radius of gyration k_O is

$$k_O = \sqrt{k_x^2 + k_y^2} = \sqrt{\frac{1}{6} h^2 + \frac{1}{2} b^2}.$$

DISCUSSION

Just as in determining the centroids of areas, you can often determine the moments of inertia of an area by using a differential element of area in the form of a strip, as we did in this example. You should consider this approach if you can conveniently express the boundaries of the area in terms of cartesian coordinates.

Example 8.2

Determine the moments of inertia and radii of gyration of the circular area in Fig. 8.6.

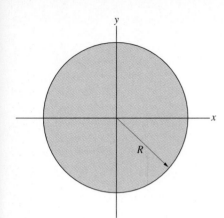

Figure 8.6

STRATEGY

We will first determine the polar moment of inertia J_O by integrating in terms of polar coordinates. We know from the symmetry of the area that $I_x = I_y$, and since $I_x + I_y = J_O$, the moments of inertia I_x and I_y are each equal to $\frac{1}{2}J_O$. We also know from the symmetry of the area that $I_{xy} = 0$.

SOLUTION

By letting r change by an amount dr, we obtain an annular element of area $dA = 2\pi r\, dr$ (Fig. a). The polar moment of inertia is

$$J_O = \int_A r^2\, dA = \int_0^R 2\pi r^3\, dr = 2\pi \left[\frac{r^4}{4}\right]_0^R = \frac{1}{2}\pi R^4,$$

and the radius of gyration about O is

$$k_O = \sqrt{\frac{J_O}{A}} = \sqrt{\frac{(1/2)\pi R^4}{\pi R^2}} = \frac{1}{\sqrt{2}}R.$$

The moments of inertia about the x and y axes are

$$I_x = I_y = \frac{1}{2}J_O = \frac{1}{4}\pi R^4,$$

and the radii of gyration about the x and y axes are

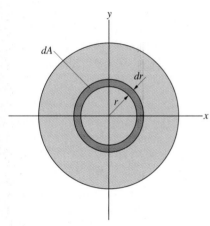

(a) An annular element dA.

$$k_x = k_y = \sqrt{\frac{I_x}{A}} = \sqrt{\frac{(1/4)\pi R^4}{\pi R^2}} = \frac{1}{2}R.$$

The product of inertia is zero:

$$I_{xy} = 0.$$

DISCUSSION

The symmetry of this example saved us from having to integrate to determine I_x, I_y, and I_{xy}. Be alert for symmetry that can shorten your work. In particular, remember that $I_{xy} = 0$ if the area is symmetric about either the x or the y axis.

Problems

Refer to P8.1 for Problems 8.1–8.3.

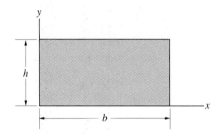

P8.1

8.1 Determine I_y and k_y.

8.2 Determine I_x and k_x by letting dA be (a) a horizontal strip of height dy; (b) a vertical strip of width dx.

8.3 Determine I_{xy}.

Refer to P8.4 for Problems 8.4 and 8.5.

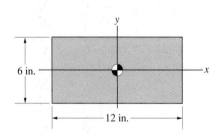

P8.4

8.4 Determine I_x, k_x, I_y, and k_y.

8.5 Determine I_{xy} and J_O.

Refer to P8.6 for Problems 8.6–8.8.

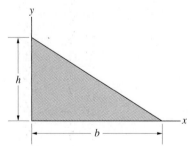

P8.6

8.6 Determine I_y and k_y.

8.7 Determine J_O and k_O.

8.8 Determine I_{xy}.

Refer to P8.9 for Problems 8.9–8.12.

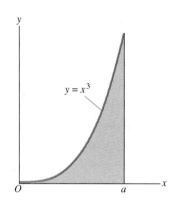

P8.9

8.9 Determine I_y and k_y.

8.10 Determine I_x and k_x.

8.11 Determine J_O and k_O.

8.12 Determine I_{xy}.

Refer to P8.13 for Problems 8.13–8.16.

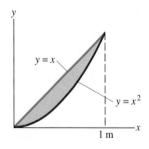

P8.13

8.13 Determine I_y and k_y.

8.14 Determine I_x and k_x.

8.15 Determine J_O and k_O.

8.16 Determine I_{xy}.

Refer to P8.17 for Problems 8.17 and 8.18.

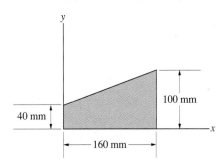

P8.17

8.17   Determine k_y.

8.18   Determine k_O.

8.19   (a) Determine I_y and k_y by letting dA be a vertical strip of width dx.
(b) The polar moment of inertia of a circular area with its center at the origin is $J_O = \frac{1}{2}\pi R^4$. Explain how you can use this information to confirm your answer to (a).

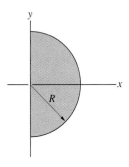

P8.19

8.20   (a) Determine I_x and k_x for the area in Problem 8.19 by letting dA be a horizontal strip of height dy.
(b) The polar moment of inertia of a circular area with its center at the origin is $J_O = \frac{1}{2}\pi R^4$. Explain how you can use this information to confirm your answer to (a).

8.21   Determine J_O and k_O.

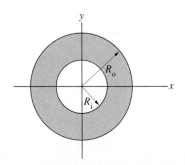

P8.21

8.22   Determine I_y and k_y for the elliptical area of the airplane's wing.

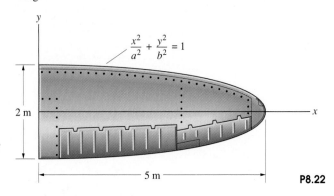

P8.22

8.23   Determine I_x and k_x for the elliptical area of the airplane's wing in Problem 8.22.

8.24   Determine I_y and k_y.

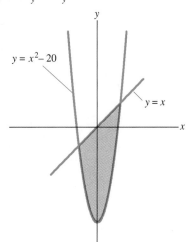

P8.24

8.25   Determine I_x and k_x for the area in Problem 8.24.

8.26   A vertical plate of area A is beneath the surface of a stationary body of water. The pressure of the water subjects each element dA of the surface of the plate to a force $(p_0 + \gamma y)\, dA$, where p_0 is the pressure at the surface of the water and γ is the weight density of the water. Show that the magnitude of the moment about the x axis due to the pressure on the front face of the plate is

$$M_{(x\,\text{axis})} = p_0 \bar{y} A + \gamma I_x,$$

where $\bar{y}$ is the y coordinate of the centroid of A and I_x is the moment of inertia of A about the x axis.

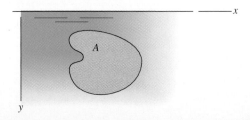

P8.26

8.2 *Parallel-Axis Theorems*

In some situations you will know the moments of inertia of an area in terms of a particular coordinate system but will need their values in terms of a different coordinate system. When the coordinate systems are parallel, you can obtain the desired moments of inertia by using the theorems we describe in this section. Furthermore, these theorems make it possible for you to determine the moments of inertia of a composite area when you know the moments of inertia of its parts.

Suppose that we know the moments of inertia of an area A in terms of a coordinate system $x'y'$ with its origin at the centroid of the area, and we wish to determine the moments of inertia in terms of a parallel coordinate system xy (Fig. 8.7a). We denote the coordinates of the centroid of A in the xy coordinate system by (d_x, d_y), and $d = \sqrt{d_x^2 + d_y^2}$ is the distance from the origin of the xy coordinate system to the centroid (Fig. 8.7b).

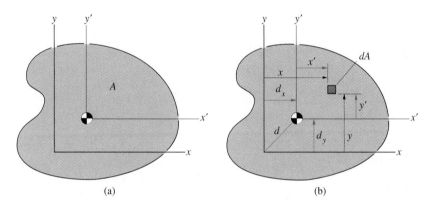

(a)   (b)

Figure 8.7

(a) The area A and the coordinate systems $x'y'$ and xy.

(b) The differential element dA.

We need to obtain two preliminary results before deriving the parallel-axis theorems. In terms of the $x'y'$ coordinate system, the coordinates of the centroid of A are

$$\bar{x}' = \frac{\int_A x' \, dA}{A}, \qquad \bar{y}' = \frac{\int_A y' \, dA}{A}.$$

But the origin of the $x'y'$ coordinate system is located at the centroid of A, so $\bar{x}' = 0$ and $\bar{y}' = 0$. Therefore

$$\int_A x' \, dA = 0, \qquad \int_A y' \, dA = 0. \tag{8.8}$$

Moment of Inertia About the x Axis   In terms of the xy coordinate system, the moment of inertia of A about the x axis is

$$I_x = \int_A y^2 \, dA, \tag{8.9}$$

where y is the coordinate of the element of area dA relative to the xy coordinate system. From Fig. 8.7(b), you can see that $y = y' + d_y$, where y' is the coordinate of dA relative to the $x'y'$ coordinate system. Substituting this expression into Eq. (8.9), we obtain

$$I_x = \int_A (y' + d_y)^2 \, dA = \int_A (y')^2 \, dA + 2d_y \int_A y' \, dA + d_y^2 \int_A dA.$$

The first integral on the right is the moment of inertia of A about the x' axis. From Eq. (8.8), the second integral on the right equals zero. Therefore we obtain

$$I_x = I_{x'} + d_y^2 A. \tag{8.10}$$

This is a **parallel-axis theorem**. It relates the moment of inertia of A about the x' axis through the centroid to the moment of inertia about the parallel axis x (Fig. 8.8).

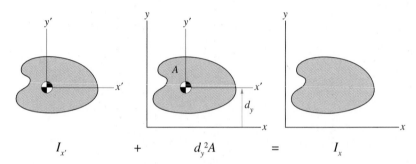

$$I_{x'} \qquad + \qquad d_y^2 A \qquad = \qquad I_x$$

Figure 8.8
The parallel-axis theorem for the moment of inertia about the x axis.

Moment of Inertia About the y Axis   In terms of the xy coordinate system, the moment of inertia of A about the y axis is

$$I_y = \int_A x^2 \, dA = \int_A (x' + d_x)^2 \, dA$$

$$= \int_A (x')^2 \, dA + 2d_x \int_A x' \, dA + d_x^2 \int_A dA.$$

From Eq. (8.8), the second integral on the right equals zero. Therefore the parallel-axis theorem that relates the moment of inertia of A about the y' axis through the centroid to the moment of inertia about the parallel axis y is

$$I_y = I_{y'} + d_x^2 A. \tag{8.11}$$

Product of Inertia   The parallel-axis theorem for the product of inertia is

$$I_{xy} = I_{x'y'} + d_x d_y A. \tag{8.12}$$

Polar Moment of Inertia   The parallel-axis theorem for the polar moment of inertia is

$$J_O = J'_O + (d_x^2 + d_y^2)A = J'_O + d^2A,$$ (8.13)

where d is the distance from the origin of the $x'y'$ coordinate system to the origin of the xy coordinate system.

How can you use the parallel-axis theorems to determine the moments of inertia of a composite area? Suppose that you want to determine the moment of inertia about the y axis of the area in Fig. 8.9(a). You can divide it into a triangle, a semicircle, and a circular cutout, which we call parts 1, 2, and 3 (Fig. 8.9b). By using the parallel-axis theorem for I_y, you can determine the moment of inertia of each part about the y axis. For example, the moment of inertia of part 2 (the semicircle) about the y axis is (Fig. 8.9c)

$$(I_y)_2 = (I_{y'})_2 + (d_x)_2^2 A_2.$$

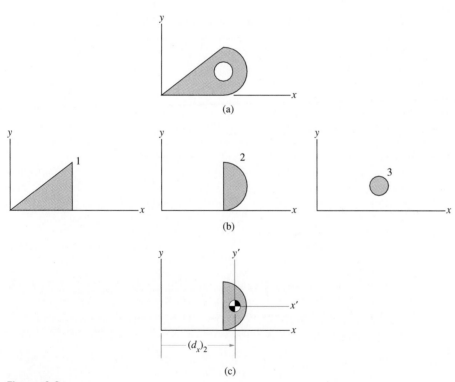

Figure 8.9

(a) A composite area.
(b) The three parts of the area.
(c) Determining $(I_y)_2$.

You must determine the values of $(I_{y'})_2$ and $(d_x)_2$. Moments of inertia and centroid locations for some simple areas are tabulated in Appendix B. Once you have carried out this procedure for each part, the moment of inertia of the composite area is

$$I_y = (I_y)_1 + (I_y)_2 - (I_y)_3.$$

Notice that you must subtract the moment of inertia of the circular cutout.

In the following example we apply the parallel-axis theorems to a simple area. We then present two examples demonstrating how to determine the moments of inertia of composite areas. Determining a moment of inertia of a composite area in terms of a given coordinate system involves three steps:

1. Choose the parts—*Try to divide the composite area into parts whose moments of inertia you know or can easily determine.*

2. Determine the moments of inertia of the parts—*Determine the moment of inertia of each part in terms of a parallel coordinate system with its origin at the centroid of the part, and then use the parallel-axis theorem to determine the moment of inertia in terms of the given coordinate system.*

3. Sum the results—*Sum the moments of inertia of the parts (or subtract in the case of a cutout) to obtain the moment of inertia of the composite area.*

Example 8.3

The moments of inertia of the rectangular area in Fig. 8.10 in terms of the $x'y'$ coordinate system are $I_{x'} = \frac{1}{12}bh^3$, $I_{y'} = \frac{1}{12}hb^3$, $I_{x'y'} = 0$, and $J'_O = \frac{1}{12}(bh^3 + hb^3)$ (see Appendix B). Determine its moments of inertia in terms of the xy coordinate system.

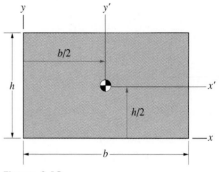

Figure 8.10

STRATEGY

The $x'y'$ coordinate system has its origin at the centroid of the area and is parallel to the xy coordinate system. We can use the parallel-axis theorems to determine the moments of inertia of A in terms of the xy coordinate system.

SOLUTION

The coordinates of the centroid in terms of the xy coordinate system are $d_x = b/2$, $d_y = h/2$. The moment of inertia about the x axis is

$$I_x = I_{x'} + d_y^2 A = \frac{1}{12}bh^3 + \left(\frac{1}{2}h\right)^2 bh = \frac{1}{3}bh^3.$$

The moment of inertia about the y axis is

$$I_y = I_{y'} + d_x^2 A = \frac{1}{12}hb^3 + \left(\frac{1}{2}b\right)^2 bh = \frac{1}{3}hb^3.$$

The product of inertia is

$$I_{xy} = I_{x'y'} + d_x d_y A = 0 + \left(\frac{1}{2}b\right)\left(\frac{1}{2}h\right) bh = \frac{1}{4}b^2h^2.$$

The polar moment of inertia is

$$J_O = J'_O + d^2 A = \frac{1}{12}(bh^3 + hb^3) + \left[\left(\frac{1}{2}b\right)^2 + \left(\frac{1}{2}h\right)^2\right] bh$$

$$= \frac{1}{3}(bh^3 + hb^3).$$

DISCUSSION

Notice that we could also have determined J_O by using the relation

$$J_O = I_x + I_y = \frac{1}{3}bh^3 + \frac{1}{3}hb^3.$$

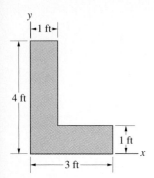

Figure 8.11

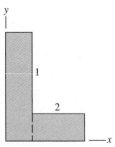

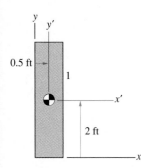

(a) Dividing the area into rectangles 1 and 2.

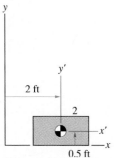

(b) Parallel coordinate systems $x'y'$ with origins at the centroids of the parts.

Example 8.4

Determine I_x, k_x, and I_{xy} for the composite area in Fig. 8.11.

SOLUTION

Choose the Parts We can determine the moments of inertia by dividing the area into the rectangular parts 1 and 2 shown in Fig. (a).

Determine the Moments of Inertia of the Parts For each part, we introduce a coordinate system $x'y'$ with its origin at the centroid of the part (Fig. b). The moments of inertia of the rectangular parts in terms of these coordinate systems are given in Appendix B. We then use the parallel-axis theorem to determine the moment of inertia of each part about the x axis (Table 8.1).

Table 8.1 Determining the moments of inertia of the parts about the x axis.

	d_y (ft)	A (ft^2)	$I_{x'}$ (ft^4)	$I_x = I_{x'} + d_y^2 A$ (ft^4)
Part 1	2	$(1)(4)$	$\frac{1}{12}(1)(4)^3$	21.33
Part 2	0.5	$(2)(1)$	$\frac{1}{12}(2)(1)^3$	0.67

Sum the Results The moment of inertia of the composite area about the x axis is

$$I_x = (I_x)_1 + (I_x)_2 = 21.33 + 0.67 = 22.00 \text{ ft}^4.$$

The sum of the areas is $A = A_1 + A_2 = 6$ ft^2, so the radius of gyration about the x axis is

$$k_x = \sqrt{\frac{I_x}{A}} = \sqrt{\frac{22}{6}} = 1.91 \text{ ft}.$$

Repeating this procedure, we determine I_{xy} for each part in Table 8.2. The product of inertia of the composite area is

$$I_{xy} = (I_{xy})_1 + (I_{xy})_2 = 4 + 2 = 6 \text{ ft}^4.$$

Table 8.2 Determining the products of inertia of the parts in terms of the xy coordinate system.

	d_x (ft)	d_y (ft)	A (ft^2)	$I_{x'y'}$	$I_{xy} = I_{x'y'} + d_x d_y A$ (ft^4)
Part 1	0.5	2	$(1)(4)$	0	4
Part 2	2	0.5	$(2)(1)$	0	2

DISCUSSION

The moments of inertia you obtain do not depend on how you divide a composite area into parts, and you will often have a choice of convenient ways to divide a given area. See Problem 8.28, in which we divide the composite area in this example in a different way.

Example 8.5

Determine I_y and k_y for the composite area in Fig. 8.12.

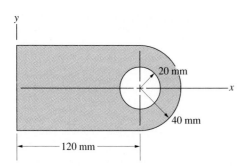

Figure 8.12

SOLUTION

Choose the Parts   We divide the area into a rectangle, a semicircle, and the circular cutout, calling them parts 1, 2, and 3, respectively (Fig. a).

Determine the Moments of Inertia of the Parts   The moments of inertia of the parts in terms of the $x'y'$ coordinate systems and the location of the centroid of the semicircular part are given in Appendix B. In Table 8.3 we use the parallel-axis theorem to determine the moment of inertia of each part about the y axis.

Table 8.3   Determining the moments of inertia of the parts.

	d_x (mm)	A (mm^2)	$I_{y'}$ (mm^4)	$I_y = I_{y'} + d_x^2 A$ (mm^4)
Part 1	60	$(120)(80)$	$\frac{1}{12}(80)(120)^3$	4.608×10^7
Part 2	$120 + \dfrac{4(40)}{3\pi}$	$\frac{1}{2}\pi(40)^2$	$\left(\dfrac{\pi}{8} - \dfrac{8}{9\pi}\right)(40)^4$	4.744×10^7
Part 3	120	$\pi(20)^2$	$\frac{1}{4}\pi(20)^4$	1.822×10^7

Sum the Results   The moment of inertia of the composite area about the y axis is

$$I_y = (I_y)_1 + (I_y)_2 - (I_y)_3 = (4.608 + 4.744 - 1.822) \times 10^7$$

$$= 7.530 \times 10^7 \text{ mm}^4.$$

The total area is

$$A = A_1 + A_2 - A_3 = (120)(80) + \frac{1}{2}\pi(40)^2 - \pi(20)^2$$

$$= 1.086 \times 10^4 \text{ mm}^2,$$

so the radius of gyration about the y axis is

$$k_y = \sqrt{\frac{I_y}{A}} = \sqrt{\frac{7.530 \times 10^7}{1.086 \times 10^4}} = 83.3 \text{ mm}.$$

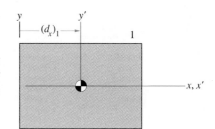
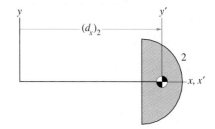
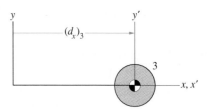

(a) Parts 1, 2, and 3.

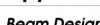

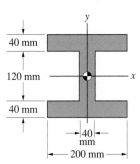

Figure 8.13

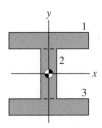

(a) Dividing the I-beam cross section into parts.

A simply supported beam.

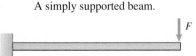

A cantilever beam.

Figure 8.14

Example 8.6

Application to Engineering

Beam Design

The equal areas in Fig. 8.13 are candidates for the cross section of a beam. (A beam with the second cross section shown is called an I-beam.) Compare their moments of inertia about the x axis.

SOLUTION

Square Cross Section From Appendix B, the moment of inertia of the square cross section about the x axis is

$$I_x = \frac{1}{12}(144.2)(144.2)^3 = 3.60 \times 10^7 \text{ mm}^4.$$

I-Beam Cross Section We can divide the area into the rectangular parts shown in Fig. (a). Introducing coordinate systems $x'y'$ with their origins at the centroids of the parts (Fig. b), we use the parallel-axis theorem to determine the moments of inertia about the x axis (Table 8.4). Their sum is

$$I_x = (I_x)_1 + (I_x)_2 + (I_x)_3 = (5.23 + 0.58 + 5.23) \times 10^7$$
$$= 11.03 \times 10^7 \text{ mm}^4.$$

The moment of inertia of the I-beam about the x axis is 3.06 times that of the square cross section of equal area.

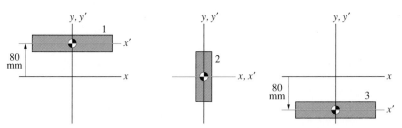

(b) Parallel coordinate systems $x'y'$ with origins at the centroids of the parts.

Table 8.4 Determining the moments of inertia of the parts about the x axis.

	d_y (mm)	A (mm^2)	$I_{x'}$ (mm^4)	$I_x = I_{x'} + d_y^2 A$ (mm^4)
Part 1	80	(200)(40)	$\frac{1}{12}(200)(40)^3$	5.23×10^7
Part 2	0	(40)(120)	$\frac{1}{12}(40)(120)^3$	0.58×10^7
Part 3	-80	(200)(40)	$\frac{1}{12}(200)(40)^3$	5.23×10^7

DESIGN ISSUES

A **beam** is a bar of material that supports lateral loads, meaning loads perpendicular to the axis of the bar. Two common types of beams are shown in Fig. 8.14 supporting a lateral load F. A beam with pinned ends is called a simply supported beam, and a beam with a single, built-in support is called a cantilever beam.

The lateral loads on a beam cause it to bend, and it must be stiff, or resistant to bending, to support them. A beam's resistance to bending depends directly on the moment of inertia of its cross-sectional area. Let's consider the beam in Fig. 8.15(a). The cross section is symmetric about the y axis and the origin of the coordinate system is placed at its centroid. If the beam consists of a homogeneous structural material such as steel and it is subjected to couples at the ends, as shown in Fig. 8.15(b), it bends into a circular arc of radius R. It can be shown that

$$R = \frac{EI_x}{M},$$

where I_x is the moment of inertia of the beam cross section about the x axis. The "elastic modulus" or "Young's modulus" E has different values for different materials. (This equation holds only if M is small enough so that the beam returns to its original shape when the couples are removed. The bending in Fig. 8.15b is exaggerated.) Thus the amount the beam bends for a given value of M depends on the material and the moment of inertia of its cross section. Increasing I_x increases the value of R, which means the resistance of the beam to bending is increased.

This explains in large part the cross sections of many of the beams you see in use—for example, in highway overpasses and in the frames of buildings. They are configured to increase their moments of inertia. The cross sections in Fig. 8.16 all have the same area. The numbers are the ratios of the moment of inertia I_x to the value of I_x for the solid square cross section.

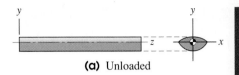

(a) Unloaded

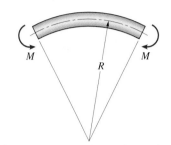

(b) Subjected to couples at the ends.

Figure 8.15

A beam with symmetrical cross section.

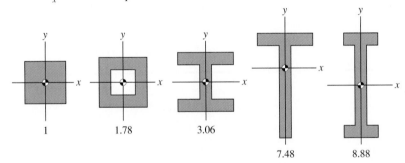

1   1.78   3.06

7.48   8.88

Figure 8.16

Typical beam cross sections and the ratio of I_x to the value for a solid square beam of equal cross-sectional area.

However, configuring the cross section of a beam to increase its moment of inertia can be carried too far. The "box" beam in Fig. 8.17(a) has a value of I_x that is four times as large as a solid square beam of the same cross-sectional area, but its walls are so thin they may "buckle," as shown in Fig. 8.17(b). The stiffness implied by the beam's large moment of inertia is not realized because it becomes geometrically unstable. One solution used by engineers to achieve a large moment of inertia in a relatively light beam while avoiding failure due to buckling is to stabilize its walls by filling the beam with a light material such as honeycombed metal or foamed plastic (Fig. 8.17c).

Figure 8.17

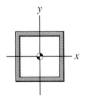

(a) A box beam with thin walls.

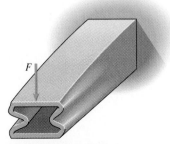

(b) Failure by buckling.

(c) Stabilizing the walls with a filler.

Problems

8.27   Determine I_y and k_y for the composite area in Example 8.4.

8.28   Determine I_x, k_x, and I_{xy} for the composite area by dividing it into rectangles 1 and 2 as shown, and compare your results to those of Example 8.4.

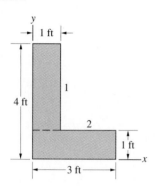

P8.28

Refer to P8.29 for Problems 8.29–8.31.

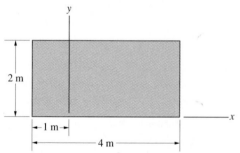

P8.29

8.29   Determine I_y and k_y.

8.30   Determine J_O and k_O.

8.31   Determine I_{xy}.

8.32   Determine I_x and k_x.

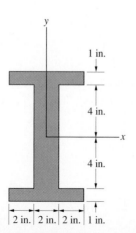

P8.32

8.33   Determine J_O and k_O for the area in Problem 8.32.

8.34   If you design the beam cross section so that $I_x = 6.4 \times 10^5$ mm^4, what are the resulting values of I_y and J_O?

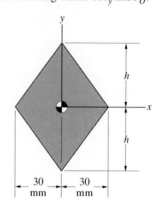

P8.34

Refer to P8.35 for Problems 8.35–8.37.

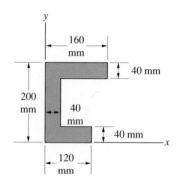

P8.35

8.35   Determine I_y and k_y.

8.36   Determine I_x and k_x.

8.37   Determine I_{xy}.

Refer to P8.38 for Problems 8.38–8.40. The origin of the coordinate system is at the centroid of the composite area.

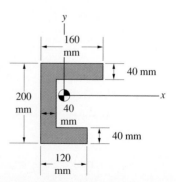

P8.38

8.38 Determine I_x and k_x.

8.39 Determine I_y and k_y.

8.40 Determine I_{xy}.

Refer to P8.41 for Problems 8.41–8.43.

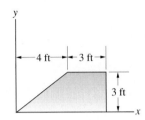

P8.41

8.41 Determine I_x and k_x.

8.42 Determine J_O and k_O.

8.43 Determine I_{xy}.

Refer to P8.44 for Problems 8.44–8.46. The origin of the coordinate system is at the centroid of the composite area.

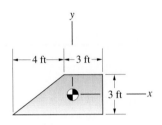

P8.44

8.44 Determine I_x and k_x.

8.45 Determine J_O and k_O.

8.46 Determine I_{xy}.

Refer to P8.47 for Problems 8.47–8.49.

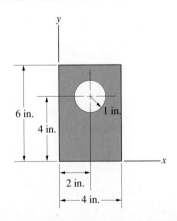

P8.47

8.47 Determine I_x and k_x.

8.48 Determine J_O and k_O.

8.49 Determine I_{xy}.

Refer to P8.50 for Problems 8.50–8.52. The origin of the coordinate system is at the centroid of the composite area.

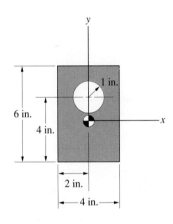

P8.50

8.50 Determine I_x and k_x.

8.51 Determine J_O and k_O.

8.52 Determine I_{xy}.

8.53 Determine I_y and k_y.

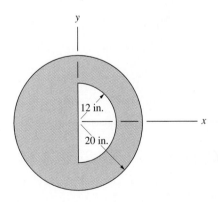

P8.53

8.54 Determine J_O and k_O for the area in Problem 8.53.

Refer to P8.55 for Problems 8.55–8.57. The origin of the coordinate system is at the centroid of the composite area.

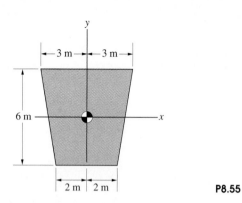

P8.55

8.55 Determine I_x and k_x.

8.56 Determine I_y and k_y.

8.57 Suppose that you want to decrease I_x 20% by placing a circular cutout with its center at the centroid of the area. What should the radius of the cutout be?

Refer to P8.58 for Problems 8.58–8.60.

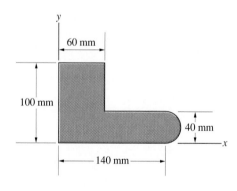

P8.58

8.58 Determine I_y and k_y.

8.59 Determine I_x and k_x.

8.60 Determine I_{xy}.

Refer to P8.61 for Problems 8.61–8.63. The origin of the coordinate system is at the centroid of the composite area.

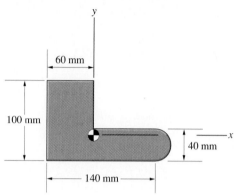

P8.61

8.61 Determine I_y and k_y.

8.62 Determine I_x and k_x.

8.63 Determine I_{xy}.

Refer to P8.64 for Problems 8.64–8.66.

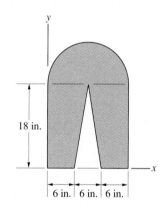

P8.64

8.64 Determine I_y and k_y.

8.65 Determine I_x and k_x.

8.66 Determine I_{xy}.

8.67 Determine I_y and k_y.

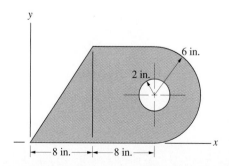

P8.67

8.68 Determine J_O and k_O for the area in Problem 8.67.

Refer to P8.69 for Problems 8.69–8.71.

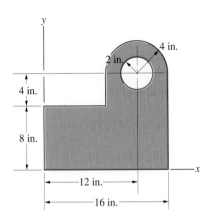

P8.69

8.69 Determine I_y and k_y.

8.70 Determine I_x and k_x.

8.71 Determine I_{xy}.

Refer to P8.72 for Problems 8.72–8.74. The origin of the coordinate system is at the centroid of the composite area.

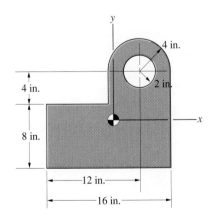

P8.72

8.72 Determine I_y and k_y.

8.73 Determine I_x and k_x.

8.74 Determine I_{xy}.

8.75 Determine I_y and k_y.

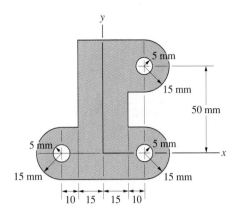

P8.75

8.76 Determine J_O and k_O for the area in Problem 8.75.

8.77 Determine I_y for the cross section of the concrete masonry unit.

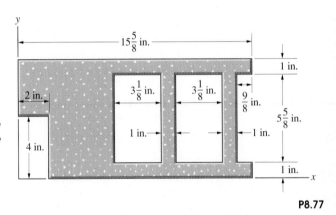

P8.77

8.78 Determine I_x for the cross section in Problem 8.77.

8.79 The area $A = 2 \times 10^4$ mm^2. Its moment of inertia about the y axis is $I_y = 3.2 \times 10^8$ mm^4. Determine its moment of inertia about the $\hat{y}$ axis.

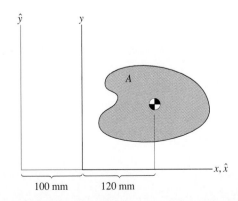

P8.79

8.80 The area $A = 100$ in^2 and it is *symmetric* about the x' axis. The moments of inertia $I_{x'} = 420$ in^4, $I_{y'} = 580$ in^4, $J_O = 11,000$ in^4, and $I_{xy} = 4800$ in^4. What are I_x and I_y?

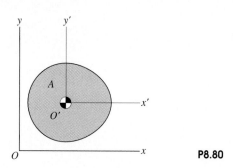

P8.80

8.81 Derive the parallel-axis theorem for the product of inertia, Eq. (8.12), by using the same procedures we used to derive Eqs. (8.10) and (8.11).

8.82 Derive the parallel-axis theorem for the polar moment of inertia, Eq. (8.13), (a) by using the same procedures we used to derive Eqs. (8.10) and (8.11); (b) by using Eqs. (8.10) and (8.11).

Problems 8.83–8.86 are related to Example 8.6.

8.83 Determine the moment of inertia of the beam cross section about the x axis. Compare your result with the moment of inertia of a solid square cross section of equal area and confirm the ratio shown in Fig. 8.16.

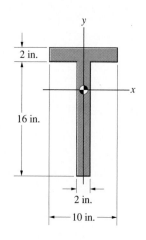

P8.83

8.84 The area of the beam cross section is 5200 mm^2. Determine the moment of inertia of the beam cross section about the x axis. Compare your result with the moment of inertia of a solid square cross section of equal area and confirm the ratio shown in Fig. 8.16.

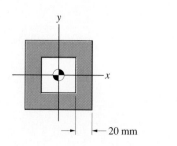

P8.84

8.85 (a) If I_x is expressed in in^4, R is in inches, and M is in in-lb, what are the dimensions of Young's modulus E?
(b) A beam with the cross section shown is subjected to couples $M = 1500$ in-lb as shown in Fig. 8.15(b), and it bends into a circular arc with radius $R = 120$ in. What is Young's modulus of the material?

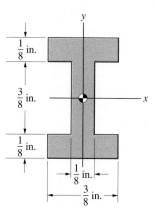

P8.85

8.86 Suppose that you want to design a beam made of material whose density is 8000 kg/m^3. The beam is to be 4 m in length and have a mass of 320 kg. Design a cross section for the beam so that $I_x = 3 \times 10^{-5}$ m^4.

8.3 *Rotated and Principal Axes*

Suppose that Fig. 8.18(a) is the cross section of a cantilever beam. If you apply a vertical force to the end of the beam, a larger vertical deflection results if the cross section is oriented as shown in Fig. 8.18(b) than if it is oriented as shown in Fig. 8.18(c). The *minimum* vertical deflection results when the beam's cross section is oriented so that the moment of inertia I_x is a maximum (Fig. 8.18d).

In many engineering applications you must determine moments of inertia of areas with various angular orientations relative to a coordinate system and also determine the orientation for which the value of a moment of inertia is a maximum or minimum. We discuss these procedures in this section.

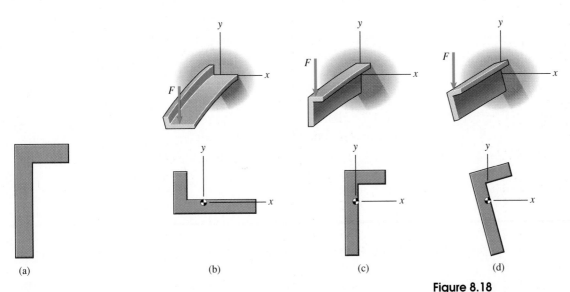

(a)　　　　　　　　　(b)　　　　　　　　　(c)　　　　　　　　　(d)

Figure 8.18

(a) A beam cross section.
(b)–(d) Applying a lateral load with different orientations of the cross section.

Rotated Axes

Let's consider an area A, a coordinate system xy, and a second coordinate system $x'y'$ that is rotated through an angle θ relative to the xy coordinate system (Fig. 8.19a). Suppose that we know the moments of inertia of A in terms of the xy coordinate system. Our objective is to determine the moments of inertia in terms of the $x'y'$ coordinate system.

In terms of the radial distance r to a differential element of area dA and the angle α in Fig. 8.19(b), the coordinates of dA in the xy coordinate system are

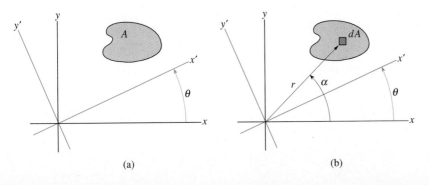

(a)　　　　　　　　　　　　　　　　　　(b)

Figure 8.19

(a) The $x'y'$ coordinate system is rotated through an angle θ relative to the xy coordinate system.
(b) A differential element of area dA.

$$x = r \cos \alpha, \tag{8.14}$$

$$y = r \sin \alpha. \tag{8.15}$$

The coordinates of dA in the $x'y'$ coordinate system are

$$x' = r \cos (\alpha - \theta) = r (\cos \alpha \cos \theta + \sin \alpha \sin \theta), \tag{8.16}$$

$$y' = r \sin (\alpha - \theta) = r (\sin \alpha \cos \theta - \cos \alpha \sin \theta). \tag{8.17}$$

In Eqs. (8.16) and (8.17), we use identities for the cosine and sine of the difference of two angles (Appendix A). By substituting Eqs. (8.14) and (8.15) into Eqs. (8.16) and (8.17), we obtain equations relating the coordinates of dA in the two coordinate systems:

$$x' = x \cos \theta + y \sin \theta, \tag{8.18}$$

$$y' = -x \sin \theta + y \cos \theta. \tag{8.19}$$

We can use these expressions to derive relations between the moments of inertia of A in terms of the xy and $x'y'$ coordinate systems:

Moment of Inertia About the x′ Axis

$$I_{x'} = \int_A (y')^2 \, dA = \int_A (-x \sin \theta + y \cos \theta)^2 \, dA$$

$$= \cos^2 \theta \int_A y^2 \, dA - 2 \sin \theta \cos \theta \int_A xy \, dA + \sin^2 \theta \int_A x^2 \, dA.$$

From this equation we obtain

$$I_{x'} = I_x \cos^2 \theta - 2I_{xy} \sin \theta \cos \theta + I_y \sin^2 \theta. \tag{8.20}$$

Moment of Inertia About the y′ Axis

$$I_{y'} = \int_A (x')^2 \, dA = \int_A (x \cos \theta + y \sin \theta)^2 \, dA$$

$$= \sin^2 \theta \int_A y^2 \, dA + 2 \sin \theta \cos \theta \int_A xy \, dA + \cos^2 \theta \int_A x^2 \, dA.$$

This equation gives us the result

$$I_{y'} = I_x \sin^2 \theta + 2I_{xy} \sin \theta \cos \theta + I_y \cos^2 \theta. \tag{8.21}$$

Product of Inertia In terms of the $x'y'$ coordinate system, the product of inertia of A is

$$I_{x'y'} = (I_x - I_y) \sin \theta \cos \theta + (\cos^2 \theta - \sin^2 \theta) I_{xy}. \tag{8.22}$$

Polar Moment of Inertia From Eqs. (8.20) and (8.21), the polar moment of inertia in terms of the $x'y'$ coordinate system is

$$J'_O = I_{x'} + I_{y'} = I_x + I_y = J_O.$$

Thus *the value of the polar moment of inertia is unchanged by a rotation of the coordinate system.*

Principal Axes

You have seen that the moments of inertia of A in terms of the $x'y'$ coordinate system depend on the angle θ in Fig. 8.19(a). Let's consider the following question: For what values of θ is the moment of inertia $I_{x'}$ a maximum or minimum?

To consider this question, it is convenient to use the identities

$$\sin 2\theta = 2 \sin \theta \cos \theta,$$

$$\cos 2\theta = \cos^2 \theta - \sin^2 \theta = 1 - 2 \sin^2 \theta = 2 \cos^2 \theta - 1.$$

With these expressions, we can write Eqs. (8.20)–(8.22) in the forms

$$I_{x'} = \frac{I_x + I_y}{2} + \frac{I_x - I_y}{2} \cos 2\theta - I_{xy} \sin 2\theta, \tag{8.23}$$

$$I_{y'} = \frac{I_x + I_y}{2} - \frac{I_x - I_y}{2} \cos 2\theta + I_{xy} \sin 2\theta, \tag{8.24}$$

$$I_{x'y'} = \frac{I_x - I_y}{2} \sin 2\theta + I_{xy} \cos 2\theta. \tag{8.25}$$

We will denote a value of θ at which $I_{x'}$ is a maximum or minimum by θ_p. To determine θ_p, we evaluate the derivative of Eq. (8.23) with respect to 2θ and equate it to zero, obtaining

$$\tan 2\theta_p = \frac{2I_{xy}}{I_y - I_x}. \tag{8.26}$$

If we set the derivative of Eq. (8.24) with respect to 2θ equal to zero to determine a value of θ for which $I_{y'}$ is a maximum or minimum, we again obtain Eq. (8.26). The second derivatives of $I_{x'}$ and $I_{y'}$ with respect to 2θ are opposite in sign,

$$\frac{d^2 I_{x'}}{d(2\theta)^2} = -\frac{d^2 I_{y'}}{d(2\theta)^2},$$

which means that at an angle θ_p for which $I_{x'}$ is a maximum, $I_{y'}$ is a minimum, and at an angle θ_p for which $I_{x'}$ is a minimum, $I_{y'}$ is a maximum.

A rotated coordinate system $x'y'$ that is oriented so that $I_{x'}$ and $I_{y'}$ have maximum or minimum values is called a set of **principal axes** of the area A. The corresponding moments of inertia $I_{x'}$ and $I_{y'}$ are called the **principal moments of inertia**. In the next section we can show that the product of inertia $I_{x'y'}$ corresponding to a set of principal axes equals zero.

Because the tangent is a periodic function, Eq. (8.26) does not yield a unique solution for the angle θ_p. We can show, however, that it does determine the orientation of the principal axes within an arbitrary multiple of 90°.

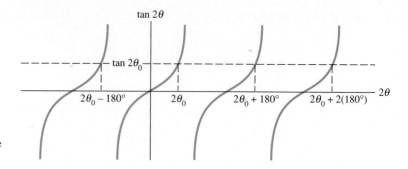

Figure 8.20

For a given value of $\tan 2\theta_0$, there are multiple roots $2\theta_0 + n(180°)$.

Observe in Fig. 8.20 that if $2\theta_0$ is a solution of Eq. (8.26), then $2\theta_0 + n(180°)$ is also a solution for any integer n. The resulting orientations of the $x'y'$ coordinate system are shown in Fig. 8.21.

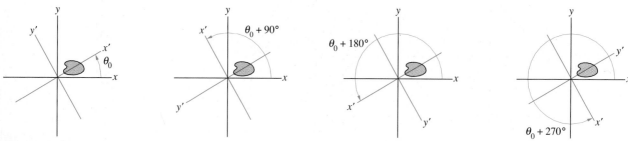

Figure 8.21

The orientation of the $x'y'$ coordinate system is determined within a multiple of 90°.

Let us now return to the example with which we began this section. The deflection of the beam in Fig. 8.18(d) will be a minimum if the x axis is the principal axis of the beam's cross section about which the moment of inertia is a maximum.

The following examples demonstrate the procedure for determining principal axes and principal moments of inertia:

1. Determine I_x, I_y, and I_{xy}—*You must determine the moments of inertia of the area in terms of the xy coordinate system.*

2. Determine θ_p—*Solve Eq. (8.26) to determine the orientation of the principal axes within an arbitrary multiple of 90°.*

3. Calculate $I_{x'}$ and $I_{y'}$—*Once you have chosen the orientation of the principal axes, you can use Eqs. (8.20) and (8.21) or Eqs. (8.23) and (8.24) to determine the principal moments of inertia.*

Example 8.7

Determine a set of principal axes and the corresponding principal moments of inertia for the triangular area in Fig. 8.22.

STRATEGY

We can obtain the moments of inertia of the triangular area from Appendix B. Then we can use Eq. (8.26) to determine the orientation of the principal axes and evaluate the principal moments of inertia with Eqs. (8.23) and (8.24).

SOLUTION

Determine I_x, I_y, and I_{xy}   The moments of inertia of the triangular area are

$$I_x = \frac{1}{12}(4)(3)^3 = 9 \text{ m}^4,$$

$$I_y = \frac{1}{4}(4)^3(3) = 48 \text{ m}^4,$$

$$I_{xy} = \frac{1}{8}(4)^2(3)^2 = 18 \text{ m}^4.$$

Determine θ_p   From Eq. (8.26),

$$\tan 2\theta_p = \frac{2I_{xy}}{I_y - I_x} = \frac{2(18)}{48 - 9} = 0.923,$$

and we obtain $\theta_p = 21.4°$. The principal axes corresponding to this value of θ_p are shown in Fig. (a).

Calculate $I_{x'}$ and $I_{y'}$ Substituting $\theta_p = 21.4°$ into Eqs. (8.23) and (8.24), we obtain

$$I_{x'} = \frac{I_x + I_y}{2} + \frac{I_x - I_y}{2} \cos 2\theta - I_{xy} \sin 2\theta$$

$$= \left(\frac{9 + 48}{2}\right) + \left(\frac{9 - 48}{2}\right) \cos[2(21.4°)] - (18) \sin[2(21.4°)] = 1.96 \text{ m}^4,$$

$$I_{y'} = \frac{I_x + I_y}{2} - \frac{I_x - I_y}{2} \cos 2\theta + I_{xy} \sin 2\theta$$

$$= \left(\frac{9 + 48}{2}\right) - \left(\frac{9 - 48}{2}\right) \cos[2(21.4°)] + (18) \sin[2(21.4°)] = 55.0 \text{ m}^4.$$

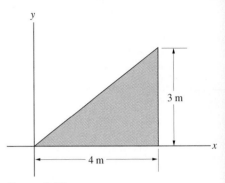

Figure 8.22

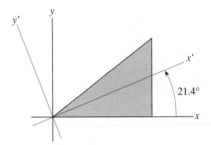

(a) The principal axes corresponding to $\theta_p = 21.4°$.

DISCUSSION

The product of inertia corresponding to a set of principal axes is zero. In this example, substituting $\theta_p = 21.4°$ into Eq. (8.25) confirms that $I_{x'y'} = 0$.

Example 8.8

The moments of inertia of the area in Fig. 8.23 in terms of the xy coordinate system shown are $I_x = 22$ ft^4, $I_y = 10$ ft^4, and $I_{xy} = 6$ ft^4. Determine $I_{x'}$, $I_{y'}$, and $I_{x'y'}$ for $\theta = 30°$; then determine a set of principal axes and the corresponding principal moments of inertia.

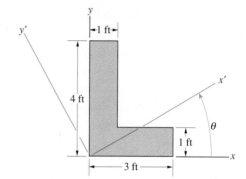

Figure 8.23

SOLUTION

Determine $I_{x'}$, $I_{y'}$, and $I_{x'y'}$ By setting $\theta = 30°$ in Eqs. (8.23)–(8.25), we obtain the moments of inertia:

$$I_{x'} = \frac{I_x + I_y}{2} + \frac{I_x - I_y}{2} \cos 2\theta - I_{xy} \sin 2\theta$$

$$= \left(\frac{22 + 10}{2}\right) + \left(\frac{22 - 10}{2}\right) \cos[2(30°)] - (6) \sin[2(30°)] = 13.8 \text{ ft}^4,$$

$$I_{y'} = \frac{I_x + I_y}{2} - \frac{I_x - I_y}{2} \cos 2\theta + I_{xy} \sin 2\theta$$

$$= \left(\frac{22 + 10}{2}\right) - \left(\frac{22 - 10}{2}\right) \cos[2(30°)] + (6) \sin[2(30°)] = 18.2 \text{ ft}^4,$$

$$I_{x'y'} = \frac{I_x - I_y}{2} \sin 2\theta + I_{xy} \cos 2\theta$$

$$= \left(\frac{22 - 10}{2}\right) \sin[2(30°)] + (6) \cos[2(30°)] = 8.2 \text{ ft}^4.$$

Determine θ_p Substituting the moments of inertia in terms of the xy coordinate system into Eq. (8.26),

$$\tan 2\theta_p = \frac{2I_{xy}}{I_y - I_x} = \frac{2(6)}{10 - 22} = -1,$$

we obtain $\theta_p = -22.5°$. The principal axes corresponding to this value of θ_p are shown in Fig. (a).

Calculate $I_{x'}$ and $I_{y'}$ We substitute $\theta_p = -22.5°$ into Eqs. (8.23) and (8.24), obtaining the principal moments of inertia:

$$I_{x'} = 24.5 \text{ ft}^4, \qquad I_{y'} = 7.5 \text{ ft}^4.$$

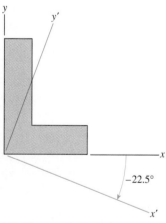

(a) The set of principal axes corresponding to $\theta_p = -22.5°$.

Mohr's Circle

Given the moments of inertia of an area in terms of a particular coordinate system, we have presented equations with which you can determine the moments of inertia in terms of a rotated coordinate system, the orientation of the principal axes, and the principal moments of inertia. You can also obtain this information by using a graphical method called **Mohr's circle**, which is very useful for visualizing the solutions of Eqs. (8.23)–(8.25).

Determining $I_{x'}$, $I_{y'}$, ***and*** $I_{x'y'}$ We first describe how to construct Mohr's circle and then explain why it works. Suppose we know the moments of inertia I_x, I_y, and I_{xy} of an area in terms of a coordinate system xy and we want to determine the moments of inertia for a rotated coordinate system $x'y'$ (Fig. 8.24). Constructing Mohr's circle involves three steps:

1. Establish a set of horizontal and vertical axes and plot two points: point 1 with coordinates (I_x, I_{xy}) and point 2 with coordinates $(I_y, -I_{xy})$, as shown in Fig. 8.25(a).

2. Draw a straight line connecting points 1 and 2. Using the intersection of the straight line with the horizontal axis as the center, draw a circle that passes through the two points (Fig. 8.25b).

3. Draw a straight line through the center of the circle at an angle 2θ measured counterclockwise from point 1. This line intersects the circle at point 1' with coordinates $(I_{x'}, I_{x'y'})$ and point 2' with coordinates $(I_{y'}, -I_{x'y'})$, as shown in Fig. 8.25(c).

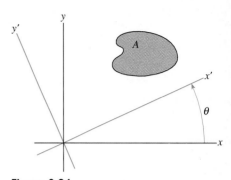

Figure 8.24

The xy coordinate system and the rotated $x'y'$ coordinate system.

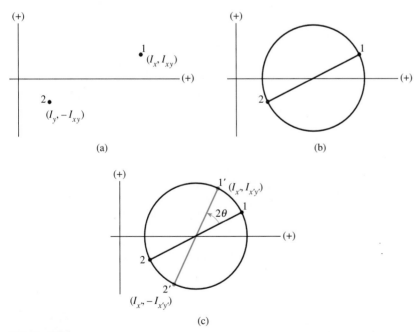

Figure 8.25

(a) Plotting the points 1 and 2.
(b) Drawing Mohr's circle. The center of the circle is the intersection of the line from 1 to 2 with the horizontal axis.
(c) Finding the points 1' and 2'.

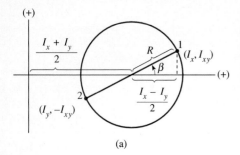

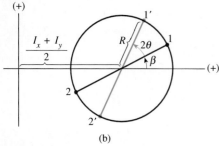

(a)

(b)

Figure 8.26

(a) The points 1 and 2 and Mohr's circle.
(b) The points 1′ and 2′.

Thus for a given angle θ, the coordinates of points 1′ and 2′ determine the moments of inertia in terms of the rotated coordinate system. Why does this graphical construction work? In Fig. 8.26(a), we show the points 1 and 2 and Mohr's circle. Notice that the horizontal coordinate of the center of the circle is $(I_x + I_y)/2$. The sine and cosine of the angle β are

$$\sin \beta = \frac{I_{xy}}{R}, \qquad \cos \beta = \frac{I_x - I_y}{2R},$$

where R, the radius of the circle, is given by

$$R = \sqrt{\left(\frac{I_x - I_y}{2}\right)^2 + (I_{xy})^2}.$$

From Fig. 8.26(b), the horizontal coordinate of point 1′ is

$$\frac{I_x + I_y}{2} + R \cos(\beta + 2\theta)$$

$$= \frac{I_x + I_y}{2} + R(\cos \beta \cos 2\theta - \sin \beta \sin 2\theta)$$

$$= \frac{I_x + I_y}{2} + \frac{I_x - I_y}{2} \cos 2\theta - I_{xy} \sin 2\theta = I_{x'},$$

and the horizontal coordinate of point 2′ is

$$\frac{I_x + I_y}{2} - R \cos(\beta + 2\theta)$$

$$= \frac{I_x + I_y}{2} - R(\cos \beta \cos 2\theta - \sin \beta \sin 2\theta)$$

$$= \frac{I_x + I_y}{2} - \frac{I_x - I_y}{2} \cos 2\theta + I_{xy} \sin 2\theta = I_{y'}.$$

The vertical coordinate of point 1′ is

$$R \sin(\beta + 2\theta) = R(\sin \beta \cos 2\theta + \cos \beta \sin 2\theta)$$

$$= I_{xy} \cos 2\theta + \frac{I_x - I_y}{2} \sin 2\theta = I_{x'y'},$$

and the vertical coordinate of point 2′ is

$$-R \sin(\beta + 2\theta) = -I_{x'y'}.$$

We have shown that the coordinates of point 1′ are $(I_{x'}, I_{x'y'})$ and the coordinates of point 2′ are $(I_{y'}, -I_{x'y'})$.

Determining Principal Axes and Principal Moments of Inertia

Because the moments of inertia $I_{x'}$ and $I_{y'}$ are the horizontal coordinates of points 1' and 2' of Mohr's circle, their maximum and minimum values occur when points 1' and 2' coincide with the intersections of the circle with the horizontal axis (Fig. 8.27). (Which intersection you designate as 1' is arbitrary. In Fig. 8.27, we have designated the minimum moment of inertia as point 1'.) You can determine the orientation of the principal axes by measuring the angle $2\theta_p$ from point 1 to point 1', and the coordinates of points 1' and 2' are the principal moments of inertia.

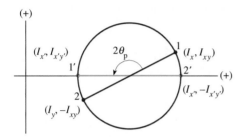

Figure 8.27

To determine the orientation of a set of principal axes, let points 1' and 2' be the points where the circle intersects the horizontal axis.

Notice that Mohr's circle demonstrates that the product of inertia $I_{x'y'}$ corresponding to a set of principal axes (the vertical coordinate of point 1' in Fig. 8.27) is always zero. Furthermore, we can use Fig. 8.26(a) to obtain an analytical expression for the horizontal coordinates of the points where the circle intersects the horizontal axis, which are the principal moments of inertia:

$$\text{Principal moments of inertia} = \frac{I_x + I_y}{2} \pm R$$

$$= \frac{I_x + I_y}{2} \pm \sqrt{\left(\frac{I_x - I_y}{2}\right)^2 + (I_{xy})^2}.$$

The following example demonstrates the use of Mohr's circle to determine moments of inertia for a rotated coordinate system, principal axes, and principal moments of inertia.

Example 8.9

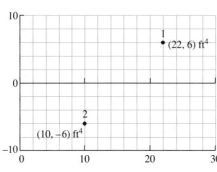

The moments of inertia of the area in Fig. 8.28 in terms of the xy coordinate system are $I_x = 22$ ft⁴, $I_y = 10$ ft⁴, and $I_{xy} = 6$ ft⁴. Use Mohr's circle to determine (a) the moments of inertia $I_{x'}$, $I_{y'}$, and $I_{x'y'}$ for $\theta = 30°$; (b) a set of principal axes and the corresponding principal moments of inertia.

Figure 8.28

SOLUTION

(a) First we plot point 1 with coordinates $(I_x, I_{xy}) = (22, 6)$ ft⁴ and point 2 with coordinates $(I_y, -I_{xy}) = (10, -6)$ ft⁴ (Fig. a). Then we draw a straight line between points 1 and 2 and, using the intersection of the line with the horizontal axis as the center, draw a circle that passes through the points (Fig. b).

(a) Plot point 1 with coordinates (I_x, I_{xy}) and point 2 with coordinates $(I_y, -I_{xy})$.

(b) Draw a line from point 1 to point 2 and construct the circle.

To determine the moments of inertia for $\theta = 30°$, we measure an angle $2\theta = 60°$ counterclockwise from point 1 (Fig. c). From the coordinates of points 1′ and 2′, we obtain

$$I_{x'} = 14 \text{ ft}^4, \qquad I_{x'y'} = 8 \text{ ft}^4, \qquad I_{y'} = 18 \text{ ft}^4.$$

(b) To determine the principal axes, we let the points 1′ and 2′ be the points where the circle intersects the horizontal axis (Fig. d). Measuring the angle from point 1 to point 1′, we determine that $2\theta_p = 135°$. From the coordinates of points 1′ and 2′, we obtain the principal moments of inertia:

$$I_{x'} = 7.5 \text{ ft}^4, \qquad I_{y'} = 24.5 \text{ ft}^4.$$

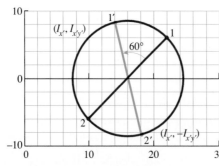

(c) Measure the angle $2\theta = 60°$ counterclockwise from point 1 to determine the points 1′ and 2′.

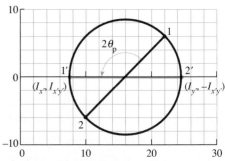

(d) Determine the principal axes by letting points 1′ and 2′ correspond to the points where the circle intersects the horizontal axis.

The principal axes are shown in Fig. (e).

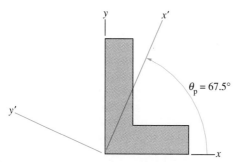

(e) The principal axes corresponding to $\theta_p = 67.5°$.

DISCUSSION

In Example 8.8 we solved this problem by using Eqs. (8.23)–(8.26). For $\theta = 30°$, we obtained $I_{x'} = 13.8$ ft^4, $I_{x'y'} = 8.2$ ft^4, and $I_{y'} = 18.2$ ft^4. The differences between these results and the ones we obtained using Mohr's circle are due to the errors inherent in measuring the answer graphically. By using Eq. (8.26) to determine the orientation of the principal axes, we obtained the principal axes shown in Fig. (a) of Example 8.8 and the principal moments of inertia $I_{x'} = 24.5$ ft^4 and $I_{y'} = 7.5$ ft^4. The difference between those results and the ones we obtained using Mohr's circle simply reflects the fact that the orientation of the principal axes can be determined only within a multiple of 90°.

Problems

8.87 Determine $I_{x'}$, $I_{y'}$, and $I_{x'y'}$. (Do not use Mohr's circle.)

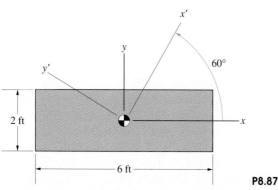

P8.87

8.88 Determine $I_{x'}$, $I_{y'}$, and $I_{x'y'}$. (Do not use Mohr's circle.)

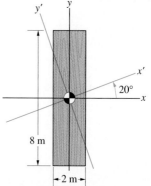

P8.88

8.89 The moments of inertia of the rectangular area are $I_x = 76.0$ m^4, $I_y = 14.7$ m^4, and $I_{xy} = 25.7$ m^4. Determine a set of principal axes and the corresponding principal moments of inertia. (Do not use Mohr's circle.)

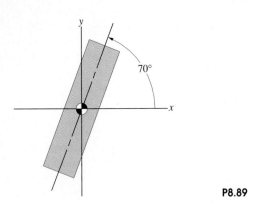

P8.89

8.90 Determine the moments of inertia $I_{x'}$, $I_{y'}$, and $I_{x'y'}$ if $\theta = 15°$. (Do not use Mohr's circle.)

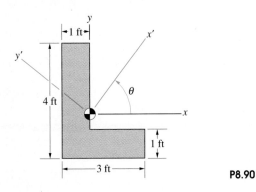

P8.90

8.91 For the area in Problem 8.90, determine a set of principal axes and the corresponding principal moments of inertia. (Do not use Mohr's circle.)

8.92 Determine a set of principal axes and the corresponding principal moments of inertia. (Do not use Mohr's circle.)

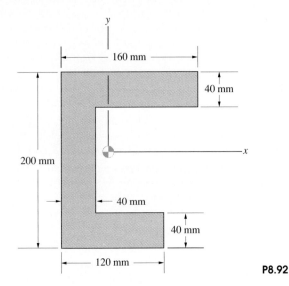

P8.92

8.93 Solve Problem 8.87 by using Mohr's circle.

8.94 Solve Problem 8.88 by using Mohr's circle.

8.95 Solve Problem 8.89 by using Mohr's circle.

8.96 Solve Problem 8.90 by using Mohr's circle.

8.97 Solve Problem 8.91 by using Mohr's circle.

8.98 Solve Problem 8.92 by using Mohr's circle.

8.99 Derive Eq. (8.22) for the product of inertia by using the same procedure we used to derive Eqs. (8.20) and (8.21).

Masses

In dynamics, the acceleration of an object that results from the forces acting on it depends on its mass. The angular acceleration, or rotational acceleration, that results from the forces and couples acting on an object depends on quantities called the mass moments of inertia of the object. In this section we discuss methods for determining mass moments of inertia of particular objects. We show that for special classes of objects, their mass moments of inertia can be expressed in terms of moments of inertia of areas, which explains how the names of those area integrals originated.

An object and a line or "axis" L_O are shown in Fig. 8.29(a). The **mass moment of inertia** of the object about the axis L_O is defined by

$$I_O = \int_m r^2 \, dm,$$ (8.27)

where r is the *perpendicular* distance from the axis to the differential element of mass dm (Fig. 8.29b). Often L_O is an axis about which the object rotates, and the value of I_O is required to determine the angular acceleration, or the rate of change of the rate of rotation, caused by a given couple about L_O.

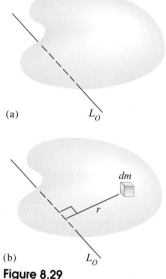

(a)

L_O

(b) L_O

Figure 8.29

(a) An object and axis L_O.
(b) A differential element of mass dm.

8.4 Simple Objects

You can determine the mass moments of inertia of complicated objects by summing the mass moments of inertia of their individual parts. We therefore begin by determining mass moments of inertia of some simple objects. Then in the next section we describe the parallel-axis theorem, which makes it possible for you to determine mass moments of inertia of objects composed of combinations of parts.

Slender Bars

Let us determine the mass moment of inertia of a straight, slender bar about a perpendicular axis L through the center of mass of the bar (Fig. 8.30a). "Slender" means that we assume that the bar's length is much greater than its width. Let the bar have length l, cross-sectional area A, and mass m. We assume that A is uniform along the length of the bar and that the material is homogeneous.

Consider a differential element of the bar of length dr at a distance r from the center of mass (Fig. 8.30b). The element's mass is equal to the product of its volume and the mass density: $dm = \rho A \, dr$. Substituting this expression into Eq. (8.27), we obtain the mass moment of inertia of the bar about a perpendicular axis through its center of mass:

$$I = \int_m r^2 \, dm = \int_{-l/2}^{l/2} \rho A r^2 \, dr = \frac{1}{12} \rho A l^3.$$

The mass of the bar equals the product of the mass density and the volume of the bar, $m = \rho A l$, so we can express the mass moment of inertia as

$$I = \frac{1}{12} m l^2.$$ (8.28)

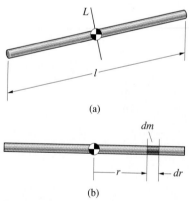

(a)

(b)

Figure 8.30

(a) A slender bar.
(b) A differential element of length dr.

We have neglected the lateral dimensions of the bar in obtaining this result. That is, we treated the differential element of mass dm as if it were concentrated on the axis of the bar. As a consequence, Eq. (8.28) is an approximation for the mass moment of inertia of a bar. Later in this section we determine the mass moments of inertia for a bar of finite lateral dimension and show that Eq. (8.28) is a good approximation when the width of the bar is small in comparison to its length.

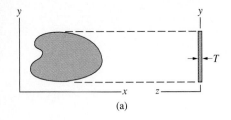

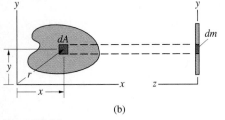

Figure 8.31

(a) A plate of arbitrary shape and uniform
   thickness T.
(b) An element of volume obtained by
   projecting an element of area dA through
   the plate.

Thin Plates

Consider a homogeneous flat plate that has mass m and uniform thickness T. We will leave the shape of the cross-sectional area of the plate unspecified. Let a cartesian coordinate system be oriented so that the plate lies in the x-y plane (Fig. 8.31a). Our objective is to determine the mass moments of inertia of the plate about the x, y, and z axes.

We can obtain a differential element of volume of the plate by projecting an element of area dA through the thickness T of the plate (Fig. 8.31b). The resulting volume is $T\,dA$. The mass of this element of volume is equal to the product of the mass density and the volume: $dm = \rho T\,dA$. Substituting this expression into Eq. (8.27), we obtain the mass moment of inertia of the plate about the z axis in the form

$$I_{(z\text{ axis})} = \int_m r^2\,dm = \rho T \int_A r^2\,dA,$$

where r is the distance from the z axis to dA. Since the mass of the plate is $m = \rho T A$, where A is the cross-sectional area of the plate, $\rho T = m/A$. The integral on the right is the polar moment of inertia J_O of the cross-sectional area of the plate. We can therefore write the mass moment of inertia of the plate about the z axis as

$$I_{(z\text{ axis})} = \frac{m}{A} J_O. \tag{8.29}$$

From Fig 8.31(b), we see that the perpendicular distance from the x axis to the element of area dA is the y coordinate of dA. Therefore the mass moment of inertia of the plate about the x axis is

$$I_{(x\text{ axis})} = \int_m y^2\,dm = \rho T \int_A y^2\,dA = \frac{m}{A} I_x, \tag{8.30}$$

where I_x is the moment of inertia of the cross-sectional area of the plate about the x axis. The mass moment of inertia of the plate about the y axis is

$$I_{(y\text{ axis})} = \int_m x^2\,dm = \rho T \int_A x^2\,dA = \frac{m}{A} I_y, \tag{8.31}$$

where I_y is the moment of inertia of the cross-sectional area of the plate about the y axis.

Thus we have expressed the mass moments of inertia of a thin homogeneous plate of uniform thickness in terms of the moments of inertia of the cross-sectional area of the plate. In fact, these results explain why the area integrals I_x, I_y, and J_O are called moments of inertia.

Since the sum of the area moments of inertia I_x and I_y is equal to the polar moment of inertia J_O, the mass moment of inertia of the thin plate about the z axis is equal to the sum of its moments of inertia about the x and y axes:

$$I_{(z\text{ axis})} = I_{(x\text{ axis})} + I_{(y\text{ axis})}. \qquad \textbf{Thin plate} \tag{8.32}$$

In the following example we use integration to determine the mass moment of inertia of an object consisting of two slender bars welded together. We then present an example that demonstrates the use of Eqs. (8.29)–(8.31) to determine the mass moments of inertia of a thin homogeneous plate with a specific cross-sectional area. Be sure to remember that the mass moment of inertia of an object about an axis depends on the position of the axis relative to the object.

Example 8.10

Two homogeneous slender bars, each of length l, mass m, and cross-sectional area A, are welded together to form the L-shaped object in Fig. 8.32. Determine the mass moment of inertia of the object about the axis L_O through point O. (The axis L_O is perpendicular to the two bars.)

STRATEGY

Using the same integration procedure we used for a single bar, we will determine the mass moment of inertia of each bar about L_O and sum the results.

SOLUTION

Our first step is to introduce a coordinate system with the z axis along L_O and the x axis collinear with bar 1 (Fig. a). The mass of the differential element of bar 1 of length dx is $dm = \rho A\, dx$. The mass moment of inertia of bar 1 about L_O is

$$(I_O)_1 = \int_m r^2 \, dm = \int_0^l \rho A x^2 \, dx = \frac{1}{3}\rho A l^3.$$

In terms of the mass of the bar, $m = \rho A l$, we can write this result as

$$(I_O)_1 = \frac{1}{3}ml^2.$$

The mass of an element of bar 2 of length dy, shown in Fig. (b), is $dm = \rho A\, dy$. From the figure we see that the perpendicular distance from L_O to the element is $r = \sqrt{l^2 + y^2}$. Therefore the mass moment of inertia of bar 2 about L_O is

$$(I_O)_2 = \int_m r^2 \, dm = \int_0^l \rho A(l^2 + y^2)\,dy = \frac{4}{3}\rho A l^3.$$

In terms of the mass of the bar, we obtain

$$(I_O)_2 = \frac{4}{3}ml^2.$$

The mass moment of inertia of the L-shaped object about L_O is

$$I_O = (I_O)_1 + (I_O)_2 = \frac{1}{3}ml^2 + \frac{4}{3}ml^2 = \frac{5}{3}ml^2.$$

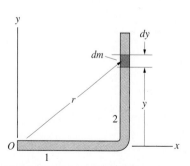

Figure 8.32

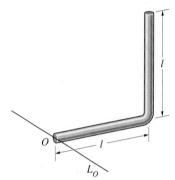

(a) Differential element of bar 1.

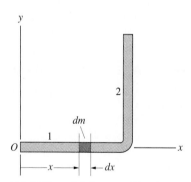

(b) Differential element of bar 2.

Example 8.11

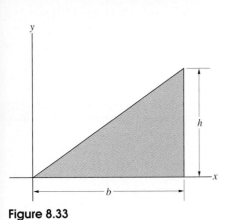

Figure 8.33

The thin homogeneous plate in Fig. 8.33 is of uniform thickness and mass m. Determine its mass moments of inertia about the x, y, and z axes.

STRATEGY

The mass moments of inertia about the x and y axes are given by Eqs. (8.30) and (8.31) in terms of the moments of inertia of the cross-sectional area of the plate. We can determine the mass moment of inertia of the plate about the z axis from Eq. (8.32).

SOLUTION

From Appendix B, the moments of inertia of the triangular area about the x and y axes are $I_x = \frac{1}{12} bh^3$ and $I_y = \frac{1}{4} hb^3$. Therefore the mass moments of inertia of the plate about the x and y axes are

$$I_{(x\,\text{axis})} = \frac{m}{A} I_x = \left(\frac{m}{\frac{1}{2}bh} \right) \left(\frac{1}{12} bh^3 \right) = \frac{1}{6} mh^2,$$

$$I_{(y\,\text{axis})} = \frac{m}{A} I_y = \left(\frac{m}{\frac{1}{2}bh} \right) \left(\frac{1}{4} hb^3 \right) = \frac{1}{2} mb^2.$$

The mass moment of inertia about the z axis is

$$I_{(z\,\text{axis})} = I_{(x\,\text{axis})} + I_{(y\,\text{axis})} = m \left(\frac{1}{6} h^2 + \frac{1}{2} b^2 \right).$$

8.5 *Parallel-Axis Theorem*

The parallel-axis theorem allows you to determine the mass moment of inertia of an object about any axis when you know the mass moment of inertia about a parallel axis through the center of mass. You can use this theorem to calculate the mass moment of inertia of a composite object about a given axis when you know the mass moments of inertia of each of its parts about axes parallel to the given axis.

Suppose that we know the mass moment of inertia I about an axis L through the center of mass of an object, and we wish to determine its mass moment of inertia I_O about a parallel axis L_O (Fig. 8.34a). To determine I_O, we introduce parallel coordinate systems xyz and $x'y'z'$ with the z axis along L_O and the z' axis along L, as shown in Fig. 8.34(b). (In this figure the axes L_O and L are perpendicular to the page.) The origin O of the xyz coordinate system is contained in the x'-y' plane. The terms d_x and d_y are the coordinates of the center of mass relative to the xyz coordinate system.

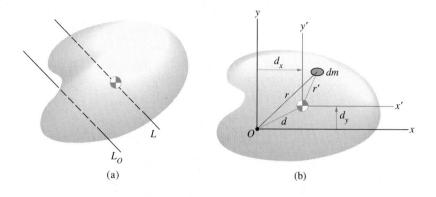

Figure 8.34

(a) An axis L through the center of mass of an object and a parallel axis L_O.

(b) The xyz and $x'y'z'$ coordinate systems.

The mass moment of inertia of the object about L_O is

$$I_O = \int_m r^2\, dm = \int_m (x^2 + y^2)\, dm, \tag{8.33}$$

where r is the perpendicular distance from L_O to the differential element of mass dm, and x, y are the coordinates of dm in the x-y plane. The coordinates of dm in the two coordinate systems are related by

$$x = x' + d_x, \qquad y = y' + d_y.$$

By substituting these expressions into Eq. (8.33), we can write it as

$$I_O = \int_m [(x')^2 + (y')^2]\, dm + 2d_x \int_m x'\, dm + 2d_y \int_m y'\, dm$$

$$+ \int_m (d_x^2 + d_y^2)\, dm. \tag{8.34}$$

Since $(x')^2 + (y')^2 = (r')^2$, where r' is the perpendicular distance from L to dm, the first integral on the right side of this equation is the mass moment of inertia I of the object about L. Recall that the x' and y' coordinates of the center of mass of the object relative to the $x'y'z'$ coordinate system are defined by

$$\bar{x}' = \frac{\displaystyle\int_m x'\, dm}{\displaystyle\int_m dm}, \qquad \bar{y}' = \frac{\displaystyle\int_m y'\, dm}{\displaystyle\int_m dm}.$$

Because the center of mass of the object is at the origin of the $x'y'z'$ system, $\bar{x}' = 0$ and $\bar{y}' = 0$. Therefore the integrals in the second and third terms on the right side of Eq. (8.34) are equal to zero. From Fig. 8.34(b), we see that $d_x^2 + d_y^2 = d^2$, where d is the perpendicular distance between the axes L and L_O. Therefore we obtain

$$\boxed{I_O = I + d^2 m,} \tag{8.35}$$

where m is the mass of the object. This is the parallel-axis theorem. If you know the mass moment of inertia of an object about a given axis, you can use this theorem to determine its mass moment of inertia about any parallel axis.

In the next two examples we use the parallel-axis theorem to determine mass moments of inertia of composite objects. Determining the mass moment of inertia about a given axis L_O typically requires three steps:

1. Choose the parts—*Try to divide the object into parts whose mass moments of inertia you know or can easily determine.*

2. Determine the mass moments of inertia of the parts—*You must first determine the mass moment of inertia of each part about the axis through its center of mass parallel to L_O. Then you can use the parallel-axis theorem to determine its mass moment of inertia about L_O.*

3. Sum the results—*Sum the mass moments of inertia of the parts (or subtract in the case of a hole or cutout) to obtain the mass moment of inertia of the composite object.*

Example 8.12

Two homogeneous slender bars, each of length l and mass m, are welded together to form the L-shaped object in Fig. 8.35. Determine the mass moment of inertia of the object about the axis L_O through point O. (The axis L_O is perpendicular to the two bars.)

SOLUTION

Choose the Parts   The parts are the two bars, which we call bar 1 and bar 2 (Fig. a).

Determine the Mass Moments of Inertia of the Parts   From Eq. (8.28), the mass moment of inertia of each bar about a perpendicular axis through its center of mass is $I = \frac{1}{12}ml^2$. The distance from L_O to the parallel axis through the center of mass of bar 1 is $\frac{1}{2}l$ (Fig. a). Therefore the mass moment of inertia of bar 1 about L_O is

$$(I_O)_1 = I + d^2m = \frac{1}{12}ml^2 + \left(\frac{1}{2}l\right)^2 m = \frac{1}{3}ml^2.$$

The distance from L_O to the parallel axis through the center of mass of bar 2 is $\left[l^2 + \left(\frac{1}{2}l\right)^2\right]^{1/2}$. The mass moment of inertia of bar 2 about L_O is

$$(I_O)_2 = I + d^2m = \frac{1}{12}ml^2 + \left[l^2 + \left(\frac{1}{2}l\right)^2\right]m = \frac{4}{3}ml^2.$$

Sum the Results   The mass moment of inertia of the L-shaped object about L_O is

$$I_O = (I_O)_1 + (I_O)_2 = \frac{1}{3}ml^2 + \frac{4}{3}ml^2 = \frac{5}{3}ml^2.$$

DISCUSSION

Compare this solution to Example 8.10, in which we used integration to determine the mass moment of inertia of this object about L_O. We obtained the result much more easily with the parallel-axis theorem, but of course we needed to know the mass moments of inertia of the bars about the axes through their centers of mass.

Figure 8.35

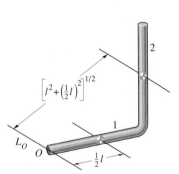

(a) The distances from L_O to parallel axes through the centers of mass of bars 1 and 2.

Example 8.13

The object in Fig. 8.36 consists of a slender, 3-kg bar welded to a thin, circular 2-kg disk. Determine its mass moment of inertia about the axis L through its center of mass. (The axis L is perpendicular to the bar and disk.)

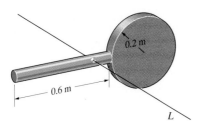

Figure 8.36

STRATEGY

We must first locate the center of mass of the composite object and then apply the parallel-axis theorem to the parts separately and sum the results.

SOLUTION

Choose the Parts   The parts are the bar and the disk. Introducing the coordinate system in Fig. (a), the x coordinate of the center of mass of the composite object is

$$\bar{x} = \frac{\bar{x}_{(bar)}m_{(bar)} + \bar{x}_{(disk)}m_{(disk)}}{} = \frac{(0.3)(3) + (0.6 + 0.2)(2)}{} = 0.5 \text{ m}.$$

Determine the Mass Moments of Inertia of the Parts   The distance from the center of mass of the bar to the center of mass of the composite object is 0.2 m (Fig. b). Therefore the mass moment of inertia of the bar about L is

$$I_{(bar)} = \frac{1}{12}(3)(0.6)^2 + (0.2)^2(3) = 0.210 \text{ kg-m}^2.$$

The distance from the center of mass of the disk to the center of mass of the composite object is 0.3 m (Fig. c). The mass moment of inertia of the disk about L is

$$I_{(disk)} = \frac{1}{2}(2)(0.2)^2 + (0.3)^2(2) = 0.220 \text{ kg-m}^2.$$

Sum the Results   The mass moment of inertia of the composite object about L is

$$I = I_{(bar)} + I_{(disk)} = 0.430 \text{ kg-m}^2.$$

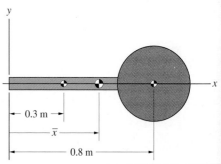

(a) The coordinate $\bar{x}$ of the center of mass of the object.

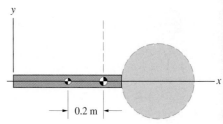

(b) Distance from L to the center of mass of the bar.

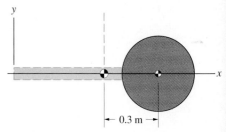

(c) Distance from L to the center of mass of the disk.

Example 8.14

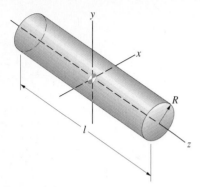

Figure 8.37

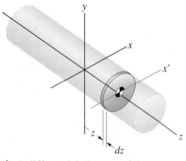

(a) A differential element of the cylinder in the form of a disk.

The homogeneous cylinder in Fig. 8.37 has mass m, length l, and radius R. Determine its mass moments of inertia about the x, y, and z axes.

STRATEGY

We first determine the mass moments of inertia about the x, y, and z axes of an infinitesimal element of the cylinder consisting of a disk of thickness dz. We then integrate the results with respect to z to obtain the mass moments of inertia of the cylinder. We must apply the parallel-axis theorem to determine the mass moments of inertia of the disk about the x and y axes.

SOLUTION

Consider an element of the cylinder of thickness dz at a distance z from the center of the cylinder (Fig. a). (You can imagine obtaining this element by "slicing" the cylinder perpendicular to its axis.) The mass of the element is equal to the product of the mass density and the volume of the element, $dm = \rho(\pi R^2 \, dz)$. We obtain the mass moments of inertia of the element by using the values for a thin circular plate given in Appendix C. The mass moment of inertia about the z axis is

$$dI_{(z \text{ axis})} = \frac{1}{2} dm \, R^2 = \frac{1}{2}(\rho \pi R^2 \, dz) R^2.$$

By integrating this result with respect to z from $-l/2$ to $l/2$, we sum the mass moments of inertia of the infinitesimal disk elements that make up the cylinder. The result is the mass moment of inertia of the cylinder about the z axis:

$$I_{(z \text{ axis})} = \int_{-l/2}^{l/2} \frac{1}{2} \rho \pi R^4 \, dz = \frac{1}{2} \rho \pi R^4 l.$$

We can write this result in terms of the mass of the cylinder, $m = \rho(\pi R^2 l)$, as

$$I_{(z \text{ axis})} = \frac{1}{2} m R^2.$$

The mass moment of inertia of the disk element about the x' axis is

$$dI_{(x' \text{ axis})} = \frac{1}{4} dm \, R^2 = \frac{1}{4}(\rho \pi R^2 \, dz) R^2.$$

We can use this result and the parallel-axis theorem to determine the mass moment of inertia of the element about the x axis:

$$dI_{(x \text{ axis})} = dI_{(x' \text{ axis})} + z^2 dm = \frac{1}{4}(\rho \pi R^2 \, dz) R^2 + z^2 (\rho \pi R^2 \, dz).$$

Integrating this expression with respect to z from $-l/2$ to $l/2$, we obtain the mass moment of inertia of the cylinder about the x axis:

$$I_{(x \text{ axis})} = \int_{-l/2}^{l/2} \left(\frac{1}{4} \rho \pi R^4 + \rho \pi R^2 z^2 \right) dz = \frac{1}{4} \rho \pi R^4 l + \frac{1}{12} \rho \pi R^2 l^3.$$

In terms of the mass of the cylinder,

$$I_{(x\text{ axis})} = \frac{1}{4}mR^2 + \frac{1}{12}ml^2.$$

Due to the symmetry of the cylinder,

$$I_{(y\text{ axis})} = I_{(x\text{ axis})}.$$

DISCUSSION

When the cylinder is very long in comparison to its width, $l \gg R$, the first term in the equation for $I_{(x\text{ axis})}$ can be neglected, and we obtain the mass moment of inertia of a slender bar about a perpendicular axis, Eq. (8.28). Conversely, when the radius of the cylinder is much greater than its length, $R \gg l$, the second term in the equation for $I_{(x\text{ axis})}$ can be neglected, and we obtain the mass moment of inertia for a thin circular disk about an axis parallel to the disk. This indicates the sizes of the terms you neglect when you use the approximate expressions for the mass moments of inertia of a "slender" bar and a "thin" disk.

Problems

8.100 The homogeneous slender bar has mass m and length l. Use integration to determine its mass moment of inertia about the perpendicular axis L_O.

Strategy: Use the same approach we used to obtain Eq. (8.28). You only need to change the limits of integration.

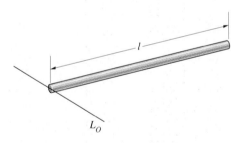

P8.100

8.101 Two homogeneous slender bars, each of mass m and length l, are welded together to form the T-shaped object. Use integration to determine the mass moment of inertia of the object about the axis through point O that is perpendicular to the bars.

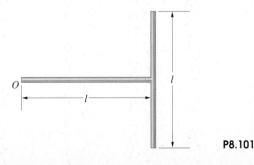

P8.101

8.102 The homogeneous slender bar has mass m and length l. Use integration to determine the mass moment of inertia of the bar about the axis L.

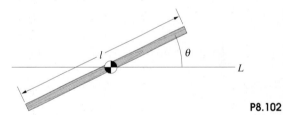

P8.102

8.103 A homogeneous slender bar is bent into a circular ring of mass m and radius R. Determine the mass moment of inertia of the ring (a) about the axis through its center of mass that is perpendicular to the ring; (b) about the axis L.

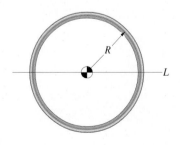

P8.103

8.104 The homogeneous thin plate is of uniform thickness and mass m. Determine its mass moments of inertia about the x, y, and z axes.

Strategy: The mass moments of inertia of a thin plate of arbitrary shape are given by Eqs. (8.30)–(8.32) in terms of the moments of inertia of the cross-sectional area of the plate. You can obtain the moments of inertia of the rectangular area from Appendix B.

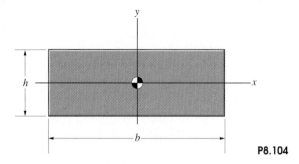

P8.104

8.105 The homogeneous thin plate is of uniform thickness and mass m.
(a) Determine its mass moments of inertia about the x and z axes.
(b) Let $R_i = 0$, and compare your results with the values given in Appendix C for a thin circular plate.
(c) Let $R_i \rightarrow R_o$, and compare your results with the solutions of Problem 8.103.

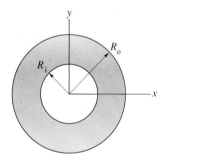

P8.105

8.106 The homogeneous thin plate is of uniform thickness and weighs 20 lb. Determine its mass moment of inertia about the y axis.

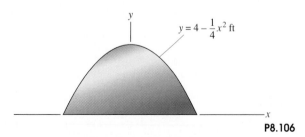

$$y = 4 - \frac{1}{4}x^2 \text{ ft}$$

P8.106

8.107 Determine the mass moment of inertia of the plate in Problem 8.106 about the x axis.

8.108 The mass of the object is 10 kg. Its mass moment of inertia about L_1 is 10 kg-m^2. What is its mass moment of inertia about L_2? (The three axes lie in the same plane.)

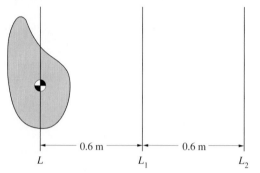

P8.108

8.109 An engineer gathering data for the design of a maneuvering unit determines that the astronaut's center of mass is at $x = 1.01$ m, $y = 0.16$ m and that his mass moment of inertia about the z axis is 105.6 kg-m^2. His mass is 81.6 kg. What is his mass moment of inertia about the z' axis through his center of mass?

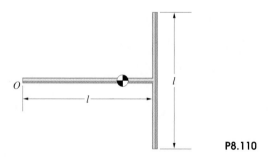

P8.109

8.110 Two homogeneous slender bars, each of mass m and length l, are welded together to form the T-shaped object. Use the parallel-axis theorem to determine the mass moment of inertia of the object about the axis through point O that is perpendicular to the bars.

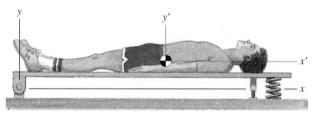

P8.110

8.111 Use the parallel-axis theorem to determine the mass moment of inertia of the T-shaped object in Problem 8.110 about the axis through the center of mass of the object that is perpendicular to the two bars.

8.112 The mass of the homogeneous slender bar is 20 kg. Determine its mass moment of inertia about the z axis.

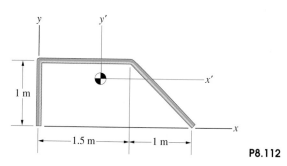

P8.112

8.113 Determine the mass moment of inertia of the bar in Problem 8.112 about the z' axis through its center of mass.

8.114 The homogeneous slender bar weighs 5 lb. Determine its mass moment of inertia about the z axis.

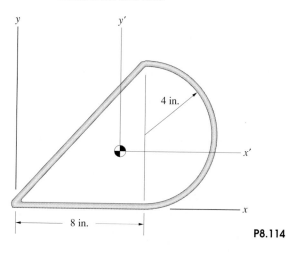

P8.114

8.115 Determine the mass moment of inertia of the bar in Problem 8.114 about the z' axis through its center of mass.

8.116 The rocket is used for atmospheric research. Its weight and its mass moment of inertia about the z axis through its center of mass (including its fuel) are 10 kip and 10,200 slug-ft^2, respectively. The rocket's fuel weighs 6000 lb, its center of mass is located at $x = -3$ ft, $y = 0$, $z = 0$, and the mass moment of inertia of the fuel about the axis through the fuel's center of mass parallel to z is 2200 slug-ft^2. When the fuel is exhausted, what is the rocket's mass moment of inertia about the axis through its new center of mass parallel to z?

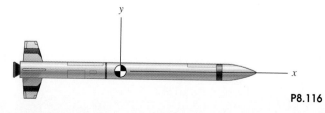

P8.116

8.117 The mass of the homogeneous thin plate is 36 kg. Determine its mass moment of inertia about the x axis.

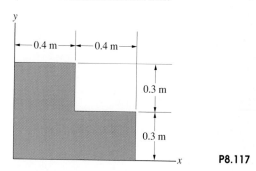

P8.117

8.118 Determine the mass moment of inertia of the plate in Problem 8.117 about the z axis.

8.119 The homogeneous thin plate weighs 10 lb. Determine its mass moment of inertia about the x axis.

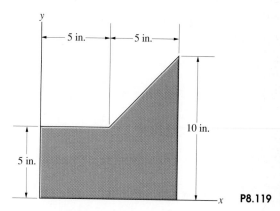

P8.119

8.120 Determine the mass moment of inertia of the plate in Problem 8.119 about the y axis.

8.121 The thermal radiator (used to eliminate excess heat from a satellite) can be modeled as a homogeneous thin rectangular plate. Its mass is 5 slugs. Determine its mass moments of inertia about the x, y, and z axes.

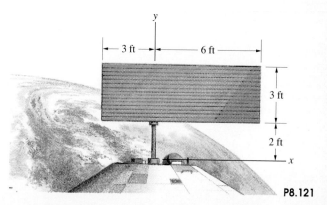

P8.121

8.122 The mass of the homogeneous thin plate is 2 kg. Determine its mass moment of inertia about the axis L_O through point O that is perpendicular to the plate.

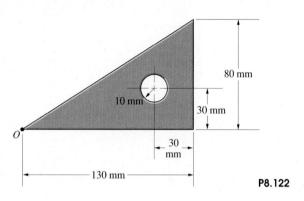

P8.122

8.123 The homogeneous cone is of mass m. Determine its mass moment of inertia about the z axis, and compare your result with the value given in Appendix C.

Strategy: Use the same approach we used in Example 8.14 to obtain the moments of inertia of a homogeneous cylinder.

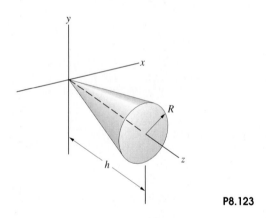

P8.123

8.124 Determine the mass moments of inertia of the homogeneous cone in Problem 8.123 about the x and y axes, and compare your results with the values given in Appendix C.

8.125 The homogeneous object has the shape of a truncated cone and consists of bronze with mass density $\rho = 8200$ kg/m^3. Determine its mass moment of inertia about the z axis.

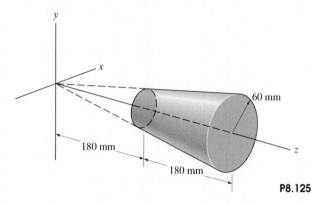

P8.125

8.126 Determine the mass moment of inertia of the object in Problem 8.125 about the x axis.

8.127 The homogeneous rectangular parallelepiped is of mass m. Determine its mass moments of inertia about the x, y, and z axes, and compare your results with the values given in Appendix C.

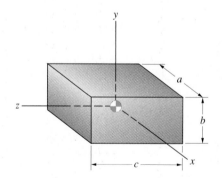

P8.127

8.128 The L-shaped machine part is composed of two homogeneous bars. Bar 1 is tungsten alloy with mass density 14,000 kg/m^3, and bar 2 is steel with mass density 7800 kg/m^3. Determine its moment of inertia about the x axis.

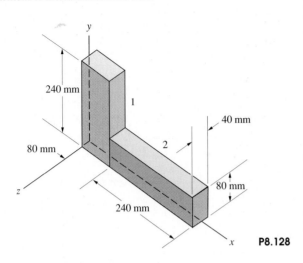

P8.128

8.129 Determine the moment of inertia of the L-shaped machine part in Problem 8.128 about the z axis.

8.130 The homogeneous ring consists of steel of density $\rho = 15$ slug/ft^3. Determine its mass moment of inertia about the axis L through its center of mass.

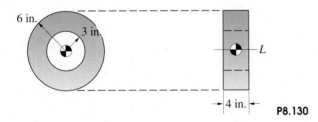

P8.130

8.131 The homogeneous half-cylinder is of mass m. Determine its mass moment of inertia about the axis L through its center of mass.

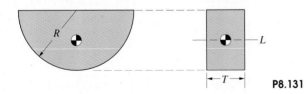

P8.131

8.132 The homogeneous machine part is made of aluminum alloy with mass density $\rho = 2800$ kg/m^3. Determine its mass moment of inertia about the z axis.

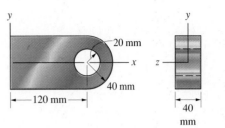

P8.132

8.133 Determine the mass moment of inertia of the machine part in Problem 8.132 about the x axis.

8.134 The object consists of steel of density $\rho = 7800$ kg/m^3. Determine its mass moment of inertia about the axis L_O.

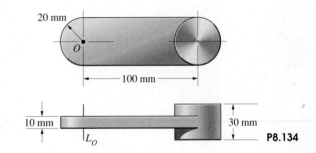

P8.134

8.135 Determine the mass moment of inertia of the object in Problem 8.134 about the axis through the center of mass of the object parallel to L_O.

8.136 The thick plate consists of steel of density $\rho = 15$ slug/ft^3. Determine its mass moment of inertia about the z axis.

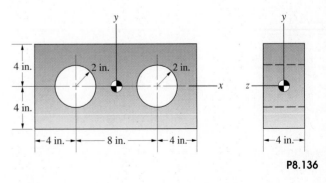

P8.136

8.137 Determine the mass moment of inertia of the plate in Problem 8.136 about the x axis.

Chapter Summary

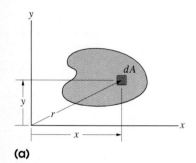

(a)

Areas

Four **area moments of inertia** are defined (Fig. a):

1. The moment of inertia about the x axis:

$$I_x = \int_A y^2 \, dA. \qquad \text{Eq. (8.1)}$$

2. The moment of inertia about the y axis:

$$I_y = \int_A x^2 \, dA. \qquad \text{Eq. (8.3)}$$

3. The product of inertia:

$$I_{xy} = \int_A xy \, dA. \qquad \text{Eq. (8.5)}$$

4. The polar moment of inertia:

$$J_O = \int_A r^2 \, dA. \qquad \text{Eq. (8.6)}$$

The **radii of gyration** about the x and y axes are defined by $k_x = \sqrt{I_x/A}$ and $k_y = \sqrt{I_y/A}$, respectively, and the radius of gyration about the origin O is defined by $k_O = \sqrt{J_O/A}$.

The polar moment of inertia is equal to the sum of the moments of inertia about the x and y axes: $J_O = I_x + I_y$. If an area is symmetric about either the x axis or the y axis, its product of inertia is zero.

Let $x'y'$ be a coordinate system with its origin at the centroid of an area A, and let xy be a parallel coordinate system. The moments of inertia of A in terms of the two systems are related by the **parallel-axis theorems** (Eqs. 8.10–8.13):

$$I_x = I_{x'} + d_y^2 A,$$
$$I_y = I_{y'} + d_x^2 A,$$
$$I_{xy} = I_{x'y'} + d_x d_y A,$$
$$J_O = J_O' + (d_x^2 + d_y^2)A = J_O' + d^2 A,$$

where d_x and d_y are the coordinates of the centroid of A in the xy coordinate system.

Masses

The mass moment of inertia of an object about an axis L_O is (Fig. b)

$$I_O = \int_m r^2 \, dm, \qquad \text{Eq. (8.27)}$$

where r is the perpendicular distance from L_O to the differential element of mass dm.

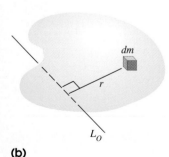

(b)

Let L be an axis through the center of mass of an object, and let L_O be a parallel axis (Fig. c). The moment of inertia I_O about L_O is given in terms of the moment of inertia I about L by the parallel-axis theorem,

$$I_O = I + d^2 m, \qquad \textbf{Eq. (8.35)}$$

where m is the mass of the object and d is the perpendicular distance between L and L_O.

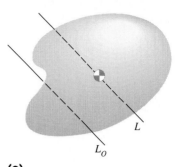

(c)

Review Problems

Refer to P8.138 for Problems 8.138–8.141.

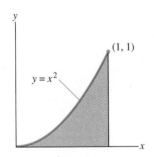

P8.138

8.138 Determine I_y and k_y.

8.139 Determine I_x and k_x.

8.140 Determine J_O and k_O.

8.141 Determine I_{xy}.

Refer to P8.142 for Problems 8.142–8.144.

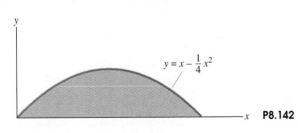

P8.142

8.142 Determine I_y and k_y.

8.143 Determine I_x and k_x.

8.144 Determine I_{xy}.

Refer to P8.145 for Problems 8.145–8.147. The origin of the $x'y'$ coordinate system is at the centroid of the area.

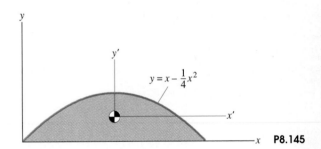

P8.145

8.145 Determine $I_{y'}$ and $k_{y'}$.

8.146 Determine $I_{x'}$ and $k_{x'}$.

8.147 Determine $I_{x'y'}$.

8.148 Determine I_y and k_y.

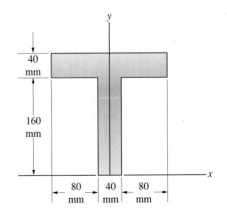

P8.148

8.149 Determine I_x and k_x for the area in Problem 8.148.

8.150 Determine I_x and k_x.

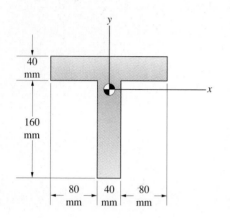

P8.150

8.151 Determine J_O and k_O for the area in Problem 8.150.

8.152 Determine I_y and k_y.

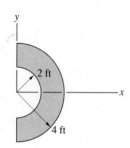

P8.152

8.153 Determine J_O and k_O for the area in Problem 8.152.

8.154 Determine I_x and k_x.

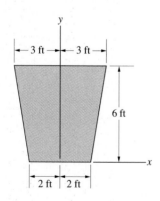

P8.154

8.155 Determine I_y and k_y for the area in Problem 8.154.

8.156 The moments of inertia of the area are $I_x = 36$ m^4, $I_y = 145$ m^4, and $I_{xy} = 44.25$ m^4. Determine a set of principal axes and the principal moments of inertia.

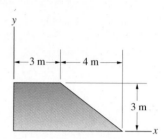

P8.156

8.157 The mass moment of inertia of the 31-oz bat about a perpendicular axis through point B is 0.093 slug-ft^2. What is the bat's mass moment of inertia about a perpendicular axis through point A? (Point A is the bat's "instantaneous center," or center of rotation, at the instant shown.)

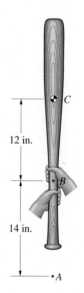

P8.157

8.158 The mass of the thin homogeneous plate is 4 kg. Determine its mass moment of inertia about the y axis.

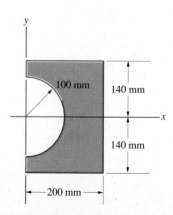

P8.158

8.159 Determine the mass moment of inertia of the plate in Problem 8.158 about the z axis.

8.160 The homogeneous pyramid is of mass m. Determine its mass moment of inertia about the z axis.

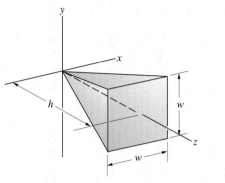

P8.160

8.161 Determine the mass moments of inertia of the homogeneous pyramid in Problem 8.160 about the x and y axes.

8.162 The homogeneous object weighs 400 lb. Determine its mass moment of inertia about the x axis.

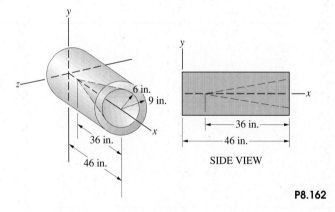

SIDE VIEW

P8.162

8.163 Determine the mass moments of inertia of the object in Problem 8.162 about the y and z axes.

8.164 Determine the mass moment of inertia of the 14-kg flywheel about the axis L.

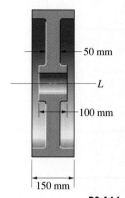

P8.164

Part of the power produced by a car's engine is transmitted by belts to pulleys that operate the alternator and coolant pump. To design a belt, and to determine the maximum torque it can exert, requires analysis of the friction forces at the surfaces of contact between the belt and the pulleys, an application we discuss in this chapter.

Chapter 9

Friction

FRICTION forces have many important effects, both desirable and unde-sirable, in engineering applications. The proper functioning of a car, for example, depends on the friction forces between the tires and the road and between the belts and pulleys in its engine, but friction between its pistons and cylinders causes wear and must be minimized by lubrication. We have seen that in some situations, friction forces can be determined with the equilibrium equations. In this chapter we discuss the Coulomb theory of friction, which allows us to estimate the maximum friction force that can be exerted by contacting surfaces and the friction force exerted by sliding surfaces. This opens the path to the analysis of important new classes of supports and machines, including wedges (shims), threaded connections, bearings, and belts.

9.1 *Theory of Dry Friction*

When you climb a ladder, it remains in place because of the friction force exerted on it by the floor (Fig. 9.1a). If you remain stationary on the ladder, the equilibrium equations determine the friction force. But an important question cannot be answered by the equilibrium equations alone: Will the ladder remain in place, or will it slip on the floor? If a truck is parked on an incline, the friction force exerted on it by the road prevents it from sliding down the incline (Fig. 9.1b). Here too there is another question: What is the *steepest* incline on which the truck can be parked?

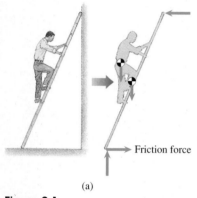

(a)

Figure 9.1

Objects supported by friction forces.

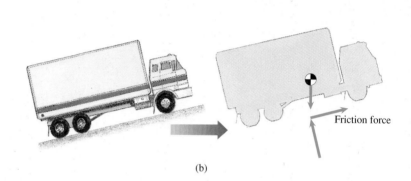

(b)

To answer these questions, we must examine the nature of friction forces in more detail. Place a book on a table and push it with a small horizontal force, as shown in Fig. 9.2(a). If the force you exert is sufficiently small, the book does not move. The free-body diagram of the book is shown in Fig. 9.2(b). The force W is the book's weight, and N is the normal force exerted by the table. The force F is the horizontal force you apply, and f is the friction force exerted by the table. Because the book is in equilibrium, $f = F$.

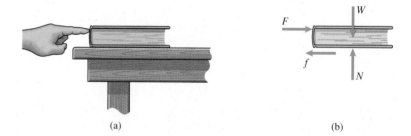

Figure 9.2

(a) Exerting a horizontal force on a book.
(b) The free-body diagram of the book.

(a)

(b)

Now slowly increase the force you apply to the book. As long as the book remains in equilibrium, the friction force must increase correspondingly, since it equals the force you apply. When the force you apply becomes too large, the book moves. It slips on the table. After reaching some maximum value, the friction force can no longer maintain the book in equilibrium. Also, notice that the force you must apply to keep the book moving on the table is smaller than the force required to cause it to slip. (You are familiar with this phenomenon if you've ever pushed a piece of furniture across a floor.)

How does the table exert a friction force on the book? Why does the book slip? Why is less force required to slide the book across the table than is required to start it moving? If the surfaces of the table and the book are magnified sufficiently, they will appear rough (Fig. 9.3). Friction forces arise in part from the interactions of the roughnesses, or **asperities**, of the contacting surfaces. On a still smaller scale, contacting surfaces tend to form atomic bonds that "glue" them together (Fig. 9.4). The fact that more force is required to start an object sliding on a surface than to keep it sliding is explained in part by the necessity to break these bonds before sliding can begin.

In the following sections we present a theory that predicts the basic phenomena we have described and has been found useful for approximating friction forces between dry surfaces in engineering applications. (Friction between lubricated surfaces is a hydrodynamic phenomenon and must be analyzed in the context of fluid mechanics.)

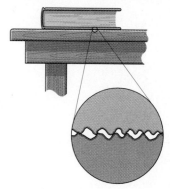

Figure 9.3

The roughnesses of the surfaces can be seen in a magnified view.

Coefficients of Friction

The theory of dry friction, or **Coulomb friction**, predicts the maximum friction forces that can be exerted by dry, contacting surfaces that are stationary relative to each other. It also predicts the friction forces exerted by the surfaces when they are in relative motion, or sliding. We first consider surfaces that are not in relative motion.

The Static Coefficient   The magnitude of the *maximum* friction force that can be exerted between two plane dry surfaces in contact is

$$f = \mu_s N, \tag{9.1}$$

where N is the normal component of the contact force between the surfaces and μ_s is a constant called the **coefficient of static friction**.

The value of μ_s is assumed to depend only on the materials of the contacting surfaces and the conditions (smoothness and degree of contamination by other materials) of the surfaces. Typical values of μ_s for various materials are shown in Table 9.1. The relatively large range of values for each pair of materials reflects the sensitivity of μ_s to the conditions of the surfaces. In engineering applications it is usually necessary to measure the value of μ_s for the actual surfaces used.

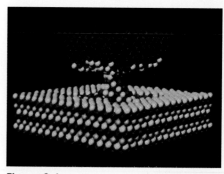

Figure 9.4

Computer simulation of a bond or "neck" of atoms formed between a nickel tip (red) and a gold surface.

Table 9.1 Typical values of the coefficient of static friction.

Materials	Coefficient of Static Friction μ_s
Metal on metal	0.15–0.20
Masonry on masonry	0.60–0.70
Wood on wood	0.25–0.50
Metal on masonry	0.30–0.70
Metal on wood	0.20–0.60
Rubber on concrete	0.50–0.90

Let's return to the example of the book on the table (Fig. 9.2). If the force F you exert on the book is small enough that the book does not move, the condition for equilibrium tells you that the friction force $f = F$. Why do you need the theory of dry friction? If you begin to increase F, the friction force f will increase until the book slips. Equation (9.1) gives the *maximum* friction force that the two surfaces can exert and thus tells you the largest force F you can apply to the book without causing it to slip. Suppose that you know the coefficient of static friction μ_s between the book and the table and the weight W of the book. Since the normal force $N = W$, the largest value of F you can apply to the book without causing it to slip is $F = f = \mu_s W$.

Equation (9.1) determines the magnitude of the maximum friction force but not its direction. The friction force is a maximum, and Eq. (9.1) is applicable, when two surfaces are on the verge of slipping relative to each other. We say that slip is **impending**, and the friction forces resist the impending motion. In Fig. 9.5(a), suppose that the lower surface is fixed and slip of the upper surface toward the right is impending. The friction force on the upper surface resists its impending motion (Fig. 9.5b). The friction force on the lower surface is in the opposite direction.

The Kinetic Coefficient According to the theory of dry friction, the magnitude of the friction force between two plane dry contacting surfaces that are in motion (sliding) relative to each other is

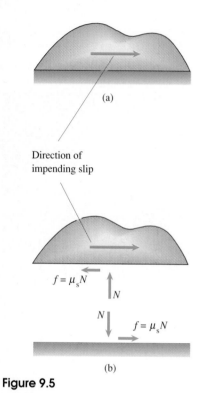

Direction of impending slip

$$f = \mu_k N, \qquad (9.2)$$

$f = \mu_s N$

N

N

$f = \mu_s N$

(b)

Figure 9.5

(a) The upper surface is on the verge of slipping to the right.
(b) Directions of the friction forces.

where N is the normal force between the surfaces and μ_k is the **coefficient of kinetic friction**. The value of μ_k is assumed to depend only on the compositions of the surfaces and their conditions. For a given pair of surfaces, its value is generally smaller than that of μ_s.

Once you have caused the book in Fig. 9.2 to begin sliding on the table, the friction force $f = \mu_k N = \mu_k W$. Therefore the force you must exert to keep the book in uniform motion is $F = f = \mu_k W$.

When two surfaces are sliding relative to each other, the friction forces resist the relative motion. In Fig. 9.6(a), suppose that the lower surface is fixed and the upper surface is moving to the right. The friction force on the upper surface acts in the direction opposite to its motion (Fig. 9.6b). The friction force on the lower surface is in the opposite direction.

Figure 9.6

(a) The upper surface is moving to the right relative to the lower surface.
(b) Directions of the friction forces.

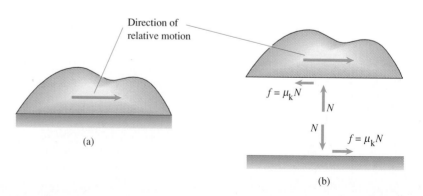

Direction of relative motion

$f = \mu_k N$

N

N

$f = \mu_k N$

(a)

(b)

Angles of Friction

Instead of resolving the reaction exerted on a surface due to its contact with another surface into the normal force N and friction force f (Fig. 9.7a), we can express it in terms of its magnitude R and the **angle of friction** θ between the force and the normal to the surface (Fig. 9.7b). The normal force and friction force are related to R and θ by

$$f = R \sin \theta, \tag{9.3}$$

$$N = R \cos \theta. \tag{9.4}$$

The value of θ when slip is impending is called the **angle of static friction** θ_s, and its value when the surfaces are sliding relative to each other is called the **angle of kinetic friction** θ_k. By using Eqs. (9.1)–(9.4), we can express the angles of static and kinetic friction in terms of the coefficients of friction:

$$\tan \theta_s = \mu_s, \tag{9.5}$$

$$\tan \theta_k = \mu_k. \tag{9.6}$$

In the following examples we discuss problems involving friction forces. An important step in the solution of such problems is deciding how to evaluate friction forces or angles of friction. When you know that slip is impending, the magnitude of the friction force is given by Eq. (9.1) and the angle of friction by Eq. (9.5). When surfaces are sliding relative to each other, the magnitude of the friction force is given by Eq. (9.2) and the angle of friction by Eq. (9.6). Otherwise, you must determine the friction force or the angle of friction from the equilibrium equations. The sequence of decisions is summarized in Fig. 9.8.

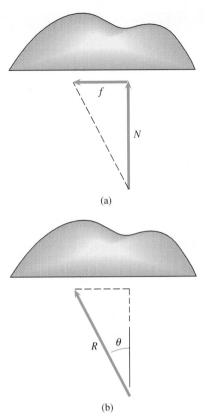

(a)

(b)

Figure 9.7
(a) The normal force N and the friction force f.
(b) The magnitude R and the angle of friction θ.

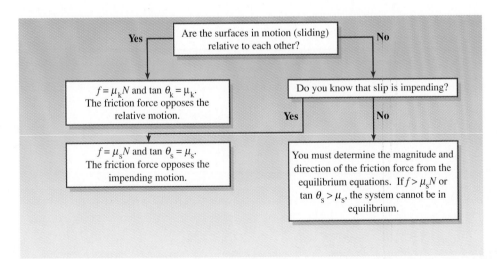

Figure 9.8
Evaluating the friction force.

Example 9.1

The arrangement in Fig. 9.9 exerts a horizontal force on the stationary 80-kg crate. The coefficient of static friction between the crate and the ramp is $\mu_s = 0.4$.
(a) If the rope exerts a 400-N force on the crate, what is the friction force exerted on the crate by the ramp?
(b) What is the largest force the rope can exert on the crate without causing it to slide up the ramp?

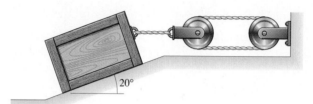

20°

Figure 9.9

STRATEGY

(a) We can follow the logic in Fig. 9.8 to decide how to evaluate the friction force. The crate is not sliding on the ramp, and we don't know whether slip is impending, so we must determine the friction force by using the equilibrium equations.
(b) We want to determine the value of the force exerted by the rope that causes the crate to be on the verge of slipping up the ramp. When slip is impending, the magnitude of the friction force is $f = \mu_s N$ and the friction force opposes the impending slip. We can use the equilibrium equations to determine the force exerted by the rope.

SOLUTION

(a) We draw the free-body diagram of the crate in Fig. (a), showing the force T exerted by the rope, the weight mg of the crate, and the normal force N and friction force f exerted by the ramp. We can choose the direction of f arbitrarily, and our solution will indicate the actual direction of the friction force. By aligning the coordinate system with the ramp as shown, we obtain the equilibrium equation

$$\Sigma F_x = f + T \cos 20° - mg \sin 20° = 0.$$

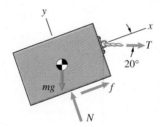

(a) Free-body diagram of the crate.

Solving for the friction force, we obtain

$$f = -T \cos 20° + mg \sin 20° = -(400) \cos 20° + (80)(9.81) \sin 20°$$

$$= -107 \text{ N}.$$

The minus sign indicates that the direction of the friction force on the crate is down the ramp.

(b) The friction force is $f = \mu_s N$, and it opposes the impending slip. To simplify our solution for T, we align the coordinate system as shown in Fig. (b), obtaining the equilibrium equations

$$\Sigma F_x = T - N \sin 20° - \mu_s N \cos 20° = 0,$$

$$\Sigma F_y = N \cos 20° - \mu_s N \sin 20° - mg = 0.$$

Solving the second equilibrium equation for N, we obtain

$$N = \frac{mg}{\cos 20° - \mu_s \sin 20°} = \frac{(80)(9.81)}{\cos 20° - (0.4) \sin 20°} = 977 \text{ N}.$$

Then from the first equilibrium equation, T is

$$T = N(\sin 20° + \mu_s \cos 20°) = (977)[\sin 20° + (0.4) \cos 20°]$$

$$= 702 \text{ N}.$$

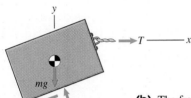

(b) The free-body diagram when slip up the ramp is impending.

Alternative Solution We can also determine T by representing the reaction exerted on the crate by the ramp as a single force (Fig. c). Because slip of the crate up the ramp is impending, R opposes the impending motion and the friction angle is $\theta_s = \arctan \mu_s = \arctan (0.4) = 21.8°$. From the triangle formed by the sum of the forces acting on the crate (Fig. d), we obtain

$$T = mg \tan (20° + \theta_s) = (80)(9.81) \tan (20° + 21.8°) = 702 \text{ N}.$$

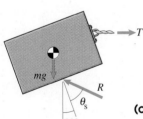

(c) Representing the reaction exerted by the ramp as a single force.

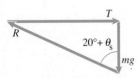

(d) The forces on the crate.

Example 9.2

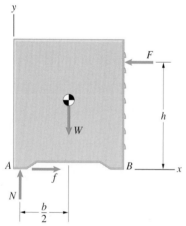

Figure 9.10

Suppose that we want to push the tool chest in Fig. 9.10 across the floor by applying the horizontal force F. If we apply the force at too great a height h, the chest will tip over before it slips. If the coefficient of static friction between the floor and the chest is μ_s, what is the largest value of h for which the chest will slip before it tips over?

STRATEGY

When the chest is on the verge of tipping over, it is in equilibrium with no reaction at B. We can use this condition to determine F in terms of h. Then, by determining the value of F that will cause the chest to slip, we will obtain the value of h that causes the chest to be on the verge of tipping over *and* on the verge of slipping.

SOLUTION

We draw the free-body diagram of the chest when it is on the verge of tipping over in Fig. (a). Summing moments about A, we obtain

$$\Sigma M_{(\text{point } A)} = Fh - W\left(\frac{1}{2}b\right) = 0.$$

Equilibrium also requires that $f = F$ and $N = W$.

When the chest is on the verge of slipping,

$$f = \mu_s N,$$

so

$$F = f = \mu_s N = \mu_s W.$$

Substituting this expression into the moment equation, we obtain

$$\mu_s W h - W\left(\frac{1}{2}b\right) = 0.$$

Solving this equation for h, we find that the chest is on the verge of tipping over *and* on the verge of slipping when

$$h = \frac{b}{2\mu_s}.$$

If h is smaller than this value, the chest will begin sliding before it tips over.

(a) The free-body diagram when the chest is on the verge of tipping over.

DISCUSSION

Notice that the largest value of h for which the chest will slip before it tips over is independent of F. Whether the chest will tip over depends only on where the force is applied, not how large it is.

Example 9.3

The motion of the disk in Fig. 9.11 is controlled by the friction force exerted at C by the brake ABC. The hydraulic actuator BE exerts a horizontal force of magnitude F on the brake at B. The coefficients of friction between the disk and the brake are μ_s and μ_k. What couple M is necessary to rotate the disk at a constant rate in the counterclockwise direction?

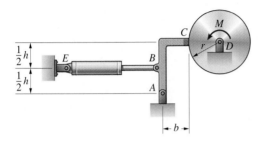

Figure 9.11

STRATEGY

We can use the free-body diagram of the disk to obtain a relation between M and the reaction exerted on the disk by the brake, then use the free-body diagram of the brake to determine the reaction in terms of F.

SOLUTION

We draw the free-body diagram of the disk in Fig. (a), representing the force exerted by the brake by a single force R. The force R opposes the counterclockwise rotation of the disk, and the friction angle is the angle of kinetic friction $\theta_k = \arctan \mu_k$. Summing moments about D, we obtain

$$\Sigma M_{(\text{point } D)} = M - (R \sin \theta_k)r = 0.$$

Then, from the free-body diagram of the brake (Fig. b), we obtain

$$\Sigma M_{(\text{point } A)} = -F\left(\frac{1}{2}h\right) + (R \cos \theta_k)h - (R \sin \theta_k)b = 0.$$

We can solve these two equations for M and R. The solution for the couple M is

$$M = \frac{(1/2)hr\, F \sin \theta_k}{h \cos \theta_k - b \sin \theta_k} = \frac{(1/2)hr\, F\mu_k}{h - b\mu_k}.$$

DISCUSSION

If μ_k is sufficiently small, then the denominator of the solution for the couple, $(h \cos \theta_k - b \sin \theta_k)$, is positive. As μ_k becomes larger, the denominator becomes smaller, because $\cos \theta_k$ decreases and $\sin \theta_k$ increases. As the denominator approaches zero, the couple required to rotate the disk approaches infinity. To understand this result, notice that the denominator equals zero when $\tan \theta_k = h/b$, which means that the line of action of R passes through point A (Fig. c). As μ_k becomes larger and the line of action of R approaches point A, the magnitude of R necessary to balance the moment of F about A approaches infinity and, as a result, M approaches infinity.

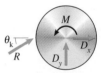

(a) The free-body diagram of the disk.

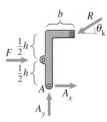

(b) The free-body diagram of the brake.

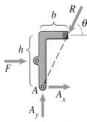

(c) The line of action of R passing through point A.

Example 9.4

The 80-kg climber at *A* in Fig. 9.12 is being helped up an icy slope by friends. The tensions in ropes *AB* and *AC* are 130 N and 220 N, respectively. The *y* axis is vertical, and the unit vector $\mathbf{e} = -0.182\mathbf{i} + 0.818\mathbf{j} + 0.545\mathbf{k}$ is perpendicular to the ground where the climber stands. What minimum coefficient of static friction between the climber's shoes and the ground is necessary to prevent him from slipping?

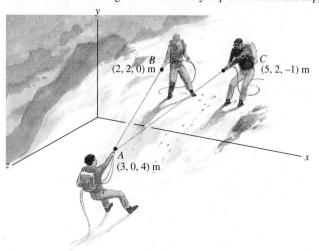

Figure 9.12

STRATEGY

We know the forces exerted on the climber by the two ropes and by his weight, so we can use equilibrium to determine the force $\mathbf{R}$ exerted on him by the ground. When slip is impending, the angle between $\mathbf{R}$ and the unit vector $\mathbf{e}$ is equal to the angle of static friction θ_s. We can use this condition to calculate the coefficient of static friction for impending slip.

SOLUTION

We draw the free-body diagram of the climber in Fig. (a), showing the forces $\mathbf{T}_{AB}$ and $\mathbf{T}_{AC}$ exerted by the ropes, the force $\mathbf{R}$ exerted by the ground, and his weight. The sum of the forces equals zero:

$$\mathbf{R} + \mathbf{T}_{AB} + \mathbf{T}_{AC} - mg\mathbf{j} = 0.$$

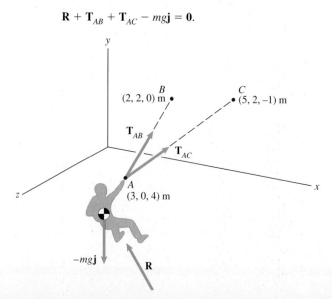

(a) Free-body diagram of the climber.

By expressing $\mathbf{T}_{AB}$ and $\mathbf{T}_{AC}$ in terms of their components, we can solve this equation for the components of $\mathbf{R}$. The force $\mathbf{T}_{AB}$ is

$$\mathbf{T}_{AB} = |\mathbf{T}_{AB}| \left[\frac{(2-3)\mathbf{i} + (2-0)\mathbf{j} + (0-4)\mathbf{k}}{\sqrt{(2-3)^2 + (2-0)^2 + (0-4)^2}} \right]$$

$$= (130)(-0.218\mathbf{i} + 0.436\mathbf{j} - 0.873\mathbf{k})$$

$$= -28.4\mathbf{i} + 56.7\mathbf{j} - 113.5\mathbf{k} \ (\text{N}),$$

and the force $\mathbf{T}_{AC}$ is

$$\mathbf{T}_{AC} = |\mathbf{T}_{AC}| \left[\frac{(5-3)\mathbf{i} + (2-0)\mathbf{j} + (-1-4)\mathbf{k}}{\sqrt{(5-3)^2 + (2-0)^2 + (-1-4)^2}} \right]$$

$$= (220)(0.348\mathbf{i} + 0.348\mathbf{j} - 0.870\mathbf{k})$$

$$= 76.6\mathbf{i} + 76.6\mathbf{j} - 191.5\mathbf{k} \ (\text{N}).$$

Substituting these expressions into the equilibrium equation and solving for $\mathbf{R}$, we obtain

$$\mathbf{R} = -48.2\mathbf{i} + 651.5\mathbf{j} + 305.0\mathbf{k} \ (\text{N}).$$

To determine the angle θ between $\mathbf{R}$ and the unit vector $\mathbf{e}$ that is normal to the surface on which the climber stands (Fig. b), we use the dot product. From the definition $\mathbf{R} \cdot \mathbf{e} = |\mathbf{R}||\mathbf{e}| \cos \theta$, we obtain

$$\cos \theta = \frac{\mathbf{R} \cdot \mathbf{e}}{|\mathbf{R}||\mathbf{e}|} = \frac{(-48.2)(-0.182) + (651.5)(0.818) + (305.0)(0.545)}{\sqrt{(-48.2)^2 + (651.5)^2 + (305.0)^2}}$$

$$= 0.982.$$

The angle $\theta = 10.9°$. Setting this angle equal to the angle of static friction, we obtain the coefficient of static friction for impending slip:

$$\mu_s = \tan \theta_s = \tan 10.9° = 0.193.$$

(b) The angle θ.

Problems

9.1 The crane's cable is attached to a stationary 4000-kg crate.
(a) If the tension in the cable is 10 kN, what is the magnitude of the friction force exerted on the crate by the ground?
(b) If slip of the crate impends when the tension in the cable is 20 kN, what is the coefficient of static friction between the crate and the ground?

    Strategy: (a) You do not know that slip is impending, so you *cannot* determine the friction force from Eq. (9.1). Draw the free-body diagram of the crate, and use the equilibrium equations to determine the friction force. (b) After using the equilibrium equations to determine the normal and friction forces, you can use Eq. (9.1) to determine μ_s.

P9.1

9.3 The truck weighs 18,000 lb.
(a) If the truck is stationary on the incline and $\alpha = 10°$, what is the magnitude of the friction force exerted on it by the road?
(b) If the coefficient of static friction between the truck's tires and the road is $\mu_s = 0.6$, what is the largest value of α for which the truck can remain stationary?
(c) If the incline is icy and $\mu_s = 0.1$, what is the largest value of α for which the truck can remain stationary?

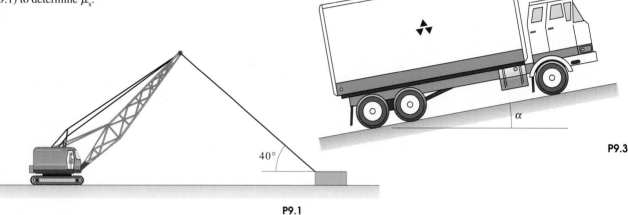

P9.3

9.2 The 10.5-kg Sojourner rover, placed on the surface of Mars by the Pathfinder Lander on July 4, 1997, was designed to negotiate a 45° slope without tipping over.
(a) What minimum static coefficient of friction between the wheels of the rover and the surface is necessary for it to rest on a 45° slope? The acceleration due to gravity at the surface of Mars is 3.69 m/s².
(b) Engineers testing the Sojourner want to confirm that it will negotiate a 45° slope without tipping over. What minimum static coefficient of friction between the wheels of the rover and the surface is necessary for it to rest on a 45° slope on Earth?

9.4 The 5-kg box is stationary on the inclined surface.
(a) What is the magnitude of the friction force on the box?
(b) What is the smallest value of the coefficient of static friction that will permit the box to remain stationary?

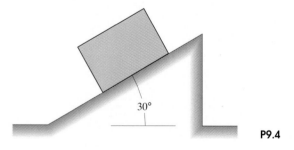

P9.4

P9.2

9.5 You can measure the coefficient of static friction between two materials as follows. Place a sheet of one of the materials on a flat board. Bond a sheet of the other material to a block A. Place the block onto the board, and slowly increase the tilt α of the board until the block slips. Show that the coefficient of static friction between the two materials is related to the angle α by $\mu_s = \tan \alpha$.

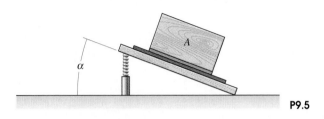

P9.5

9.6 The mechanism shown is designed to position pieces of luggage on a ramp. The suitcase S weighs 20 lb. The mechanism exerts a force on the suitcase in the direction parallel to the ramp. The coefficients of friction between the suitcase and the ramp are $\mu_s = 0.20$ and $\mu_k = 0.18$.
(a) Will the suitcase remain stationary on the ramp when the mechanism exerts no force on it?
(b) Determine the force the mechanism must exert on the suitcase to start it moving up the ramp.
(c) Determine the force the mechanism must exert on the suitcase to keep it moving up the ramp once it has started moving.

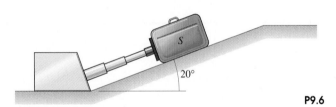

P9.6

9.7 The mass of the stationary crate is 40 kg. The length of the spring is 180 mm, its unstretched length is 200 mm, and the spring constant is $k = 2500$ N/m. The coefficient of static friction between the crate and the inclined surface is $\mu_s = 0.6$. Determine the magnitude of the friction force exerted on the crate.

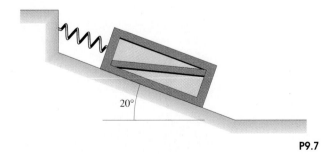

P9.7

9.8 In Problem 9.7, what is the minimum value of the coefficient of static friction between the crate and the inclined surface for which the crate will remain stationary?

9.9 The mass of the box is 3 kg. The force F is horizontal. The coefficient of static friction between the box and the inclined surface is $\mu_s = 0.2$. For what range of values of F will the box remain stationary?

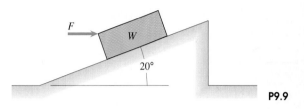

P9.9

9.10 Box A weighs 100 lb, and box B weighs 30 lb. The coefficients of friction between box A and the ramp are $\mu_s = 0.30$ and $\mu_k = 0.28$. What is the magnitude of the friction force exerted on box A by the ramp?

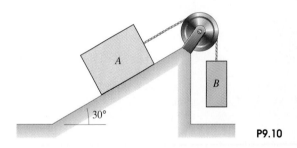

P9.10

9.11 In Problem 9.10, box A weighs 100 lb, and the coefficients of friction between box A and the ramp are $\mu_s = 0.30$ and $\mu_k = 0.28$. For what range of weights of the box B will the system remain stationary?

9.12 The mass of the box on the left is 30 kg, and the mass of the box on the right is 40 kg. The coefficient of static friction between each box and the inclined surface is $\mu_s = 0.2$. Determine the minimum angle α for which the boxes will remain stationary.

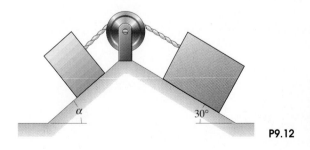

P9.12

9.13 In Problem 9.12, determine the maximum angle α for which the boxes will remain stationary.

9.14 The box is stationary on the inclined surface. The coefficient of static friction is μ_s.
(a) Determine the force T necessary to start the box sliding up the surface.
(b) Show that the force T necessary to start the box sliding up the surface is a minimum when $\tan \beta = \mu_s$.

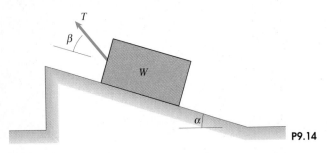

P9.14

9.15 To explain observations of ship launchings at the port of Rochefort in 1779, Coulomb analyzed the system in Problem 9.14 to determine the minimum force T necessary to hold the box stationary on the inclined surface. Show that the result is $T = W(\sin \alpha - \mu_s \cos \alpha)/(\cos \beta - \mu_s \sin \beta)$.

9.16 Two sheets of plywood A and B lie on the bed of the truck. They have the same weight W, and the coefficient of static friction between the two sheets of wood and between sheet B and the truck bed is μ_s.
(a) If you apply a horizontal force to sheet A and apply no force to sheet B, can you slide sheet A off the truck without causing sheet B to move? What force is necessary to cause sheet A to start moving?
(b) If you prevent sheet A from moving by exerting a horizontal force on it, what horizontal force on sheet B is necessary to start it moving?

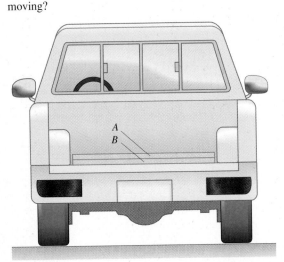

P9.16

9.17 Suppose that the truck in Problem 9.16 is loaded with N sheets of plywood of the same weight W, labeled (from the top) sheets 1, 2, . . . , N. The coefficient of static friction between the sheets of wood and between the bottom sheet and the truck bed is

μ_s. If you apply a horizontal force to the sheets above it to prevent them from moving, can you pull out the ith sheet, $1 \le i \le N$, without causing any of the sheets below it to move? What force must you apply to cause it to start moving?

9.18 The weights of the two boxes are $W_1 = 100$ lb and $W_2 = 50$ lb. The coefficients of friction between the left box and the inclined surface are $\mu_s = 0.12$ and $\mu_k = 0.10$. Determine the tension the man must exert on the rope to pull the boxes upward at a constant rate.

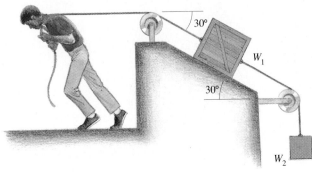

P9.18

9.19 In Problem 9.18, for what range of tensions exerted on the rope by the man will the boxes remain stationary?

9.20 The coefficient of static friction between the two boxes and between the lower box and the inclined surface is μ_s. What is the largest angle α for which the lower box will not slip?

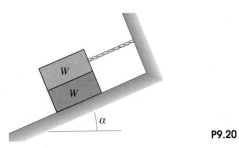

P9.20

9.21 The coefficient of static friction between the two boxes and between the lower box and the inclined surface is μ_s. What is the largest force F that will not cause the boxes to slip?

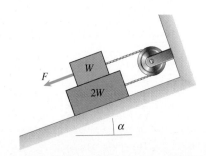

P9.21

9.22 Consider the system shown in Problem 9.21. The coefficient of static friction between the two boxes and between the lower box and the inclined surface is μ_s. If $F = 0$, the lower box will slip down the inclined surface. What is the smallest force F for which the boxes will not slip?

9.23 A sander consists of a rotating cylinder with sandpaper bonded to the outer surface. The normal force exerted on the workpiece A by the sander is 30 lb. The workpiece A weighs 50 lb. The coefficients of friction between the sander and the workpiece A are $\mu_s = 0.65$ and $\mu_k = 0.60$. The coefficients of friction between the workpiece A and the table are $\mu_s = 0.35$ and $\mu_k = 0.30$. Will the workpiece remain stationary while it is being sanded?

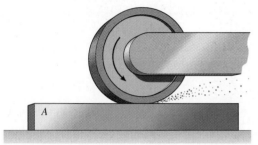

P9.23

9.24 Suppose that you want the bar of length L to act as a simple brake that will allow the workpiece A to slide to the left but will not allow it to slide to the right no matter how large a horizontal force is applied to it. The weight of the bar is W, and the coefficient of static friction between it and the workpiece A is μ_s. You can neglect friction between the workpiece and the surface it rests on.
(a) What is the largest angle α for which the bar will prevent the workpiece from moving to the right?
(b) If α has the value determined in (a), what horizontal force is necessary to slide the workpiece A toward the left at a constant rate?

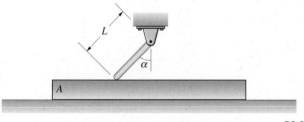

P9.24

9.25 The coefficient of static friction between the 20-lb bar and the floor is $\mu_s = 0.3$. Neglect friction between the bar and the wall.
(a) If $\alpha = 20°$, what is the magnitude of the friction force exerted on the bar by the floor?

(b) What is the maximum value of α for which the bar will not slip?

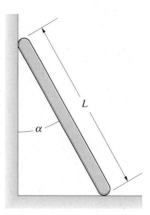

P9.25

9.26 The masses of the ladder and the person are 18 kg and 90 kg, respectively. The center of mass of the 4-m ladder is at its midpoint. If $\alpha = 30°$, what is the minimum coefficient of static friction between the ladder and the floor necessary for the person to climb to the top of the ladder? Neglect friction between the ladder and the wall.

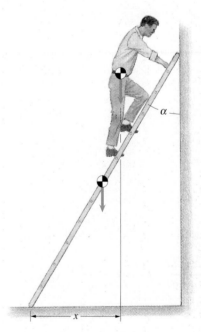

P9.26

9.27 In Problem 9.26, the coefficient of static friction between the ladder and the floor is $\mu_s = 0.6$. The masses of the ladder and the person are 18 kg and 100 kg, respectively. The center of mass of the 4-m ladder is at its midpoint. What is the maximum value of α for which the person can climb to the top of the ladder? Neglect friction between the ladder and the wall.

9.28 In Problem 9.26, the coefficient of static friction between the ladder and the floor is $\mu_s = 0.6$, and $\alpha = 35°$. The center of mass of the 4-m ladder is at its midpoint, and its mass is 18 kg.
(a) If a football player with a mass of 140 kg attempts to climb the ladder, what maximum value of x will he reach? Neglect friction between the ladder and the wall.
(b) What minimum friction coefficient would be required for him to reach the top of the ladder?

9.29 The disk weighs 50 lb. Neglect the weight of the bar. The coefficients of friction between the disk and the floor are $\mu_s = 0.6$ and $\mu_k = 0.4$.
(a) What is the largest couple M that can be applied to the stationary disk without causing it to start rotating?
(b) What couple M is necessary to rotate the disk at a constant rate?

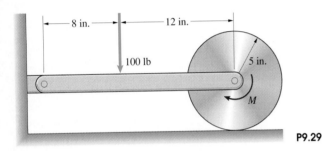

P9.29

9.30 The cylinder has weight W. The coefficient of static friction between the cylinder and the floor and between the cylinder and the wall is μ_s. What is the largest couple M that can be applied to the stationary cylinder without causing it to rotate?

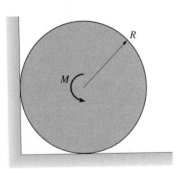

P9.30

9.31 The cylinder has weight W. The coefficient of static friction between the cylinder and the floor and between the cylinder and the wall is μ_s. What is the largest couple M that can be applied to the stationary cylinder without causing it to rotate?

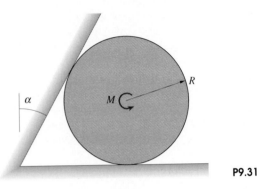

P9.31

9.32 Suppose that $\alpha = 30°$ in Problem 9.31 and that a couple $M = 0.5RW$ is required to turn the cylinder at a constant rate. What is the coefficient of kinetic friction?

9.33 The disk of weight W and radius R is held in equilibrium on the circular surface by a couple M. The coefficient of static friction between the disk and the surface is μ_s. Show that the largest value M can have without causing the disk to slip is

$$M = \frac{\mu_s RW}{\sqrt{1 + \mu_s^2}}.$$

P9.33

9.34 The coefficient of static friction between the jaws of the pliers and the gripped object is μ_s. What is the largest value of the angle α for which the gripped object will not slip out? (Neglect the object's weight.)
Strategy: Draw the free-body diagram of the gripped object, and assume that slip is impending.

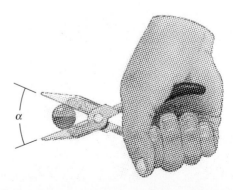

P9.34

9.35 The stationary disk, of 300-mm radius, is attached to a pin support at D. The disk is held in place by the brake ABC in contact with the disk at C. The hydraulic actuator BE exerts a horizontal 400-N force on the brake at B. The coefficients of friction between the disk and the brake are $\mu_s = 0.6$ and $\mu_k = 0.5$. What couple must be applied to the stationary disk to cause it to slip in the counterclockwise direction?

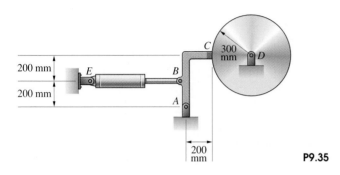

P9.35

9.36 What couple must be applied to the stationary disk in Problem 9.35 to cause it to slip in the clockwise direction?

9.37 The mass of block B is 8 kg. The coefficient of static friction between the surfaces of the clamp and the block is $\mu_s = 0.2$. When the clamp is aligned as shown, what minimum force must the spring exert to prevent the block from slipping out?

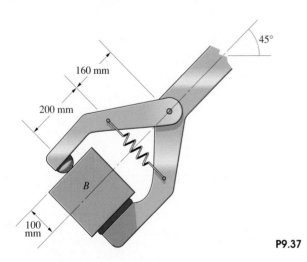

P9.37

9.38 By altering its dimensions, redesign the clamp in Problem 9.37 so that the minimum force the spring must exert to prevent the block from slipping out is 180 N. Draw a sketch of your new design.

9.39 The horizontal bar is attached to a collar that slides on the smooth vertical bar. The collar at P slides on the smooth horizontal bar. The total mass of the horizontal bar and the two collars is 12 kg. The system is held in place by the pin in the circular slot. The pin contacts only the lower surface of the slot, and the coefficient of static friction between the pin and the slot is 0.8. If the system is in equilibrium and $y = 260$ mm, what is the magnitude of the friction force exerted on the pin by the slot?

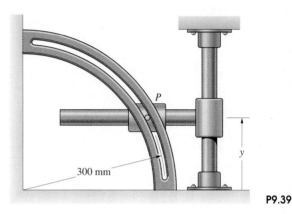

P9.39

9.40 In Problem 9.39, what is the minimum height y at which the system can be in equilibrium?

9.41 The rectangular 100-lb plate is supported by the pins A and B. If friction can be neglected at A and the coefficient of static friction between the pin at B and the slot is $\mu_s = 0.4$, what is the largest angle α for which the plate will not slip?

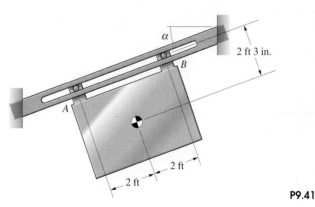

P9.41

9.42 If you can neglect friction at B in Problem 9.41 and the coefficient of static friction between the pin at A and the slot is $\mu_s = 0.4$, what is the largest angle α for which the plate will not slip?

9.43   The airplane's weight is $W = 2400$ lb. Its brakes keep the rear wheels locked, and the coefficient of static friction between the wheels and the runway is $\mu_s = 0.6$. The front (nose) wheel can turn freely and so exerts only a normal force on the runway. Determine the largest horizontal thrust force T the plane's propeller can generate without causing the rear wheels to slip.

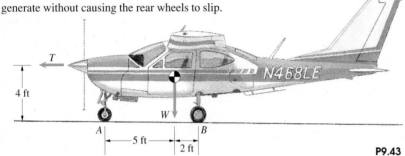

P9.43

9.44   The refrigerator weighs 350 lb. The distances $h = 60$ in. and $b = 14$ in. The coefficient of static friction at A and B is $\mu_s = 0.24$.
(a) What force F is necessary for impending slip?
(b) Will the refrigerator tip over before it slips?

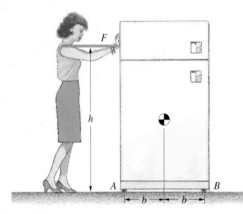

P9.44

9.45   If you want the refrigerator in Problem 9.44 to slip before it tips over, what is the maximum height h at which you can push it?

9.46   The refrigerator in Problem 9.44 weighs 350 lb. The distances $h = 45$ in. and $b = 15$ in. The coefficients of static friction at A and B are 0.24 and 0.30, respectively.
(a) What force F is necessary for impending slip?
(b) Will the refrigerator tip over before it slips?

9.47   The mass of the Mercedes-Benz 600SL is 2020 kg, and all four wheels are locked. The coefficient of static friction between the tires and the road is $\mu_s = 0.8$. Determine the force F that must be exerted on the rear bumper to cause the car to slip.

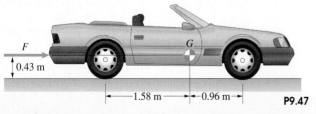

P9.47

9.48   Determine the force F that must be exerted on the car in Problem 9.47 to cause it to slip if (a) the rear wheels are locked but the front wheels are free to roll; (b) the front wheels are locked but the rear wheels are free to roll. (The wheels that are free to roll exert no friction force.)

9.49   The coefficient of static friction between the 3000-lb car's tires and the road is $\mu_s = 0.5$. Determine the steepest grade (the largest value of the angle α) the car can drive up at constant speed if the car has (a) rear-wheel drive; (b) front-wheel drive; (c) four-wheel drive.

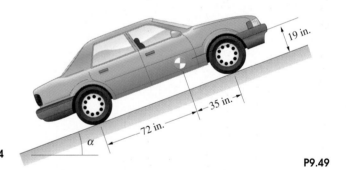

P9.49

9.50   The stationary cabinet has weight W. Determine the force F that must be exerted to cause it to move if (a) the coefficient of static friction at A and at B is μ_s; (b) the coefficient of static friction at A is μ_{sA} and the coefficient of static friction at B is μ_{sB}.

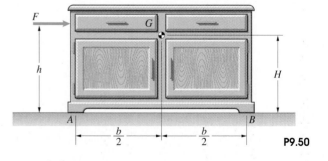

P9.50

9.51 The mass of the 3-m bar is 20 kg. It will slip if the angle α is larger than 15°. What is the coefficient of static friction between the ends of the bar and the circular surface?

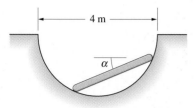

P9.51

9.52 The coefficient of static friction between the right bar and the surface at A is $\mu_s = 0.6$. Neglect the weights of the bars. If $\alpha = 20°$, what is the magnitude of the friction force exerted at A?

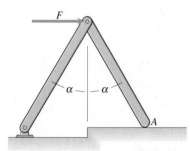

P9.52

9.53 Consider the system in Problem 9.52. The coefficient of static friction between the right bar and the surface at A is $\mu_s = 0.6$. Neglect the weights of the bars. What is the largest angle α at which the truss will remain stationary without slipping?

9.54 Each of the uniform 2-ft bars weighs 4 lb. Neglect the weight of the collar at P. The coefficient of static friction between the collar and the horizontal bar is $\mu_s = 0.6$. If the system is in equilibrium and the angle $\theta = 45°$, what is the magnitude of the friction force exerted on the collar by the horizontal bar?

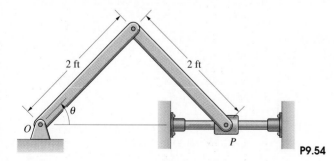

P9.54

9.55 In Problem 9.54, what is the minimum coefficient of static friction between the collar P and the horizontal bar necessary for the system to be in equilibrium when $\theta = 45°$?

9.56 The coefficient of static friction between the box and the floor is μ_s. Neglect the weights of the bars. What force F will cause the box to slip?

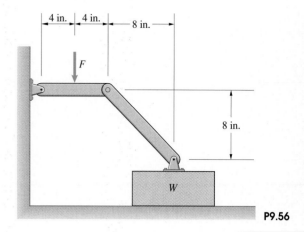

P9.56

9.57 The structure is supported at B by the normal and friction forces exerted on the plate by the wall. Neglect the weights of the bars.
(a) What is the magnitude of the friction force?
(b) What is the minimum coefficient of static friction necessary for the structure to remain in place?

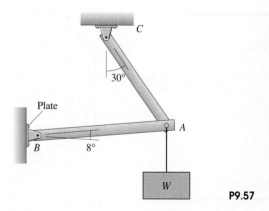

P9.57

9.58 Suppose that bars AB and AC in Problem 9.57 are of equal length L and that each has a weight of $W/2$.
(a) What is the magnitude of the friction force at B?
(b) What is the minimum coefficient of static friction necessary for the structure to remain in place?

9.59 The frame is supported by the normal and friction forces exerted on the plates at A and G by the fixed surfaces. The coefficient of static friction at A is $\mu_s = 0.6$. Will the frame slip at A when it is subjected to the loads shown?

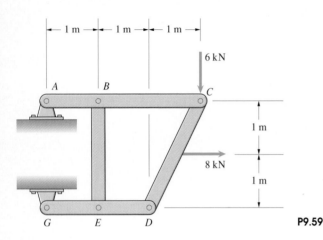

P9.59

9.60 The frame in Problem 9.59 is supported by the normal and friction forces exerted on the plates at A and G by the fixed surfaces. The coefficient of static friction at G is $\mu_s = 0.25$. Will the frame slip at G when it is subjected to the loads shown?

9.61 The direction cosines of the crane's cable are $\cos \theta_x = 0.588$, $\cos \theta_y = 0.766$, $\cos \theta_z = 0.260$. The y axis is vertical. The stationary caisson to which the cable is attached weighs 2000 lb and rests on horizontal ground. If the coefficient of static friction between the caisson and the ground is $\mu_s = 0.4$, what tension in the cable is necessary to cause the caisson to slip?

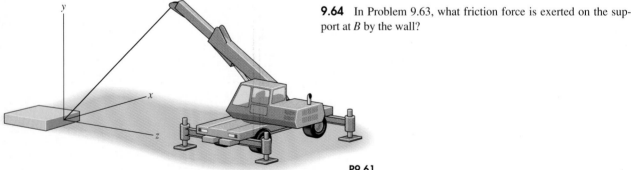

P9.61

9.62 The 10-lb metal disk A is at the center of the inclined surface. The tension in the string AB is 5 lb. What minimum coefficient of static friction between the disk and the surface is necessary to keep the disk from slipping?

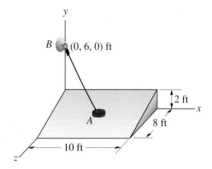

P9.62

9.63 The suspended weight $W = 600$ lb. The bars AB and AC have ball and socket supports at each end. Suppose that you want the ball and socket at B to be held in place by the normal and friction forces between the support and the wall. What minimum coefficient of friction is required?

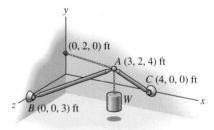

P9.63

9.64 In Problem 9.63, what friction force is exerted on the support at B by the wall?

9.2 *Applications*

Effects of friction forces, such as wear, loss of energy, and generation of heat, are often undesirable. But many devices cannot function properly without friction forces and may actually be designed to create them. A car's brakes work by exerting friction forces on the rotating wheels, and its tires are designed to maximize the friction forces they exert on the road under various weather conditions. In this section we analyze several types of devices in which friction forces play important roles.

Wedges

A **wedge** is a bifacial tool with the faces set at a small acute angle (Figs. 9.13a and b). When a wedge is pushed forward, the faces exert large lateral forces as a result of the small angle between them (Fig. 9.13c). In various forms, wedges are used in many engineering applications.

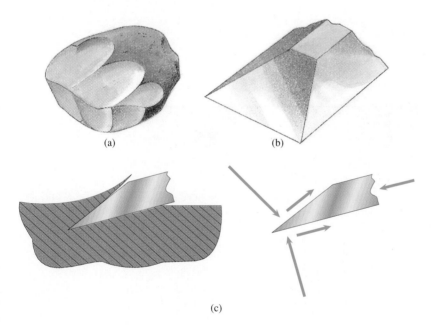

(a)

(b)

(c)

Figure 9.13

(a) An early wedge tool—a bifacial "chopper" from the Rift Valley in eastern Africa.
(b) A modern chisel blade.
(c) The faces of a wedge exert large lateral forces.

The large lateral force generated by a wedge can be used to lift a load (Fig. 9.14a). Let W_L be the weight of the load and W_W the weight of the wedge. To determine the force F necessary to start raising the load, we assume that slip of the load and wedge are impending (Fig. 9.14b). From the free-body diagram of the load, we obtain the equilibrium equations

$$\Sigma F_x = Q - N \sin \alpha - \mu_s N \cos \alpha = 0,$$
$$\Sigma F_y = N \cos \alpha - \mu_s N \sin \alpha - \mu_s Q - W_L = 0.$$

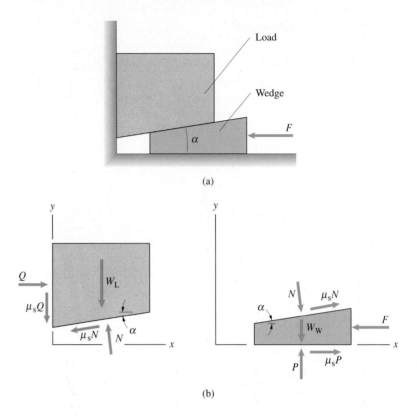

Figure 9.14

(a) Raising a load with a wedge.
(b) Free-body diagrams of the load and the wedge when slip is impending.

From the free-body diagram of the wedge, we obtain the equations

$$\Sigma F_x = N \sin \alpha + \mu_s N \cos \alpha + \mu_s P - F = 0,$$

$$\Sigma F_y = P - N \cos \alpha + \mu_s N \sin \alpha - W_W = 0.$$

These four equations determine the three normal forces Q, N, and P and the force F. The solution for F is

$$F = \mu_s W_W + \left[\frac{(1 - \mu_s^2) \tan \alpha + 2\mu_s}{(1 - \mu_s^2) - 2\mu_s \tan \alpha} \right] W_L.$$

Suppose that $W_W = 0.2W_L$ and $\alpha = 10°$. If $\mu_s = 0$, the force necessary to lift the load is only $0.176W_L$. But if $\mu_s = 0.2$, the force becomes $0.680W_L$, and if $\mu_s = 0.4$, it becomes $1.44W_L$. From this standpoint, friction is undesirable. But if there were no friction, the wedge would not remain in place when the force F is removed.

Example 9.5

Splitting a log must have been one of the first applications of the wedge (Fig. 9.15). Although it is a dynamic process—the wedge is hammered into the wood—you can get an idea of the forces involved from a static analysis. Suppose that $\alpha = 10°$ and the coefficients of friction between the surfaces of the wedge and the log are $\mu_s = 0.22$ and $\mu_k = 0.20$. Neglect the weight of the wedge.
(a) If the wedge is driven into the log at a constant rate by a vertical force F, what are the magnitudes of the normal forces exerted on the log by the wedge?
(b) Will the wedge remain in place in the log when the force is removed?

STRATEGY

(a) The friction forces resist the motion of the wedge into the log and are equal to $\mu_k N$, where N is the normal force the log exerts on the faces. We can use equilibrium to determine N in terms of F.
(b) By assuming that the wedge is on the verge of slipping out of the log, we can determine the minimum value of μ_s necessary for the wedge to stay in place.

SOLUTION

(a) In Fig. (a), we draw the free-body diagram of the wedge as it is pushed into the log by a force F. The faces of the wedge are subjected to normal forces and friction forces by the log. The friction forces resist the motion of the wedge. From the equilibrium equation

$$2N \sin\left(\frac{\alpha}{2}\right) + 2\mu_k N \cos\left(\frac{\alpha}{2}\right) - F = 0,$$

we obtain the normal force N:

$$N = \frac{F}{2[\sin(\alpha/2) + \mu_k \cos(\alpha/2)]} = \frac{F}{2[\sin(10°/2) + (0.20)\cos(10°/2)]}$$

$$= 1.75F.$$

(b) In Fig. (b), we draw the free-body diagram when $F = 0$ and the wedge is on the verge of slipping out. From the equilibrium equation

$$2N \sin\left(\frac{\alpha}{2}\right) - 2\mu_s N \cos\left(\frac{\alpha}{2}\right) = 0,$$

we obtain the minimum coefficient of friction necessary for the wedge to remain in place:

$$\mu_s = \tan\left(\frac{\alpha}{2}\right) = \tan\left(\frac{10°}{2}\right) = 0.087.$$

We can also obtain this result by representing the reaction exerted on the wedge by the log as a single force (Fig. c). When the wedge is on the verge of slipping out, the friction angle is the angle of static friction θ_s. The sum of the forces in the vertical direction is zero only if

$$\theta_s = \arctan(\mu_s) = \frac{\alpha}{2} = 5°,$$

so $\mu_s = \tan 5° = 0.087$. Thus we conclude that the wedge will remain in place.

Figure 9.15

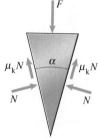

(a) Free-body diagram of the wedge with a vertical force F applied to it.

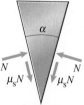

(b) Free-body diagram of the wedge when it is on the verge of slipping out.

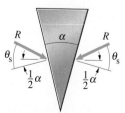

(c) Representing the reactions by a single force.

Threads

You are familiar with threads from their use on wood screws, machine screws, and other machine elements. We show a shaft with square threads in Fig. 9.16(a). The axial distance p from one thread to the next is called the **pitch** of the thread, and the angle α is its **slope**. We will consider only the case in which the shaft has a single continuous thread, so the relation between the pitch and slope is

$$\tan \alpha = \frac{p}{2\pi r}, \tag{9.7}$$

where r is the mean radius of the thread.

Suppose that the threaded shaft is enclosed in a fixed sleeve with a mating groove and is subjected to an axial load F (Fig. 9.16b). Applying a couple M in the direction shown will tend to cause the shaft to start rotating and moving in the axial direction opposite to F. Our objective is to determine the couple M necessary to cause the shaft to start rotating.

We draw the free-body diagram of a differential element of the thread of length dL in Fig. 9.16(c), representing the reaction exerted by the mating groove by the force dR. If the shaft is on the verge of rotating, dR resists the impending motion and the friction angle is the angle of static friction θ_s. The vertical component of the reaction on the element is $dR \cos(\theta_s + \alpha)$. To determine the total vertical force on the thread, we must integrate this expression over the length L of the thread. For equilibrium, the result must equal the axial force F acting on the shaft:

$$\cos(\theta_s + \alpha) \int_L dR = F. \tag{9.8}$$

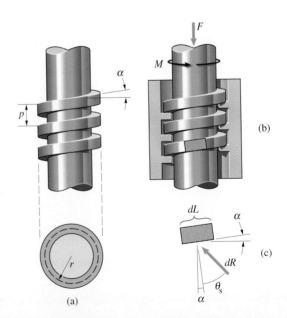

Figure 9.16

(a) A shaft with a square thread.
(b) The shaft within a sleeve with a mating groove and the direction of M that can cause the shaft to start moving in the axial direction opposite to F.
(c) A differential element of the thread when slip is impending.

The moment about the center of the shaft due to the reaction on the element is $r\,dR \sin(\theta_\mathrm{s} + \alpha)$. The total moment must equal the couple M exerted on the shaft:

$$r \sin(\theta_\mathrm{s} + \alpha) \int_L dR = M.$$

Dividing this equation by Eq. (9.8), we obtain the couple M necessary for the shaft to be on the verge of rotating and moving in the axial direction opposite to F:

$$M = rF \tan(\theta_\mathrm{s} + \alpha). \tag{9.9}$$

Replacing the angle of static friction θ_s in this expression with the angle of kinetic friction θ_k gives the couple required to cause the shaft to rotate at a constant rate.

If the couple M is applied to the shaft in the opposite direction (Fig. 9.17a), the shaft tends to start rotating and moving in the axial direction of the load F. Figure 9.17(b) shows the reaction on a differential element of the thread of length dL when slip is impending. The direction of the reaction opposes the rotation of the shaft. In this case, the vertical component of the reaction on the element is $dR \cos(\theta_\mathrm{s} - \alpha)$. Equilibrium requires that

$$\cos(\theta_\mathrm{s} - \alpha) \int_L dR = F. \tag{9.10}$$

The moment about the center of the shaft due to the reaction is $r\,dR \sin(\theta_\mathrm{s} - \alpha)$, so

$$r \sin(\theta_\mathrm{s} - \alpha) \int_L dR = M.$$

Dividing this equation by Eq. (9.10), we obtain the couple M necessary for the shaft to be on the verge of rotating and moving in the direction of the force F:

$$M = rF \tan(\theta_\mathrm{s} - \alpha). \tag{9.11}$$

Replacing θ_s with θ_k in this expression gives the couple necessary to rotate the shaft at a constant rate.

Notice in Eq. (9.11) that the couple required for impending motion is zero when $\theta_\mathrm{s} = \alpha$. When the angle of static friction is less than this value, the shaft will rotate and move in the direction of the force F with no couple applied.

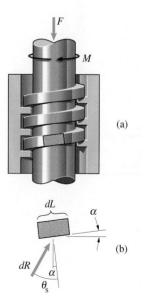

(a)

(b)

Figure 9.17

(a) The direction of M that can cause the shaft to move in the axial direction of F.

(b) A differential element of the thread when slip is impending.

Example 9.6

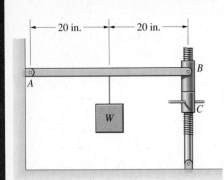

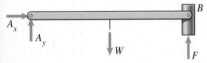

Figure 9.18

The right end of bar AB in Fig. 9.18 is pinned to an unthreaded collar B that rests on the collar C. Collar C has a groove that mates with the thread on the shaft. The mean radius of the thread is $r = 1.5$ in., and its pitch is $p = 0.2$ in. The coefficient of static friction between the collar C and the threaded shaft is $\mu_s = 0.25$. The 400-lb weight W can be raised or lowered by turning the collar C.

(a) What couple must be applied to the collar C to start the weight moving upward?

(b) Will the weight remain in the position shown when no couple is exerted on the collar C?

STRATEGY

(a) By drawing the free-body diagram of the bar and collar B, we can determine the axial force exerted on collar C. Then we can use Eq. (9.9) to determine the required couple.

(b) The question is whether the threaded collar C will move downward under the action of the axial load exerted on it by collar B when no couple is exerted on collar C. We can determine whether it will from Eq. (9.11).

SOLUTION

(a) Free-body diagram of bar AB and the collar B.

(b) The threaded shaft and the collar C.

(a) We draw the free-body diagram of the bar and collar B in Fig. (a), where F is the force exerted on collar B by collar C. From the equilibrium equation

$$\Sigma M_{(\text{point } A)} = (40)F - (20)W = 0,$$

we obtain $F = \frac{1}{2}W = 200$ lb. This is the axial load exerted on collar C (Fig. b). From Eq. (9.9), we know that the couple necessary for impending upward motion of collar C is

$$M = rF \tan (\theta_s + \alpha).$$

The angle α is related to the pitch and mean radius of the thread by Eq. (9.7):

$$\tan \alpha = \frac{p}{2\pi r} = \frac{0.2}{2\pi (1.5)} = 0.0212.$$

Thus $\alpha = \arctan (0.0212) = 1.22°$. The angle of static friction is

$$\theta_s = \arctan (\mu_s) = \arctan (0.25) = 14.04°.$$

Using these values, the moment is

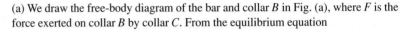

$$M = rF \tan (\theta_s + \alpha) = \left(\frac{1.5}{12}\right)(200) \tan (14.04° + 1.22°) = 6.82 \text{ ft-lb}.$$

(b) Equation (9.11) indicates that the couple required for impending motion is zero when $\theta_s = \alpha$. When the angle of static friction is less than this value, the shaft will rotate and move in the direction of the force F with no couple applied. In this example, $\theta_s = 14.04°$ is greater than $\alpha = 1.22°$, so the weight will remain in place when no couple is applied to collar C.

Problems

9.65 A force $F = 200$ N is necessary to raise A at a constant rate. Between all of the contacting surfaces, $\mu_s = 0.25$ and $\mu_k = 0.20$. The weight of the wedge is negligible. What is the mass of A?

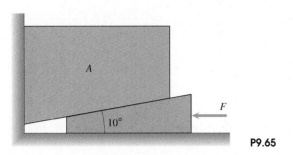

P9.65

9.66 Suppose that in Problem 9.65, A weighs 1000 lb. Between all contacting surfaces, $\mu_s = 0.25$ and $\mu_k = 0.20$. Neglect the weight of the wedge.
(a) What force F is necessary to start the wedge moving to the left?
(b) Will the wedge remain in place if F is removed?
  Strategy: To answer (b), assume that slip of the wedge to the right is impending and determine the minimum value of μ_s necessary for equilibrium.

9.67 Suppose that in Problem 9.65, A weighs 1000 lb and the wedge weighs 200 lb. Between all contacting surfaces, $\mu_s = 0.25$ and $\mu_k = 0.20$. What force F is necessary to raise A at a constant rate?

9.68 The masses of A and B are 42 kg and 50 kg, respectively. Between all contacting surfaces, $\mu_s = 0.05$. What force F is required to start A moving to the right?

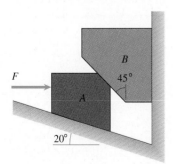

P9.68

9.69 In Problem 9.68, what is the minimum force F necessary for the blocks to remain stationary?

9.70 The stationary blocks A, B, and C each have a mass of 200 kg. Between all contacting surfaces, $\mu_s = 0.6$. What force F is necessary to start B moving downward?

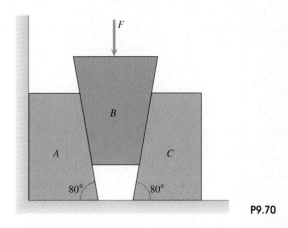

P9.70

9.71 Suppose that in Problem 9.70, the force $F = 1$ kN and A, B, and C each have a mass of 200 kg. The coefficient of static friction between all of the contacting surfaces is μ_s. What is the smallest value of μ_s for which the blocks will remain stationary?

9.72 The object A is held in place by two wedges. The coefficient of static friction between all of the contacting surfaces is μ_s. Neglect the weights of the wedges. What is the largest angle α for which the wedges will remain in place?

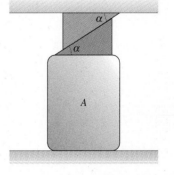

P9.72

9.73 The crate A weighs 600 lb. Between all contacting surfaces, $\mu_s = 0.32$ and $\mu_k = 0.30$. Neglect the weights of the wedges. What force F is required to move A to the right at a constant rate?

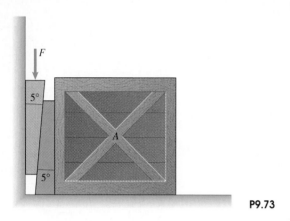

P9.73

9.74 Suppose that between all contacting surfaces in Problem 9.73, $\mu_s = 0.32$ and $\mu_k = 0.30$. Neglect the weights of the 5° wedges. If a force $F = 800$ N is required to move A to the right at a constant rate, what is the mass of A?

9.75 The box A has a mass of 80 kg, and the wedge B has a mass of 40 kg. Between all contacting surfaces, $\mu_s = 0.15$ and $\mu_k = 0.12$. What force F is required to raise A at a constant rate?

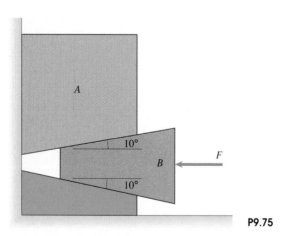

P9.75

9.76 Suppose that in Problem 9.75, A weighs 800 lb and B weighs 400 lb. The coefficients of friction between all of the contacting surfaces are $\mu_s = 0.15$ and $\mu_k = 0.12$. Will B remain in place if the force F is removed?

9.77 Between A and B, $\mu_s = 0.20$, and between B and C, $\mu_s = 0.18$. Between C and the wall, $\mu_s = 0.30$. The weights $W_B = 20$ lb and $W_C = 80$ lb. What force F is required to start C moving upward?

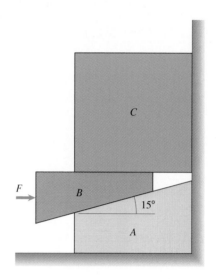

P9.77

9.78 The masses of A, B, and C are 8 kg, 12 kg, and 80 kg, respectively. Between all contacting surfaces, $\mu_s = 0.4$. What force F is required to start C moving upward?

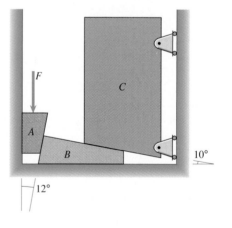

P9.78

9.79 The vertical threaded shaft fits into a mating groove in the tube C. The pitch of the threaded shaft is $p = 0.1$ in., and the mean radius of the thread is $r = 0.5$ in. The coefficients of friction between the thread and the mating groove are $\mu_s = 0.15$ and $\mu_k = 0.10$. The weight $W = 200$ lb. Neglect the weight of the threaded shaft.
(a) Will the stationary threaded shaft support the weight if no couple is applied to the shaft?

(b) What couple must be applied to the threaded shaft to raise the weight at a constant rate?

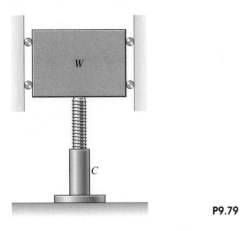

P9.79

9.80 Suppose that in Problem 9.79, the pitch of the threaded shaft is $p = 2$ mm and the mean radius of the thread is $r = 20$ mm. The coefficients of friction between the thread and the mating groove are $\mu_s = 0.22$ and $\mu_k = 0.20$. The weight $W = 500$ N. Neglect the weight of the threaded shaft. What couple must be applied to the threaded shaft to lower the weight at a constant rate?

9.81 The position of the horizontal beam can be adjusted by turning the machine screw A. Neglect the weight of the beam. The pitch of the screw is $p = 1$ mm, and the mean radius of the thread is $r = 4$ mm. The coefficients of friction between the thread and the mating groove are $\mu_s = 0.20$ and $\mu_k = 0.18$. If the system is initially stationary, determine the couple that must be applied to the screw to cause the beam to start moving (a) upward; (b) downward.

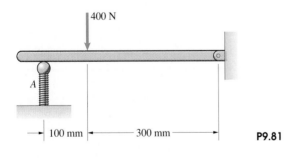

P9.81

9.82 Suppose that in Problem 9.81, the pitch of the machine screw is $p = 1$ mm and the mean radius of the thread is $r = 4$ mm. What minimum value of the coefficient of static friction between the thread and the mating groove is necessary for the beam to remain in the position shown with no couple applied to the screw?

9.83 The mass of block A is 60 kg. Neglect the weight of the 5° wedge. The coefficient of kinetic friction between the contacting surfaces of the block A, the wedge, the table, and the wall is $\mu_k = 0.4$. The pitch of the threaded shaft is 5 mm, the mean radius of the thread is 15 mm, and the coefficient of kinetic friction between the thread and the mating groove is 0.2. What couple must be exerted on the threaded shaft to raise the block A at a constant rate?

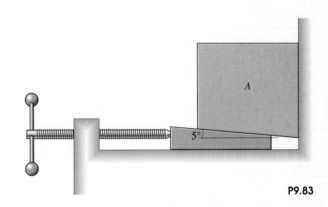

P9.83

9.84 The vise exerts 80-lb forces on A. The threaded shafts are subjected only to axial loads by the jaws of the vise. The pitch of their threads is $p = 1/8$ in., the mean radius of the threads is $r = 1$ in., and the coefficient of static friction between the threads and the mating grooves is 0.2. Suppose that you want to loosen the vise by turning one of the shafts. Determine the couple you must apply (a) to shaft B; (b) to shaft C.

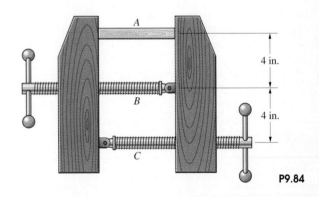

P9.84

9.85 Suppose that you want to tighten the vise in Problem 9.84 by turning one of the shafts. Determine the couple you must apply (a) to shaft B; (b) to shaft C.

9.86 The threaded shaft has a ball and socket support at B. The 400-lb load A can be raised or lowered by rotating the threaded shaft, causing the threaded collar at C to move relative to the shaft. Neglect the weights of the members. The pitch of the shaft is $p = \frac{1}{4}$ in., the mean radius of the thread is $r = 1$ in., and the coefficient of static friction between the thread and the mating groove is 0.24. If the system is stationary in the position shown, what couple is necessary to start the shaft rotating to raise the load?

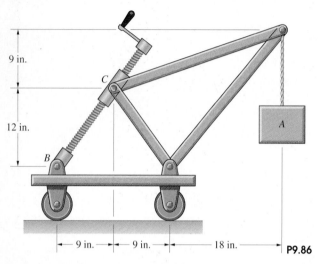

P9.86

9.87 In Problem 9.86, if the system is stationary in the position shown, what couple is necessary to start the shaft rotating to lower the load?

9.88 The car jack is operated by turning the threaded shaft at A. The threaded shaft fits into a mating groove in the collar at B, causing the collar to move relative to the shaft as the shaft turns. As a result, points B and D move closer together or farther apart, causing point C (where the jack is in contact with the car) to move up or down. The pitch of the threaded shaft is $p = 5$ mm, the mean radius of the thread is $r = 10$ mm, and the coefficient of kinetic friction between the thread and the mating groove is 0.15. What couple is necessary to turn the shaft at a constant rate and raise the jack when it is in the position shown if $F = 6.5$ kN?

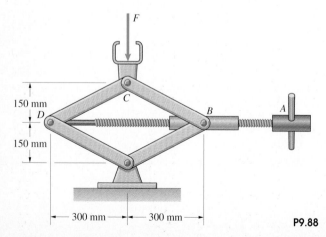

P9.88

9.89 In Problem 9.88, what couple is necessary to turn the threaded shaft at a constant rate and lower the jack when it is in the position shown if the force $F = 6.5$ kN?

9.90 A turnbuckle is used to adjust the tension in a bar or cable. The central link is threaded at both ends. Rotating the turnbuckle draws the threaded ends of the bar together or moves them apart. Suppose that the pitch of the threads is $p = \frac{1}{8}$ in., their mean radius is $r = 1$ in., and the coefficient of static friction between the threads and the mating grooves is 0.24. If $T = 200$ lb, what couple must be exerted on the turnbuckle to start tightening it?

P9.90

9.91 In Problem 9.90, if $T = 200$ lb, what couple must be exerted on the turnbuckle to start loosening it?

9.92 Member BE of the frame has a turnbuckle. (See Problem 9.90.) The threads have pitch $p = \frac{1}{32}$ in., the mean radius of the threads is $r = \frac{1}{4}$ in., and the coefficient of static friction between the threads and the mating grooves is 0.2. What couple must be exerted on the turnbuckle to start loosening it?

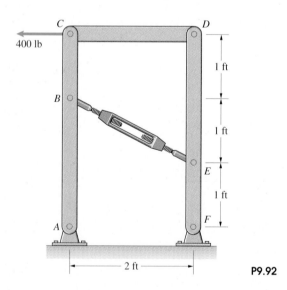

P9.92

9.93 In Problem 9.92, what couple must be exerted on the turnbuckle to start tightening it?

9.94 Members CD and DG of the truss have turnbuckles. (See Problem 9.90.) The pitch of the threads is $p = 4$ mm, their mean radius is $r = 10$ mm, and the coefficient of static friction between the threads and the mating grooves is 0.18. What couple must be exerted on the turnbuckle of member *CD* to start loosening it?

shown, what couple must be exerted to turn the threaded shaft at a constant rate, raising the load?

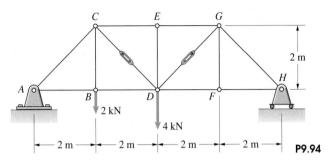

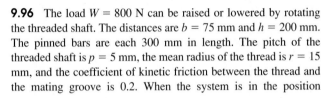

P9.94

9.95 In Problem 9.94, what couple must be exerted on the turnbuckle of member *DG* to start loosening it?

9.96 The load $W = 800$ N can be raised or lowered by rotating the threaded shaft. The distances are $b = 75$ mm and $h = 200$ mm. The pinned bars are each 300 mm in length. The pitch of the threaded shaft is $p = 5$ mm, the mean radius of the thread is $r = 15$ mm, and the coefficient of kinetic friction between the thread and the mating groove is 0.2. When the system is in the position

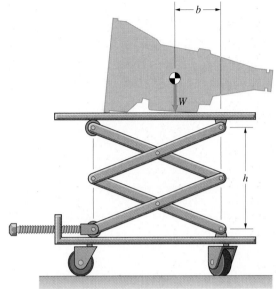

P9.96

Journal Bearings

A **bearing** is a support. This term usually refers to supports designed to allow the supported object to move. For example, in Fig. 9.19(a), a horizontal shaft is supported by two **journal bearings**, which allow the shaft to rotate. The shaft can then be used to support a load perpendicular to its axis, such as that subjected by a pulley (Fig. 9.19b).

Here we analyze journal bearings consisting of brackets with holes through which the shaft passes. The radius of the shaft is slightly smaller than the radius of the holes in the bearings. Our objective is to determine the couple that must be applied to the shaft to cause it to rotate in the bearings. Let F be the total load supported by the shaft including the weight of the shaft itself. When no couple is exerted on the shaft, the force F presses it against the bearings as shown in Fig. 9.19(c). When a couple M is exerted on the shaft, it rolls up the surfaces of the bearings (Fig. 9.19d). The term α is the angle from the original point of contact of the shaft to its point of contact when M is applied.

Figure 9.19

(a) A shaft supported by journal bearings.
(b) A pulley supported by the shaft.
(c) The shaft and bearing when no couple is applied to the shaft.
(d) A couple causes the shaft to roll within the bearing.
(e) Free-body diagram of the shaft.
(f) The two forces on the shaft must be equal and opposite.

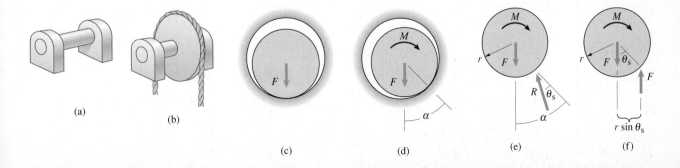

(a)

(b)

(c)

(d)

(e)

(f)

In Fig. 9.19(e), we draw the free-body diagram of the shaft when M is sufficiently large that slip is impending. The force R is the total reaction exerted on the shaft by the two bearings. Since R and F are the only forces acting on the shaft, equilibrium requires that $\alpha = \theta_s$ and $R = F$ (Fig. 9.19f). The reaction exerted on the shaft by the bearings is displaced a distance $r \sin \theta_s$ from the vertical line through the center of the shaft. By summing moments about the center of the shaft, we obtain the couple M that causes the shaft to be on the verge of slipping:

$$M = rF \sin \theta_s. \tag{9.12}$$

This is the largest couple that can be exerted on the shaft without causing it to start rotating. Replacing θ_s in this expression by the angle of kinetic friction θ_k gives the couple necessary to rotate the shaft at a constant rate.

The simple type of journal bearing we have described is too primitive for most applications. The surfaces where the shaft and bearing are in contact would quickly become worn. Designers usually incorporate "ball" or "roller" bearings in journal bearings to minimize friction (Fig. 9.20).

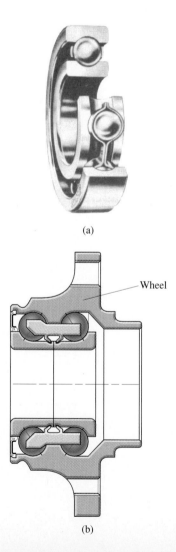

(a)

Wheel

(b)

Figure 9.20

(a) A journal bearing with one row of balls.
(b) Journal bearing assembly of the wheel of a car. There are two rows of balls between the rotating wheel and the fixed inner cylinder.

Example 9.7

The weight of the load in Fig. 9.21 is $W = 1000$ lb. The pulley P has a 6-in. radius and is rigidly attached to a horizontal shaft supported by journal bearings. The radius of the shaft is $\frac{1}{2}$ in., and the coefficient of kinetic friction between the shaft and the bearings is $\mu_k = 0.2$. Neglect the weights of the pulley and the shaft. What tension must the power winch A exert on the cable to raise the load at a constant rate?

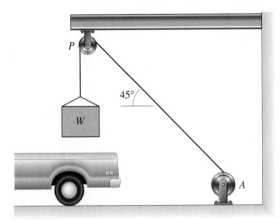

Figure 9.21

STRATEGY

Equation (9.12) with θ_s replaced by θ_k relates the couple M required to turn the pulley at a constant rate to the total force F on the shaft. By expressing M and F in terms of the load and the tension exerted by the winch, we can obtain an equation for the tension.

SOLUTION

Let T be the tension exerted by the winch (Fig. a). By calculating the magnitude of the sum of the forces exerted by the tension and the load, we obtain an expression for the total force F on the shaft supporting the pulley (Fig. b):

$$F = \sqrt{(W + T \sin 45°)^2 + (T \cos 45°)^2}.$$

The couple exerted on the pulley by the tension and the load is

$$M = (0.5 \text{ ft})T - (0.5 \text{ ft}) W.$$

The radius of the shaft is $r = \frac{1}{2}$ in. $= 0.0417$ ft, and the angle of kinetic friction is $\theta_k = \arctan(\mu_k) = \arctan(0.2) = 11.31°$. From Eq. (9.12), we know that the couple M necessary to turn the pulley at a constant rate is given in terms of F by

$$M = rF \sin \theta_k.$$

Substituting our expressions for F and M in terms of the tension T and the load W into this equation, we obtain

$$(0.5)T - (0.5)W = (0.0417) \sin(11.31°)\sqrt{(W + T \sin 45°)^2 + (T \cos 45°)^2}.$$

Solving this equation for the tension, we obtain $T = 1030$ lb.

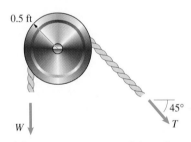

(a) Free-body diagram of the pulley.

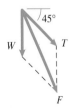

(b) The total force F on the shaft.

Thrust Bearings and Clutches

A **thrust bearing** supports a rotating shaft that is subjected to an axial load. In the type shown in Figs. 9.22(a) and (b), the conical end of the shaft is pressed against the mating conical cavity by an axial load F. Let us determine the couple M necessary to rotate the shaft.

The differential element of area dA in Fig. 9.22(c) is

$$dA = 2\pi r \, ds = 2\pi r \left(\frac{dr}{\cos \alpha} \right).$$

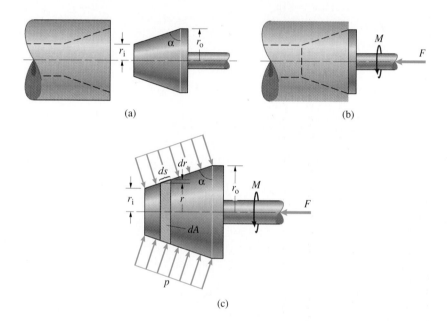

Figure 9.22

(a), (b) A thrust bearing supports a shaft subjected to an axial load.
(c) The differential element dA and the uniform pressure p exerted by the cavity.

Integrating this expression from $r = r_i$ to $r = r_o$, we obtain the area of contact:

$$A = \frac{\pi(r_o^2 - r_i^2)}{\cos \alpha}.$$

If we assume that the mating surface exerts a uniform pressure p, the axial component of the total force due to p must equal F: $pA \cos \alpha = F$. Therefore the pressure is

$$p = \frac{F}{A \cos \alpha} = \frac{F}{\pi(r_o^2 - r_i^2)}.$$

As the shaft rotates about its axis, the moment about the axis due to the friction force on the element dA is $r\mu_k (p \, dA)$. The total moment equals M:

$$M = \int_A \mu_k rp \, dA = \int_{r_i}^{r_o} \mu_k r \left[\frac{F}{\pi(r_o^2 - r_i^2)} \right] \left(\frac{2\pi r \, dr}{\cos \alpha} \right).$$

Integrating, we obtain the couple M necessary to rotate the shaft at a constant rate:

$$M = \frac{2\mu_{\mathrm{k}}F}{3\cos\alpha}\left(\frac{r_{\mathrm{o}}^3 - r_{\mathrm{i}}^3}{r_{\mathrm{o}}^2 - r_{\mathrm{i}}^2}\right). \tag{9.13}$$

A simpler thrust bearing is shown in Figs. 9.23(a) and (b). The bracket supports the flat end of a shaft of radius r that is subjected to an axial load F. We can obtain the couple necessary to rotate the shaft at a constant rate from Eq. (9.13) by setting $\alpha = 0$, $r_{\mathrm{i}} = 0$, and $r_{\mathrm{o}} = r$:

$$M = \frac{2}{3}\mu_{\mathrm{k}}Fr. \tag{9.14}$$

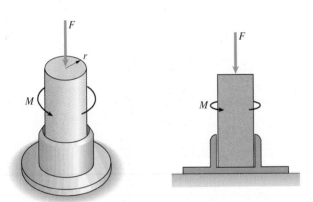

Figure 9.23

A thrust bearing that supports a flat-ended shaft.

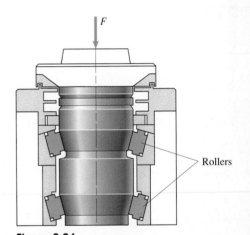

Figure 9.24

A thrust bearing with two rows of cylindrical rollers between the shaft and the fixed support.

Although they are good examples of the analysis of friction forces, the thrust bearings we have described would become worn too quickly to be used in most applications. The designer of the thrust bearing in Fig. 9.24 minimizes friction by incorporating "roller" bearings.

A **clutch** is a device used to connect and disconnect two coaxial rotating shafts. The type shown in Figs. 9.25(a) and (b) consists of disks of radius r attached to the ends of the shafts. When the disks are separated (Fig. 9.25a), the clutch is **disengaged**, and the shafts can rotate freely relative to each other. When the clutch is engaged by pressing the disks together with axial forces F (Fig. 9.25b), the shafts can support a couple M due to the friction forces between the disks. If the couple M becomes too large, the clutch slips.

The friction forces exerted on one face of the clutch by the other face are identical to the friction forces exerted on the flat-ended shaft by the bracket in Fig. 9.23. We can therefore determine the largest couple the clutch can support without slipping by replacing μ_{k} by μ_{s} in Eq. (9.14):

$$M = \frac{2}{3}\mu_{\mathrm{s}}Fr. \tag{9.15}$$

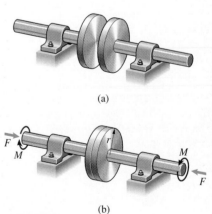

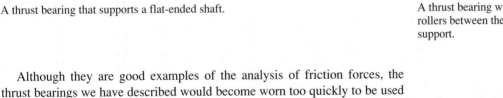

Figure 9.25

A clutch.
(a) Disengaged position.
(b) Engaged position.

Example 9.8

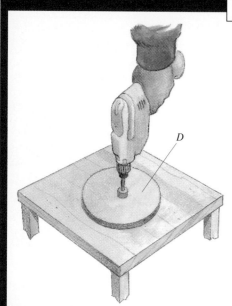

The handheld sander in Fig. 9.26 has a rotating disk D of 4-in. radius with sandpaper bonded to it. The total downward force exerted by the operator and the weight of the sander is 15 lb. The coefficient of kinetic friction between the sandpaper and the surface is $\mu_k = 0.6$. What couple (torque) M must the motor exert to turn the sander at a constant rate?

STRATEGY

As the disk D rotates, it is subjected to friction forces analogous to the friction forces exerted on the flat-ended shaft by the bracket in Fig. 9.23. We can determine the couple required to turn the disk D at a constant rate from Eq. (9.14).

SOLUTION

The couple required to turn the disk at a constant rate is

$$ M = \frac{2}{3}\mu_k rF = \frac{2}{3}(0.6)\left(\frac{4}{12}\right)(15) = 2 \text{ ft-lb.} $$

Figure 9.26

Problems

9.97 The horizontal shaft is supported by two journal bearings. The coefficient of kinetic friction between the shaft and the bearings is $\mu_k = 0.2$. The radius of the shaft is 20 mm, and its mass is 5 kg. Determine the couple M necessary to rotate the shaft at a constant rate.

Strategy: You can obtain the couple necessary to rotate the shaft at a constant rate by replacing θ_s by θ_k in Eq. (9.12).

P9.97

9.98 The horizontal shaft is supported by two journal bearings. The coefficient of static friction between the shaft and the bearings is $\mu_s = 0.3$. The radius of the shaft is 20 mm, and its mass is 5 kg. Determine the largest mass m that can be suspended as shown without causing the stationary shaft to slip in the bearings.

P9.98

9.99 Suppose that in Problem 9.98 the mass $m = 8$ kg and the coefficient of kinetic friction between the shaft and the bearings is $\mu_k = 0.26$. What couple must be applied to the shaft to raise the mass at a constant rate?

9.100 The pulley is mounted on a horizontal shaft supported by journal bearings. The coefficient of kinetic friction between the shaft and the bearings is $\mu_k = 0.3$. The radius of the shaft is 20 mm, and the radius of the pulley is 150 mm. The mass $m = 10$ kg. Neglect the masses of the pulley and shaft. What force T must be applied to the cable to move the mass upward at a constant rate?

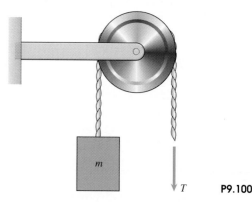

P9.100

9.101 In Problem 9.100, what force T must be applied to the cable to lower the mass at a constant rate?

9.102 The pulley of 8-in. radius is mounted on a shaft of 1-in. radius. The shaft is supported by two journal bearings. The coefficient of static friction between the bearings and the shaft is $\mu_s = 0.15$. Neglect the weights of the pulley and shaft. The 50-lb block A rests on the floor. If sand is slowly added to the bucket B, what do the bucket and sand weigh when the shaft slips in the bearings?

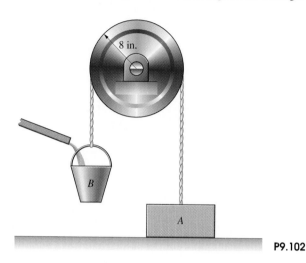

P9.102

9.103 The pulley of 50-mm radius is mounted on a shaft of 10-mm radius. The shaft is supported by two journal bearings. The mass of the block A is 8 kg. Neglect the weights of the pulley and shaft. If a force $T = 84$ N is necessary to raise block A at a constant rate, what is the coefficient of kinetic friction between the shaft and the bearings?

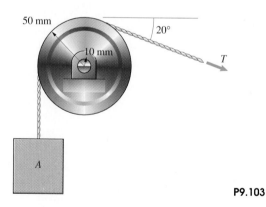

P9.103

9.104 Suppose that in Problem 9.103, the coefficient of kinetic friction between the shaft and the bearings is $\mu_k = 0.18$. What force T is necessary to raise the block A at a constant rate?

9.105 Suppose that in Problem 9.103, the coefficient of kinetic friction between the shaft and the bearings is $\mu_k = 0.24$. What force T is necessary to lower the block A at a constant rate?

9.106 The radius of the pulley is 200 mm, and it is mounted on a shaft of 20-mm radius. The coefficient of static friction between

the pulley and shaft is $\mu_s = 0.18$. If $F_A = 200$ N, what is the largest force F_B that can be applied without causing the pulley to turn? Neglect the weight of the pulley.

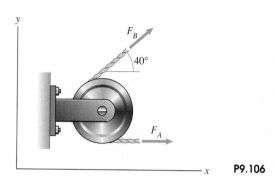

P9.106

9.107 The mass of the pulley in Problem 9.106 is 4 kg. The force $F_A = 200$ N. Including the effect of the weight of the pulley, determine the largest force F_B that can be applied without causing the pulley to turn, and compare your answer to that of Problem 9.106.

9.108 The two pulleys have a radius of 4 in. and are mounted on shafts of 1-in. radius supported by journal bearings. Neglect the weights of the pulleys and shafts. The tension in the spring is 40 lb. The coefficient of kinetic friction between the shafts and the bearings is $\mu_k = 0.3$. What couple M is required to turn the left pulley at a constant rate?

P9.108

9.109 The weights of the two boxes are $W_1 = 100$ lb and $W_2 = 50$ lb. The coefficient of kinetic friction between the left box and the inclined surface is $\mu_k = 0.14$. Each pulley has a 6-in. radius and is mounted on a shaft of $\frac{1}{2}$-in. radius. The coefficient of kinetic friction between each pulley and its shaft is $\mu_k = 0.12$. Determine the tension the man must exert on the rope to pull the boxes upward at a constant rate.

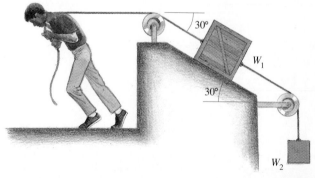

P9.109

9.110 Each pulley has a radius of 100 mm and a mass of 2 kg. Both are mounted on shafts of 5-mm radius supported by journal bearings. The coefficient of kinetic friction between the shafts and the bearings is $\mu_k = 0.18$. The mass of A is 14 kg. What force T is required to raise A at a constant rate?

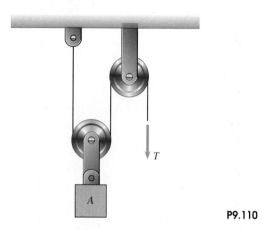

P9.110

9.111 The circular flat-ended shaft is pressed into the thrust bearing by an axial load of 100 N. Neglect the weight of the shaft. The coefficients of friction between the end of the shaft and the bearing are $\mu_s = 0.20$ and $\mu_k = 0.15$. What is the largest couple M that can be applied to the stationary shaft without causing it to rotate in the bearing?

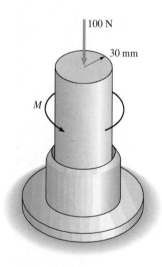

P9.111

9.112 In Problem 9.111, what couple M is required to rotate the shaft at a constant rate?

9.113 Suppose that the end of the shaft in Problem 9.111 is supported by a thrust bearing of the type shown in Fig. 9.22, where $r_o = 30$ mm, $r_i = 10$ mm, $\alpha = 30°$, and $\mu_k = 0.15$. What couple M is required to rotate the shaft at a constant rate?

9.114 The disk D is rigidly attached to the vertical shaft. The shaft has flat ends supported by thrust bearings. The disk and the shaft together weigh 500 lb, and the diameter of the shaft is 2 in. The vertical force exerted on the end of the shaft by the upper thrust bearing is 100 lb. The coefficient of kinetic friction between the ends of the shaft and the bearings is $\mu_k = 0.25$. What couple M is required to rotate the shaft at a constant rate?

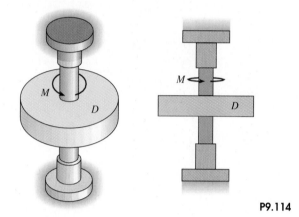

P9.114

9.115 Suppose that the ends of the shaft in Problem 9.114 are supported by thrust bearings of the type shown in Fig. 9.22, where $r_o = 1$ in., $r_i = \frac{1}{4}$ in., $\alpha = 45°$, and $\mu_k = 0.25$. What couple M is required to rotate the shaft at a constant rate?

9.116 The shaft is supported by thrust bearings that subject it to an axial load of 800 N. The coefficients of kinetic friction between the shaft and the left and right bearings are 0.20 and 0.26, respectively. What couple is required to rotate the shaft at a constant rate?

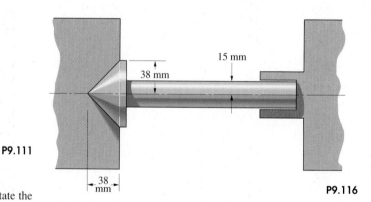

P9.116

9.117 A motor is used to rotate a paddle for mixing chemicals. The shaft of the motor is coupled to the paddle using a friction clutch of the type shown in Fig. 9.25. The radius of the disks of the clutch is 120 mm, and the coefficient of static friction between the disks is 0.6. If the motor transmits a maximum torque of 15 N-m to the paddle, what minimum normal force between the plates of the clutch is necessary to prevent slipping?

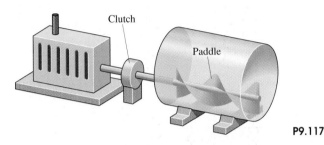

P9.117

9.118 The thrust bearing is supported by contact of the collar C with a fixed plate. The area of contact is an annulus with an inside diameter $D_1 = 40$ mm and an outside diameter $D_2 = 120$ mm. The coefficient of kinetic friction between the collar and the plate is $\mu_k = 0.3$. The force $F = 400$ N. What couple M is required to rotate the shaft at a constant rate?

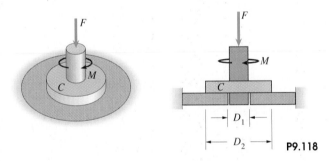

P9.118

9.119 An experimental automobile brake design works by pressing the red annular plate against the rotating wheel. If $\mu_k = 0.6$, what force F pressing the plate against the wheel is necessary to exert a couple of 200 N-m on the wheel?

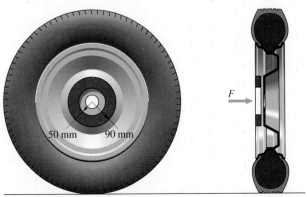

P9.119

9.120 In Problem 9.119, suppose that $\mu_k = 0.65$ and the force pressing the plate against the wheel is $F = 2$ kN.
(a) What couple is exerted on the wheel?
(b) What percentage increase in the couple exerted on the wheel is obtained if the outer radius of the brake is increased from 90 mm to 100 mm?

9.121 The coefficient of static friction between the plates of the car's clutch is $\mu_s = 0.8$. If the plates are pressed together with a force $F = 600$ lb, what is the maximum torque the clutch will support without slipping?

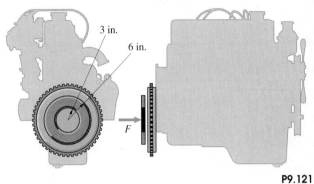

P9.121

9.122 The "Morse taper" is used to support the workpiece on a machinist's lathe. The taper is driven into the spindle and is held in place by friction. If the spindle exerts a uniform pressure $p = 15$ psi on the taper and $\mu_s = 0.2$, what couple must be exerted about the axis of the taper to loosen it?

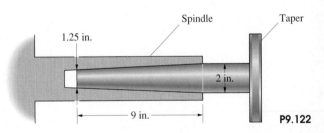

P9.122

Belt Friction

You know that if a rope is wrapped around a fixed post as shown in Fig. 9.27, a large force T_2 exerted on one end can be supported by a relatively small force T_1 applied to the other end. In this section we analyze this familiar phenomenon. It is referred to as **belt friction** because a similar approach can be used to analyze belts used in machines, such as the belts that drive alternators and other devices in a car.

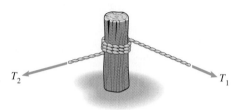

Figure 9.27

A rope wrapped around a post.

Let's consider a rope wrapped through an angle β around a fixed cylinder (Fig. 9.28a). We will assume that the tension T_1 is known. Our objective is to determine the largest force T_2 that can be applied to the other end of the rope without causing the rope to slip.

We begin by drawing the free-body diagram of an element of the rope whose boundaries are at angles α and $\alpha + \Delta\alpha$ from the point where the rope comes into contact with the cylinder (Figs. 9.28b and c). The force T is the tension in the rope at the position defined by the angle α. We know that the tension in the rope varies with position, because it increases from T_1 at $\alpha = 0$ to T_2 at $\alpha = \beta$. We therefore write the tension in the rope at the position $\alpha + \Delta\alpha$ as $T + \Delta T$. The force ΔN is the normal force exerted on the element by the cylinder. Because we want to determine the largest value of T_2 that will not cause the rope to slip, we assume that the friction force is equal to its maximum possible value $\mu_s \Delta N$, where μ_s is the coefficient of static friction between the rope and the cylinder.

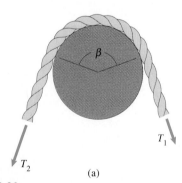

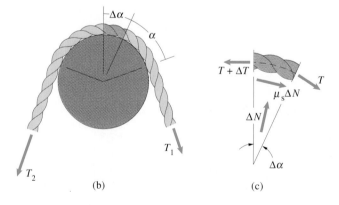

(a)

(b)

(c)

Figure 9.28

(a) A rope wrapped around a fixed cylinder.
(b) A differential element with boundaries at angles α and $\alpha + \Delta\alpha$.
(c) Free-body diagram of the element.

The equilibrium equations in the directions tangential to and normal to the centerline of the rope are

$$\Sigma F_{\text{(tangential)}} = \mu_s \Delta N + T \cos\left(\frac{\Delta\alpha}{2}\right) - (T + \Delta T) \cos\left(\frac{\Delta\alpha}{2}\right) = 0,$$

$$\text{(9.16)}$$

$$\Sigma F_{\text{(normal)}} = \Delta N - (T + \Delta T) \sin\left(\frac{\Delta\alpha}{2}\right) - T \sin\left(\frac{\Delta\alpha}{2}\right) = 0.$$

Eliminating ΔN, we can write the resulting equation as

$$\left[\cos\left(\frac{\Delta\alpha}{2}\right) - \mu_s \sin\left(\frac{\Delta\alpha}{2}\right)\right]\frac{\Delta T}{\Delta\alpha} - \mu_s T \frac{\sin(\Delta\alpha/2)}{(\Delta\alpha/2)} = 0.$$

Evaluating the limit of this equation as $\Delta\alpha \to 0$ and observing that

$$\frac{\sin(\Delta\alpha/2)}{(\Delta\alpha/2)} \to 1,$$

we obtain

$$\frac{dT}{d\alpha} - \mu_s T = 0.$$

This differential equation governs the variation of the tension in the rope. By separating variables,

$$\frac{dT}{T} = \mu_s \, d\alpha,$$

we can integrate to determine the tension T_2 in terms of the tension T_1 and the angle β:

$$\int_{T_1}^{T_2} \frac{dT}{T} = \int_0^\beta \mu_s \, d\alpha.$$

Thus we obtain the largest force T_2 that can be applied without causing the rope to slip when the force on the other end is T_1:

$$\boxed{T_2 = T_1 e^{\mu_s \beta}.}$$

$$\text{(9.17)}$$

The angle β in this equation must be expressed in radians. Replacing μ_s by the coefficient of kinetic friction μ_k gives the force T_2 required to cause the rope to slide at a constant rate.

Equation (9.17) explains why a large force can be supported by a relatively small force when a rope is wrapped around a fixed support. The force required to cause the rope to slip increases exponentially as a function of the angle through which the rope is wrapped. Suppose that $\mu_s = 0.3$. When the rope is wrapped one complete turn around the post ($\beta = 2\pi$), the ratio $T_2/T_1 = 6.59$. When the rope is wrapped four complete turns around the post ($\beta = 8\pi$), the ratio $T_2/T_1 = 1880$.

Example 9.9

The 50-kg crate in Fig. 9.29 is suspended from a rope that passes over two fixed cylinders. The coefficient of static friction is 0.2 between the rope and the left cylinder and 0.4 between the rope and the right cylinder. What is the smallest force the woman can exert and support the crate?

Figure 9.29

STRATEGY

She exerts the smallest possible force when slip of the rope is impending on both cylinders. Because we know the weight of the crate, we can use Eq. (9.17) to determine the tension in the rope between the two cylinders and then use Eq. (9.17) again to determine the force she exerts.

SOLUTION

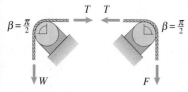

(a) The tensions in the rope.

The weight of the crate is $W = (50)(9.81) = 491$ N. Let T be the tension in the rope between the two cylinders (Fig. a). The rope is wrapped around the left cylinder through an angle $\beta = \pi/2$ rad. The tension T necessary to prevent the rope from slipping on the left cylinder is related to W by

$$W = Te^{\mu_s\beta} = Te^{(0.2)(\pi/2)}.$$

Solving for T, we obtain

$$T = We^{-(0.2)(\pi/2)} = (491)e^{-(0.2)(\pi/2)} = 358 \text{ N.}$$

The rope is also wrapped around the right cylinder through an angle $\beta = \pi/2$ rad. The force F the woman must exert to prevent the rope from slipping on the right cylinder is related to T by

$$T = Fe^{\mu_s\beta} = Fe^{(0.4)(\pi/2)}.$$

The solution for F is

$$F = Te^{-(0.4)(\pi/2)} = (358)e^{-(0.4)(\pi/2)} = 191 \text{ N.}$$

Example 9.10

Application to Engineering

Belts and Pulleys

The pulleys in Fig. 9.30 turn at a constant rate. The large pulley is attached to a fixed support. The small pulley is supported by a smooth horizontal slot and is pulled to the right by the force $F = 200$ N. The coefficient of static friction between the pulleys and the belt is $\mu_s = 0.8$, the dimension $b = 500$ mm, and the radii of the pulleys are $R_A = 200$ mm and $R_B = 100$ mm. What are the largest values of the couples M_A and M_B for which the belt will not slip?

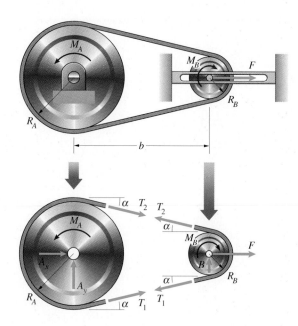

(a) Free-body diagram of the large pulley.

(b) Free-body diagram of the small pulley.

Figure 9.30

STRATEGY

By drawing free-body diagrams of the pulleys, we can use the equilibrium equations to relate the tensions in the belt to M_A and M_B and obtain a relation between the tensions in the belt and the force F. When slip is impending, the tensions are also related by Eq. (9.17). From these equations we can determine M_A and M_B.

SOLUTION

From the free-body diagram of the large pulley (Fig. a), we obtain the equilibrium equation

$$M_A = R_A(T_2 - T_1), \tag{9.18}$$

and from the free-body diagram of the small pulley (Fig. b), we obtain

$$F = (T_1 + T_2) \cos \alpha, \tag{9.19}$$

$$M_B = R_B(T_2 - T_1). \tag{9.20}$$

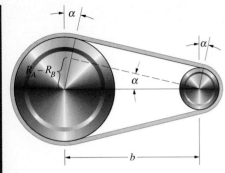

(c) Determining the angle α.

The belt is in contact with the small pulley through the angle $\pi - 2\alpha$ (Fig. c). From the dashed line parallel to the belt, we see that the angle α satisfies the relation

$$\sin \alpha = \frac{R_A - R_B}{b} = \frac{200 - 100}{500} = 0.2.$$

Therefore $\alpha = 11.5° = 0.201$ rad. If we assume that slip impends between the small pulley and the belt, Eq. (9.17) states that

$$T_2 = T_1 e^{\mu_s \beta} = T_1 e^{0.8(\pi - 2\alpha)} = 8.95 T_1.$$

We solve this equation together with Eq. (9.19) for the two tensions, obtaining $T_1 = 20.5$ N and $T_2 = 183.6$ N. Then from Eqs. (9.18) and (9.20), the couples are $M_A = 32.6$ N-m and $M_B = 16.3$ N-m.

If we assume that slip impends between the large pulley and the belt, we obtain $M_A = 36.3$ N-m and $M_B = 18.1$ N-m, so the belt slips on the small pulley at smaller values of the couples.

DESIGN ISSUES

Belts and pulleys are used to transfer power in cars and many other types of machines, including printing presses, farming equipment, and industrial robots. Because two pulleys of different diameters connected by a belt are subjected to different torques and have different rates of rotation, they can be used as a mechanical "transformer" to alter torque or rotation rate.

In this example we assumed that the belt was flat, but "V-belts" that fit into matching grooves in the pulleys are often used in applications (Fig. 9.31a). This configuration keeps the belt in place on the pulleys and also decreases the tendency of the belt to slip. Suppose that a V-belt is wrapped through an angle β around a pulley (Fig. 9.31b). If the tension T_1 is known, what is the largest tension T_2 that can be applied to the other end of the belt without causing it to slip relative to the pulley?

In Fig. 9.31(c), we draw the free-body diagram of an element of the belt whose boundaries are at angles α and $\alpha + \Delta\alpha$ from the point where the belt comes into contact with the pulley. (Compare this figure with Fig. 9.28c.) The equilibrium equations in the directions tangential to and normal to the centerline of the belt are

$$\Sigma F_{(\text{tangential})} = 2\,\mu_s \Delta N + T \cos\left(\frac{\Delta\alpha}{2}\right) - (T + \Delta T) \cos\left(\frac{\Delta\alpha}{2}\right) = 0,$$

$$\text{(9.21)}$$

$$\Sigma F_{(\text{normal})} = 2\Delta N \sin\left(\frac{\gamma}{2}\right) - (T + \Delta T) \sin\left(\frac{\Delta\alpha}{2}\right) - T \sin\left(\frac{\Delta\alpha}{2}\right) = 0.$$

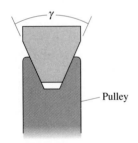

(a) Cross-sectional view of a V-belt and pulley.

(b) V-belt wrapped around a pulley.

Figure 9.31

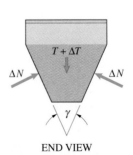

END VIEW

(c) Free-body diagram of an element of the belt.

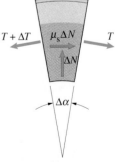

SIDE VIEW

By the same steps leading from Eqs. (9.16) to Eq. (9.17), it can be shown that

$$T_2 = T_1 e^{\mu_s \beta / \sin(\gamma/2)}. \tag{9.22}$$

Thus using a V-belt effectively increases the coefficient of friction between the belt and pulley by the factor $1/\sin(\gamma/2)$.

When it is essential that the belt not slip relative to the pulley, a belt with cogs and a matching pulley (Fig. 9.32a) or a chain and sprocket wheel (Fig. 9.32b) can be used. The chains and sprocket wheels in bicycles and motorcycles are examples.

Figure 9.32
Designs that prevent slip of the belt relative to the pulley.

Problems

9.123 Suppose that you want to lift a 50-lb crate off the ground by using a rope looped over a tree limb as shown. The coefficient of static friction between the rope and the limb is 0.4, and the rope is wound 120° around the limb. What force must you exert to lift the crate?

Strategy: The tension necessary to cause impending slip of the rope on the limb is given by Eq. (9.17), with $T_1 = 50$ lb, $\mu_s = 0.4$, and $\beta = (\pi/180)(120)$ rad.

P9.123

9.124 In Problem 9.123, once you have lifted the crate off the ground, what is the minimum force you must exert on the rope to keep the crate suspended?

9.125 *Winches* are used on sailboats to help support the ropes (*sheets*) that position the sails. The simple winch shown is a post that will rotate in the clockwise direction (as seen from above) but not in the counterclockwise direction. The sail exerts an 800-N force on the sheet. The sheet is wrapped three complete turns around the winch. The coefficient of static friction between the sheet and the winch is $\mu_s = 0.2$. Determine the force that the crew member must exert on the end of the sheet to prevent it from slipping on the winch.

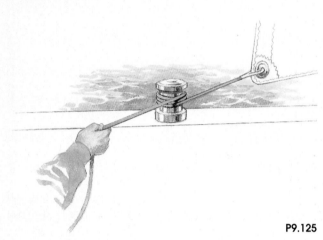

P9.125

9.126 The block *A* weighs 30 lb. The rope is wrapped one and one-fourth turns around the fixed post. The coefficients of friction between the fixed post and the rope are $\mu_s = 0.2$ and $\mu_k = 0.15$. What is the smallest force the person must exert to support the stationary block?

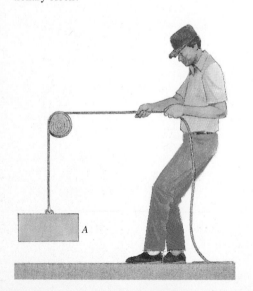

P9.126

9.127 In Problem 9.126, what force would the person have to exert to raise the block at a constant rate?

9.128 The weight of block *A* is *W*. The disk is supported by a smooth bearing. The coefficient of kinetic friction between the disk and the belt is μ_k. What couple *M* is necessary to turn the disk at a constant rate?

P9.128

9.129 The couple required to turn the wheel of the exercise bicycle is adjusted by changing the weight *W*. The coefficient of kinetic friction between the wheel and the belt is μ_k. Assume the wheel turns clockwise.
(a) Show that the couple *M* required to turn the wheel is $M = WR(1 - e^{-3.4\mu_k})$.
(b) If $W = 40$ lb and $\mu_k = 0.2$, what force will the scale S indicate when the bicycle is in use?

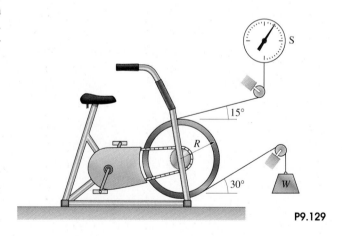

P9.129

9.130 The box *B* weighs 50 lb. The coefficients of friction between the cable and the fixed round supports are $\mu_s = 0.4$ and $\mu_k = 0.3$.
(a) What is the minimum force *F* required to support the box?
(b) What force *F* is required to move the box upward at a constant rate?

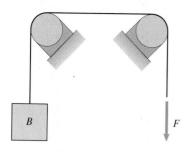

B   F

P9.130

9.131 The 20-kg box *A* is held in equilibrium on the inclined surface by the force *T* acting on the rope wrapped over the fixed cylinder. The coefficient of static friction between the box and the inclined surface is 0.1. The coefficient of static friction between the rope and the cylinder is 0.05. Determine the largest value of *T* that will not cause the box to slip up the inclined surface.

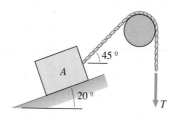

45 º

A

20 º

T   **P9.131**

9.132 In Problem 9.131, determine the smallest value of *T* necessary to hold the box in equilibrium on the inclined surface.

9.133 The block *A* weighs 30 lb. The coefficient of kinetic friction between the rope and the cylinder is 0.2. If the cylinder is rotated at a constant rate, first in the counterclockwise direction and then in the clockwise direction, the difference in the height of block *A* is 1 ft. What is the spring constant *k*?

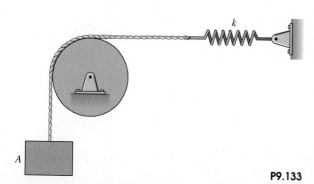

k

A

P9.133

Problems 9.134–9.138 are related to Example 9.10.

9.134 If the force *F* in Example 9.10 is increased to 400 N, what are the largest values of the couples M_A and M_B for which the belt will not slip?

9.135 If the belt in Example 9.10 is a V-belt with angle $\gamma = 45°$, what are the largest values of the couples M_A and M_B for which the belt will not slip?

9.136 The spring exerts a 320-N force on the left pulley. The coefficient of static friction between the flat belt and the pulleys is $\mu_s = 0.5$. The right pulley cannot rotate. What is the largest couple *M* that can be exerted on the left pulley without causing the belt to slip?

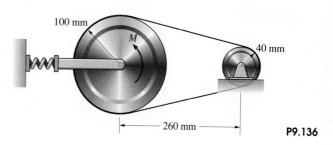

100 mm

M

40 mm

260 mm   **P9.136**

9.137 Suppose that the belt in Problem 9.136 is a V-belt with angle $\gamma = 30°$. What is the largest couple *M* that can be exerted on the left pulley without causing the belt to slip?

9.138 Beginning with Eqs. (9.21), derive Eq. (9.22):

$$T_2 = T_1 e^{\mu_s \beta / \sin (\gamma/2)}.$$

Computational Mechanics

The following example and problems are designed for the use of a programmable calculator or computer.

Example 9.11

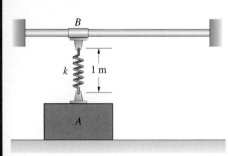

Figure 9.33

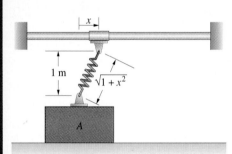

(a) Moving the slider to the right a distance x.

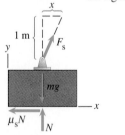

(b) Free-body diagram of the block when slip is impending.

The mass of the block A in Fig. 9.33 is 20 kg, and the coefficient of static friction between the block and the floor is $\mu_s = 0.3$. The spring constant $k = 1$ kN/m, and the spring is unstretched. How far can the slider B be moved to the right without causing the block to slip?

SOLUTION

Suppose that moving the slider B a distance x to the right causes impending slip of the block (Fig. a). The resulting stretch of the spring is $\sqrt{1 + x^2} - 1$ m, so the magnitude of the force exerted on the block by the spring is

$$F_s = k\left(\sqrt{1 + x^2} - 1\right).$$

(9.23)

From the free-body diagram of the block (Fig. b), we obtain the equilibrium equations

$$\Sigma F_x = \left(\frac{x}{\sqrt{1 + x^2}}\right) F_s - \mu_s N = 0,$$

$$\Sigma F_y = \left(\frac{1}{\sqrt{1 + x^2}}\right) F_s + N - mg = 0.$$

Substituting Eq. (9.23) into these two equations and then eliminating N, we can write the resulting equation in the form

$$h(x) = k(x + \mu_s)\left(\sqrt{1 + x^2} - 1\right) - \mu_s mg \sqrt{1 + x^2} = 0.$$

We must obtain the root of this function to determine the value of x corresponding to impending slip of the block. From the graph of h(x) in Fig. 9.34, we estimate that $h(x) = 0$ at $x = 0.43$ m. By examining computed results near this value of x,

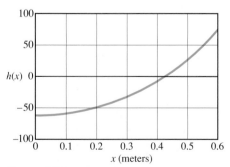

Figure 9.34

Graph of the function h(x).

x (m)	$h(x)$
0.4281	−0.1128
0.4282	−0.0777
0.4283	−0.0425
0.4284	−0.0074
0.4285	0.0278
0.4286	0.0629
0.4287	0.0981

we see that $h(x) = 0$, and slip is impending, when x is approximately 0.4284 m.

Problems

9.139 The mass of the block A is 20 kg, and the coefficient of static friction between the block and the floor is $\mu_s = 0.3$. The spring constant $k = 1$ kN/m, and the spring is unstretched. How far can the slider B be moved to the right without causing the block to slip?

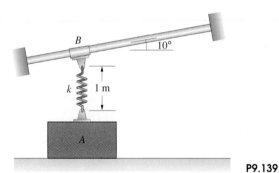

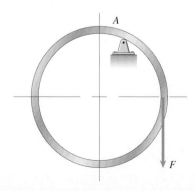

P9.139

9.140 The slender circular ring of weight W is supported by normal and friction forces at A. If slip is impending when the vertical force $F = 0.4W$, what is the coefficient of static friction between the ring and the support?

P9.140

9.141 Suppose that the vertical force on the ring in Problem 9.140 is $F = KW$ and slip is impending. Draw a graph of K as a function of the coefficient of static friction between the ring and the support for $0 \le \mu_s \le 1$.

9.142 The mass of the 3-m bar is 20 kg, and the coefficient of static friction between the ends of the bar and the circular surface is $\mu_s = 0.3$. What is the largest value of the angle α for which the bar will not slip?

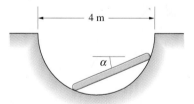

P9.142

9.143 The load $W = 800$ N can be raised or lowered by rotating the threaded shaft. The distance $b = 75$ mm, and the pinned bars are each 300 mm in length. The pitch of the threaded shaft is $p = 5$ mm, the mean radius of the thread is $r = 15$ mm, and the coefficient of kinetic friction between the thread and the mating groove is 0.2. Draw a graph of the moment that must be exerted to turn the threaded shaft at a constant rate, raising the load, as a function of the height h from $h = 100$ mm to $h = 400$ mm.

friction between the caisson and the ground is $\mu_s = 0.4$, what is the largest tension in the cable that will not cause the caisson to slip?

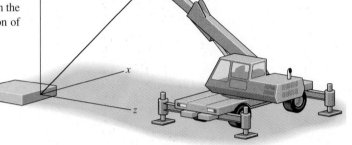

P9.145

9.146 The thrust bearing is supported by contact of the collar C with a fixed plate. The area of contact is an annulus with inside diameter D_1 and outside diameter D_2. Suppose that because of thermal constraints, you want the area of contact to be 0.02 m². The coefficient of kinetic friction between the collar and the plate is $\mu_k = 0.3$. The force $F = 600$ N, and the couple M required to rotate the shaft at a constant rate is 10 N-m. What are the diameters D_1 and D_2?

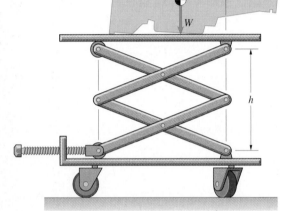

P9.143

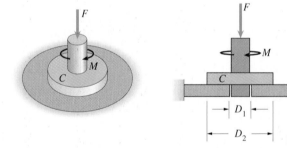

P9.146

9.144 The 10-lb metal disk A is at the center of the inclined surface. The coefficient of static friction between the disk and the surface is 0.3. What is the largest tension in the string AB that will not cause the disk to slip?

9.147 The block A weighs 30 lb, and the spring constant $k = 30$ lb/ft. If the cylinder is rotated at a constant rate, first in the counterclockwise direction and then in the clockwise direction, the difference in the height of block A is 2 ft. What is the coefficient of kinetic friction between the rope and the cylinder?

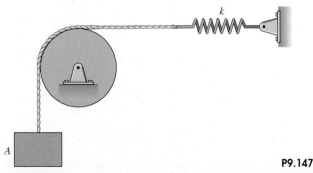

P9.144

9.145 The direction cosines of the crane's cable are $\cos \theta_x = 0.588$, $\cos \theta_y = 0.766$, and $\cos \theta_z = 0.260$. The y axis is vertical. The stationary caisson to which the cable is attached weighs 2000 lb. The unit vector $\mathbf{e} = 0.260\mathbf{i} + 0.940\mathbf{j} - 0.221\mathbf{k}$ is perpendicular to the ground where the caisson rests. If the coefficient of static

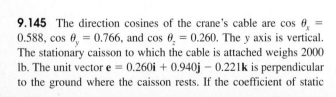

P9.147

9.148 The coefficient of static friction between the 1-kg slider and the vertical bar is $\mu_s = 0.6$. The constant of the spring is $k = 20$ N/m, and its unstretched length is 1 m. Determine the range of values of y at which the slider will remain stationary on the bar.

9.149 The axial force on the thrust bearing is $F = 200$ lb, and the dimension $b = 6$ in. The uniform pressure exerted by the mating surface is $p = 7$ psi, and the coefficient of kinetic friction is $\mu_k = 0.28$. If a couple $M = 360$ in-lb is required to turn the shaft, what are the dimensions D_o and D_i?

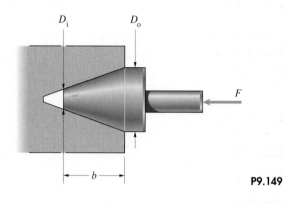

P9.149

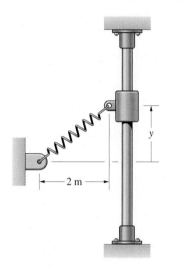

P9.148

Chapter Summary

Dry Friction

The forces resulting from the contact of two plane surfaces can be expressed in terms of the normal force N and friction force f (Fig. a) or the magnitude R and angle of friction θ (Fig. b).

If slip is impending, the magnitude of the friction force is

$$f = \mu_s N, \qquad \text{Eq. (9.1)}$$

and its direction opposes the impending slip. The angle of friction equals the angle of static friction $\theta_s = \arctan(\mu_s)$.

If the surfaces are sliding, the magnitude of the friction force is

$$f = \mu_k N, \qquad \text{Eq. (9.2)}$$

and its direction opposes the relative motion. The angle of friction equals the angle of kinetic friction $\theta_k = \arctan(\mu_k)$.

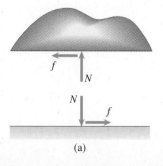

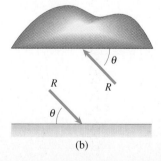

(a)   (b)

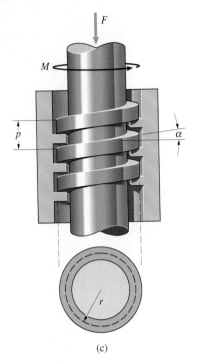

(c)

(d)

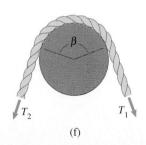

(f)

Threads

The slope α of the thread (Fig. c) is related to its pitch p by

$$\tan \alpha = \frac{p}{2\pi r}. \qquad \textbf{Eq. (9.7)}$$

The couple required for impending rotation and axial motion opposite to the direction of F is

$$M = rF \tan (\theta_s + \alpha), \qquad \textbf{Eq. (9.9)}$$

and the couple required for impending rotation and axial motion of the shaft in the direction of F is

$$M = rF \tan (\theta_s - \alpha). \qquad \textbf{Eq. (9.11)}$$

When $\theta_s < \alpha$, the shaft will rotate and move in the direction of the force F with no couple applied.

Journal Bearings

The couple required for impending slip of the circular shaft (Fig. d) is

$$M = rF \sin \theta_s, \qquad \textbf{Eq. (9.12)}$$

where F is the total load on the shaft.

Thrust Bearings and Clutches

The couple required to rotate the shaft at a constant rate (Fig. e) is

$$M = \frac{2\mu_k F}{3 \cos \alpha} \left(\frac{r_o^3 - r_i^3}{r_o^2 - r_i^2} \right). \qquad \textbf{Eq. (9.13)}$$

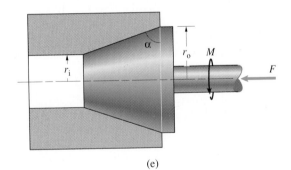

(e)

Belt Friction

The force T_2 required for impending slip in the direction of T_2 (Fig. f) is

$$T_2 = T_1 e^{\mu_s \beta}, \qquad \textbf{Eq. (9.17)}$$

where β is in radians.

Review Problems

9.150 The weight of the box is $W = 30$ lb, and the force F is perpendicular to the inclined surface. The coefficient of static friction between the box and the inclined surface is $\mu_s = 0.2$.
(a) If $F = 30$ lb, what is the magnitude of the friction force exerted on the stationary box?
(b) If $F = 10$ lb, show that the box cannot remain at rest on the inclined surface.

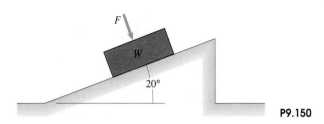

P9.150

9.151 In Problem 9.150, what is the smallest force F necessary to hold the box stationary on the inclined surface?

9.152 Blocks A and B are connected by a horizontal bar. The coefficient of static friction between the inclined surface and the 400-lb block A is 0.3. The coefficient of static friction between the surface and the 300-lb block B is 0.5. What is the smallest force F that will prevent the blocks from slipping down the surface?

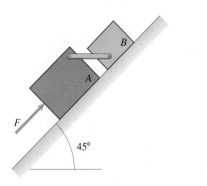

P9.152

9.153 What force F is necessary to cause the blocks in Problem 9.152 to start sliding up the plane?

9.154 The masses of crates A and B are 25 kg and 30 kg, respectively. The coefficient of static friction between the contacting surfaces is $\mu_s = 0.34$. What is the largest value of α for which the crates will remain in equilibrium?

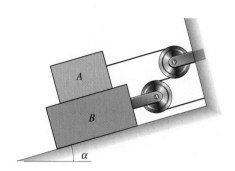

P9.154

9.155 The side of a soil embankment has a 45° slope (Fig. a). If the coefficient of static friction of soil on soil is $\mu_s = 0.6$, will the embankment be stable or will it collapse? If it will collapse, what is the smallest slope that can be stable?

Strategy: Draw a free-body diagram by isolating part of the embankment as shown in Fig. (b).

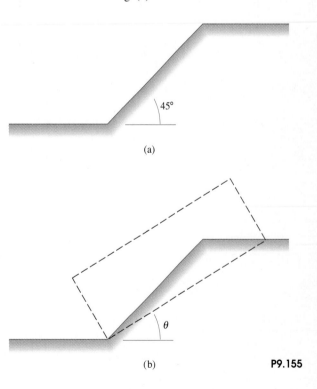

(a)

(b)

P9.155

9.156 The mass of the van is 2250 kg, and the coefficient of static friction between its tires and the road is 0.6. If its front wheels are locked and its rear wheels can turn freely, what is the largest value of α for which it can remain in equilibrium?

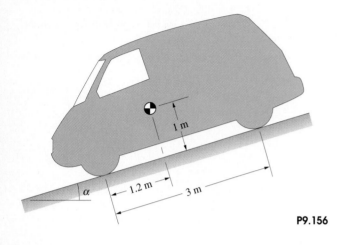

P9.156

9.157 In Problem 9.156, what is the largest value of α for which the van can remain in equilibrium if it points up the slope?

9.158 The shelf is designed so that it can be placed at any height on the vertical beam. The shelf is supported by friction between the two horizontal cylinders and the vertical beam. The combined weight of the shelf and camera is W. If the coefficient of static friction between the vertical beam and the horizontal cylinders is μ_s, what is the minimum distance b necessary for the shelf to stay in place?

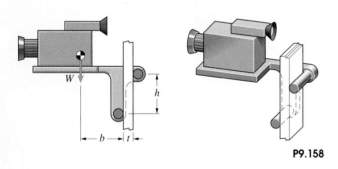

P9.158

9.159 The 20-lb homogeneous object is supported at A and B. The distance $h = 4$ in., friction can be neglected at B, and the coefficient of static friction at A is 0.4. Determine the largest force F that can be exerted without causing the object to slip.

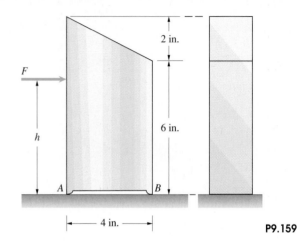

P9.159

9.160 In Problem 9.159, suppose that the coefficient of static friction at B is 0.36. What is the largest value of h for which the object will slip before it tips over?

9.161 The 180-lb climber is supported in the "chimney" by the normal and friction forces exerted on his shoes and back. The static coefficients of friction between his shoes and the wall and between his back and the wall are 0.8 and 0.6, respectively. What is the minimum normal force his shoes must exert?

P9.161

9.162 The sides of the 200-lb door fit loosely into grooves in the walls. Cables at A and B raise the door at a constant rate. The coefficient of kinetic friction between the door and the grooves is $\mu_k = 0.3$. What force must the cable at A exert to continue raising the door at a constant rate if the cable at B breaks?

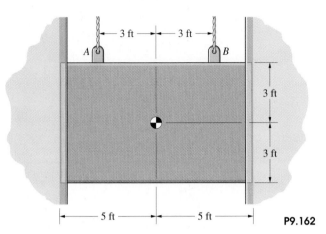

P9.162

9.163 The coefficients of static friction between the tires of the 1000-kg tractor and the ground and between the 450-kg crate and the ground are 0.8 and 0.3, respectively. Starting from rest, what torque must the tractor's engine exert on the rear wheels to cause the crate to move? (The front wheels can turn freely.)

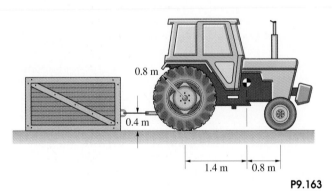

P9.163

9.164 In Problem 9.163, what is the most massive crate the tractor can cause to move from rest if its engine can exert sufficient torque? What torque is necessary?

9.165 The mass of the vehicle is 900 kg, it has rear-wheel drive, and the coefficient of static friction between its tires and the surface is 0.65. The coefficient of static friction between the crate and the surface is 0.4. If the vehicle attempts to pull the crate up the

incline, what is the largest value of the mass of the crate for which it will slip up the incline before the vehicle's tires slip?

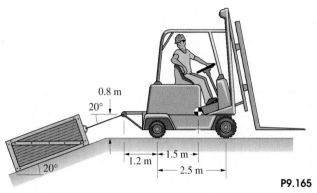

P9.165

9.166 Each of the uniform 1-m bars has a mass of 4 kg. The coefficient of static friction between the bar and the surface at B is 0.2. If the system is in equilibrium, what is the magnitude of the friction force exerted on the bar at B?

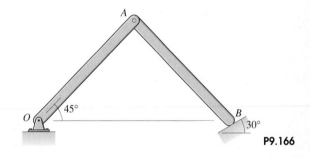

P9.166

9.167 In Problem 9.166, what is the minimum coefficient of static friction between the bar and the surface at B necessary for the system to be in equilibrium?

9.168 The collars A and B each have a mass of 2 kg. If friction between collar B and the bar can be neglected, what minimum coefficient of static friction between collar A and the bar is necessary for the collars to remain in equilibrium in the position shown?

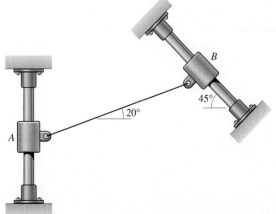

P9.168

9.169 In Problem 9.168, if the coefficient of static friction has the same value μ_s between collars A and B and the bars, what minimum value of μ_s is necessary for the collars to remain in equilibrium in the position shown? (Assume that slip impends at A and B.)

9.170 The frame is supported by the normal and friction forces exerted on the plate at A by the fixed surface. The coefficient of static friction between the plate and wall is $\mu_s = 0.5$. Will the frame slip at A when it is subjected to the 600-lb load?

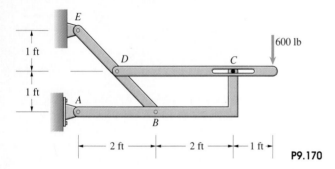

P9.170

9.171 The clamp presses two pieces of wood together. The pitch of the threads is $p = 2$ mm, the mean radius of the thread is $r = 8$ mm, and the coefficient of kinetic friction between the thread and the mating groove is 0.24. What couple must be exerted on the threaded shaft to press the pieces of wood together with a force of 200 N?

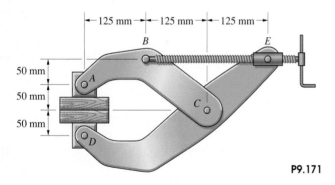

P9.171

9.172 The axles of the tram are supported by journal bearings. The radius of the wheels is 75 mm, the radius of the axles is 15 mm, and the coefficient of kinetic friction between the axles and the bearings is $\mu_k = 0.14$. The mass of the tram and its load is 160 kg. If the weight of the tram and its load is evenly divided between the axles, what force P is necessary to push the tram at a constant speed?

P9.172

9.173 The two pulleys have a radius of 6 in. and are mounted on shafts of 1-in. radius supported by journal bearings. Neglect the weights of the pulleys and shafts. The coefficient of kinetic friction between the shafts and the bearings is $\mu_k = 0.2$. If a force $T = 200$ lb is required to raise the man at a constant rate, what is his weight?

P9.173

9.174 If the man in Problem 9.173 weighs 160 lb, what force T is necessary to lower him at a constant rate?

9.175 If the two cylinders are held fixed, what is the range of W for which the two weights will remain stationary?

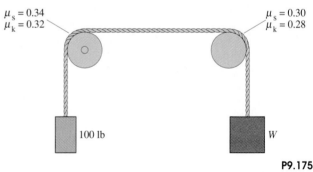

P9.175

9.176 In Problem 9.175, if the system is initially stationary and the left cylinder is slowly rotated, determine the largest weight W that can be (a) raised; (b) lowered.

9.177 The chinning bar is secured in the door frame by turning the threaded collar C. The pitch of the threaded shaft is $p = 0.1$ in., and the mean radius of the thread is $r = 0.5$ in. The coefficients of friction between the thread and the mating groove are $\mu_s = 0.30$ and $\mu_k = 0.24$. The coefficient of static friction between the door frame and the ends of the bar is 0.8. What couple must be applied to the collar C for the bar to be able to support a 400-lb vertical load without slipping in the door frame?

P9.177

9.178 In Problem 9.177, what couple must be applied to the threaded collar C to loosen it?

Project 9.1 Use the procedure described in Problem 9.5 to measure the coefficient of static friction for the following pairs of materials:

(a) Paper on paper
(b) Metal on metal
(c) Wood on wood
(d) Metal on wood

Write a brief report describing your procedure, comparing your results with the typical values in Table 9.1, and commenting on possible sources of error.

The Verrazano-Narrows Bridge in New York City had the longest suspension span in the world, 4260 ft, at the time of its completion in 1965. The enormous weight of the deck is supported by 7200-ft-long steel cables. By modeling the forces exerted on the main cable of a suspension bridge as a distributed load, we can determine the shape of the cable and the variation of the tension—the internal force within the cable—along its length.

Chapter 10

Internal Forces and Moments

WE began our study of equilibrium by drawing free-body diagrams of individual objects to determine unknown forces and moments acting on them. Then, by conceptually "disassembling" structures and drawing free-body diagrams of their members, we determined forces and moments acting on the members. In this chapter we carry this process one step further and draw free-body diagrams of parts of individual objects to determine internal forces and moments exerted by the objects themselves. In doing so, we arrive at the central concern of the design engineer: It is the internal forces within an object that determine whether it will support the loads to which it is subjected.

You are already familiar with two examples of internal forces: the axial forces in two-force members and the tensions in ropes and cables. Here we first discuss the internal forces and moments in beams subjected to arbitrary loads. We then analyze ropes and cables subjected to loads along their lengths. Finally, we discuss the pressure distributions that describe the internal forces in a liquid at rest.

Beams

10.1 *Axial Force, Shear Force, and Bending Moment*

To ensure that a structural member will not fail (break or collapse) due to the forces and moments acting on it, the design engineer must know not only the external loads and reactions acting on it but also the forces and moments acting *within* the member.

Let's consider a beam subjected to an external load and reactions (Fig. 10.1a). How can we determine the forces and moments within the beam? In Fig. 10.1(b), we "cut" the beam by a plane at an arbitrary cross section and isolate part of it. You can see that the isolated part cannot be in equilibrium unless it is subjected to some system of forces and moments at the plane where it joins the other part of the beam. These are the internal forces and moments we seek.

In Chapter 4 we demonstrated that *any* system of forces and moments can be represented by an equivalent system consisting of a force and a couple. Since the system of external loads and reactions on the beam is two-dimensional, we can represent the internal forces and moments by an equivalent system consisting of two components of force and a couple (Fig. 10.1c). The component P parallel to the beam's axis is called the **axial force**. The component V normal to the beam's axis is called the **shear force**, and the couple M is called the **bending moment**. The axial force, shear force, and bending moment on the free-body diagram of the other part of the beam are shown

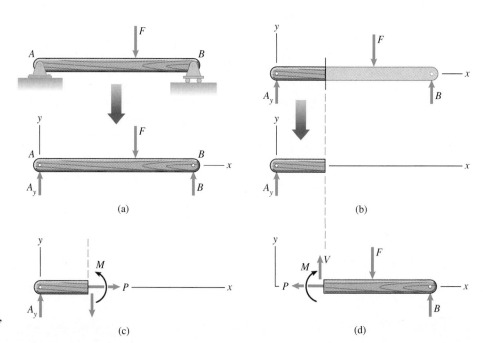

Figure 10.1

(a) A beam subjected to a load and reactions.
(b) Isolating a part of the beam.
(c), (d) The axial force, shear force, and bending moment.

in Fig. 10.1(d). Notice that they are equal in magnitude but opposite in direction to the internal forces and moment on the free-body diagram in Fig. 10.1(c).

The directions of the axial force, shear force, and bending moment in Figs. 10.1(c) and (d) are the established definitions of the positive directions of these quantities. A positive axial force P subjects the beam to tension. A positive shear force V tends to rotate the axis of the beam clockwise (Fig. 10.2a). Bending moments are defined to be positive when they tend to bend the axis of the beam upward (Fig. 10.2b). In terms of the coordinate system we use, "upward" means in the direction of the positive y axis.

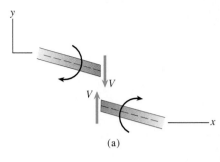
(a)

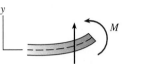
(b)

Figure 10.2
(a) Positive shear forces tend to rotate the axis of the beam clockwise.
(b) Positive bending moments tend to bend the axis of the beam upward.

The following examples demonstrate how to determine the internal forces and moment at a particular cross section of a beam. This typically involves three steps:

1. Determine the external forces and moments—*Draw the free-body diagram of the beam, and determine the reactions at its supports. If the beam is a member of a structure, you must analyze the structure.*

2. Draw the free-body diagram of part of the beam—*Cut the beam at the point at which you want to determine the internal forces and moment, and draw the free-body diagram of one of the resulting parts. You can choose the part with the simplest free-body diagram. If your cut divides a distributed load, don't represent the distributed load by an equivalent force until after you have obtained your free-body diagram.*

3. Apply the equilibrium equations—*Use the equilibrium equations to determine P, V, and M.*

Example 10.1

For the beam in Fig. 10.3, determine the internal forces and moment at C.

SOLUTION

Determine the External Forces and Moments   We begin by drawing the free-body diagram of the beam and determining the reactions at its supports; the results are shown in Fig. (a).

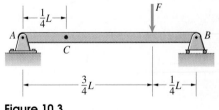

Figure 10.3

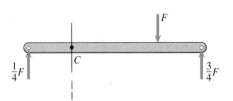

(a) The free-body diagram of the beam and a plane through point C.

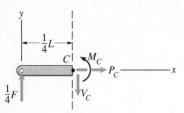

(b) The free-body diagram of the part of the beam to the left of the plane through point C.

(c) The free-body diagram of the part of the beam to the right of the plane through point C.

Draw the Free-Body Diagram of Part of the Beam
We cut the beam at C (Fig. a) and draw the free-body diagram of the left part, including the internal forces and moment in their defined positive directions (Fig. b).

Apply the Equilibrium Equations
From the equilibrium equations

$$\Sigma F_x = P_C = 0,$$

$$\Sigma F_y = \frac{1}{4}F - V_C = 0,$$

$$\Sigma M_{(\text{point } C)} = M_C - \left(\frac{1}{4}L\right)\left(\frac{1}{4}F\right) = 0,$$

we obtain $P_C = 0$, $V_C = \frac{1}{4}F$, and $M_C = \frac{1}{16}LF$.

DISCUSSION

We should check our results with the free-body diagram of the other part of the beam (Fig. c). The equilibrium equations are

$$\Sigma F_x = -P_C = 0,$$

$$\Sigma F_y = V_C - F + \frac{3}{4}F = 0,$$

$$\Sigma M_{(\text{point } C)} = -M_C - \left(\frac{1}{2}L\right)F + \left(\frac{3}{4}L\right)\left(\frac{3}{4}F\right) = 0,$$

confirming that $P_C = 0$, $V_C = \frac{1}{4}F$, and $M_C = \frac{1}{16}LF$.

Example 10.2

For the beam in Fig. 10.4, determine the internal forces and moment (a) at B; (b) at C.

Figure 10.4

SOLUTION

Determine the External Forces and Moments
We draw the free-body diagram of the beam and represent the distributed load by an equivalent force in Fig. (a). The equilibrium equations are

$$\Sigma F_x = A_x = 0,$$

$$\Sigma F_y = A_y + D - 180 = 0,$$

$$\Sigma M_{(\text{point } A)} = 12D - (4)(180) = 0.$$

Solving them, we obtain $A_x = 0$, $A_y = 120$ N, and $D = 60$ N.

(a) Free-body diagram of the entire beam with the distributed load represented by an equivalent force.

Draw the Free-Body Diagram of Part of the Beam   We cut the beam at B, obtaining the free-body diagram in Fig. (b). Because point B is at the midpoint of the triangular distributed load, the value of the distributed load at B is 30 N/m. By representing the distributed load in Fig. (b) by an equivalent force, we obtain the free-body diagram in Fig. (c). From the equilibrium equations

$$\Sigma F_x = P_B = 0,$$

$$\Sigma F_y = 120 - 45 - V_B = 0,$$

$$\Sigma M_{(\text{point } B)} = M_B + (1)(45) - (3)(120) = 0,$$

we obtain $P_B = 0$, $V_B = 75$ N, and $M_B = 315$ N-m.

To determine the internal forces and moment at C, we obtain the simplest free-body diagram by isolating the part of the beam to the right of C (Fig. d). From the equilibrium equations

$$\Sigma F_x = -P_C = 0,$$

$$\Sigma F_y = V_C + 60 = 0,$$

$$\Sigma M_{(\text{point } C)} = -M_C + (3)(60) = 0,$$

we obtain $P_C = 0$, $V_C = -60$ N, and $M_C = 180$ N-m.

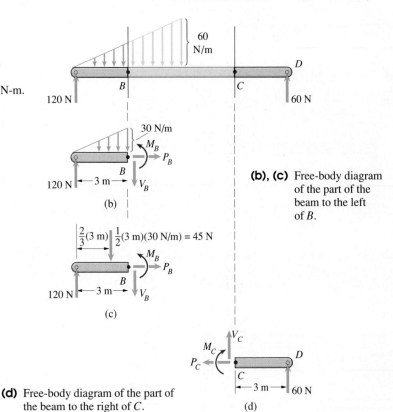

(b), (c) Free-body diagram of the part of the beam to the left of B.

(b)

(c)

(d) Free-body diagram of the part of the beam to the right of C.

(d)

DISCUSSION

If you attempt to determine the internal forces and moment at B by cutting the free-body diagram in Fig. (a) at B, you do *not* obtain correct results. (You can confirm that the resulting free-body diagram of the part of the beam to the left of B gives $P_B = 0$, $V_B = 120$ N, and $M_B = 360$ N-m.) The reason is that you do not properly account for the effect of the distributed load on your free-body diagram. You must wait until *after* you have obtained the free-body diagram of part of the beam before representing distributed loads by equivalent forces.

Problems

10.1 Determine the internal forces and moment at A, B, and C.

Strategy: In this case you don't need to determine the reactions at the built-in support. Cut the beam at the point where you want to determine the internal forces and moment, and draw the free-body diagram of the part of the beam to the left of your cut. Remember that P, V, and M must be in their defined positive directions in your free-body diagrams.

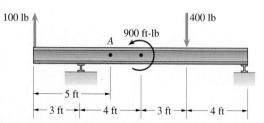

P10.1

10.2 Determine the internal forces and moment at A.

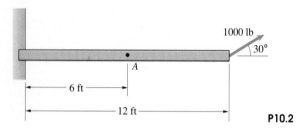

P10.2

10.3 Determine the internal forces and moment at A.

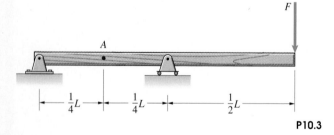

P10.3

10.4 Determine the internal forces and moment at A.

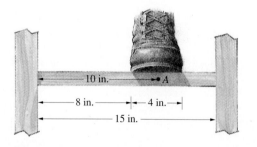

P10.4

10.5 Determine the internal forces and moment at A.

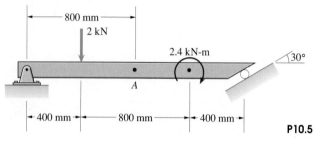

P10.5

10.6 Determine the internal forces and moment at A for each loading.

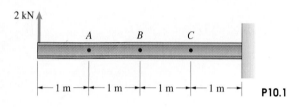

P10.6

10.7 Model the ladder rung as a simply supported (pin-supported) beam, and assume that the 200-lb load exerted by the person's shoe is uniformly distributed. Determine the internal forces and moment at A.

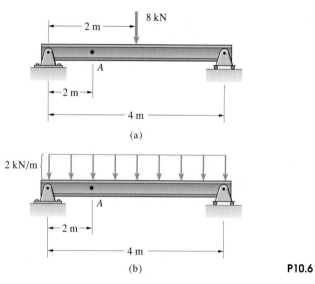

P10.7

10.8   Determine the internal forces and moment at *A*.

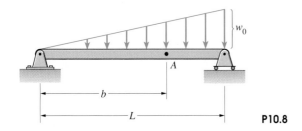

P10.8

10.9   If *x* = 3 m, what are the internal forces and moment at *A*?

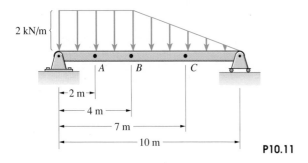

P10.9

10.10   If *x* = 8 m in Problem 10.9, what are the internal forces and moment at *A*?

10.11   Determine the internal forces and moment at *A*.

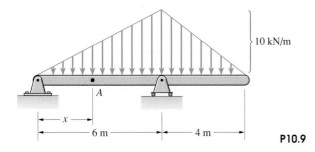

P10.11

10.12   Determine the internal forces and moment at point *B* of the beam in Problem 10.11.

10.13   Determine the internal forces and moment at point *C* of the beam in Problem 10.11.

10.14   Determine the internal forces and moment at *A*.

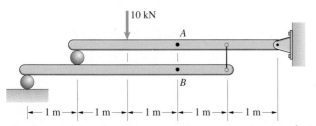

P10.14

10.15   Determine the internal forces and moment at point *B* in Problem 10.14.

10.16   Determine the internal forces and moment at *A*.

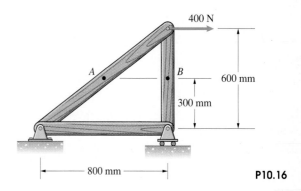

P10.16

10.17   Determine the internal forces and moment at point *B* of the truss in Problem 10.16.

10.18   The tension in the cable is 500 lb. Determine the internal forces and moment at *A*.

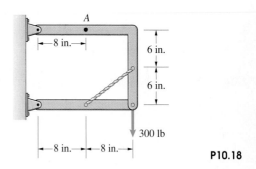

P10.18

10.19   Determine the internal forces and moment at *A*.

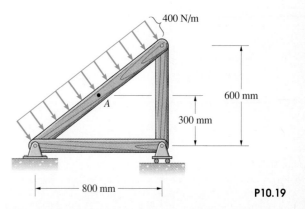

P10.19

10.20 Determine the internal forces and moment at A.

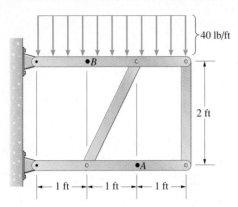

P10.20

10.21 Determine the internal forces and moment at point B of the frame in Problem 10.20.

10.2 Shear Force and Bending Moment Diagrams

To design a beam, an engineer must know the internal forces and moments throughout its length. Of special concern are the maximum and minimum values of the shear force and bending moment and where they occur. In this section we show how the values of P, V, and M can be determined as functions of x and introduce shear force and bending moment diagrams.

Let's consider a simply supported beam loaded by a force (Fig. 10.5a). Instead of cutting the beam at a specific cross section to determine the internal forces and moment, we cut it at an arbitrary position x between the left end of the beam and the load F (Fig. 10.5b). Applying the equilibrium equations to this free-body diagram, we obtain

$$\left. \begin{array}{l} P = 0 \\[1.5em] V = \dfrac{1}{3}F \\[1.5em] M = \dfrac{1}{3}Fx \end{array} \right\} \quad 0 < x < \dfrac{2}{3}L.$$

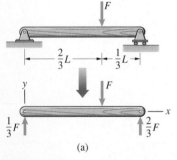

(a)

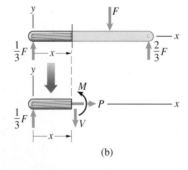

(b)

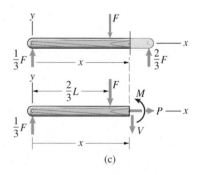

(c)

Figure 10.5

(a) A beam loaded by a force F and its free-body diagram.
(b) Cutting the beam at an arbitrary position x to the left of F.
(c) Cutting the beam at an arbitrary position x to the right of F.

To determine the internal forces and moment for values of x greater than $\frac{2}{3}L$, we obtain a free-body diagram by cutting the beam at an arbitrary position x between the load F and the right end of the beam (Fig. 10.5c). The results are

$$
\left.
\begin{array}{l}
P = 0 \\[2mm]
V = -\dfrac{2}{3}F \\[2mm]
M = \dfrac{2}{3}F\,(L - x)
\end{array}
\right\} \quad \frac{2}{3}L < x < L.
$$

The **shear force and bending moment diagrams** are simply the graphs of V and M, respectively, as functions of x (Fig. 10.6). They permit you to see the changes in the shear force and bending moment that occur along the beam's length as well as their maximum and minimum values. (By *maximum* we mean the least upper bound of the shear force or bending moment, and by *minimum* we mean the greatest lower bound.)

Thus you can determine the distributions of the internal forces and moment in a beam by considering a plane at an arbitrary distance x from the end of the beam and solving for P, V, and M as functions of x. Depending on the complexity of the loading, you may have to draw several free-body diagrams to determine the distributions over the entire length of the beam.

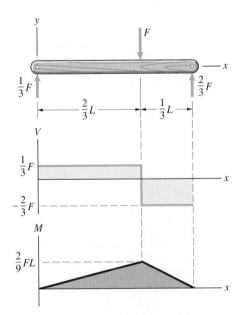

Figure 10.6

The shear force and bending moment diagrams indicating the maximum and minimum values of V and M.

The following example demonstrates how to obtain the shear force and bending moment diagrams for a beam. Once you have determined the external reactions on the beam, your procedure is to cut it at an arbitrary position x, draw the free-body diagram of the part of the beam on either side of the cut (choose the simplest part), and determine the internal forces and moment. You may need to repeat this process several times to obtain results that are valid over the entire length of the beam.

Example 10.3

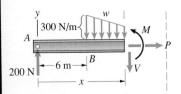

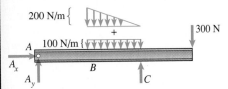

Figure 10.7

For the beam in Fig. 10.7, (a) draw the shear force and bending moment diagrams; (b) determine the locations and values of the maximum and minimum shear forces and bending moments.

STRATEGY

To determine the internal forces and moment as functions of x for the entire beam, we must use three free-body diagrams: one for the range $0 < x < 6$ m, one for $6 < x < 12$ m, and one for $12 < x < 18$ m.

SOLUTION

(a) We begin by drawing the free-body diagram of the beam, treating the distributed load as the sum of uniform and triangular distributed loads (Fig. a). We then represent these distributed loads by equivalent forces (Fig. b). From the equilibrium equations

$$\Sigma F_x = A_x = 0,$$

$$\Sigma F_y = A_y + C - 600 - 600 - 300 = 0,$$

$$\Sigma M_{(\text{point } A)} = 12C - (8)(600) - (9)(600) - (18)(300) = 0,$$

we obtain the reactions $A_x = 0$, $A_y = 200$ N, and $C = 1300$ N.

We draw the free-body diagram for the range $0 < x < 6$ m in Fig. (c). From the equilibrium equations

$$\Sigma F_x = P = 0,$$

$$\Sigma F_y = 200 - V = 0,$$

$$\Sigma M_{(\text{right end})} = M - 200x = 0,$$

we obtain

$$\left. \begin{array}{l} P = 0 \\ V = 200 \text{ N} \\ M = 200x \text{ N-m} \end{array} \right\} \quad 0 < x < 6 \text{ m}.$$

(a) Free-body diagram of the entire beam.

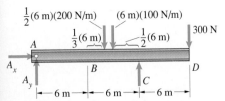

(b) Representing the distributed loads by equivalent forces.

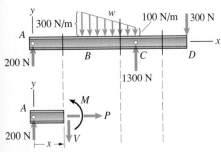

(c) Free-body diagram for $0 < x < 6$ m.

(d) Free-body diagram for $6 < x < 12$ m.

We draw the free-body diagram for the range $6 < x < 12$ m in Fig. (d). To obtain the equilibrium equations, we determine w as a function of x and integrate to determine the force and moment exerted by the distributed load. We can express w in the form $w = cx + d$, where c and d are constants. Using the conditions $w = 300$ N/m at $x = 6$ m and $w = 100$ N/m at $x = 12$ m, we obtain the equation $w = -(100/3)x + 500$ N/m. The downward force on the free body in Fig. (d) due to the distributed load is

$$F = \int_L^x w \, dx = \int_6^x \left(-\frac{100}{3}x + 500 \right) dx = -\frac{50}{3}x^2 + 500x - 2400 \text{ N}.$$

The clockwise moment about the origin (point A) due to the distributed load is

$$\int_L^x xw \, dx = \int_6^x \left(-\frac{100}{3}x^2 + 500x \right) dx = -\frac{100}{9}x^3 + 250x^2 - 6600 \text{ N-m}.$$

The equilibrium equations are

$$\Sigma F_x = P = 0,$$

$$\Sigma F_y = 200 - V + \frac{50}{3}x^2 - 500x + 2400 = 0,$$

$$\Sigma M_{(\text{point } A)} = M - Vx + \frac{100}{9}x^3 - 250x^2 + 6600 = 0.$$

Solving them, we obtain

$$P = 0$$

$$V = \frac{50}{3}x^2 - 500x + 2600 \text{ N}$$

$$M = \frac{50}{9}x^3 - 250x^2 + 2600x - 6600 \text{ N-m}$$

$$\left.\right\} \quad 6 < x < 12 \text{ m.}$$

For the range $12 < x < 18$ m, we obtain a very simple free-body diagram by using the part of the beam on the right of the cut (Fig. e). From the equilibrium equations

$$\Sigma F_x = -P = 0,$$

$$\Sigma F_y = V - 300 = 0,$$

$$\Sigma M_{(\text{left end})} = -M - 300(18 - x) = 0,$$

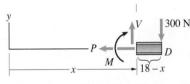

(e) Free-body diagram for $12 < x < 18$ m.

we obtain

$$P = 0$$

$$V = 300 \text{ N}$$

$$M = 300x - 5400 \text{ N-m}$$

$$\left.\right\} \quad 12 < x < 18 \text{ m.}$$

The shear force and bending moment diagrams obtained by plotting the equations for V and M for the three ranges of x are shown in Fig. 10.8.
(b) From the shear force diagram, the minimum shear force is -1000 N at $x = 12$ m, and its maximum value is 300 N over the range $12 < x < 18$ m. The minimum bending moment is -1800 N-m at $x = 12$ m. The bending moment has its maximum value in the range $6 < x < 12$ m. It occurs where $dM/dx = 0$. Using the equation for M as a function of x in the range $6 < x < 12$ m, we obtain

$$\frac{dM}{dx} = \frac{150}{9}x^2 - 500x + 2600 = 0.$$

The applicable root is $x = 6.69$ m. Substituting it into the equation for M, we determine that the value of the maximum bending moment is 1270 N-m.

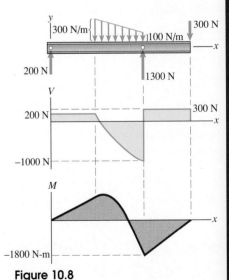

Figure 10.8

The shear force and bending moment diagrams.

Problems

10.22 (a) Determine the internal forces and moment as functions of x.

(b) Draw the shear force and bending moment diagrams.

 Strategy: You don't need to determine the reactions at the built-in support. Cut the beam at an arbitrary position x, and draw the free-body diagram of the part on the left.

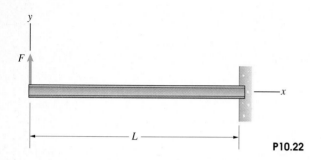

P10.22

10.23 (a) Determine the internal forces and moment as functions of x.

(b) Show that the equations for V and M as functions of x satisfy the equation $V = dM/dx$.

(c) Draw the shear force and bending moment diagrams.

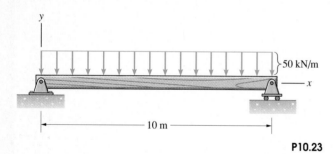

P10.23

10.24 (a) Determine the internal forces and moment as functions of x.

(b) Show that the equations for V and M as functions of x satisfy the equation $V = dM/dx$.

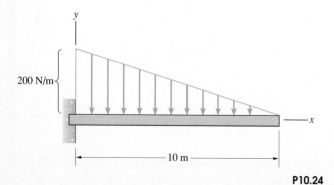

P10.24

10.25 Draw the shear force and bending moment diagrams for the beam in Problem 10.24.

10.26 Determine the internal forces and moment as functions of x.

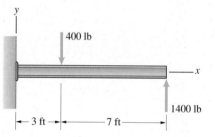

P10.26

10.27 Draw the shear force and bending moment diagrams for the beam in Problem 10.26.

10.28 (a) Determine the internal forces and moment as functions of x.

(b) Draw the shear force and bending moment diagrams.

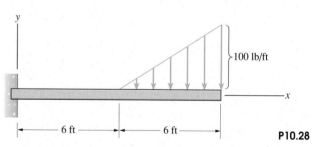

P10.28

10.29 The loads $F = 200$ N and $C = 800$ N-m.

(a) Determine the internal forces and moment as functions of x.

(b) Draw the shear force and bending moment diagrams.

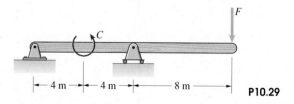

P10.29

10.30 The beam in Problem 10.29 will safely support shear forces and bending moments of magnitudes 2 kN and 6.5 kN-m, respectively. On the basis of this criterion, can it safely be subjected to the loads $F = 1$ kN, $C = 1.6$ kN-m?

10.31 Model the ladder rung as a simply supported (pin-supported) beam, and assume that the 200-lb load exerted by the person's shoe is uniformly distributed. Draw the shear force and bending moment diagrams.

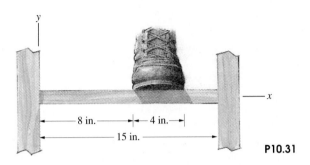

P10.31

10.32 What is the maximum bending moment in the ladder rung in Problem 10.31, and where does it occur?

10.33 Assume that the surface the beam rests on exerts a uniformly distributed load. Draw the shear force and bending moment diagrams.

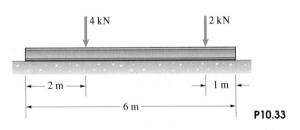

P10.33

10.34 The homogeneous beams *AB* and *CD* weigh 600 lb and 500 lb, respectively. Draw the shear force and bending moment diagrams for beam *CD*.

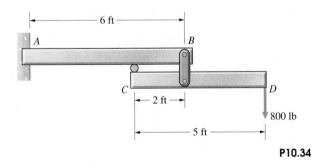

P10.34

10.35 Draw the shear force and bending moment diagrams for beam *AB* in Problem 10.34.

10.36 The beam is subjected to a distributed load $w = 100x - cx^2$ lb/ft, where c is a constant. The moment about A due to the distributed load is zero. Determine the internal forces and moment as functions of x.

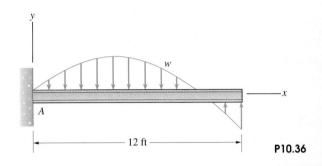

P10.36

10.37 Draw the shear force and bending moment diagrams for the beam in Problem 10.36.

10.38 The load $F = 4650$ lb. Draw the shear force and bending moment diagrams.

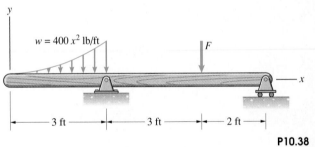

P10.38

10.39 If the load $F = 2150$ lb in Problem 10.38, what are the maximum and minimum shear forces and bending moments, and at what values of x do they occur?

10.40 Draw the shear force and bending moment diagrams.

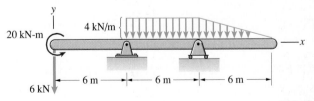

P10.40

10.3 *Relations Between Distributed Load, Shear Force, and Bending Moment*

The shear force and bending moment in a beam subjected to a distributed load are governed by simple differential equations. In this section we derive these equations and show that they provide an interesting and enlightening way to obtain shear force and bending moment diagrams. These equations are also useful for determining the deflections of beams.

Suppose that a portion of a beam is subjected to a distributed load w (Fig. 10.9a). In Fig. 10.9(b), we obtain a free-body diagram by cutting the beam at x and at $x + \Delta x$. The terms ΔP, ΔV, and ΔM are the changes in the axial force, shear force, and bending moment, respectively, from x to $x + \Delta x$. From this free-body diagram we obtain the equilibrium equations

$$\Sigma F_x = P + \Delta P - P = 0,$$

$$\Sigma F_y = V - V - \Delta V - w\Delta x - O(\Delta x^2) = 0,$$

$$\Sigma M_{(\text{point } Q)} = M + \Delta M - M - (V + \Delta V)\Delta x - wO(\Delta x^2) = 0,$$

where the notation $O(\Delta x^2)$ means a term of order Δx^2. Dividing these equations by Δx and taking the limit as $\Delta x \to 0$, we obtain

$$\frac{dP}{dx} = 0, \tag{10.1}$$

$$\frac{dV}{dx} = -w, \tag{10.2}$$

$$\frac{dM}{dx} = V. \tag{10.3}$$

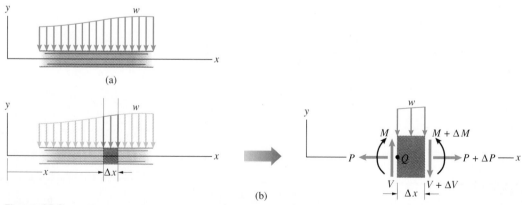

(b)

Figure 10.9
(a) A portion of a beam subjected to a distributed force w.
(b) Obtaining the free-body diagram of an element of the beam.

Equation (10.1) simply states that the axial force does not depend on x in a portion of a beam subjected only to a lateral distributed load. But notice that you can integrate Eq. (10.2) to determine V as a function of x if you know w, and then you can integrate Eq. (10.3) to determine M as a function of x.

We derived Eqs. (10.2) and (10.3) for a portion of beam subjected only to a distributed load. When you use these equations to determine shear force and bending moment diagrams, you must also account for the effects of forces and couples. Let's determine what happens to the shear force and bending moment diagrams where a beam is subjected to a force F in the positive y direction (Fig. 10.10a). By cutting the beam just to the left and just to the right of the force, we obtain the free-body diagram in Fig 10.10(b), where the subscripts $-$ and $+$ denote values to the left and right of the force, respectively. Equilibrium requires that

$$V_+ - V_- = F, \qquad (10.4)$$

$$M_+ - M_- = 0. \qquad (10.5)$$

The shear force diagram undergoes a jump discontinuity of magnitude F (Fig. 10.10c), but the bending moment diagram is continuous (Fig. 10.10d). The jump in the shear force is positive if the force is in the positive y direction.

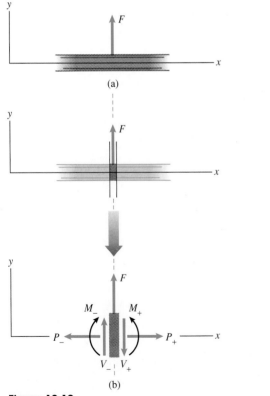

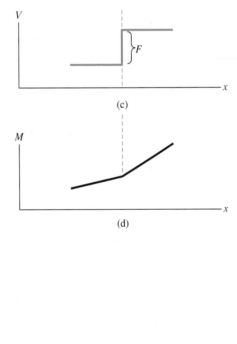

Figure 10.10

(a) A portion of a beam subjected to a distributed force F in the positive y direction.
(b) Obtaining a free-body diagram by cutting the beam to the left and right of F.
(c) The shear force diagram undergoes a positive jump of magnitude F.
(d) The bending moment diagram is continuous.

Now we consider what happens to the shear force and bending moment diagrams when a beam is subjected to a counterclockwise couple C (Fig. 10.11a). Cutting the beam just to the left and just to the right of the couple (Fig.10.11b), we determine that

$$V_+ - V_- = 0, \tag{10.6}$$

$$M_+ - M_- = -C. \tag{10.7}$$

The shear force diagram is continuous (Fig. 10.11c), but the bending moment diagram undergoes a jump discontinuity of magnitude C (Fig. 10.11d) where a beam is subjected to a couple. The jump in the bending moment is *negative* if the couple is in the counterclockwise direction.

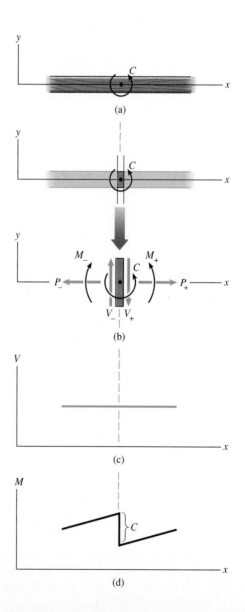

(a)

(b)

(c)

(d)

Figure 10.11

(a) A portion of a beam subjected to a counterclockwise couple C.
(b) Obtaining a free-body diagram by cutting the beam to the left and right of C.
(c) The shear force diagram is continuous.
(d) The bending moment diagram undergoes a *negative* jump of magnitude C.

We illustrate the application of these results by considering the simply sup-ported beam in Fig. 10.12(a). To determine its shear force diagram, we begin at $x = 0$, where the upward reaction at A results in a positive value of V of magnitude $\frac{1}{3}F$. Since there is no load between A and B, Eq. (10.2) states that $dV/dx = 0$. The shear force remains constant between A and B (Fig. 10.12b). At B, the downward load F causes a negative jump in V of magnitude F. There is no load between B and C, so the shear force remains constant between B and C (Fig. 10.12c).

Now that we have completed the shear force diagram, we begin again at $x = 0$ to determine the bending moment diagram. There is no couple at $x = 0$, so the bending moment is zero there. Between A and B, $V = \frac{1}{3}F$. Integrating Eq. (10.3) from $x = 0$ to an arbitrary value of x between A and B,

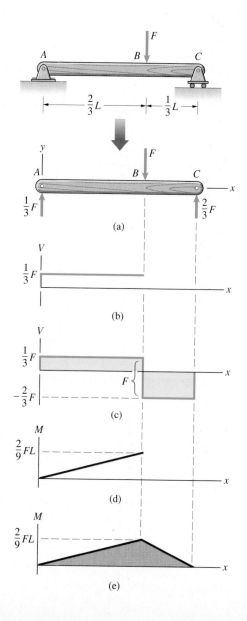

(a)

(b)

(c)

(d)

(e)

Figure 10.12

(a) A beam loaded by a force F.
(b) The shear force diagram from A to B.
(c) The complete shear force diagram.
(d) The bending moment diagram from A to B.
(e) The complete bending moment diagram.

$$\int_0^M dM = \int_0^x V \, dx = \int_0^x \frac{1}{3} F \, dx,$$

we determine M as a function of x from A to B:

$$M = \frac{1}{3} Fx, \qquad 0 < x < \frac{2}{3} L.$$

The bending moment diagram from A to B is shown in Fig. 10.12(d). The value of the bending moment at B is $M_B = \frac{2}{9} FL$.

Between B and C, $V = -\frac{2}{3} F$. Integrating Eq. (10.3) from $x = \frac{2}{3} L$ to an arbitrary value of x between B and C,

$$\int_{M_B}^M dM = \int_{(2/3)L}^x V \, dx = \int_{(2/3)L}^x -\frac{2}{3} F \, dx,$$

we obtain M as a function of x from B to C:

$$M = M_B - \frac{2}{3} F \left(x - \frac{2}{3} L \right) = \frac{2}{3} F (L - x), \qquad \frac{2}{3} L < x < L.$$

The completed bending moment diagram is shown in Fig. 10.12(e). Compare the shear and bending moment diagrams in Figs. 10.12(c) and (e) with those in Fig. 10.6, which we obtained by cutting the beam and solving equilibrium equations.

The following examples demonstrate the use of Eqs. (10.2) through (10.7) to determine shear force and bending moment diagrams of beams.

1. Shear force diagram—*For segments of the beam that are unloaded or are subjected to a distributed load, you can integrate Eq. (10.2) to determine V as a function of x. In addition, you must use Eq. (10.4) to determine the effects of forces on V.*

2. Bending moment diagram—*Once you have determined V as a function of x, integrate Eq. (10.3) to determine M as a function of x. Use Eq. (10.7) to determine the effects of couples on M.*

Example 10.4

Determine the shear force and bending moment diagrams for the beam in Fig. 10.13.

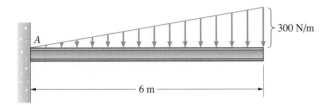

300 N/m

Figure 10.13

6 m

SOLUTION

We must first determine the reactions at A. The results are shown on the free-body diagram of the beam in Fig. (a). The equation describing the distributed load as a function of x is $w = (x/6)300 = 50x$ N/m.

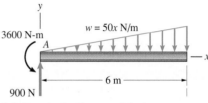

y

3600 N-m

$w = 50x$ N/m

A

x

6 m

900 N

(a) Free-body diagram of the beam.

Shear Force Diagram   The upward force at A causes a positive value of V of 900-N magnitude, so that $V_A = 900$ N. Integrating Eq. (10.2) from $x = 0$ to an arbitrary value of x,

$$\int_{V_A}^{V} dV = \int_0^x -w \, dx = \int_0^x -50x \, dx,$$

we obtain V as a function of x:

$$V = V_A - 25x^2 = 900 - 25x^2.$$

The shear force diagram is shown in Fig. (b).

Bending Moment Diagram   The counterclockwise couple at A causes a *negative* value of M of 3600 N-m magnitude, so that $M_A = -3600$ N-m. Integrating Eq. (10.3) from $x = 0$ to an arbitrary value of x,

$$\int_{M_A}^{M} dM = \int_0^x V \, dx = \int_0^x (900 - 25x^2) \, dx,$$

we obtain

$$M = M_A + 900x - \frac{25}{3}x^3 = -3600 + 900x - \frac{25}{3}x^3.$$

The bending moment diagram is shown in Fig. (c).

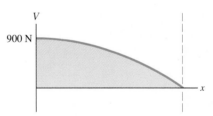

V

900 N

x

(b) Shear force diagram.

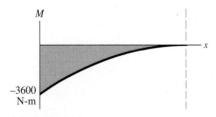

M

x

−3600 N-m

(c) Bending moment diagram.

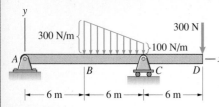

Figure 10.14

Example 10.5

Determine the shear force and bending moment diagrams for the beam in Fig. 10.14.

SOLUTION

The first step, determining the reactions at the supports, was carried out for this beam and loading in Example 10.3. The results are shown in Fig. (a).

Shear Force Diagram   *From A to B.* There is no load between A and B, so the shear force increases by 200 N at A and then remains constant from A to B:

$$V = 200 \text{ N}, \qquad 0 < x < 6 \text{ m}.$$

From B to C. We can express the distributed load w between B and C in the form $w = cx + d$, where c and d are constants. Using the conditions $w = 300$ N/m at $x = 6$ m and $w = 100$ N/m at $x = 12$ m, we obtain the equation $w = -(100/3)x + 500$ N/m. Integrating Eq. (10.2) from $x = 6$ m to an arbitrary value of x between B and C,

$$\int_{V_B}^{V} dV = \int_{6}^{x} -w \, dx = \int_{6}^{x} \left(\frac{100}{3}x - 500 \right) dx,$$

we obtain an equation for V between B and C:

$$V = \frac{50}{3} x^2 - 500x + 2600 \text{ N}, \qquad 6 < x < 12 \text{ m}.$$

At $x = 12$ m, $V = -1000$ N.

From C to D. At C, V undergoes a positive jump of 1300-N magnitude, so that its value becomes $-1000 + 1300 = 300$ N. There is no loading between C and D, so V remains constant from C to D:

$$V = 300 \text{ N}, \qquad 12 < x < 18 \text{ m}.$$

The shear force diagram is shown in Fig. (b).

Bending Moment Diagram   *From A to B.* There is no couple at $x = 0$, so the bending moment is zero there. Integrating Eq. (10.3) from $x = 0$ to an arbitrary value of x between A and B,

$$\int_{0}^{M} dM = \int_{0}^{x} V \, dx = \int_{0}^{x} 200 \, dx,$$

we obtain

$$M = 200x \text{ N-m}, \qquad 0 < x < 6 \text{ m}.$$

At $x = 6$ m, $M_B = 1200$ N-m.

From B to C. Integrating from $x = 6$ m to an arbitrary value of x between B and C,

$$\int_{M_B}^{M} dM = \int_{6}^{x} V \, dx = \int_{6}^{x} \left(\frac{50}{3} x^2 - 500x + 2600 \right) dx,$$

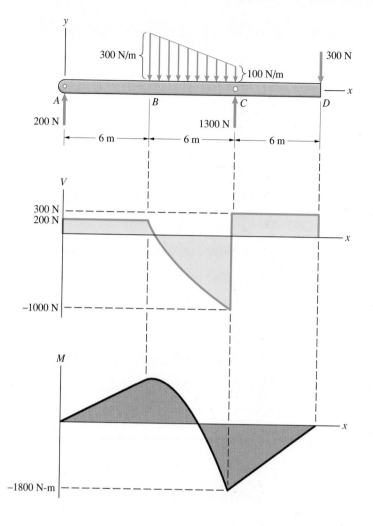

(a) Free-body diagram of the beam.

(b) Shear force diagram.

(c) Bending moment diagram.

we obtain

$$M = \frac{50}{9}x^3 - 250x^2 + 2600x - 6600 \text{ N-m}, \qquad 6 < x < 12 \text{ m}.$$

At $x = 12$ m, $M_C = -1800$ N-m.

From C to D. Integrating from $x = 12$ m to an arbitrary value of x between C and D,

$$\int_{M_C}^{M} dM = \int_{12}^{x} V \, dx = \int_{12}^{x} 300 \, dx,$$

we obtain

$$M = 300x - 5400 \text{ N-m}, \qquad 12 < x < 18 \text{ m}.$$

The bending moment diagram is shown in Fig. (c).

Problems

The following problems are to be solved using Eqs. (10.2–10.7).

10.41 Determine V and M as functions of x.

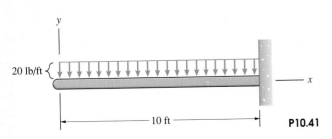

P10.41

10.42 (a) Determine V and M as functions of x.
(b) Use the equilibrium equations to determine the reactions at the built-in support, and use the results of (a) to check your answers.

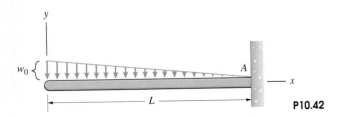

P10.42

10.43 Determine V and M as functions of x.

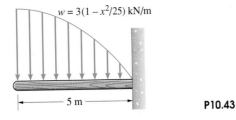

$w = 3(1 - x^2/25)$ kN/m

P10.43

10.44 Determine V and M as functions of x.

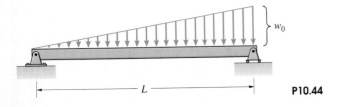

P10.44

10.45 Determine V and M as functions of x.

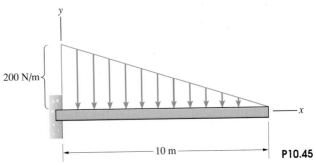

P10.45

10.46 Determine V and M as functions of x for $0 < x < 3$ ft.

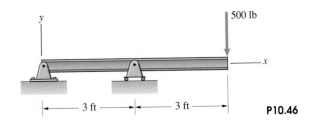

P10.46

10.47 For the beam in Problem 10.46, determine V and M as functions of x for $3 < x < 6$ ft.

10.48 Draw the shear force and bending moment diagrams.

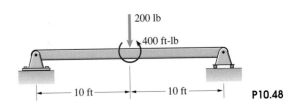

P10.48

10.49 Determine V and M as functions of x.

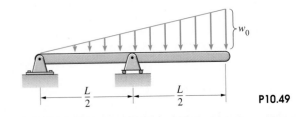

P10.49

Cables

Because of their unique combination of strength, lightness, and flexibility, ropes and cables are often used to support loads and transmit forces in structures, machines, and vehicles. The great suspension bridges are supported by enormous steel cables. Architectural engineers use cables to create aesthetic structures with open interior spaces (Fig. 10.15). In the following sections we determine the tensions in cables subjected to distributed and discrete loads.

Figure 10.15

The use of cables to suspend the roof of this sports stadium provides spectators with a view unencumbered by supporting columns.

10.4 *Loads Distributed Uniformly Along Straight Lines*

The main cable of a suspension bridge is the classic example of a cable subjected to a load uniformly distributed along a straight line (Fig. 10.16). The weight of the bridge is (approximately) uniformly distributed horizontally. The load, transmitted to the main cable by the large number of vertical cables, can be modeled as a distributed load. In this section we determine the shape and the variation in the tension of a cable loaded in this way.

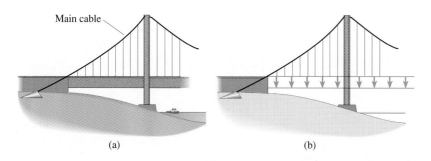

(a)   (b)

Figure 10.16

(a) Main cable of a suspension bridge.
(b) The load is distributed horizontally.

Consider a suspended cable subjected to a load distributed uniformly along a horizontal line (Fig. 10.17a). We neglect the weight of the cable. The origin of the coordinate system is located at the cable's lowest point. Let the function $y(x)$ be the curve described by the cable in the x-y plane. Our objective is to determine the curve $y(x)$ and the tension in the cable.

Shape of the Cable

We obtain a free-body diagram by cutting the cable at its lowest point and at an arbitrary position x (Fig. 10.17b). The term T_0 is the tension in the cable at its lowest point, and T is the tension at x. The downward force exerted by the distributed load is wx. From this free-body diagram we obtain the equilibrium equations

$$T \cos \theta = T_0,$$

$$T \sin \theta = wx. \qquad (10.8)$$

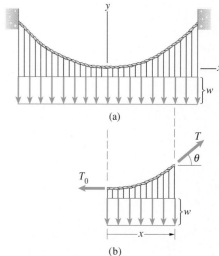

Figure 10.17

(a) A cable subjected to a load uniformly distributed along a horizontal line.
(b) Free-body diagram of the cable between $x = 0$ and an arbitrary position x.

We eliminate the tension T by dividing the second equation by the first one, obtaining

$$\tan \theta = \frac{w}{T_0} x = ax,$$

where we define the constant $a = w/T_0$. The slope of the cable at x is $dy/dx = \tan \theta$, so we obtain a differential equation governing the curve described by the cable:

$$\frac{dy}{dx} = ax. \tag{10.9}$$

We have chosen the coordinate system so that $y = 0$ at $x = 0$. Integrating Eq. (10.9),

$$\int_0^y dy = \int_0^x ax \, dx,$$

we find that the curve described by the cable is the parabola

$$\boxed{y = \frac{1}{2} ax^2.} \tag{10.10}$$

Tension of the Cable

To determine the distribution of the tension in the cable, we square both sides of Eqs. (10.8) and then sum them, obtaining

$$\boxed{T = T_0 \sqrt{1 + a^2 x^2}.} \tag{10.11}$$

The tension is a minimum at the lowest point of the cable and increases monotonically with distance from the lowest point.

Length of the Cable

In some applications it is useful to have an expression for the length of the cable in terms of x. We can write the relation $ds^2 = dx^2 + dy^2$, where ds is an element of length of the cable (Fig. 10.18), in the form

$$ds = \sqrt{1 + \left(\frac{dy}{dx}\right)^2} \, dx.$$

Substituting Eq. (10.9) into this expression and integrating, we obtain an equation for the length s of the cable in the horizontal interval from 0 to x:

$$\boxed{s = \frac{1}{2}\left\{ x\sqrt{1 + a^2 x^2} + \frac{1}{a} \ln\left[ax + \sqrt{1 + a^2 x^2} \right]\right\}.} \tag{10.12}$$

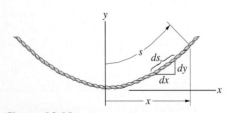

Figure 10.18

The length s of the cable in the horizontal interval from 0 to x.

Example 10.6

The horizontal distance between the supporting towers of the Golden Gate Bridge in San Francisco, California, is 1280 m (Fig. 10.19). The tops of the towers are 160 m above the lowest point of the main supporting cables. Obtain the equation for the curve described by the cables.

Figure 10.19

STRATEGY

We know the coordinates of the cables' attachment points relative to their lowest points. By substituting the coordinates into Eq. (10.10), we can determine a. Once a is known, Eq. (10.10) describes the shapes of the cables.

SOLUTION

The coordinates of the top of the right supporting tower relative to the lowest point of the support cables are $x_R = 640$ m, $y_R = 160$ m (Fig. a). By substituting these values into Eq. (10.10),

$$160 = \frac{1}{2}a(640)^2,$$

we obtain

$$a = 7.81 \times 10^{-4}\,\text{m}^{-1}.$$

The curve described by the supporting cables is

$$y = \frac{1}{2}ax^2 = (3.91 \times 10^{-4})\,x^2.$$

Fig. (a) compares this parabola with a photograph of the supporting cables.

$$y = (3.91 \times 10^{-4})x^2\,\text{m}$$

(a) The theoretical curve superimposed on a photograph of the supporting cable.

Example 10.7

The cable in Fig. 10.20 supports a distributed load of 100 lb/ft. What is the maximum tension in the cable?

STRATEGY

We are given the vertical coordinate of each attachment point, but we are told only the total horizontal span. However, the coordinates of each attachment point relative to a coordinate system with its origin at the lowest point of the cable must satisfy Eq. (10.10). This permits us to determine the horizontal coordinates of the attachment points. Once we know them, we can use Eq. (10.10) to determine $a = w/T_0$, which tells us the tension at the lowest point, and then use Eq. (10.11) to obtain the maximum tension.

SOLUTION

We introduce a coordinate system with its origin at the lowest point of the cable, denoting the coordinates of the left and right attachment points by (x_L, y_L) and (x_R, y_R), respectively (Fig. a). Equation (10.10) must be satisfied for both of these points:

$$y_L = 40 \text{ ft} = \frac{1}{2} a x_L^2,$$

$$y_R = 20 \text{ ft} = \frac{1}{2} a x_R^2. \tag{10.13}$$

We don't know a, but we can eliminate it by dividing the first equation by the second one, obtaining

$$\frac{x_L^2}{x_R^2} = 2.$$

We also know that

$$x_R - x_L = 40 \text{ ft}.$$

(The reason for the minus sign is that x_L is negative.) We therefore have two equations we can solve for x_L and x_R; the results are $x_L = -23.4$ ft and $x_R = 16.6$ ft.

We can now use either of Eqs. (10.13) to determine a. We obtain $a = 0.146 \text{ ft}^{-1}$, so the tension T_0 at the lowest point of the cable is

$$T_0 = \frac{w}{a} = \frac{100}{0.146} = 686 \text{ lb}.$$

From Eq. (10.11), we know that the maximum tension in the cable occurs at the maximum horizontal distance from its lowest point, which in this example is the left attachment point. The maximum tension is therefore

$$T_{max} = T_0 \sqrt{1 + a^2 x_L^2} = 686 \sqrt{1 + (0.146)^2 (-23.4)^2} = 2440 \text{ lb}.$$

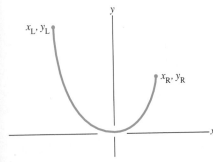

Figure 10.20

(a) A coordinate system with its origin at the lowest point and the coordinates of the left and right attachment points.

40 ft

20 ft

100 lb/ft

40 ft

10.5 *Loads Distributed Uniformly Along Cables*

A cable's own weight subjects it to a load that is distributed uniformly along its length. If a cable is subjected to equal, parallel forces spaced uniformly along its length, the load on the cable can often be modeled as a load distributed uniformly along its length. In this section we show how to determine both the cable's resulting shape and the variation in its tension.

Suppose that a cable is acted on by a distributed load that subjects each element ds of its length to a force $w\,ds$, where w is constant. In Fig. 10.21 we show the free-body diagram obtained by cutting the cable at its lowest point and at a point a distance s along its length. The terms T_0 and T are the tensions at the lowest point and at s, respectively. The distributed load exerts a downward force ws. The origin of the coordinate system is located at the lowest point of the cable. Let the function $y(x)$ be the curve described by the cable in the x-y plane. Our objective is to determine $y(x)$ and the tension T.

Shape of the Cable

From the free-body diagram in Fig. 10.21 we obtain the equilibrium equations

$$T \sin \theta = ws, \tag{10.14}$$

$$T \cos \theta = T_0. \tag{10.15}$$

Dividing the first equation by the second one, we obtain

$$\tan \theta = \frac{w}{T_0} s = as, \tag{10.16}$$

where we define the constant $a = w/T_0$. The slope of the cable $dy/dx = \tan \theta$, so

$$\frac{dy}{dx} = as.$$

The derivative of this equation with respect to x is

$$\frac{d}{dx}\left(\frac{dy}{dx}\right) = a\frac{ds}{dx}. \tag{10.17}$$

By using the relation

$$ds^2 = dx^2 + dy^2,$$

we can write the derivative of s with respect to x as

$$\frac{ds}{dx} = \sqrt{1 + \left(\frac{dy}{dx}\right)^2} = \sqrt{1 + \sigma^2}, \tag{10.18}$$

where we define σ to be the slope:

$$\sigma = \frac{dy}{dx} = \tan \theta.$$

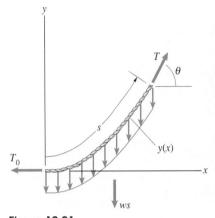

Figure 10.21

A cable subjected to a load distributed uniformly along its length.

With Eq. (10.18), we can write Eq. (10.17) as

$$\frac{d\sigma}{\sqrt{1 + \sigma^2}} = a \, dx.$$

The slope $\sigma = 0$ at $x = 0$. Integrating this equation,

$$\int_0^\sigma \frac{d\sigma}{\sqrt{1 + \sigma^2}} = \int_0^x a \, dx,$$

we determine the slope as a function of x:

$$\sigma = \frac{dy}{dx} = \frac{1}{2}(e^{ax} - e^{-ax}) = \sinh ax. \tag{10.19}$$

Then by integrating this equation with respect to x, we obtain the curve described by the cable, which is called a **catenary**:

$$y = \frac{1}{2a}(e^{ax} + e^{-ax} - 2) = \frac{1}{a}(\cosh ax - 1). \tag{10.20}$$

Tension of the Cable

Using Eq. (10.15) and the relation $dx = \cos \theta \, ds$, we obtain

$$T = \frac{T_0}{\cos \theta} = T_0 \frac{ds}{dx}.$$

Substituting Eq. (10.18) into this expression and using Eq. (10.19) yields the tension in the cable as a function of x:

$$T = T_0 \sqrt{1 + \frac{1}{4}(e^{ax} - e^{-ax})^2} = T_0 \cosh ax. \tag{10.21}$$

Length of the Cable

From Eq. (10.16), the length s is

$$s = \frac{1}{a} \tan \theta = \frac{\sigma}{a}.$$

Substituting Eq. (10.19) into this equation, we obtain an expression for the length s of the cable in the horizontal interval from its lowest point to x:

$$s = \frac{1}{2a}(e^{ax} - e^{-ax}) = \frac{\sinh ax}{a}. \tag{10.22}$$

Example 10.8

The mass per unit length of the cable in Fig. 10.22 is 1 kg/m. The tension at its lowest point is 50 N. Determine the distance h and the maximum tension in the cable.

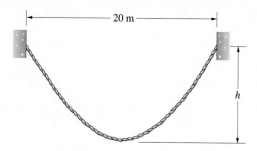

Figure 10.22

STRATEGY

The cable is subjected to a load $w = (9.81 \text{ m/s}^2)(1 \text{ kg/m}) = 9.81 \text{ N/m}$ distributed uniformly along its length. Since we know w and T_0, we can determine $a = w/T_0$. Then we can determine h from Eq. (10.20). Since the maximum tension occurs at the greatest distance from the lowest point of the cable, we can determine it by letting $x = 10$ m in Eq. (10.21).

SOLUTION

The parameter a is

$$a = \frac{w}{T_0} = \frac{9.81}{50} = 0.196 \text{ m}^{-1}.$$

In terms of a coordinate system with its origin at the lowest point of the cable (Fig. a), the coordinates of the right attachment point are $x = 10$ m, $y = h$. From Eq. (10.20),

$$h = \frac{1}{a}(\cosh ax - 1) = \frac{1}{0.196}\{\cosh[(0.196)(10)] - 1\} = 13.4 \text{ m}.$$

From Eq. (10.21), the maximum tension is

$$T_{\max} = T_0 \sqrt{1 + \sinh^2 ax} = 50 \sqrt{1 + \sinh^2[(0.196)(10)]} = 181 \text{ N}.$$

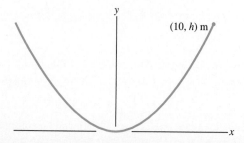

(a) A coordinate system with its origin at the lowest point of the cable.

Problems

10.50 The cable supports a uniformly distributed load $w = 1$ kN/m.
(a) What is the maximum tension in the cable?
(b) What is the length of the cable?
  Strategy: You know the coordinates of the attachment points of the cable relative to its lowest point, so you can use Eq. (10.10) to determine the coefficient a and then use $a = w/T_0$ to determine the tension at the lowest point.

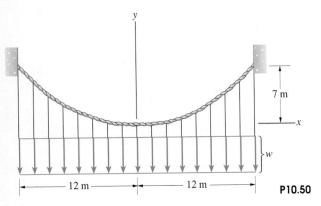

P10.50

10.51 The cable in Problem 10.50 will safely support a tension of 40 kN. On the basis of this criterion, what is the largest value of the distributed load w?

10.52 Engineers use a cable to suspend a pipeline above a river. The towers are 120 ft apart. The lowest point of the cable is 10 ft below the tops of the towers. The total weight of the suspended pipe is 6000 lb.
(a) Determine the maximum tension in the suspending cable.
(b) Determine the length of the suspending cable.

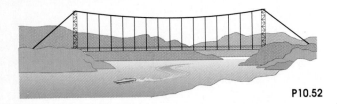

P10.52

10.53 In Problem 10.52, let the lowest point of the cable be a distance h below the tops of the towers.
(a) If the cable supporting the pipeline will safely support a tension of 15 kip, what is the minimum safe value of h?
(b) If h has the value determined in (a), what is the length of the cable?

10.54 The cable supports a uniformly distributed load $w = 100$ lb/ft. The lowest point of the cable is 20 ft below the attachment points C and D. Determine the axial loads in the truss members AC and BC.

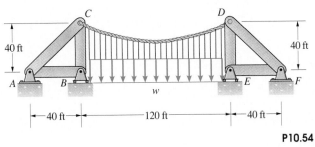

P10.54

10.55 The cable supports a railway bridge between two tunnels. The distributed load is $w = 1$ MN/m, and $h = 40$ m.
(a) What is the maximum tension in the cable?
(b) What is the length of the cable?

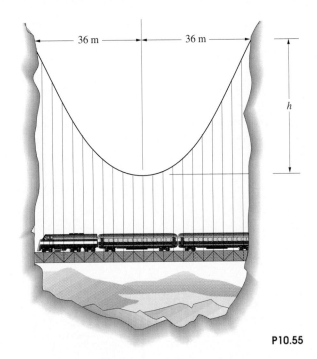

P10.55

10.56 The cable in Problem 10.55 will safely support a tension of 40 MN. What is the shortest cable that can be used, and what is the corresponding value of h?

10.57 An oceanographic research ship tows an instrument package from a cable. Hydrodynamic drag subjects the cable to a uniformly distributed force $w = 2$ lb/ft. The tensions in the cable at 1 and 2 are 800 lb and 1300 lb, respectively. Determine the distance h.

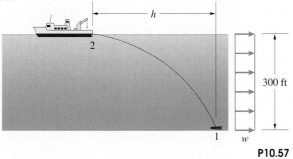

P10.57

10.58 Draw a graph of the shape of the cable in Problem 10.57.

10.59 The rope weighs 1 N/m. The tension at its lowest point is 5 N.
(a) What is the maximum tension in the rope?
(b) What is the length of the rope?
  Strategy: You are given the information needed to evaluate the coefficient $a = w/T_0$. Since the rope is loaded only by its own weight, the tension is given as a function of x by Eq. (10.21), and the length of the rope in the horizontal interval from $x = 0$ to x is given by Eq. (10.22).

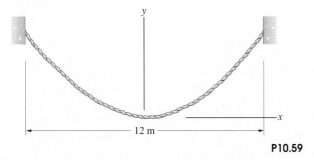

P10.59

10.60 The stationary balloon's tether is horizontal at point O where it is attached to the truck. The mass per unit length of the tether is 0.45 kg/m. The tether exerts a 50-N horizontal force on the truck. The horizontal distance from point O to point A where the tether is attached to the balloon is 20 m. What is the height of point A relative to point O?

P10.60

10.61 In Problem 10.60, determine the magnitudes of the horizontal and vertical components of the force exerted on the balloon at A by the tether.

10.62 The mass per unit length of lines AB and BC is 2 kg/m. The tension at the lowest point of cable AB is 1.8 kN. The two lines exert equal horizontal forces at B.
(a) Determine the sags h_1 and h_2.
(b) Determine the maximum tensions in the two lines.

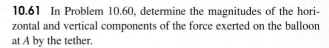
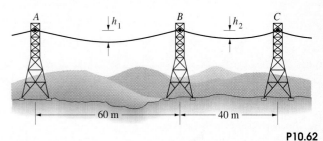

P10.62

10.63 The rope is loaded by 10-lb weights suspended at 1-ft intervals along its length. Neglect the weight of the rope. If the tension at the lowest point of the rope is 45 lb, determine h, the maximum tension in the rope, and the length of the rope.
  Strategy: Obtain an approximate answer by modeling the discrete loads exerted on the rope as a uniformly distributed load.

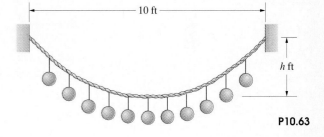

P10.63

10.6 *Discrete Loads*

Your first applications of equilibrium in Chapter 3 involved determining the tensions in cables supporting suspended objects. In this section we consider the case of an arbitrary number N of objects suspended from a cable (Fig. 10.23a). We assume that the weight of the cable can be neglected in comparison to the suspended weights and that the cable is sufficiently flexible that we can approximate its shape by a series of straight segments.

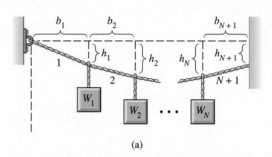

(a)

Figure 10.23

(a) N weights suspended from a cable.
(b) The first free-body diagram.
(c) The second free-body diagram.

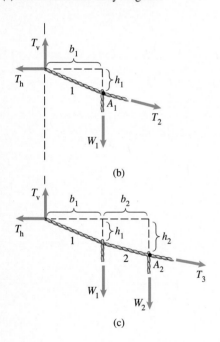

(b)

(c)

Determining the Configuration and Tensions

Suppose that the horizontal distances $b_1, b_2, \ldots, b_{N+1}$ are known and that the vertical distance h_{N+1} specifying the position of the cable's right attachment point is known. We have two objectives: (1) to determine the configuration (shape) of the cable by solving for the vertical distances $h_1, h_2, \ldots, h_N$ specifying the positions of the attachment points of the weights and (2) to determine the tensions in the segments $1, 2, \ldots, N+1$ of the cable.

We begin by drawing a free-body diagram, cutting the cable at its left attachment point and just to the right of the weight W_1 (Fig. 10.23b). We resolve the tension in the cable at the left attachment point into its horizontal and vertical components T_h and T_v. Summing moments about the attachment point A_1, we obtain the equation

$$\Sigma M_{(\text{point } A_1)} = h_1 T_h - b_1 T_v = 0.$$

Our next step is to obtain a free-body diagram by cutting the cable at its left attachment point and just to the right of the weight W_2 (Fig. 10.23c). Summing moments about A_2, we obtain

$$\Sigma M_{(\text{point } A_2)} = h_2 T_h - (b_1 + b_2) T_v + b_2 W_1 = 0.$$

Proceeding in this way, cutting the cable just to the right of each of the N weights, we obtain N equations. We can also draw a free-body diagram by cutting the cable at its left and right attachment points and sum moments about the right attachment point. In this way, we obtain $N + 1$ equations in terms of $N + 2$ unknowns: the two components of the tension T_h and T_v and the vertical positions of the attachment points $h_1, h_2, \ldots, h_N$. If the vertical position of just one attachment point is also specified, we can solve the system of equations for the vertical positions of the other attachment points, determining the configuration of the cable.

Once we know the configuration of the cable and the force T_h, we can determine the tension in any segment by cutting the cable at the left attachment point and within the segment and summing forces in the horizontal direction.

Continuous and Discrete Models

By comparing cables subjected to distributed and discrete loads, we can make some observations about how continuous and discrete systems are modeled in engineering. Consider a cable subjected to a horizontally distributed load w (Fig. 10.24a). The total force exerted on it is wL. Since the cable passes through the point $x = L/2$, $y = L/2$, we find from Eq. (10.10) that $a = 4/L$, so the equation for the curve described by the cable is $y = (2/L)x^2$.

In Fig. 10.24(b), we compare the shape of the cable with the distributed load to that of a cable of negligible weight subjected to three discrete loads $W = wL/3$ with equal horizontal spacing. (We chose the dimensions of the cable with discrete loads so that the heights of the two cables would be equal at their midpoints.) In Fig. 10.24(c), we compare the shape of the cable with the distributed load to that of a cable subjected to five discrete loads $W = wL/5$ with equal horizontal spacing. In Figs. 10.25(a) and (b), we compare the tension in the cable subjected to the distributed load to those in the cables subjected to three and five discrete loads.

The shape and the tension in the cable with a distributed load are approximated by the shapes and tensions in the cables with discrete loads. Although the approximation of the tension is less impressive than the approximation of the shape, it is clear that the former can be improved by increasing the number of discrete loads.

This approach—approximating a continuous distribution by a discrete model—is extremely important in engineering. It is the starting point of the finite difference and finite element methods. The opposite approach—modeling discrete systems by continuous models—is also widely used, for example, when the forces exerted on a bridge by traffic are modeled as a distributed load.

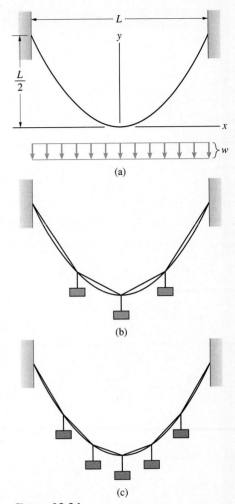

(a)

(b)

(c)

Figure 10.24

(a) A cable subjected to a continuous load.
(b) A cable with three discrete loads.
(c) A cable with five discrete loads.

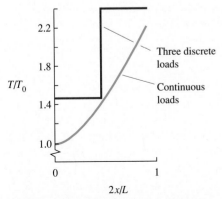

(a) The tension in a cable with a continuous load compared to the cable with three discrete loads.

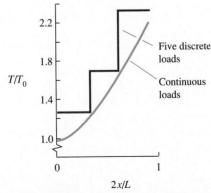

(b) The tension in a cable with a continuous load compared to the cable with five discrete loads.

Figure 10.25

Example 10.9

Two weights $W_1 = 100$ N and $W_2 = 200$ N are suspended from the cable in Fig. 10.26.

(a) Determine the vertical distance h_2.

(b) Determine the tension in cable segment 2.

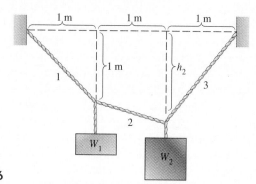

Figure 10.26

SOLUTION

(a) First we cut the cable at the left attachment point and just to the right of W_1 and express the tension at the left attachment point in terms of horizontal and vertical components (Fig. a). Summing moments about A_1, we obtain

$$\Sigma M_{(\text{point } A_1)} = (1)T_h - (1)T_v = 0.$$

Next we cut the cable just to the right of W_2 (Fig. b) and sum moments about A_2:

$$\Sigma M_{(\text{point } A_2)} = h_2 T_h - (2)T_v + (1)W_1 = 0.$$

The last step is to cut the cable at the right attachment point (Fig. c) and sum moments about A_3:

$$\Sigma M_{(\text{point } A_3)} = -(3)T_v + (2)W_1 + (1)W_2 = 0.$$

There are three equations in terms of the unknowns T_h, T_v, and h_2. Solving them, we obtain $T_h = 133$ N, $T_v = 133$ N, and $h_2 = 1.25$ m.

(b) To determine the tension in segment 2, we use the free-body diagram in Fig. (a). The angle between the force T_2 and the horizontal is

$$\arctan\left(\frac{h_2 - 1}{1}\right) = 14.0°.$$

Summing forces in the horizontal direction,

$$T_2 \cos 14.0° - T_h = 0.$$

Solving for T_2, we obtain

$$T_2 = \frac{T_h}{\cos 14.0°} = \frac{133}{\cos 14.0°} = 137 \text{ N}.$$

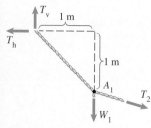

(a) First free-body diagram.

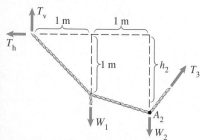

(b) Second free-body diagram.

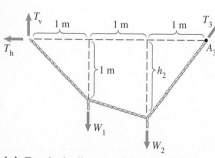

(c) Free-body diagram of the entire cable.

Problems

10.64 In Example 10.9, what are the tensions in cable segments 1 and 3?

10.65 If the weights in Example 10.9 are $W_1 = 240$ N and $W_2 = 400$ N, what is the vertical distance h_2 and what is the tension in cable segment 3?

10.66 Two weights, $W_1 = W_2 = 50$ lb, are suspended from a cable. The vertical distance $h_1 = 4$ ft.
(a) Determine the vertical distance h_2.
(b) What is the maximum tension in the cable?

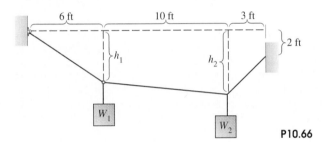

P10.66

10.67 In Problem 10.66, $W_1 = 50$ lb, $W_2 = 100$ lb, and the vertical distance $h_1 = 4$ ft.
(a) Determine the vertical distance h_2.
(b) What is the maximum tension in the cable?

10.68 In Problem 10.66, $W_1 = 50$ lb, $h_1 = 4$ ft, and $h_2 = 8$ ft.
(a) Determine the weight W_2.
(b) What is the maximum tension in the cable?

10.69 Three weights are suspended from a cable. Their weight $W = 10$ N.
(a) Determine the vertical distances h_1 and h_3, and sketch the configuration of the cable.
(b) What are the tensions in cable segments 1 and 2?

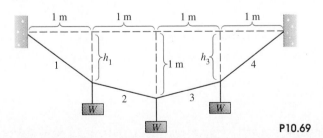

P10.69

10.70 Three weights are suspended from a cable. The weight $W = 100$ N, and the vertical distance $h_1 = 400$ mm. Determine the vertical distances h_2 and h_3.

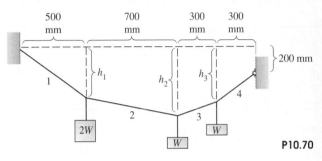

P10.70

10.71 In Problem 10.70, what is the maximum tension in the cable, and in what segment does it occur?

10.72 The cable in Problem 10.70 will safely support a tension of 15 kN. Is $W = 2$ kN a safe load if $h_1 = 200$ mm?

Computational Mechanics

The following example and problems are designed to be solved using a programmable calculator or computer.

Example 10.10

As the first step in constructing a suspended pedestrian bridge, a cable is suspended across the span from attachment points of equal height (Fig. 10.27). The cable weighs 5 lb/ft and is 42 ft long. Determine the maximum tension in the cable and the vertical distance from the attachment points to the cable's lowest point.

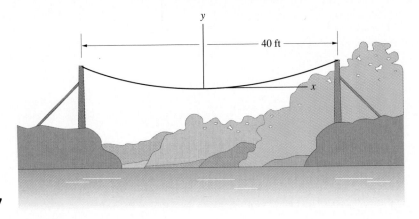

Figure 10.27

STRATEGY

Equation (10.22) gives the length s of the cable as a function of the horizontal distance x from the cable's lowest point and the parameter $a = w/T_0$. The term w is the weight per unit length, and T_0 is the tension in the cable at its lowest point. We know that the half-span of the cable is 20 ft, so we can draw a graph of s as a function of a and estimate the value of a for which $s = 21$ ft. Then we can determine the maximum tension from Eq. (10.21) and the vertical distance to the cable's lowest point from Eq. (10.20).

SOLUTION

Setting $x = 20$ ft in Eq. (10.22),

$$s = \frac{\sinh 20a}{a},$$

we compute s as a function of a (Fig. 10.28). The length $s = 21$ ft when the parameter a is approximately 0.027. By examining the computed results near $a = 0.027$,

a (ft^{-1})	s (ft)
0.0269	20.9789
0.0270	20.9863
0.0271	20.9937
0.0272	21.0012
0.0273	21.0086
0.0274	21.0162
0.0275	21.0237

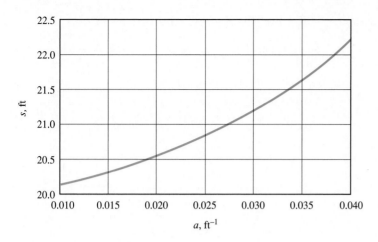

Figure 10.28

Graph of the length s as a function of the parameter a.

we see that s is approximately 21 ft when $a = 0.0272$ ft^{-1}. Therefore the tension at the cable's lowest point is

$$T_0 = \frac{w}{a} = \frac{5}{0.0272} = 184 \text{ lb},$$

and the maximum tension is

$$T_{max} = T_0 \cosh ax = 184 \cosh[(0.0272)(20)] = 212 \text{ lb.}$$

From Eq. (10.20), the vertical distance from the cable's lowest point to the attachment points is

$$y_{max} = \frac{1}{a}(\cosh ax - 1) = \frac{1}{0.0272}\{\cosh[(0.0272)(20)] - 1\} = 5.58 \text{ ft.}$$

Problems

10.73 The cable is 10 m long and is subjected to a distributed load $w = 200$ N/m. What is the maximum tension in the cable?

Strategy: By setting $x = 4$ m in Eq. (10.12) and obtaining a graph of s as a function of the parameter a, you can determine the value of a corresponding to $s = 5$ m and so determine $T_0 = w/a$. Then you can determine the maximum tension from Eq. (10.11).

10.74 The rope weighs 1 N/m and is 16 m in length.
(a) What is the maximum tension?
(b) What is the vertical distance from the attachment points to the lowest point of the rope?

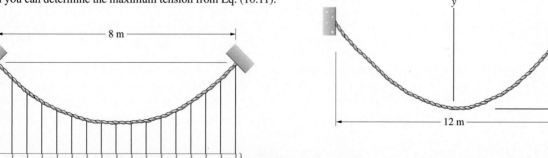

P10.74

P10.73

10.75 A chain weighs 20 lb and is 20 ft long. It is suspended from two points of equal height that are 10 ft apart.
(a) Determine the maximum tension in the chain.
(b) Draw a sketch of the shape of the chain.

10.76 An engineer wants to suspend high-voltage power lines between poles 200 m apart. Each line has a mass of 2 kg/m.
(a) If the engineer wants to subject the lines to a tension no greater than 10 kN, what should be the maximum allowable sag between poles? That is, what is the largest allowable vertical distance between the attachment points and the lowest point of the line?
(b) What is the length of each line?

10.77 The mass per unit length of lines AB and BC is 2 kg/m. The length of line AB is 62 m. The two lines exert equal horizontal forces at B.
(a) Determine the sags h_1 and h_2.
(b) Determine the maximum tensions in the two lines.

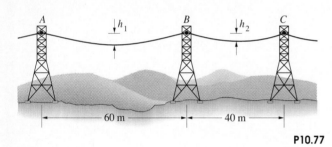

P10.77

10.78 The mass per unit length of the lines AB and BC in Problem 10.77 is 2 kg/m. The sag $h_1 = 4.5$ m, but the length of line AB is unknown. The two lines exert equal horizontal forces at B.
(a) Determine the sag h_2.
(b) Determine the maximum tensions in the two lines.

10.79 Two 30-ft cables A and B are suspended from points of equal height that are 20 ft apart. Cable A is subjected to a 200-lb load uniformly distributed horizontally. Cable B is subjected to a 200-lb load distributed uniformly along its length. What are the maximum tensions in the two cables?

10.80 Draw a graph of the two cables in Problem 10.79, comparing their shapes.

10.81 The weights $W_1 = 100$ N and $W_2 = 200$ N, and the total length L of the three segments of the rope is 5 m.
(a) What are h_1 and h_2?
(b) What is the maximum tension in the rope?
    Strategy: If you choose a value of h_1, you can determine h_2 and then L. By obtaining a graph of L as a function of h_1, you can determine the value of h_1 that corresponds to $L = 5$ m.

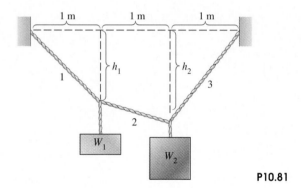

P10.81

Liquids and Gases

Wind forces on buildings and aerodynamic forces on cars and airplanes are examples of forces that are distributed over *areas*. The downward force exerted on the bed of a dump truck by a load of gravel is distributed over the area of the bed. The upward force that supports a building is distributed over the area of its foundation. Loads distributed over the roofs of buildings by accumulated snow can be hazardous. Many forces of concern in engineering are distributed over areas. In this section we analyze the most familiar example, the force exerted by the pressure of a gas or liquid.

10.7 *Pressure and the Center of Pressure*

A surface immersed in a gas or liquid is subjected to forces exerted by molecular impacts. If the gas or liquid is stationary, the load can be described by a function p, the **pressure**, defined such that the normal force exerted on a differential element dA of the surface is $p\ dA$ (Figs. 10.29a and b). (Notice the parallel between the pressure and a load w distributed along a line, which is defined such that the force on a differential element dx of the line is $w\ dx$.)

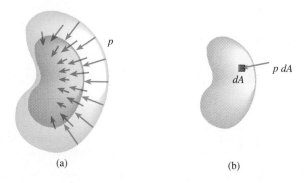

(a) (b)

Figure 10.29

(a) The pressure on an area.
(b) The force on an element dA is $p\ dA$.

The dimensions of p are (force)/(area). In U.S. Customary units, pressure can be expressed in pounds per square foot or pounds per square inch (psi). In SI units, pressure can be expressed in newtons per square meter, which are called pascals (Pa).

In some applications, it is convenient to use the **gage pressure**

$$p_{\mathrm{g}} = p - p_{\mathrm{atm}}, \tag{10.23}$$

where p_{atm} is the pressure of the atmosphere. Atmospheric pressure varies with location and climatic conditions. At sea level, its value is approximately 1×10^5 Pa in SI units and 14.7 psi or 2120 lb/ft^2 in U.S. Customary units.

If the distributed force due to pressure on a surface is represented by an equivalent force, the point at which the line of action of the force intersects the surface is called the **center of pressure**. Let's consider a *plane* area A subjected to a pressure p and introduce a coordinate system such that the area lies in the *x-y* plane (Fig. 10.30a). The normal force on each differential element of area dA is $p\ dA$ (Fig. 10.30b), so the total normal force on A is

$$F = \int_A p\ dA. \tag{10.24}$$

Now we will determine the coordinates (x_p, y_p) of the center of pressure (Fig. 10.30c). Equating the moment of F about the origin to the total moment due to the pressure about the origin,

$$(x_p\mathbf{i} + y_p\mathbf{j}) \times (-F\mathbf{k}) = \int_A (x\mathbf{i} + y\mathbf{j}) \times (-p\ dA\mathbf{k}),$$

and using Eq. (10.24), we obtain

$$x_p = \frac{\int_A xp \, dA}{\int_A p \, dA}, \qquad y_p = \frac{\int_A yp \, dA}{\int_A p \, dA}. \qquad (10.25)$$

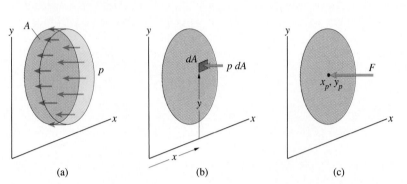

Figure 10.30

(a) A plane area subjected to pressure.
(b) The force on a differential element dA.
(c) The total force acting at the center of pressure.

(a) (b) (c)

These equations determine the position of the center of pressure when the pressure p is known. If the pressure p is uniform, the total normal force is $F = pA$ and Eqs. (10.25) indicate that the center of pressure is the centroid of A.

In Chapter 7 we showed that if you calculate the "area" defined by a load distributed along a line and place the resulting force at its centroid, the force is equivalent to the distributed load. A similar result holds for a pressure distributed on a plane area. The term $p \, dA$ in Eq. (10.24) is equal to a differential element dV of the "volume" between the surface defined by the pressure distribution and the area A (Fig. 10.31a). The total force exerted by the pressure is therefore equal to this "volume":

$$F = \int_V dV = V.$$

Substituting $p \, dA = dV$ into Eqs. (10.25), we obtain

$$x_p = \frac{\int_V x \, dV}{\int_V dV}, \qquad y_p = \frac{\int_V y \, dV}{\int_V dV}.$$

The center of pressure coincides with the x and y coordinates of the centroid of the "volume" (Fig. 10.31b).

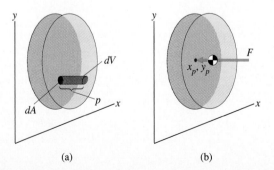

Figure 10.31

(a) The differential element $dV = p \, dA$.
(b) The line of action of F passes through the centroid of V.

(a) (b)

10.8 *Pressure in a Stationary Liquid*

Designers of pressure vessels and piping, ships, dams, and other submerged structures must be concerned with forces and moments exerted by water pressure. If you swim toward the bottom of a swimming pool, you can feel the pressure on your ears increase—the pressure in a liquid at rest increases with depth. We can determine the dependence of pressure on depth by using a simple free-body diagram.

Introducing a coordinate system with its origin at the surface of the liquid and the positive x axis downward (Fig. 10.32a), we draw a free-body diagram of a cylinder of liquid that extends from the surface to a depth x (Fig. 10.32b). The top of the cylinder is subjected to the pressure at the surface, which we call p_0. The sides and bottom of the cylinder are subjected to pressure by the surrounding liquid, which increases from p_0 at the surface to a value p at the depth x. The volume of the cylinder is Ax, where A is its cross-sectional area. Therefore its weight is $W = \gamma Ax$, where γ is the weight density of the liquid. (Recall that the weight and mass densities are related by $\gamma = \rho g$.)

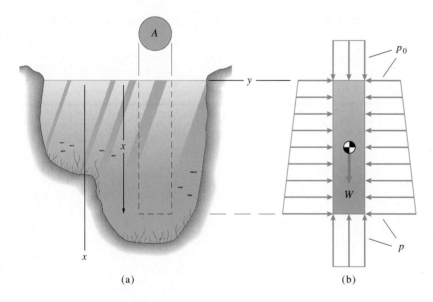

(a)   (b)

Figure 10.32

(a) A cylindrical volume that extends to a depth x in a body of stationary liquid.
(b) Free-body diagram of the cylinder.

Since the liquid is stationary, the cylinder is in equilibrium. From the equilibrium equation

$$\Sigma F_x = p_0 A - pA + \gamma Ax = 0,$$

we obtain a simple expression for the pressure p in the liquid at depth x:

$$\boxed{p = p_0 + \gamma x.} \tag{10.26}$$

Thus the pressure increases linearly with depth, and the derivation we have used illustrates why: The pressure at a given depth literally holds up the liquid above that depth.

If the surface of the liquid is open to the atmosphere, $p_0 = p_{atm}$ and we can write Eq. (10.26) in terms of the gage pressure $p_g = p - p_{atm}$ as

$$p_g = \gamma x. \tag{10.27}$$

In SI units, the mass density of water at sea level conditions is $\rho = 1000$ kg/m^3, so its weight density is approximately $\gamma = \rho g = 9.81$ kN/m^3. In U.S. Customary units, the weight density of water is approximately $\gamma = 62.4$ lb/ft^3.

The following examples involve determining forces and moments due to pressures on submerged objects. You can determine the forces and moments on a plane area A in two ways:

1. Integration—*Integrate Eq. (10.26) or Eq. (10.27).*

2. Volume analogy—*Determine the total force by calculating the "volume" between the surface defined by the pressure distribution and the area A. The center of pressure coincides with the x and y coordinates of the centroid of the volume.*

Example 10.12

An engineer making preliminary design studies for a canal lock needs to determine the total pressure force on a submerged rectangular plate (Fig. 10.33) and the location of the center of pressure. The top of the plate is 6 m below the surface. Atmospheric pressure is $p_{atm} = 1 \times 10^5$ Pa, and the weight density of the water is $\gamma = 9.81$ kN/m^3.

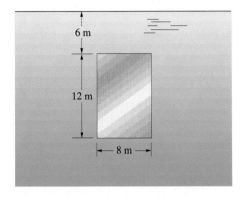

6 m

12 m

8 m

Figure 10.33

STRATEGY

We will determine the pressure force on a differential element of area of the plate in the form of a horizontal strip and integrate to determine the total force and moment exerted by the pressure.

SOLUTION

In terms of a coordinate system with its origin at the surface and the positive x axis downward (Fig. a), the pressure of the water is $p = p_{atm} + \gamma x$. The horizontal strip $dA = 8\ dx$. Therefore the total force exerted on the face of the plate by the pressure is

$$F = \int_A p\ dA = \int_6^{18} (p_{atm} + \gamma x)(8\ dx) = 96p_{atm} + 1150\gamma$$

$$= (96)(1 \times 10^5) + (1150)(9810) = 20.9 \text{ MN}.$$

The moment about the y axis due to the pressure on the plate is

$$M = \int_A xp\ dA = \int_6^{18} x(p_{atm} + \gamma x)(8\ dx) = 262 \text{ MN-m}.$$

The force F acting at the center of pressure (Fig. b) exerts a moment about the y axis equal to M:

$$x_p F = M.$$

Therefore the location of the center of pressure is

$$x_p = \frac{M}{F} = \frac{262 \text{ MN-m}}{20.9 \text{ MN}} = 12.5 \text{ m}.$$

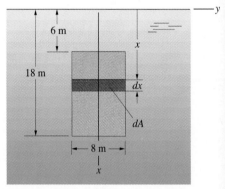

(a) An element of area in the form of a horizontal strip.

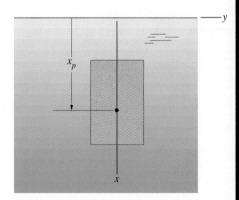

(b) The center of pressure.

Example 10.12

The gate AB in Fig. 10.34 has water of 2-ft depth on the right side. The width of the gate (the dimension into the page) is 3 ft, and its weight is 100 lb. The weight density of the water is $\gamma = 62.4$ lb/ft^3. Determine the reactions on the gate at the supports A and B.

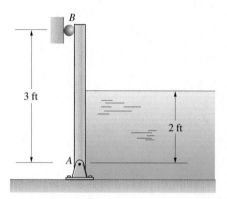

Figure 10.34

STRATEGY

The left face of the gate and the right face above the level of the water are exposed to atmospheric pressure. From Eqs. (10.23) and (10.26), the pressure in the water is the sum of atmospheric pressure and the gage pressure $p_g = \gamma x$, where x is measured downward from the surface of the water. The effects of atmospheric pressure cancel (Fig. a), so we need to consider only the forces and moments exerted on the gate by the gage pressure. We will determine them by integrating and also by calculating the "volume" of the pressure distribution.

SOLUTION

Integration   The face of the gate is shown in Fig. (b). In terms of the differential element of area dA, the force exerted on the gate by the gage pressure is

$$F = \int_A p_g \, dA = \int_0^2 (\gamma x)3 \, dx = 374 \text{ lb,}$$

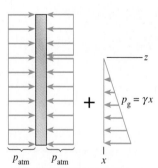

(a) The pressures acting on the faces of the gate.

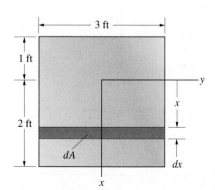

(b) The face of the gate and the differential element dA.

and the moment about the y axis is

$$M = \int_A xp_g \, dA = \int_0^2 x(\gamma x)3 \, dx = 499 \text{ ft-lb.}$$

The position of the center of pressure is

$$x_p = \frac{M}{F} = \frac{499 \text{ ft-lb}}{374 \text{ lb}} = 1.33 \text{ ft.}$$

Volume Analogy   The gage pressure at the bottom of the gate is $p_g = (2 \text{ ft})\gamma$ (Fig. c), so the "volume" of the pressure distribution is

$$F = \frac{1}{2}[2 \text{ ft}][(2 \text{ ft})(62.4 \text{ lb/ft}^3)][3 \text{ ft}] = 374 \text{ lb.}$$

The x coordinate of the centroid of the triangular distribution, which is the center of pressure, is $\frac{2}{3}(2) = 1.33$ ft.

Determining the Reactions   We draw the free-body diagram of the gate in Fig. (d). From the equilibrium equations,

$$\Sigma F_x = A_x + 100 = 0,$$

$$\Sigma F_z = A_z + B - 374 = 0,$$

$$\Sigma M_{(y \text{ axis})} = (1)B - (2)A_z + (1.33)(374) = 0,$$

we obtain $A_x = -100$ lb, $A_z = 291.2$ lb, and $B = 83.2$ lb.

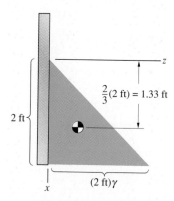

(c) Determining the "volume" of the pressure distribution and its centroid.

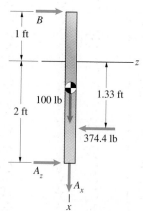

(d) Free-body diagram of the gate.

Example 10.13

The container in Fig. 10.35 is filled with a liquid with weight density γ. Determine the force exerted by the pressure of the liquid on the cylindrical wall AB.

Figure 10.35

STRATEGY

The pressure of the liquid on the cylindrical wall varies with depth (Fig. a). It is the force exerted by this pressure distribution we want to determine. We could determine it by integrating over the cylindrical surface, but we can avoid that by drawing a free-body diagram of the quarter-cylinder of liquid to the right of A.

(a) The pressure of the liquid on the wall AB.

SOLUTION

We draw the free-body diagram of the quarter-cylinder of liquid in Fig. (b). The pressure distribution on the cylindrical surface of the liquid is the same one that acts on the cylindrical wall. If we denote the force exerted on the liquid by this pressure distribution by $\mathbf{F}_p$, the force exerted by the liquid on the cylindrical wall is $-\mathbf{F}_p$.

The other forces parallel to the x-y plane that act on the quarter-cylinder of liquid are its weight, atmospheric pressure at the upper surface, and the pressure distribution of the liquid on the left side. The volume of liquid is $(\frac{1}{4}\pi R^2)b$, so the force exerted on the free-body diagram by the weight of the liquid is $\frac{1}{4}\gamma\pi R^2 b\mathbf{i}$. The force exerted on the upper surface by atmospheric pressure is $Rbp_{\text{atm}}\mathbf{i}$.

We can integrate to determine the force exerted by the pressure on the left side of the free-body diagram. Its magnitude is

$$\int_A p\,dA = \int_0^R (p_{\text{atm}} + \gamma x)b\,dx = Rb\left(p_{\text{atm}} + \frac{1}{2}\gamma R\right).$$

From the equilibrium equation

$$\Sigma \mathbf{F} = \frac{1}{4}\gamma\pi R^2 b\mathbf{i} + Rbp_{\text{atm}}\mathbf{i} + Rb\left(p_{\text{atm}} + \frac{1}{2}\gamma R\right)\mathbf{j} + \mathbf{F}_p = \mathbf{0},$$

we obtain the force exerted on the wall AB by the pressure of the liquid:

$$-\mathbf{F}_p = Rb\left(p_{\text{atm}} + \frac{\pi}{4}\gamma R\right)\mathbf{i} + Rb\left(p_{\text{atm}} + \frac{1}{2}\gamma R\right)\mathbf{j}.$$

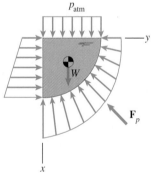

(b) Free-body diagram of the liquid to the right of A.

Problems

10.82   A deep submersible research vehicle operates near the bottom of the Indian Ocean at a depth of 1 mi. The weight density of the water is approximately $\gamma = 64$ lb/ft^3, and atmospheric pressure is $p_{atm} = 2120$ lb/ft^2. Determine the pressure on the vehicle's surface (a) in pounds per square inch (psi); (b) in pascals (Pa).

10.83   An engineer planning a water system for a new community estimates that at maximum expected usage, the pressure drop between the central system and the farthest planned fire hydrant will be 25 psi. Firefighting personnel indicate that a gage pressure of 40 psi at the fire hydrant is required. The weight density of the water is $\gamma = 62.4$ lb/ft^3. How tall would a water tower at the central system have to be to provide the needed pressure?

10.84   A cube of material is suspended below the surface of a liquid of weight density γ. By calculating the forces exerted on the faces of the cube by pressure, show that their sum is an upward force of magnitude γb^3.

P10.84

10.85   The area shown is subjected to a *uniform* pressure $p_{atm} = 1 \times 10^5$ Pa.
(a) What is the total force exerted on the area by the pressure?
(b) What is the moment about the y axis due to the pressure on the area?

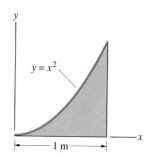

$y = x^2$

1 m

P10.85

10.86   Determine the coordinates of the center of pressure in Problem 10.85.

10.87   The area shown is subjected to a *uniform* pressure $p_{atm} = 14.7$ psi.
(a) What is the total force exerted on the area by the pressure?
(b) What is the moment about the y axis due to the pressure on the area?

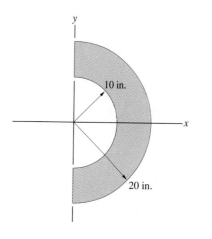

10 in.

20 in.

P10.87

10.88   Determine the coordinates of the center of pressure in Problem 10.87.

10.89   The top of the rectangular plate is 2 ft below the surface of a lake. Atmospheric pressure is $p_{atm} = 2120$ lb/ft^2, and the weight density of the water is $\gamma = 62.4$ lb/ft^3.
(a) What is the maximum pressure exerted on the plate by the water?
(b) Determine the force exerted on a face of the plate by the pressure of the water.

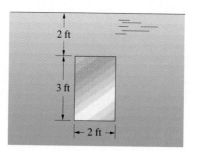

2 ft

3 ft

2 ft

P10.89

10.90   How far down from the top of the plate in Problem 10.89 is the center of pressure located?

10.91 The width of the dam (the dimension into the page) is 100 m. The mass density of the water is $\rho = 1000$ kg/m³. Determine the force exerted on the dam by the gage pressure of the water (a) by integration; (b) by calculating the "volume" of the pressure distribution.

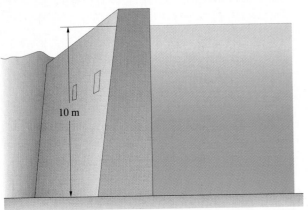

10 m

P10.91

10.92 In Problem 10.91, how far down from the surface of the water is the center of pressure due to the gage pressure of the water on the dam?

10.93 The width of the gate (the dimension into the page) is 4 ft, and its weight is 200 lb. The weight density of the water is $\gamma = 62.4$ lb/ft³, and atmospheric pressure is $p_{atm} = 2120$ lb/ft². Determine the reactions on the gate at A.

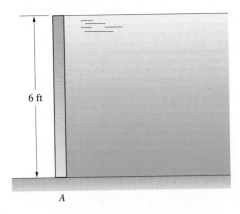

6 ft

A

P10.93

10.94 The homogeneous gate weighs 100 lb, and its width (the dimension into the page) is 3 ft. The weight density of the water is $\gamma = 62.4$ lb/ft³, and atmospheric pressure is $p_{atm} = 2120$ lb/ft². Determine the reactions at A and B.

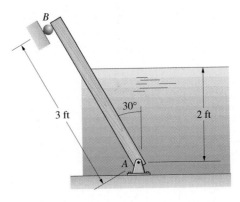

B

3 ft   30°   2 ft

A

P10.94

10.95 The gate has water of depth $d = 1$ m on one side. The width of the gate (the dimension into the page) is 2 m, and its weight is 800 N. The mass density of the water is $\rho = 1000$ kg/m³, and atmospheric pressure is $p_{atm} = 1 \times 10^5$ Pa. Determine the reactions on the gate at A and B. (The support at B exerts only a horizontal reaction on the gate.)

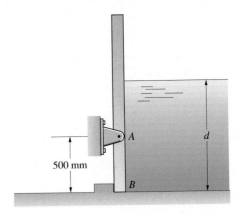

A

d

500 mm

B

P10.95

10.96 When the water depth d in Problem 10.95 exceeds a certain value, the gate will rotate, releasing the water. What is the value?

10.97 The dam has water of depth 4 ft on one side. The width of the dam (the dimension into the page) is 8 ft. The weight density of the water is $\gamma = 62.4$ lb/ft^3, and atmospheric pressure $p_{atm} = 2120$ lb/ft^2. If you neglect the weight of the dam, what are the reactions at A and B?

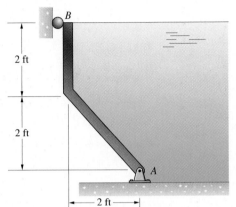

P10.97

10.98 A spherical tank of 400-mm inner radius is full of water ($\rho = 1000$ kg/m^3). The pressure of the water at the top of the tank is 4×10^5 Pa.
(a) What is the pressure of the water at the bottom of the tank?
(b) What is the total force exerted on the inner surface of the tank by the pressure of the water?

Strategy: For (b), draw a free-body diagram of the sphere of water in the tank.

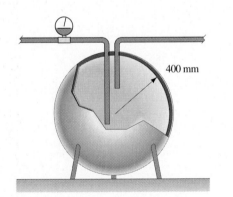

P10.98

10.99 Consider a plane, vertical area A below the surface of a liquid. Let p_0 be the pressure at the surface.
(a) Show that the force exerted by pressure on the area is $F = \bar{p}A$, where $\bar{p} = p_0 + \gamma\bar{x}$ is the pressure of the liquid at the centroid of the area.
(b) Show that the x coordinate of the center of pressure is

$$x_p = \bar{x} + \frac{\gamma I_{y'}}{\bar{p}A},$$

where $I_{y'}$ is the moment of inertia of the area about the y' axis through its centroid.

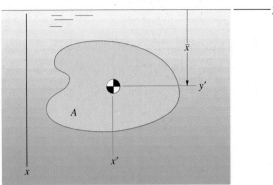

P10.99

10.100 A circular plate of 1-m radius is below the surface of a stationary pool of water. Atmospheric pressure is $p_{atm} = 1 \times 10^5$ Pa, and the mass density of the water is $\rho = 1000$ kg/m^3. Determine (a) the force exerted on a face of the plate by the pressure of the water; (b) the x coordinate of the center of pressure. (See Problem 10.99.)

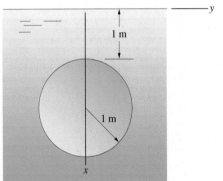

P10.100

10.101 A tank consists of a cylinder with hemispherical ends. It is filled with water ($\gamma = 62.4$ lb/ft^3), and the pressure of the water at the top of the tank is 3000 lb/ft^2. Determine the magnitude of the force exerted by the pressure of the water on one of the hemispherical ends of the tank.

Strategy: Draw a free-body diagram of the water to the right of the dashed line in the figure. See Example 10.13.

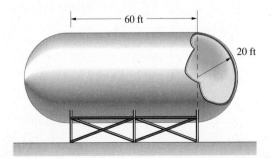

P10.101

10.102   An object of volume V and weight W is suspended below the surface of a stationary liquid of weight density γ (Fig. a). Show that the tension in the cord is $W - V\gamma$. In other words, show that the pressure distribution on the surface of the object exerts an upward force equal to the product of the object's volume and the weight density of the water. This result is due to Archimedes (287–212 B.C.).

Strategy: Draw the free-body diagram of a volume of liquid that has the same shape and position as the object (Fig. b).

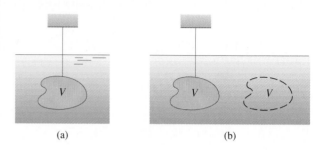

(a)   (b)

Chapter Summary

Beams

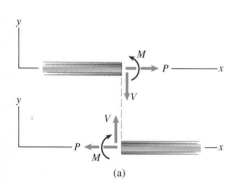

(a)

The internal forces and moment in a beam are expressed as the **axial force** P, **shear force** V, and **bending moment** M. Their positive directions are defined in Fig. (a).

By cutting a beam at an arbitrary position x, the axial load P, shear force V, and bending moment M can be determined as functions of x. Depending on the loading and supports of the beam, it may be necessary to draw several free-body diagrams to determine the distributions for the entire beam. The graphs of V and M as functions of x are the **shear force and bending moment diagrams**.

The distributed load, shear force, and bending moment in a portion of a beam subjected only to a distributed load satisfy the relations

$$\frac{dV}{dx} = -w, \qquad \textbf{Eq. (10.2)}$$

$$\frac{dM}{dx} = V. \qquad \textbf{Eq. (10.3)}$$

For segments of a beam that are unloaded or are subjected to a distributed load, these equations can be integrated to determine V and M as functions of x. To obtain the complete shear force and bending moment diagrams, forces and couples must also be accounted for.

Cables

Loads Distributed Uniformly Along a Straight Line   If a suspended cable is subjected to a horizontally distributed load w (Fig. b), the curve described by the cable is the parabola

$$y = \frac{1}{2}ax^2, \qquad \textbf{Eq. (10.10)}$$

where $a = w/T_0$ and T_0 is the tension in the cable at $x = 0$. The tension in the cable at a position x is

$$T = T_0\sqrt{1 + a^2 x^2}, \qquad \textbf{Eq. (10.11)}$$

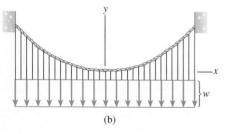

(b)

and the length of the cable in the horizontal interval from 0 to x is

$$s = \frac{1}{2}\left\{ x\sqrt{1 + a^2x^2} + \frac{1}{a}\ln\left[ax + \sqrt{1 + a^2x^2} \right] \right\}. \qquad \textbf{Eq. (10.12)}$$

Loads Distributed Uniformly Along a Cable   If a suspended cable is subjected to a load w distributed along its length, the curve described by the cable is the catenary

$$y = \frac{1}{2a}(e^{ax} + e^{-ax} - 2) = \frac{1}{a}(\cosh ax - 1), \qquad \textbf{Eq. (10.20)}$$

where $a = w/T_0$ and T_0 is the tension in the cable at $x = 0$. The tension in the cable at a position x is

$$T = T_0\sqrt{1 + \frac{1}{4}(e^{ax} - e^{-ax})^2} = T_0\cosh ax, \qquad \textbf{Eq. (10.21)}$$

and the length of the cable in the horizontal interval from 0 to x is

$$s = \frac{1}{2a}(e^{ax} - e^{-ax}) = \frac{\sinh ax}{a}. \qquad \textbf{Eq. (10.22)}$$

Discrete Loads   If N known weights are suspended from a cable and the positions of the attachment points of the cable, the horizontal positions of the attachment points of the weights, and the vertical position of the attachment point of one of the weights are known, the configuration of the cable and the tension in each of its segments can be determined.

Liquids and Gases

The pressure p on a surface is defined so that the normal force exerted on an element dA of the surface is $p\,dA$. The total normal force exerted by pressure on a *plane area A* is

$$F = \int_A p\,dA. \qquad \textbf{Eq. (10.24)}$$

The **center of pressure** is the point on A at which F must be placed to be equivalent to the pressure on A. The coordinates of the center of pressure are

$$x_p = \frac{\int_A xp\,dA}{A}, \qquad y_p = \frac{\int_A yp\,dA}{A}. \qquad \textbf{Eq. (10.25)}$$

The pressure in a stationary liquid is

$$p = p_0 + \gamma x, \qquad \textbf{Eq. (10.26)}$$

where p_0 is the pressure at the surface, γ is the weight density of the liquid, and x is the depth. If the surface of the liquid is open to the atmosphere, $p_0 = p_{atm}$, the atmospheric pressure.

Review Problems

10.103 Determine the internal forces and moment at B (a) if $x = 250$ mm; (b) if $x = 750$ mm.

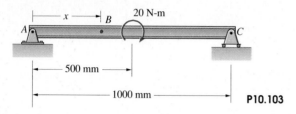

P10.103

10.104 Determine the internal forces and moment (a) at B; (b) at C.

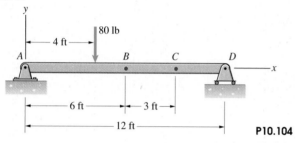

P10.104

10.105 (a) Determine the maximum bending moment in the beam and the value of x where it occurs.
(b) Show that the equations for V and M as functions of x satisfy the equation $V = dM/dx$.

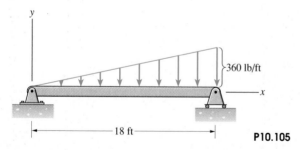

P10.105

10.106 Draw the shear and bending moment diagrams for the beam in Problem 10.105.

10.107 Determine the shear force and bending moment diagrams for the beam.

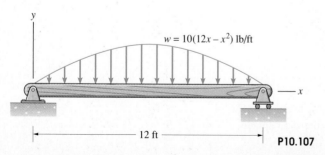

P10.107

10.108 Draw the shear force and bending moment diagrams for beam ABC.

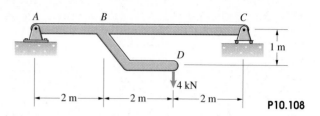

P10.108

10.109 Draw the shear force and bending moment diagrams for beam ABC.

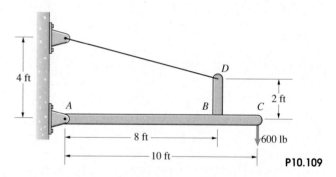

P10.109

10.110 Determine the internal forces and moments at A.

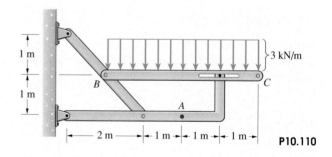

P10.110

10.111 Draw the shear force and bending moment diagrams of beam BC in Problem 10.110.

10.112 Determine the internal forces and moment at B (a) if $x = 250$ mm; (b) if $x = 750$ mm.

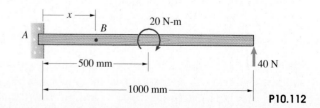

P10.112

10.113 Draw the shear force and bending moment diagrams for the beam in Problem 10.112.

10.114 The homogeneous beam weighs 1000 lb. What are the internal forces and bending moment at its midpoint?

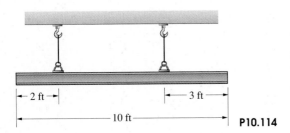

P10.114

10.115 Draw the shear force and bending moment diagrams for the beam in Problem 10.114.

10.116 At A the main cable of the suspension bridge is horizontal and its tension is 1×10^8 lb.
(a) Determine the distributed load acting on the cable.
(b) What is the tension at B?

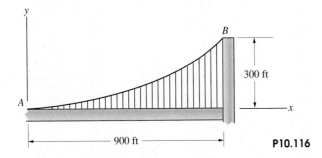

P10.116

10.117 The power line has a mass of 1.4 kg/m. If the line will safely support a tension of 5 kN, determine whether it will safely support an ice accumulation of 4 kg/m.

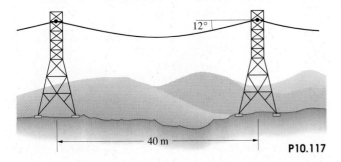

P10.117

10.118 The water depth at the center of the elliptical aquarium window is 20 ft. Determine the magnitude of the net force exerted on the window by the pressure of the seawater ($\gamma = 64$ lb/ft³) and

the atmospheric pressure of the air on the opposite side. (See Problem 10.99.)

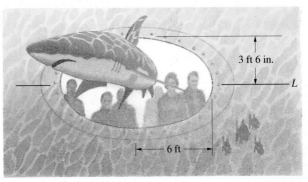

P10.118

10.119 In Problem 10.118, determine the magnitude of the net moment exerted on the window about the horizontal axis L by the pressure of the seawater ($\gamma = 64$ lb/ft³) and the atmospheric pressure of the air on the opposite side. (See Problem 10.99.)

10.120 The gate has water of 2-m depth on one side. The width of the gate (the dimension into the page) is 4 m, and its mass is 160 kg. The mass density of the water is $\rho = 1000$ kg/m³, and atmospheric pressure is $p_{atm} = 1 \times 10^5$ Pa. Determine the reactions on the gate at A and B. (The support at B exerts only a horizontal reaction on the gate.)

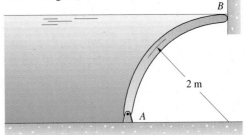

P10.120

10.121 The dam has water of depth 4 ft on one side. The width of the dam (the dimension into the page) is 8 ft. The weight density of the water is $\gamma = 62.4$ lb/ft³, and atmospheric pressure is $p_{atm} = 2120$ lb/ft². If you neglect the weight of the dam, what are the reactions at A and B?

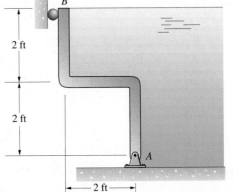

P10.121

A nalyzing structures with many members, such as this extensible platform, can be tedious and time-consuming. In certain cases, you can analyze structures in equilibrium more easily by subjecting them to hypothetical motions and calculating the work done by external forces and couples. This method, which is called the principle of virtual work, is the subject of this chapter.

Chapter 11

Virtual Work and Potential Energy

WHEN you stretch a spring, the work you perform is stored in the spring as potential energy. Raising an extensible platform increases its gravitational potential energy. In this chapter we define work and potential energy and introduce a general and powerful result called the principle of virtual work. In some instances, you can use this principle to analyze objects and structures in equilibrium more easily than by drawing free-body diagrams and applying the equilibrium equations. But beyond this, the principle of virtual work is used extensively in advanced applications of mechanics, especially in the analysis of deformable members and structures.

11.1 *Virtual Work*

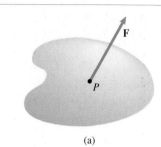

(a)

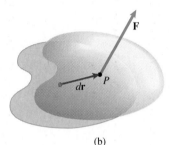

(b)

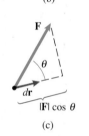

(c)

Figure 11.1

(a) A force **F** acting on an object.
(b) A displacement $d\mathbf{r}$ of P.
(c) The work $dU = (|\mathbf{F}| \cos \theta) |d\mathbf{r}|$.

The principle of virtual work is a statement about work done by forces and couples when an object or structure is subjected to various hypothetical motions. Before we can introduce this principle, we must define work.

Work

Consider a force acting on an object at a point P (Fig. 11.1a). Suppose that the object undergoes an infinitesimal motion, so that P has a differential displacement $d\mathbf{r}$ (Fig. 11.1b). The **work** dU done by **F** as a result of the displacement $d\mathbf{r}$ is defined to be

$$dU = \mathbf{F} \cdot d\mathbf{r}. \tag{11.1}$$

From the definition of the dot product, $dU = (|\mathbf{F}| \cos \theta) |d\mathbf{r}|$, where θ is the angle between **F** and $d\mathbf{r}$ (Fig. 11.1c). The work is equal to the product of the component of **F** in the direction of $d\mathbf{r}$ and the magnitude of $d\mathbf{r}$. Notice that if the component of **F** parallel to $d\mathbf{r}$ points in the direction opposite to $d\mathbf{r}$, the work is negative. Also notice that if **F** is perpendicular to $d\mathbf{r}$, the work is zero. The dimensions of work are (force) × (length).

Now consider a couple acting on an object (Fig. 11.2a). The moment due to the couple is $M = Fh$ in the counterclockwise direction. If the object rotates through an infinitesimal counterclockwise angle $d\alpha$ (Fig. 11.2b), the points of application of the forces are displaced through differential distances $\frac{1}{2}h\,d\alpha$. Consequently, the total work done is $dU = F(\frac{1}{2}h\,d\alpha) + F(\frac{1}{2}h\,d\alpha) = M\,d\alpha$.

We see that when an object acted on by a couple M is rotated through an angle $d\alpha$ in the same direction as the couple (Fig. 11.2c), the resulting work is

$$dU = M\,d\alpha. \tag{11.2}$$

If the direction of the couple is opposite to the direction of $d\alpha$, the work is negative.

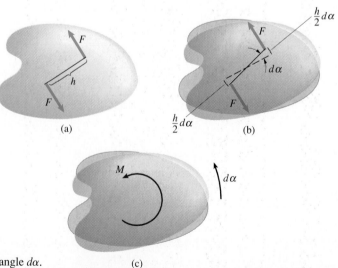

Figure 11.2

(a) A couple acting on an object.
(b) An infinitesimal rotation of the object.
(c) An object acted on by a couple M rotating through an angle $d\alpha$.

Principle of Virtual Work

Now that we have defined the work done by forces and couples, we can introduce the principle of virtual work. Before stating it, we first discuss an example to give you context for understanding the principle.

The homogeneous bar in Fig. 11.3(a) is supported by the wall and by the pin support at A and is loaded by a couple M. The free-body diagram of the bar is shown in Fig. 11.3(b). The equilibrium equations are

$$\Sigma F_x = A_x - N = 0, \tag{11.3}$$

$$\Sigma F_y = A_y - W = 0, \tag{11.4}$$

$$\Sigma M_{\text{(point } A)} = NL \sin \alpha - W\frac{1}{2}L \cos \alpha - M = 0. \tag{11.5}$$

We can solve these three equations for the reactions A_x, A_y, and N. However, we have a different objective.

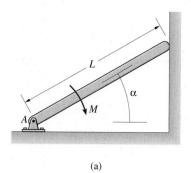
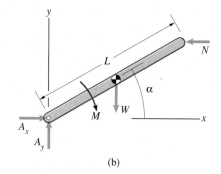

(a)   (b)

Figure 11.3

(a) A bar subjected to a couple M.
(b) Free-body diagram of the bar.

Let's ask the following question: If the bar is acted on by the forces and couple in Fig. 11.3(b) and we subject it to a *hypothetical* infinitesimal translation in the x direction, as shown in Fig. 11.4, what work is done? The hypothetical displacement δx is called a **virtual displacement** of the bar, and the resulting work δU is called the **virtual work**. The pin support and the wall prevent the bar from actually moving in the x direction; the virtual displacement is a theoretical artifice. Our objective is to calculate the resulting virtual work:

$$\delta U = A_x\,\delta x + (-N)\,\delta x = (A_x - N)\,\delta x. \tag{11.6}$$

The forces A_y and W do no work because they are perpendicular to the displacements of their points of application. The couple M also does no work, because the bar does not rotate. Comparing this equation with Eq. (11.3), we find that *the virtual work equals zero*.

Next we give the bar a virtual translation in the y direction (Fig. 11.5). The resulting virtual work is

$$\delta U = A_y\,\delta y + (-W)\,\delta y = (A_y - W)\,\delta y. \tag{11.7}$$

From Eq. (11.4), the virtual work again equals zero.

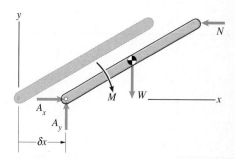

Figure 11.4

A virtual displacement δx.

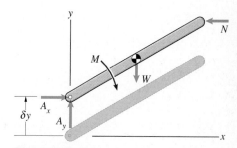

Figure 11.5

A virtual displacement δy.

Finally, we give the bar a virtual rotation while holding point A fixed (Fig. 11.6a). The forces A_x and A_y do no work because their point of application does not move. The work done by the couple M is $-M\,\delta\alpha$, because its direction is opposite to that of the rotation. The displacements of the points of application of the forces N and W are shown in Fig. 11.6(b), and the components of the forces in the direction of the displacements are shown in Fig. 11.6(c). The work done by N is $(N\sin\alpha)(L\,\delta\alpha)$, and the work done by W is $(-W\cos\alpha)(\frac{1}{2}L\,\delta\alpha)$. The total work is

$$\delta U = (N\sin\alpha)(L\,\delta\alpha) + (-W\cos\alpha)\left(\frac{1}{2}L\,\delta\alpha\right) - M\,\delta\alpha$$

$$= \left(NL\sin\alpha - W\frac{1}{2}L\cos\alpha - M\right)\delta\alpha. \tag{11.8}$$

From Eq. (11.5), the virtual work resulting from the virtual rotation is also zero.

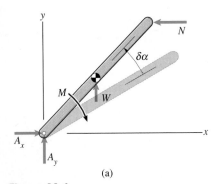

(a)

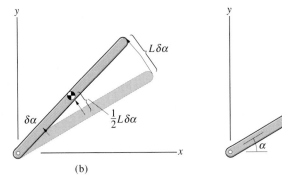

(b)　　　　　　　(c)

Figure 11.6

(a) A virtual rotation $\delta\alpha$.
(b) Displacements of the points of application of N and W.
(c) Components of N and W in the direction of the displacements.

We have shown that for three virtual motions of the bar, the virtual work is zero. These results are examples of a form of the principle of virtual work: *If an object is in equilibrium, the virtual work done by the external forces and couples acting on it is zero for any virtual translation or rotation:*

$$\delta U = 0. \tag{11.9}$$

As our example illustrates, this principle can be used to derive the equilibrium equations for an object. Subjecting the bar to virtual translations δx and δy and a virtual rotation $\delta\alpha$ results in Eqs. (11.6)–(11.8). Because the virtual work must be zero in each case, we obtain Eqs. (11.3)–(11.5). But there is no advantage to this approach compared to simply drawing the free-body diagram of the object and writing the equations of equilibrium in the usual way. The advantages of the principle of virtual work become evident when we consider structures.

Application to Structures

The principle of virtual work stated in the preceding section applies to each member of a structure. By subjecting certain types of structures in equilibrium to virtual motions and calculating the total virtual work, we can determine unknown reactions at their supports as well as internal forces in their members. The procedure involves finding virtual motions that result in virtual work being done both by known loads and by unknown forces and couples.

Suppose that we want to determine the axial load in member BD of the truss in Fig. 11.7(a). The other members of the truss are subjected to the 4-kN load and the forces exerted on them by member BD (Fig. 11.7b). If we give the structure a virtual rotation $\delta\alpha$ as shown in Fig. 11.7(c), virtual work is done by the force T_{BD} acting at B and by the 4-kN load at C. Furthermore, the virtual work done by these two forces is the total virtual work done on the members of the structure, because the virtual work done by the internal forces they exert on each other cancels out. For example, consider joint C (Fig. 11.7d). The force T_{BC} is the axial load in member BC. The virtual work done at C on member BC is $T_{BC}(1.4\ \text{m})\ \delta\alpha$, and the work done at C on member CD is $(4\ \text{kN} - T_{BC})(1.4\ \text{m})\ \delta\alpha$. When we add up the virtual work done on the members to obtain the total virtual work on the structure, the virtual work due to the internal force T_{BC} cancels out. (If the members exerted an internal *couple* on each other at C—for example, as a result of friction in the pin support—the virtual work would not cancel out.)

Therefore we can ignore internal forces in calculating the total virtual work on the structure:

$$\delta U = (T_{BD} \cos\theta)(1.4\ \text{m})\ \delta\alpha + (4\ \text{kN})(1.4\ \text{m})\ \delta\alpha = 0.$$

The angle $\theta = \arctan(1.4/1) = 54.5°$. Solving this equation, we obtain $T_{BD} = -6.88$ kN.

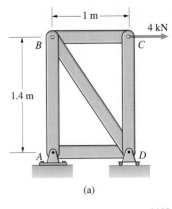

(a)

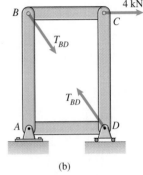

(b)

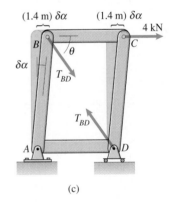

(c)

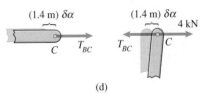

(d)

Figure 11.7
(a) A truss with a 4-kN load.
(b) Forces exerted by member BD.
(c) A virtual motion of the structure.
(d) Calculating the virtual work on members BC and CD at the joint C.

In following examples we use virtual work to determine reactions on members of structures. This method typically involves two steps:

1. Choose a virtual motion—*Identify a virtual motion that results in virtual work being done by known loads and by an unknown force or couple you want to determine.*

2. Determine the virtual work—*Calculate the total virtual work resulting from the virtual motion to obtain an equation for the unknown force or couple.*

Example 11.1

For the structure in Fig. 11.8, use the principle of virtual work to determine the horizontal reaction at C.

Figure 11.8

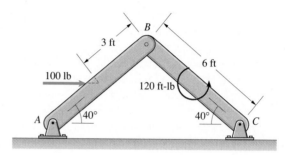

SOLUTION

Choose a Virtual Motion   We draw the free-body diagram of the structure in Fig. (a). Our objective is to determine C_x. If we hold point A fixed and subject bar AB to a virtual rotation $\delta\alpha$ while requiring point C to move horizontally (Fig. b), virtual work is done only by the external loads on the structure and by C_x. The reactions A_x and A_y do no work because A does not move, and the reaction C_y does no work because it is perpendicular to the virtual displacement of point C.

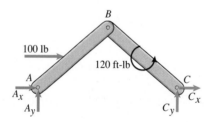

(a) Free-body diagram of the structure.

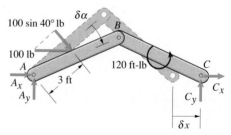

(b) A virtual displacement in which A remains fixed and C moves horizontally.

Determine the Virtual Work   The virtual work done by the 100-lb force is $(100 \sin 40° \text{ lb})(3 \text{ ft}) \, \delta\alpha$. The bar BC undergoes a virtual rotation $\delta\alpha$ in the counterclockwise direction, so the work done by the couple is $(120 \text{ ft-lb}) \, \delta\alpha$. In terms of the virtual displacement δx of point C, the work done by C_x is $C_x \, \delta x$. The total virtual work is

$$\delta U = (100 \sin 40°)(3) \, \delta\alpha + (120) \, \delta\alpha + C_x \, \delta x = 0.$$

To obtain C_x from this equation, we must determine the relation between $\delta\alpha$ and δx. From the geometry of the structure (Fig. c), the relation between the angle α and the distance x from A to C is

$$x = 2(6 \cos \alpha).$$

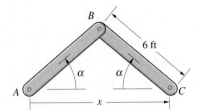

(c) The geometry for determining the relation between $\delta\alpha$ and δx.

The derivative of this equation with respect to α is

$$\frac{dx}{d\alpha} = -12 \sin \alpha.$$

Thus an infinitesimal change in x is related to an infinitesimal change in α by

$$dx = -12 \sin \alpha \, d\alpha.$$

Because the virtual rotation $\delta\alpha$ in Fig. (b) is a decrease in α, we conclude that δx is related to $\delta\alpha$ by

$$\delta x = 12 \sin 40° \, \delta\alpha.$$

Therefore the total virtual work is

$$\delta U = [300 \sin 40° + 120 + C_x(12 \sin 40°)] \, \delta\alpha = 0,$$

and we obtain $C_x = -40.6$ lb.

DISCUSSION

Notice that we ignored the internal forces the members exert on each other at B. The virtual work done by these internal forces cancels out. To obtain the solution, we needed to determine the relationship between the virtual displacements δx and $\delta\alpha$. Determining the geometrical relationships between virtual displacements is often the most challenging aspect of applying the principle of virtual work.

Example 11.2

The extensible platform in Fig. 11.9 is raised and lowered by the hydraulic cylinder *BC*. The total weight of the platform and men is *W*. The weights of the beams supporting the platform can be neglected. What axial force must the hydraulic cylinder exert to hold the platform in equilibrium in the position shown?

STRATEGY

We can use a virtual motion that coincides with the actual motion of the platform and beams when the length of the hydraulic cylinder changes. By calculating the virtual work done by the hydraulic cylinder and by the weight of the men and platform, we can determine the force exerted by the hydraulic cylinder.

SOLUTION

Choose a Virtual Motion We draw the free-body diagram of the platform and beams in Fig. (a). Our objective is to determine the force *F* exerted by the hydraulic cylinder. If we hold point *A* fixed and subject point *C* to a horizontal virtual displacement δx, the only external forces that do virtual work are *F* and the weight *W*. (The reaction due to the roller support at *C* is perpendicular to the virtual displacement.)

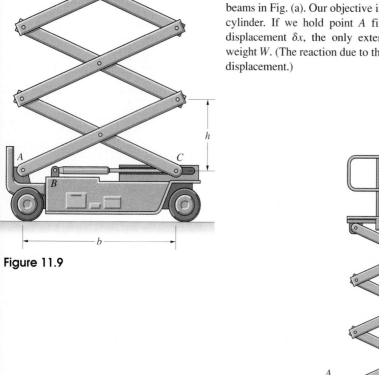

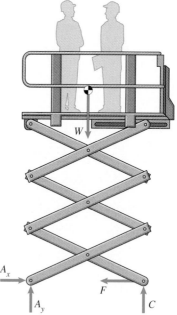

Figure 11.9

(a) Free-body diagram of the platform and supporting beams.

Determine the Virtual Work The virtual work done by the force *F* as point *C* undergoes a virtual displacement δx to the right (Fig. b) is $-F\,\delta x$. To determine the virtual work done by the weight *W*, we must determine the vertical displacement of point *D* in Fig. (b) when point *C* moves to the right a distance δx. The dimensions *b* and *h* are related by

$$b^2 + h^2 = L^2,$$

where L is the length of the beam AD. Taking the derivative of this equation with respect to b, we obtain

$$2b + 2h\frac{dh}{db} = 0,$$

which we can solve for dh in terms of db:

$$dh = -\frac{b}{h}\,db.$$

Thus when b increases an amount δx, the dimension h *decreases* an amount $(b/h)\,\delta x$. Because there are three pairs of beams, the platform moves downward a distance $(3b/h)\,\delta x$, and the virtual work done by the weight is $(3b/h)W\,\delta x$. The total virtual work is

$$\delta U = \left[-F + \left(\frac{3b}{h}\right)W\right]\delta x = 0,$$

and we obtain $F = (3b/h)W$.

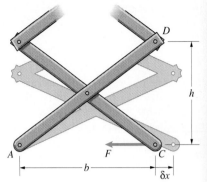

(b) A virtual displacement in which A remains fixed and C moves horizontally.

Problems

The following problems are to be solved using the principle of virtual work.

11.1   Determine the reaction at B.
  Strategy: Subject the beam to a virtual rotation about A.

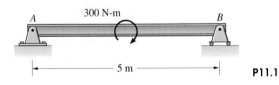

P11.1

11.2 (a) Determine the virtual work done by the 2-kN force and the 2.4 kN-m couple when the beam is rotated through a counterclockwise angle $\delta\theta$ about point A.
(b) Use the result of (a) to determine the reaction at B.

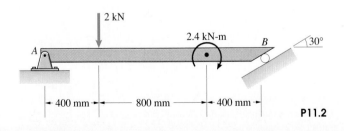

P11.2

11.3   Determine the tension in the cable.

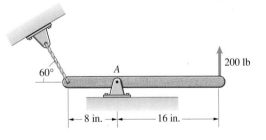

P11.3

11.4   The L-shaped bar is in equilibrium. Determine F.

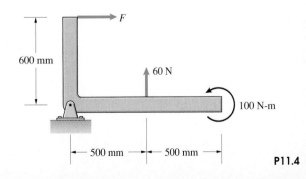

P11.4

11.5 (a) Determine the virtual work done by the distributed load when the beam is rotated through a counterclockwise angle $\delta\theta$ about point A.

(b) Use the result of (a) to determine the reaction at B.

 Strategy: In (a), remember that the force exerted on an element of the beam of length dx is $w \, dx$, where w is the value of the distributed load at x. You can calculate the virtual work done by the force $w \, dx$ and then integrate to determine the virtual work done by the distributed load. See if you obtain the same result by first representing the distributed load by an equivalent force and then calculating the virtual work done by the equivalent force.

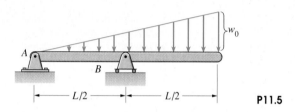

P11.5

11.6 Expecting to keep a new engineer occupied for several hours, her supervisor gave her the mechanism shown and asked her to estimate the force F that must be applied to the type platen at N to hold the system in equilibrium when the 2-N force acts on the finger pad at A. The supervisor was amazed when the engineer returned five minutes later with the answer. She had moved the finger pad downward 1 mm and measured the upward motion of the type platen, determining it to be approximately 14 mm, and then applied the principle of virtual work. What is the approximate value of F?

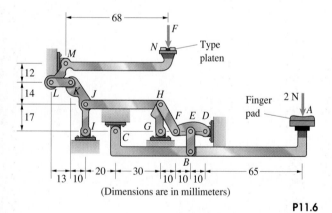

(Dimensions are in millimeters)

P11.6

11.7 The mechanism is in equilibrium. Determine the force R in terms of F.

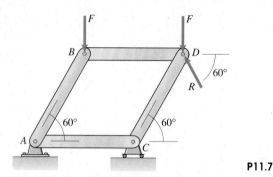

P11.7

11.8 Determine the reaction at the roller support.

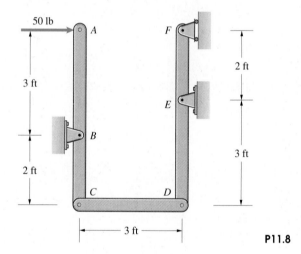

P11.8

11.9 Determine the force P necessary for the mechanism to be in equilibrium.

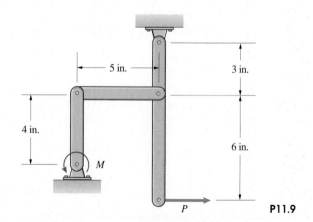

P11.9

11.10 The system is in equilibrium. The total mass of the suspended load and assembly A is 120 kg.

(a) By using equilibrium, determine the force F.

(b) Using the result of (a) and the principle of virtual work, determine the distance the suspended load rises if the cable is pulled upward 300 mm at B.

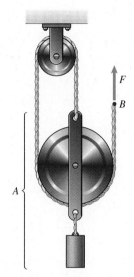

P11.10

11.11 Determine the force P necessary for the mechanism to be in equilibrium.

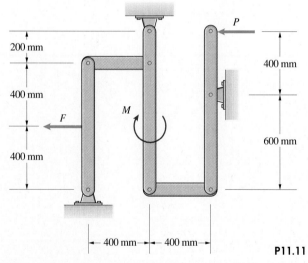

P11.11

11.12 The system is in equilibrium, the weights of the bars are negligible, and the angle $\alpha = 20°$. Determine the magnitude of the friction force exerted on the bar at A.

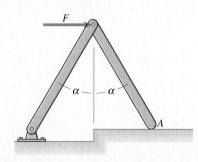

P11.12

11.13 Determine the magnitude of the force exerted on the wall by the block at A.

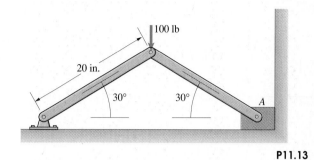

P11.13

11.14 Show that δx is related to $\delta \alpha$ by

$$\delta x = \frac{L_1 x \sin \alpha}{x - L_1 \cos \alpha} \delta \alpha.$$

Strategy: Write the law of cosines in terms of α and take the derivative of the resulting equation with respect to α. (See Example 11.2.)

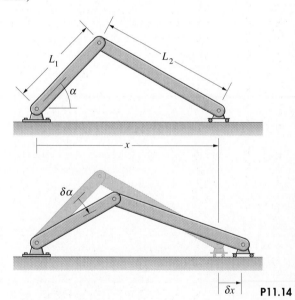

P11.14

11.15 The linkage is in equilibrium. What is the force F?

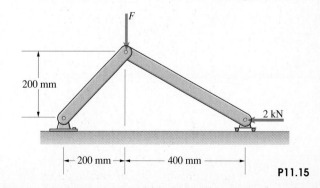

P11.15

11.16 The linkage is in equilibrium. What is the force F?

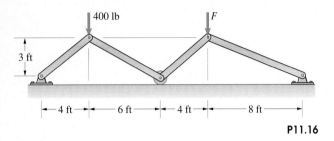

P11.16

11.17 Bar AC is connected to bar BD by a pin that fits in the smooth vertical slot. The weights of the bars are negligible. If $M_A = 20$ in-lb, what couple M_B is necessary for the system to be in equilibrium?

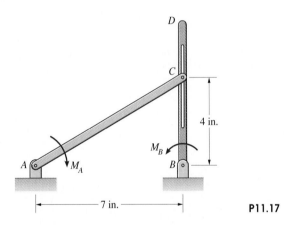

P11.17

11.18 The angle $\alpha = 20°$, and the force exerted on the stationary piston by pressure is 4 kN toward the left. What couple M is necessary to keep the system in equilibrium?

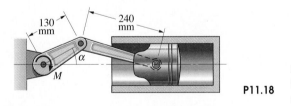

P11.18

11.19 The structure is subjected to a 400-N load and is held in place by a horizontal cable. Determine the tension in the cable.

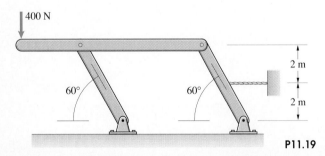

P11.19

11.20 If the car jack is subjected to a force $F = 6.5$ kN, what is the tension in the threaded shaft between B and D?

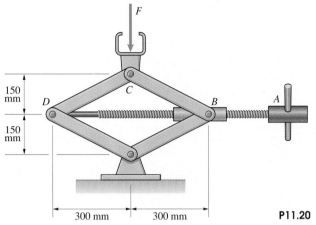

P11.20

11.21 What are the reactions at A and B?

Strategy: Use the equilibrium equations to determine the horizontal components of the reactions, and use the principle of virtual work to determine the vertical components.

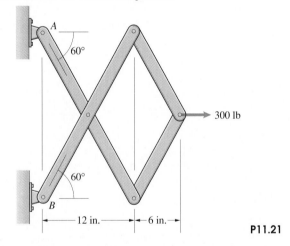

P11.21

11.22 This device raises a load W by extending the hydraulic actuator DE. The bars AD and BC are each 2 m long, and the distances $b = 1.4$ m and $h = 0.8$ m. If $W = 4$ kN, what force must the actuator exert to hold the load in equilibrium?

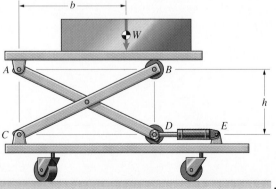

P11.22

11.23 Determine the force P necessary for the mechanism to be in equilibrium.

11.24 The collar A weighs 100 lb, and friction is negligible. Determine the tension in the cable AB.

Strategy: Let s be the distance along the bar from C to the collar, and let $\mathbf{e}_{CD}$ be a unit vector that points from C toward D. To apply the principle of virtual work, let the collar undergo a virtual displacement $\delta s\, \mathbf{e}_{CD}$.

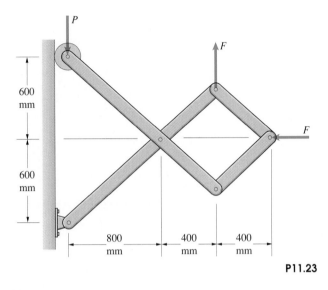

P11.23

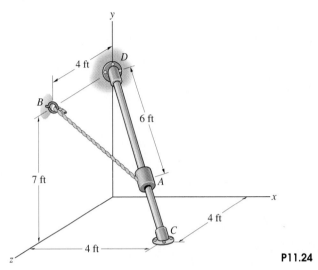

P11.24

11.2 *Potential Energy*

The work of a force $\mathbf{F}$ due to a differential displacement of its point of application is

$$dU = \mathbf{F} \cdot d\mathbf{r}.$$

If a function of position V exists such that for any $d\mathbf{r}$,

$$dU = \mathbf{F} \cdot d\mathbf{r} = -dV, \qquad (11.10)$$

the function V is called the **potential energy** associated with the force $\mathbf{F}$, and $\mathbf{F}$ is said to be **conservative**. (The negative sign in this equation is in keeping with the interpretation of V as "potential" energy. Positive work results from a decrease in V.) If the forces that do work on a system are conservative, we will show that you can use the total potential energy of the system to determine its equilibrium positions.

Examples of Conservative Forces

Weights of objects and the forces exerted by linear springs are conservative. In the following sections we derive the potential energies associated with these forces.

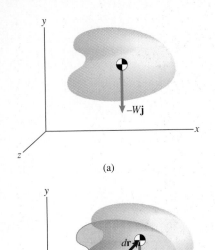

(a)

(b)

Figure 11.10

(a) Force exerted by the weight of an object.
(b) A differential displacement.

Weight   In terms of a coordinate system with its y axis upward, the force exerted by the weight of an object is $\mathbf{F} = -W\mathbf{j}$ (Fig. 11.10a). If we give the object an arbitrary displacement $d\mathbf{r} = dx\,\mathbf{i} + dy\,\mathbf{j} + dz\,\mathbf{k}$ (Fig. 11.10b), the work done by its weight is

$$dU = \mathbf{F} \cdot d\mathbf{r} = (-W\mathbf{j}) \cdot (dx\,\mathbf{i} + dy\,\mathbf{j} + dz\,\mathbf{k}) = -W\,dy.$$

We seek a potential energy V such that

$$dU = -W\,dy = -dV, \tag{11.11}$$

or

$$\frac{dV}{dy} = W.$$

If we neglect the variation in the weight with height and integrate, we obtain

$$V = Wy + C.$$

The constant C is arbitrary, since this function satisfies Eq. (11.11) for any value of C, and we will let $C = 0$. The position of the origin of the coordinate system can also be chosen arbitrarily. Thus the potential energy associated with the weight of an object is

$$V = Wy, \tag{11.12}$$

where y is the height of the object above some chosen reference level, or **datum**.

Springs   Consider a linear spring connecting an object to a fixed support (Fig. 11.11a). In terms of the stretch $S = r - r_0$, where r is the length of the spring and r_0 is its unstretched length, the force exerted on the object is kS (Fig. 11.11b). If the point at which the spring is attached to the object undergoes a differential displacement $d\mathbf{r}$ (Fig. 11.11c), the work done by the force on the object is

$$dU = -kS\,dS,$$

where dS is the increase in the stretch of the spring resulting from the displacement (Fig. 11.11d). We seek a potential energy V such that

$$dU = -kS\,dS = -dV, \tag{11.13}$$

or

$$\frac{dV}{dS} = kS.$$

Integrating this equation and letting the integration constant be zero, we obtain the potential energy associated with the force exerted by a linear spring:

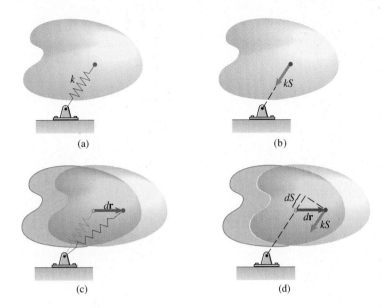

Figure 11.11

(a) A spring connected to an object.
(b) The force exerted on the object.
(c) A differential displacement of the object.
(d) The work done by the force is $dU = -kS\,dS$.

$$V = \frac{1}{2}kS^2. \tag{11.14}$$

Notice that V is positive if the spring is either stretched (S is positive) or compressed (S is negative). Potential energy (the potential to do work) is stored in a spring by either stretching or compressing it.

Principle of Virtual Work for Conservative Forces

Because the work done by a conservative force is expressed in terms of its potential energy through Eq. (11.10), we can give an alternative statement of the principle of virtual work when an object is subjected to conservative forces: *Suppose that an object is in equilibrium. If the forces that do work as the result of a virtual translation or rotation are conservative, the change in the total potential energy is zero:*

$$\delta V = 0. \tag{11.15}$$

We emphasize that it is not necessary that all of the forces acting on the object be conservative for this result to hold; it is necessary only that the forces that do work be conservative. This principle also applies to a system of interconnected objects if the external forces that do work are conservative and the internal forces at the connections between objects either do no work or are conservative. Such a system is called a **conservative system**.

If the position of a system can be specified by a single coordinate q, the system is said to have one **degree of freedom**. The total potential energy of a conservative, one-degree-of-freedom system can be expressed in terms of q, and we can write Eq. (11.15) as

$$\delta V = \frac{dV}{dq}\delta q = 0.$$

Thus when the object or system is in equilibrium, the derivative of its total potential energy with respect to q is zero:

$$\frac{dV}{dq} = 0. \tag{11.16}$$

You can use this equation to determine the values of q at which the system is in equilibrium.

Stability of Equilibrium

Suppose that a homogeneous bar of weight W and length L is pinned at one end. In terms of the angle α shown in Fig. 11.12(a), the height of the center of mass relative to the pinned end is $-\frac{1}{2}L\cos\alpha$. Choosing the level of the pin support as the datum, we can therefore express the potential energy associated with the weight of the bar as

$$V = -\frac{1}{2}WL\cos\alpha.$$

When the bar is in equilibrium,

$$\frac{dV}{d\alpha} = \frac{1}{2}WL\sin\alpha = 0.$$

This condition is satisfied when $\alpha = 0$ (Fig. 11.12b) and also when $\alpha = 180°$ (Fig. 11.12c).

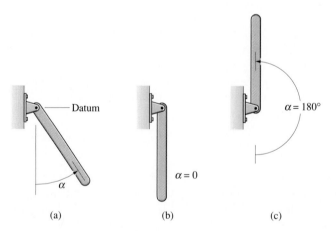

Figure 11.12

(a) A bar suspended from one end.
(b) The equilibrium position $\alpha = 0$.
(c) The equilibrium position $\alpha = 180°$.

There is a fundamental difference between the two equilibrium positions of the bar. In the position shown in Fig. 11.12(b), if we displace the bar slightly from its equilibrium position and release it, the bar will remain near the equilibrium position. We say that this equilibrium position is **stable**. When the bar is in the position shown in Fig. 11.12(c), if we displace it slightly and release it, the bar will move away from the equilibrium position. This equilibrium position is **unstable**.

The graph of the bar's potential energy V as a function of α is shown in Fig. 11.13(a). The potential energy is a minimum at the stable equilibrium position $\alpha = 0$ and a maximum at the unstable equilibrium position $\alpha = 180°$. The derivative of V (Fig. 11.13b) equals zero at both equilibrium positions. The second derivative of V (Fig. 11.13c) is positive at the stable equilibrium position $\alpha = 0$ and negative at the unstable equilibrium position $\alpha = 180°$.

If a conservative, one-degree-of-freedom system is in equilibrium and the second derivative of V evaluated at the equilibrium position is positive, the equilibrium position is stable. If the second derivative of V is negative, it is unstable (Fig. 11.14).

$$\frac{dV}{dq} = 0, \qquad \frac{d^2V}{dq^2} > 0: \qquad \text{Stable equilibrium}$$

$$\frac{dV}{dq} = 0, \qquad \frac{d^2V}{dq^2} < 0: \qquad \text{Unstable equilibrium}$$

Proving these results requires analyzing the motion of the system near an equilibrium position.

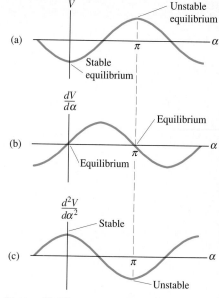

Figure 11.13

Graphs of V, $dV/d\alpha$, and $d^2V/d\alpha^2$.

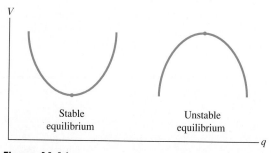

Figure 11.14

Graphs of the potential energy V as a function of the coordinate q that exhibit stable and unstable equilibrium positions.

In the following examples we use potential energy to analyze the equilibrium of one-degree-of-freedom systems. This typically involves three steps:

1. Determine the potential energy—*Express the total potential energy in terms of a single coordinate that specifies the position of the system.*

2. Find the equilibrium positions—*By calculating the first derivative of the potential energy, determine the equilibrium position or positions.*

3. Examine the stability—*Use the sign of the second derivative of the potential energy to determine whether the equilibrium positions are stable.*

Example 11.3

In Fig. 11.15 a crate of weight W is suspended from the ceiling by a wire modeled as a linear spring with constant k. The coordinate x measures the position of the center of mass of the crate relative to its position when the wire is unstretched. Find the equilibrium position of the crate, and determine whether it is stable or unstable.

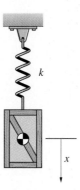

Figure 11.15

STRATEGY

The forces acting on the crate—its weight and the force exerted by the spring—are conservative. Therefore the system is conservative, and we can use the potential energy to determine both the equilibrium position and whether the equilibrium position is stable.

SOLUTION

Determine the Potential Energy   We can use $x = 0$ as the datum for the potential energy associated with the weight. Because the coordinate x is positive downward, the potential energy is $-Wx$. The stretch of the spring equals x, so the potential energy associated with the force of the spring is $\frac{1}{2}kx^2$. The total potential energy is

$$V = \frac{1}{2}kx^2 - Wx.$$

Find the Equilibrium Positions   When the crate is in equilibrium,

$$\frac{dV}{dx} = kx - W = 0.$$

The equilibrium position is $x = W/k$.

Examine the Stability   The second derivative of the potential energy is

$$\frac{d^2V}{dx^2} = k.$$

The equilibrium position is stable.

Example 11.4

The homogeneous hemisphere in Fig. 11.16 is at rest on the plane surface. Show that it is in equilibrium in the position shown. Is the equilibrium position stable?

Figure 11.16

STRATEGY

To determine whether the hemisphere is in equilibrium and whether its equilibrium is stable, we must introduce a coordinate that specifies its orientation and express its potential energy in terms of that coordinate. We can use as the coordinate the angle of rotation of the hemisphere relative to the position shown.

SOLUTION

Determine the Potential Energy   Suppose that the hemisphere is rotated through an angle α relative to its original position (Fig. a). Using the datum shown, the potential energy associated with the weight W of the hemisphere is

$$V = -\frac{3}{8}RW \cos \alpha.$$

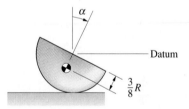

(a) The hemisphere rotated through an angle α.

Find the Equilibrium Positions   When the hemisphere is in equilibrium,

$$\frac{dV}{d\alpha} = \frac{3}{8}RW \sin \alpha = 0,$$

which confirms that $\alpha = 0$ is an equilibrium position.

Examine the Stability   The second derivative of the potential energy is

$$\frac{d^2V}{d\alpha^2} = \frac{3}{8}RW \cos \alpha.$$

This expression is positive at $\alpha = 0$, so the equilibrium position is stable.

DISCUSSION

Notice that we ignored the normal force exerted on the hemisphere by the plane surface. This force does no work and so does not affect the potential energy.

Example 11.5

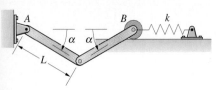

Figure 11.17

The pinned bars in Fig. 11.17 are held in place by the linear spring. Each bar has weight W and length L. The spring is unstretched when $\alpha = 0$, and the bars are in equilibrium when $\alpha = 60°$. Determine the spring constant k, and determine whether the equilibrium position is stable or unstable.

STRATEGY

The only forces that do work on the bars are their weights and the force exerted by the spring. By expressing the total potential energy in terms of α and using Eq. (11.16), we will obtain an equation we can solve for the spring constant k.

SOLUTION

Determine the Potential Energy   If we use the datum shown in Fig. (a), the potential energy associated with the weights of the two bars is

$$W\left(-\frac{1}{2}L \sin \alpha\right) + W\left(-\frac{1}{2}L \sin \alpha\right) = -WL \sin \alpha.$$

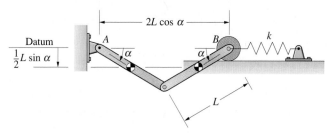

(a) Determining the total potential energy.

The spring is unstretched when $\alpha = 0$ and the distance between points A and B is $2L \cos \alpha$ (Fig. a), so the stretch of the spring is $2L - 2L \cos \alpha$. Therefore the potential energy associated with the spring is $\frac{1}{2}k(2L - 2L \cos \alpha)^2$, and the total potential energy is

$$V = -WL \sin \alpha + 2kL^2(1 - \cos \alpha)^2.$$

When the system is in equilibrium,

$$\frac{dV}{d\alpha} = -WL \cos \alpha + 4kL^2 \sin \alpha(1 - \cos \alpha) = 0.$$

Because the system is in equilibrium when $\alpha = 60°$, we can solve this equation for the spring constant in terms of W and L:

$$k = \frac{W \cos \alpha}{4L \sin \alpha(1 - \cos \alpha)} = \frac{W \cos 60°}{4L \sin 60°(1 - \cos 60°)} = \frac{0.289W}{L}.$$

Examine the Stability   The second derivative of the potential energy is

$$\frac{d^2V}{d\alpha^2} = WL \sin \alpha + 4kL^2(\cos \alpha - \cos^2 \alpha + \sin^2 \alpha)$$

$$= WL \sin 60° + 4kL^2(\cos 60° - \cos^2 60° + \sin^2 60°)$$

$$= 0.866\ WL + 4kL^2.$$

This is a positive number, so the equilibrium position is stable.

Problems

11.25 The potential energy of a conservative system is given by $V = 2x^3 + 3x^2 - 12x$.
(a) For what values of x is the system in equilibrium?
(b) Determine whether the equilibrium positions you found in (a) are stable or unstable.

11.26 The potential energy of a conservative system is given by $V = 2q^3 - 21q^2 + 72q$.
(a) For what values of q is the system in equilibrium?
(b) Determine whether the equilibrium positions you found in (a) are stable or unstable.

11.27 The weight $W = 100$ lb, and the spring constant $k = 50$ lb/ft. The spring is unstretched when $x = 0$.
(a) Determine the value of x for which the weight is in equilibrium.
(b) Determine whether the equilibrium position is stable or unstable.

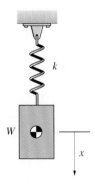

P11.27

11.28 The *nonlinear* spring exerts a force $-kx + \varepsilon x^3$ on the mass, where k and ε are constants. Determine the potential energy V associated with the force exerted on the mass by the spring.

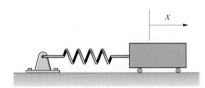

P11.28

11.29 The 1-kg mass is suspended from the nonlinear spring described in Problem 11.28. The constants $k = 10$ and $\varepsilon = 1$, where x is in meters.
(a) Show that the mass is in equilibrium when $x = 1.12$ m and when $x = 2.45$ m.

(b) Determine whether the equilibrium positions are stable or unstable.

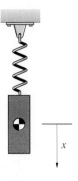

P11.29

11.30 The two straight segments of the bar are each of weight W and length L. Determine whether the equilibrium position shown is stable if (a) $0 < \alpha_0 < 90°$; (b) $90° < \alpha_0 < 180°$.

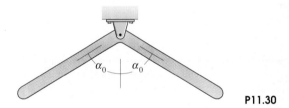

P11.30

11.31 The homogeneous composite object consists of a hemisphere and a cylinder. It is at rest on the plane surface. Show that this equilibrium position is stable only if $L < R/\sqrt{2}$.

P11.31

11.32 The homogeneous composite object consists of a hemisphere and a cone. It is at rest on the plane surface. Show that this equilibrium position is stable only if $h < \sqrt{3}R$.

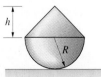

P11.32

11.33 The homogeneous bar has weight W, and the spring is unstretched when the bar is vertical ($\alpha = 0$).
(a) Use potential energy to show that the bar is in equilibrium when $\alpha = 0$.
(b) Show that the equilibrium position $\alpha = 0$ is stable only if $2kL > W$.

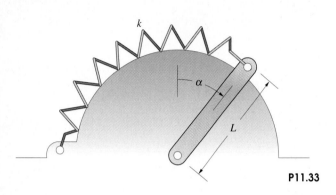

P11.33

11.34 Suppose that the bar in Problem 11.33 is in equilibrium when $\alpha = 20°$.
(a) Show that the spring constant $k = 0.490 \, W/L$.
(b) Determine whether the equilibrium position is stable.

11.35 The bar AB has weight W and length L. The spring is unstretched when the bar is vertical ($\alpha = 0$). The light collar C slides on the smooth vertical bar so that the spring remains horizontal. Show that the equilibrium position $\alpha = 0$ is stable only if $2kL > W$.

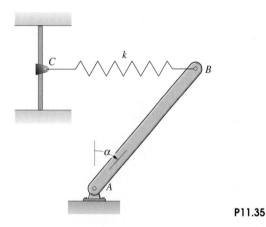

P11.35

11.36 The bar AB described in Problem 11.35 weighs 20 lb, its length is 4 ft, and the spring constant is $k = 5$ lb/ft.
(a) Determine the value of α in the range $0 < \alpha < 90°$ for which the bar is in equilibrium.
(b) Is the equilibrium position determined in (a) stable?

11.37 The bar AB has weight W and length L. The spring is unstretched when the bar is vertical ($\alpha = 0$). The light collar C slides on the smooth horizontal bar so that the spring remains vertical. Show that the equilibrium position $\alpha = 0$ is unstable.

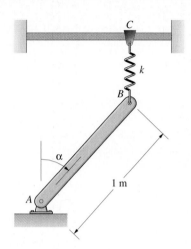

P11.37

11.38 The bar AB described in Problem 11.37 has a mass of 2 kg, and the spring constant is $k = 80$ N/m.
(a) Determine the value of α in the range $0 < \alpha < 90°$ for which the bar is in equilibrium.
(b) Is the equilibrium position determined in (a) stable?

11.39 Each homogeneous bar is of mass m and length L. The spring is unstretched when $\alpha = 0$. If $mg = kL$, determine the value of α in the range $0 < \alpha < 90°$ for which the system is in equilibrium.

P11.39

11.40 Determine whether the equilibrium position found in Problem 11.39 is stable or unstable.

11.41 The spring is unstretched when $\alpha = 90°$. If $mg = bk/2$, determine the value of α in the range $0 < \alpha < 90°$ for which the system is in equilibrium.

11.44 Determine whether the equilibrium positions of the bar in Problem 11.43 are stable or unstable.

11.45 Each bar is of weight W, and the spring is unstretched when $\alpha = 90°$.
(a) Show that the system is in equilibrium when
$\alpha = \arcsin (W/4kL)$.
(b) Is the equilibrium position described in (a) stable?

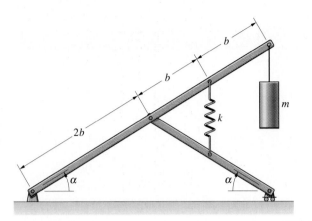

P11.41

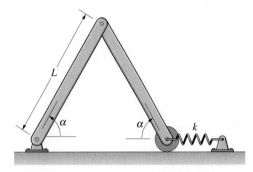

P11.45

11.42 Determine whether the equilibrium position found in Problem 11.41 is stable or unstable.

11.43 The bar weighs 15 lb. The spring is unstretched when $\alpha = 0$. The bar is in equilibrium when $\alpha = 30°$. Determine the spring constant k.

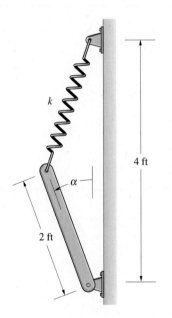

P11.43

Computational Mechanics

The following example and problems are designed for the use of a programmable calculator or computer.

Example 11.6

The two bars in Fig. 11.18 are held in place by the linear spring. Each bar has weight W and length L. The spring is unstretched when $\alpha = 0$. If $W = kL$, what is the value of α for which the bars are in equilibrium? Is the equilibrium position stable?

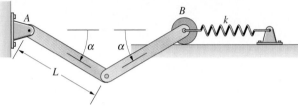

Figure 11.18

STRATEGY

By obtaining a graph of the derivative of the total potential energy as a function of α, we can estimate the value of α corresponding to equilibrium and determine whether the equilibrium position is stable.

SOLUTION

We derived the total potential energy of the system and determined its derivative with respect to α in Example 11.5, obtaining

$$\frac{dV}{d\alpha} = -WL \cos \alpha + 4kL^2 \sin \alpha(1 - \cos \alpha).$$

Substituting $W = kL$, we obtain

$$\frac{dV}{d\alpha} = kL^2[-\cos \alpha + 4 \sin \alpha(1 - \cos \alpha)].$$

From the graph of this function (Fig. 11.19), we estimate that the system is in equilibrium when $\alpha = 43°$.

The slope of $dV/d\alpha$, which is the second derivative of V, is positive at $\alpha = 43°$. The equilibrium position is therefore stable.

Figure 11.19

Graph of the derivative of V.

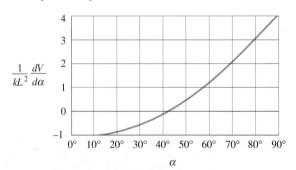

Problems

11.46 The 1-kg mass is suspended from a *nonlinear* spring that exerts a force $-10x + x^3$, where x is in meters.
(a) Draw a graph of the total potential energy of the system as a function of x from $x = 0$ to $x = 4$ m.
(b) Use your graph to estimate the equilibrium positions of the mass.
(c) Determine whether the equilibrium positions you obtained in (b) are stable or unstable.

P11.46

11.47 Suppose that the homogeneous bar in Problem 11.33 weighs 20 lb and has length $L = 2$ ft, and that $k = 4$ lb/ft.
(a) Determine the value of α in the range $0 < \alpha < 90°$ for which the bar is in equilibrium.
(b) Is the equilibrium position found in (a) stable?

11.48 The bar in Problem 11.43 weighs 15 lb, and the spring is unstretched when $\alpha = 0$. The spring constant is $k = 6$ lb/ft.
(a) Determine the value of α in the range $0 < \alpha < 90°$ for which the bar is in equilibrium.
(b) Is the equilibrium position found in (a) stable?

11.49 The homogeneous bar has length L and mass $4m$.
(a) Determine the value of α in the range $0 < \alpha < 90°$ for which the bar is in equilibrium.
(b) Is the equilibrium position found in (a) stable?

P11.49

11.50 The 2-m long, 10-kg homogeneous bar is pinned at A and at its midpoint B to light collars that slide on a smooth bar. The spring attached at A is unstretched when $\alpha = 0$, and its constant is $k = 1.2$ kN/m.
(a) Determine the value of α when the bar is in equilibrium.
(b) Determine whether the equilibrium position found in (a) is stable.

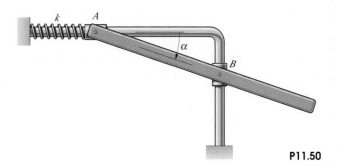

P11.50

Chapter Summary

Work

The **work** done by a force $\mathbf{F}$ as a result of a displacement $d\mathbf{r}$ of its point of application is defined by

$$dU = \mathbf{F} \cdot d\mathbf{r}. \qquad \text{Eq. (11.1)}$$

The work done by a counterclockwise couple M due to a counterclockwise rotation $d\alpha$ is

$$dU = M \, d\alpha. \qquad \text{Eq. (11.2)}$$

Principle of Virtual Work

If an object is in equilibrium, the **virtual work** done by the external forces and couples acting on it is zero for any **virtual translation** or **rotation**:

$$\delta U = 0. \qquad \text{Eq. (11.9)}$$

Potential Energy

If a function of position V exists such that for any displacement $d\mathbf{r}$, the work done by a force $\mathbf{F}$ is

$$dU = \mathbf{F} \cdot d\mathbf{r} = -dV, \qquad \text{Eq. (11.10)}$$

V is called the **potential energy** associated with the force, and $\mathbf{F}$ is said to be **conservative**.

The potential energy associated with the weight W of an object is

$$V = Wy, \qquad \text{Eq. (11.12)}$$

where y is the height of the center of mass above some reference level, or **datum**.

The potential energy associated with the force exerted by a linear spring is

$$V = \frac{1}{2} kS^2, \qquad \text{Eq. (11.14)}$$

where k is the spring constant and S is the stretch of the spring.

Principle of Virtual Work for Conservative Forces

An object or a system of interconnected objects is **conservative** if the external forces and couples that do work are conservative and internal forces at the connections between objects either do no work or are conservative. The change in the total potential energy resulting from any virtual motion of a conservative object or system is zero:

$$\delta V = 0. \qquad \text{Eq. (11.15)}$$

If the position of an object or a system can be specified by a single coordinate q, it is said to have one **degree of freedom**. When a conservative, one-degree-of-freedom object or system is in equilibrium,

$$\frac{dV}{dq} = 0. \qquad \text{Eq. (11.16)}$$

If the second derivative of V is positive, the equilibrium position is stable, and if the second derivative of V is negative, it is unstable.

Review Problems

11.51 (a) Determine the couple exerted on the beam at A.
(b) Determine the vertical force exerted on the beam at A.

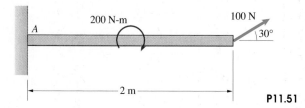

200 N-m

100 N

A

30°

2 m

P11.51

11.52 The structure is subjected to a 20 kN-m couple. Determine the horizontal reaction at C.

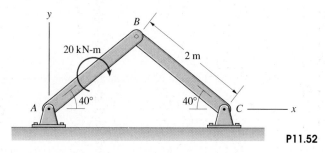

y

B

20 kN-m

2 m

A

40°

40°

C

x

P11.52

11.53 The "rack and pinion" mechanism is used to exert a vertical force on a sample at A for a stamping operation. If a force $F = 30$ lb is exerted on the handle, use the principle of virtual work to determine the force exerted on the sample.

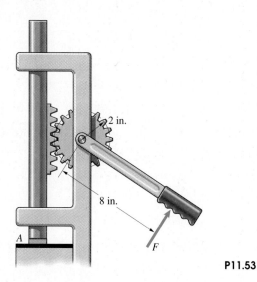

2 in.

8 in.

A

F

P11.53

11.54 If you were assigned to calculate the force exerted on the bolt by the pliers when the grips are subjected to forces F as shown in Fig. (a), you could carefully measure the dimensions, draw free-body diagrams, and use the equilibrium equations. But another approach would be to measure the change in the distance between the jaws when the distance between the handles is changed by a small amount. If your measurements indicate that the distance d in Fig. (b) decreases by 1 mm when D is decreased 8 mm, what is the approximate value of the force exerted on the bolt by each jaw when the forces F are applied?

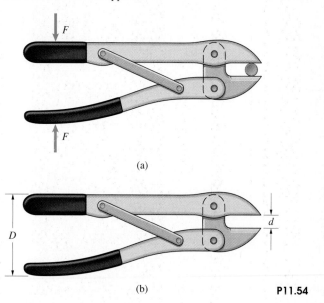

F

F

(a)

D

d

(b)

P11.54

11.55 The system is in equilibrium. The total weight of the suspended load and assembly A is 300 lb.
(a) By using equilibrium, determine the force F.
(b) Using the result of (a) and the principle of virtual work, determine the distance the suspended load rises if the cable is pulled downward 1 ft at B.

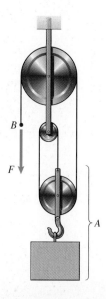

B

F

A

P11.55

11.56   The system is in equilibrium.
(a) By drawing free-body diagrams and using equilibrium equations, determine the couple M.
(b) Using the result of (a) and the principle of virtual work, determine the angle through which pulley B rotates if pulley A rotates through an angle α.

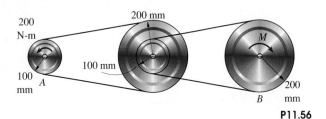

P11.56

11.57   The mechanism is in equilibrium. Neglect friction between the horizontal bar and the collar. Determine M in terms of F, α, and L.

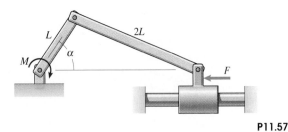

P11.57

11.58   In an injection casting machine, a couple M applied to arm AB exerts a force on the injection piston at C. Given that the horizontal component of the force exerted at C is 4 kN, use the principle of virtual work to determine M.

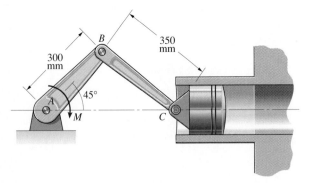

P11.58

11.59   Show that if bar AB is subjected to a clockwise virtual rotation $\delta\alpha$, bar CD undergoes a counterclockwise virtual rotation $(b/a)\,\delta\alpha$.

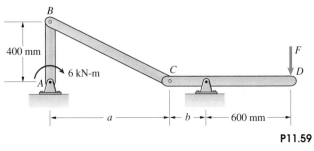

P11.59

11.60   The system in Problem 11.59 is in equilibrium, $a = 800$ mm, and $b = 400$ mm. Use the principle of virtual work to determine the force F.

11.61   Show that if bar AB is subjected to a clockwise virtual rotation $\delta\alpha$, bar CD undergoes a clockwise virtual rotation $[ad/(ac + bc - bd)]\,\delta\alpha$.

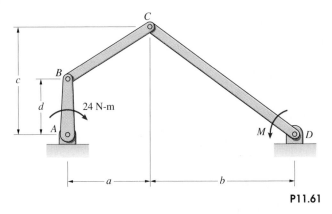

P11.61

11.62   The system in Problem 11.61 is in equilibrium, $a = 300$ mm, $b = 350$ mm, $c = 350$ mm, and $d = 200$ mm. Use the principle of virtual work to determine the couple M.

11.63 The mass of the bar is 10 kg, and it is 1 m in length. Neglect the masses of the two collars. The spring is unstretched when the bar is vertical ($\alpha = 0$), and the spring constant is $k = 100$ N/m. Determine the values of α at which the bar is in equilibrium.

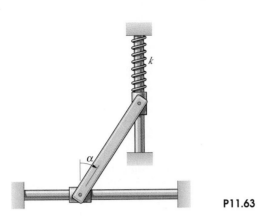

P11.63

11.64 Determine whether the equilibrium positions of the bar in Problem 11.63 are stable or unstable.

11.65 The spring is unstretched when $\alpha = 90°$. Determine the value of α in the range $0 < \alpha < 90°$ for which the system is in equilibrium.

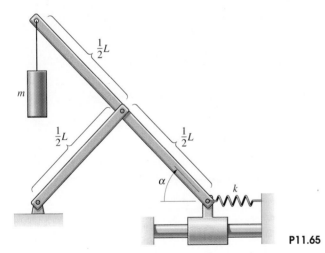

P11.65

11.66 Determine whether the equilibrium position found in Problem 11.65 is stable or unstable.

11.67 The hydraulic cylinder C exerts a horizontal force at A, raising the weight W. Determine the magnitude of the force the hydraulic cylinder must exert to support the weight in terms of W and α.

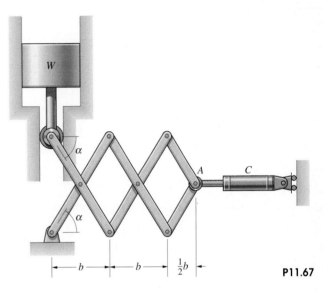

P11.67

Appendix A

Review of Mathematics

A.1 Algebra

Quadratic Equations

The solutions of the quadratic equation

$$ax^2 + bx + c = 0$$

are

$$x = \frac{-b \pm \sqrt{b^2 - 4ac}}{2a}.$$

Natural Logarithms

The natural logarithm of a positive real number x is denoted by $\ln x$. It is defined to be the number such that

$$e^{\ln x} = x,$$

where $e = 2.7182 \ldots$ is the base of natural logarithms. Logarithms have the following properties:

$$\ln (xy) = \ln x + \ln y,$$

$$\ln (x/y) = \ln x - \ln y,$$

$$\ln y^x = x \ln y.$$

A.2 Trigonometry

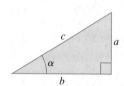

The trigonometric functions for a right triangle are

$$\sin \alpha = \frac{1}{\csc \alpha} = \frac{a}{c}, \qquad \cos \alpha = \frac{1}{\sec \alpha} = \frac{b}{c}, \qquad \tan \alpha = \frac{1}{\cot \alpha} = \frac{a}{b}.$$

The sine and cosine satisfy the relation

$$\sin^2 \alpha + \cos^2 \alpha = 1,$$

and the sine and cosine of the sum and difference of two angles satisfy

$$\sin (\alpha + \beta) = \sin \alpha \cos \beta + \cos \alpha \sin \beta,$$

$$\sin (\alpha - \beta) = \sin \alpha \cos \beta - \cos \alpha \sin \beta,$$

$$\cos (\alpha + \beta) = \cos \alpha \cos \beta - \sin \alpha \sin \beta,$$

$$\cos (\alpha - \beta) = \cos \alpha \cos \beta + \sin \alpha \sin \beta.$$

The **law of cosines** for an arbitrary triangle is

$$c^2 = a^2 + b^2 - 2ab \cos \alpha_c,$$

and the **law of sines** is

$$\frac{\sin \alpha_a}{a} = \frac{\sin \alpha_b}{b} = \frac{\sin \alpha_c}{c}.$$

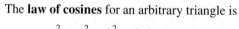
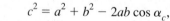

A.3 Derivatives

$$\frac{d}{dx} x^n = nx^{n-1}$$

$$\frac{d}{dx} e^x = e^x$$

$$\frac{d}{dx} \ln x = \frac{1}{x}$$

$$\frac{d}{dx} \sin x = \cos x$$

$$\frac{d}{dx} \cos x = -\sin x$$

$$\frac{d}{dx} \tan x = \frac{1}{\cos^2 x}$$

$$\frac{d}{dx} \sinh x = \cosh x$$

$$\frac{d}{dx} \cosh x = \sinh x$$

$$\frac{d}{dx} \tanh x = \frac{1}{\cosh^2 x}$$

A.4 Integrals

$$\int x^n \, dx = \frac{x^{n+1}}{n+1} \qquad (n \neq -1)$$

$$\int x^{-1} \, dx = \ln x$$

$$\int (a + bx)^{1/2} \, dx = \frac{2}{3b} (a + bx)^{3/2}$$

$$\int x(a + bx)^{1/2} \, dx = -\frac{2(2a - 3bx)(a + bx)^{3/2}}{15b^2}$$

$$\int (1 + a^2x^2)^{1/2} \, dx = \frac{1}{2}\left\{ x(1 + a^2x^2)^{1/2} + \frac{1}{a} \ln\left[x + \left(\frac{1}{a^2} + x^2 \right)^{1/2} \right] \right\}$$

$$\int x(1 + a^2x^2)^{1/2} \, dx = \frac{a}{3}\left(\frac{1}{a^2} + x^2 \right)^{3/2}$$

$$\int x^2(1 + a^2x^2)^{1/2} \, dx = \frac{1}{4} ax\left(\frac{1}{a^2} + x^2 \right)^{3/2} - \frac{1}{8a^2} x(1 + a^2x^2)^{1/2}$$
$$- \frac{1}{8a^3} \ln\left[x + \left(\frac{1}{a^2} + x^2 \right)^{1/2} \right]$$

$$\int (1 - a^2x^2)^{1/2} \, dx = \frac{1}{2}\left[x(1 - a^2x^2)^{1/2} + \frac{1}{a} \arcsin ax \right]$$

$$\int x(1 - a^2x^2)^{1/2} \, dx = -\frac{a}{3}\left(\frac{1}{a^2} - x^2 \right)^{3/2}$$

$$\int x^2(a^2 - x^2)^{1/2} \, dx = -\frac{1}{4} x(a^2 - x^2)^{3/2}$$
$$+ \frac{1}{8} a^2\left[x(a^2 - x^2)^{1/2} + a^2 \arcsin \frac{x}{a} \right]$$

$$\int \frac{dx}{(1 + a^2x^2)^{1/2}} = \frac{1}{a} \ln\left[x + \left(\frac{1}{a^2} + x^2 \right)^{1/2} \right]$$

$$\int \frac{dx}{(1 - a^2x^2)^{1/2}} = \frac{1}{a} \arcsin ax, \quad \text{or} \quad -\frac{1}{a} \arccos ax$$

$$\int \sin x \, dx = -\cos x$$

$$\int \cos x \, dx = \sin x$$

$$\int \sin^2 x \, dx = -\frac{1}{2} \sin x \cos x + \frac{1}{2} x$$

$$\int \cos^2 x \, dx = \frac{1}{2} \sin x \cos x + \frac{1}{2} x$$

$$\int \sin^3 x \, dx = -\frac{1}{3} \cos x (\sin^2 x + 2)$$

$$\int \cos^3 x \, dx = \frac{1}{3} \sin x (\cos^2 x + 2)$$

$$\int \cos^4 x \, dx = \frac{3}{8} x + \frac{1}{4} \sin 2x + \frac{1}{32} \sin 4x$$

$$\int \sin^n x \cos x \, dx = \frac{(\sin x)^{n+1}}{n+1} \qquad (n \neq -1)$$

$$\int \sinh x \, dx = \cosh x$$

$$\int \cosh x \, dx = \sinh x$$

$$\int \tanh x \, dx = \ln \cosh x$$

$$\int e^{ax} \, dx = \frac{e^{ax}}{a}$$

$$\int x \, e^{ax} \, dx = \frac{e^{ax}}{a^2}(ax - 1)$$

A.5 Taylor Series

The Taylor series of a function $f(x)$ is

$$f(a + x) = f(a) + f'(a)x + \frac{1}{2!}f''(a)x^2 + \frac{1}{3!}f'''(a)x^3 + \cdots,$$

where the primes indicate derivatives.

Some useful Taylor series are

$$e^x = 1 + x + \frac{x^2}{2!} + \frac{x^3}{3!} + \cdots,$$

$$\sin(a + x) = \sin a + (\cos a)x - \frac{1}{2}(\sin a)x^2 - \frac{1}{6}(\cos a)x^3 + \cdots,$$

$$\cos(a + x) = \cos a - (\sin a)x - \frac{1}{2}(\cos a)x^2 + \frac{1}{6}(\sin a)x^3 + \cdots,$$

$$\tan(a + x) = \tan a + \left(\frac{1}{\cos^2 a}\right)x + \left(\frac{\sin a}{\cos^3 a}\right)x^2$$

$$+ \left(\frac{\sin^2 a}{\cos^4 a} + \frac{1}{3\cos^2 a}\right)x^3 + \cdots.$$

A.6 Vector Analysis

Cartesian Coordinates

The gradient of a scalar field ψ is

$$\nabla\psi = \frac{\partial\psi}{\partial x}\mathbf{i} + \frac{\partial\psi}{\partial y}\mathbf{j} + \frac{\partial\psi}{\partial z}\mathbf{k}.$$

The divergence and curl of a vector field $\mathbf{v} = v_x\mathbf{i} + v_y\mathbf{j} + v_z\mathbf{k}$ are

$$\nabla\cdot\mathbf{v} = \frac{\partial v_x}{\partial x} + \frac{\partial v_y}{\partial y} + \frac{\partial v_z}{\partial z},$$

$$\nabla\times\mathbf{v} = \begin{vmatrix} \mathbf{i} & \mathbf{j} & \mathbf{k} \\ \dfrac{\partial}{\partial x} & \dfrac{\partial}{\partial y} & \dfrac{\partial}{\partial z} \\ v_x & v_y & v_z \end{vmatrix}.$$

Cylindrical Coordinates

The gradient of a scalar field ψ is

$$\nabla\psi = \frac{\partial\psi}{\partial r}\mathbf{e}_r + \frac{1}{r}\frac{\partial\psi}{\partial\theta}\mathbf{e}_\theta + \frac{\partial\psi}{\partial z}\mathbf{e}_z.$$

The divergence and curl of a vector field $\mathbf{v} = v_r\mathbf{e}_r + v_\theta\mathbf{e}_\theta + v_z\mathbf{e}_z$ are

$$\nabla\cdot\mathbf{v} = \frac{\partial v_r}{\partial r} + \frac{v_r}{r} + \frac{1}{r}\frac{\partial v_\theta}{\partial\theta} + \frac{\partial v_z}{\partial z},$$

$$\nabla\times\mathbf{v} = \frac{1}{r}\begin{vmatrix} \mathbf{e}_r & r\mathbf{e}_\theta & \mathbf{e}_z \\ \dfrac{\partial}{\partial r} & \dfrac{\partial}{\partial\theta} & \dfrac{\partial}{\partial z} \\ v_r & rv_\theta & v_z \end{vmatrix}.$$

Properties of Areas and Lines

B.1 Areas

The coordinates of the centroid of the area A are

$$\bar{x} = \frac{\int_A x \, dA}{\int_A dA}, \qquad \bar{y} = \frac{\int_A y \, dA}{\int_A dA}.$$

The moment of inertia about the x axis I_x, the moment of inertia about the y axis I_y, and the product of inertia I_{xy} are

$$I_x = \int_A y^2 \, dA, \qquad I_y = \int_A x^2 \, dA, \qquad I_{xy} = \int_A xy \, dA.$$

The polar moment of inertia about O is

$$J_O = \int_A r^2 \, dA = \int_A (x^2 + y^2) \, dA = I_x + I_y.$$

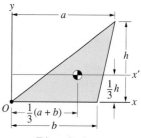

Triangular Area

$$\text{Area} = \frac{1}{2} bh$$

$$I_x = \frac{1}{12} bh^3, \qquad I_{x'} = \frac{1}{36} bh^3$$

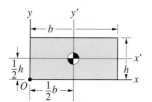

Rectangular Area

$$\text{Area} = bh$$

$$I_x = \frac{1}{3} bh^3, \qquad I_y = \frac{1}{3} hb^3, \qquad I_{xy} = \frac{1}{4} b^2 h^2$$

$$I_{x'} = \frac{1}{12} bh^3, \qquad I_{y'} = \frac{1}{12} hb^3, \qquad I_{x'y'} = 0$$

Circular Area

$$\text{Area} = \pi R^2 \qquad I_{x'} = I_{y'} = \frac{1}{4} \pi R^4, \qquad I_{x'y'} = 0$$

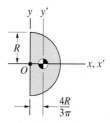

Semicircular Area

$$\text{Area} = \frac{1}{2} \pi R^2$$

$$I_x = I_y = \frac{1}{8} \pi R^4, \qquad I_{xy} = 0$$

$$I_{x'} = \frac{1}{8} \pi R^4, \qquad I_{y'} = \left(\frac{\pi}{8} - \frac{8}{9\pi}\right) R^4, \qquad I_{x'y'} = 0$$

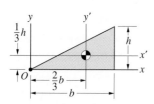

Triangular Area

$$\text{Area} = \frac{1}{2} bh$$

$$I_x = \frac{1}{12} bh^3, \qquad I_y = \frac{1}{4} hb^3, \qquad I_{xy} = \frac{1}{8} b^2 h^2$$

$$I_{x'} = \frac{1}{36} bh^3, \qquad I_{y'} = \frac{1}{36} hb^3, \qquad I_{x'y'} = \frac{1}{72} b^2 h^2$$

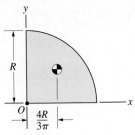

Quarter-Circular Area

$$\text{Area} = \frac{1}{4}\pi R^2$$

$$I_x = I_y = \frac{1}{16}\pi R^4, \qquad I_{xy} = \frac{1}{8}R^4$$

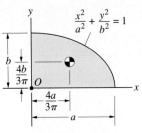

Quarter-Elliptical Area

$$\text{Area} = \frac{1}{4}\pi ab$$

$$I_x = \frac{1}{16}\pi ab^3, \qquad I_y = \frac{1}{16}\pi a^3 b, \qquad I_{xy} = \frac{1}{8}a^2 b^2$$

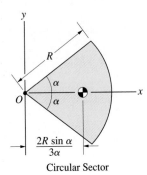

Circular Sector

$$\text{Area} = \alpha R^2$$

$$I_x = \frac{1}{4}R^4\left(\alpha - \frac{1}{2}\sin 2\alpha\right), \qquad I_y = \frac{1}{4}R^4\left(\alpha + \frac{1}{2}\sin 2\alpha\right),$$

$$I_{xy} = 0$$

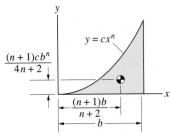

Spandrel

$$\text{Area} = \frac{cb^{n+1}}{n+1}$$

$$I_x = \frac{c^3 b^{3n+1}}{9n+3}, \qquad I_y = \frac{cb^{n+3}}{n+3}, \qquad I_{xy} = \frac{c^2 b^{2n+2}}{4n+4}$$

B.2 Lines

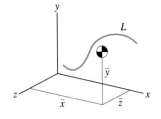

The coordinates of the centroid of the line L are

$$\bar{x} = \frac{\int_L x\,dL}{\int_L dL}, \qquad \bar{y} = \frac{\int_L y\,dL}{\int_L dL}, \qquad \bar{z} = \frac{\int_L z\,dL}{\int_L dL}.$$

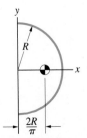

Semicircular Arc

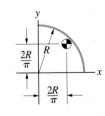

Quarter-Circular Arc

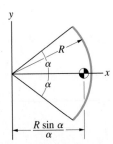

Circular Arc

Properties of Volumes and Homogeneous Objects

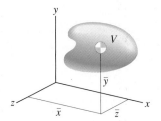

The coordinates of the centroid of the volume V are

$$\bar{x} = \frac{\int_V x \, dV}{V}, \qquad \bar{y} = \frac{\int_V y \, dV}{V}, \qquad \bar{z} = \frac{\int_V z \, dV}{V}.$$

(The center of mass of a homogeneous object coincides with the centroid of its volume.)

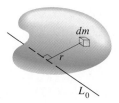

The mass moment of inertia of the object about the axis L_0 is

$$I_0 = \int_m r^2 \, dm.$$

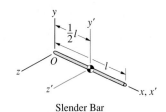

Slender Bar

$$I_{(x \text{ axis})} = 0, \qquad I_{(y \text{ axis})} = I_{(z \text{ axis})} = \frac{1}{3} ml^2$$

$$I_{(x' \text{ axis})} = 0, \qquad I_{(y' \text{ axis})} = I_{(z' \text{ axis})} = \frac{1}{12} ml^2$$

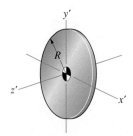

Thin Circular Plate

$$I_{(x' \text{ axis})} = I_{(y' \text{ axis})} = \frac{1}{4} mR^2, \qquad I_{(z' \text{ axis})} = \frac{1}{2} mR^2$$

$$I_{(x \text{ axis})} = \frac{1}{3} mh^2, \qquad I_{(y \text{ axis})} = \frac{1}{3} mb^2, \qquad I_{(z \text{ axis})} = \frac{1}{3} m(b^2 + h^2)$$

$$I_{(x' \text{ axis})} = \frac{1}{12} mh^2, \qquad I_{(y' \text{ axis})} = \frac{1}{12} mb^2, \qquad I_{(z' \text{ axis})} = \frac{1}{12} m(b^2 + h^2)$$

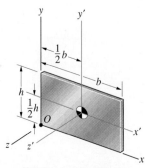

Thin Rectangular Plate

585

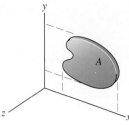

Thin Plate

$$I_{(x\text{ axis})} = \frac{m}{A} I_x^A, \quad I_{(y\text{ axis})} = \frac{m}{A} I_y^A, \quad I_{(z\text{ axis})} = I_{(x\text{ axis})} + I_{(y\text{ axis})}$$

(The superscripts A denote moments of inertia of the plate's cross-sectional area A.)

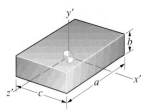

Rectangular Prism

Volume $= abc$

$$I_{(x'\text{ axis})} = \frac{1}{12} m(a^2 + b^2), \quad I_{(y'\text{ axis})} = \frac{1}{12} m(a^2 + c^2),$$

$$I_{(z'\text{ axis})} = \frac{1}{12} m(b^2 + c^2)$$

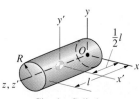

Circular Cylinder

Volume $= \pi R^2 l$

$$I_{(x\text{ axis})} = I_{(y\text{ axis})} = m\left(\frac{1}{3} l^2 + \frac{1}{4} R^2\right), \quad I_{(z\text{ axis})} = \frac{1}{2} mR^2$$

$$I_{(x'\text{ axis})} = I_{(y'\text{ axis})} = m\left(\frac{1}{12} l^2 + \frac{1}{4} R^2\right), \quad I_{(z'\text{ axis})} = \frac{1}{2} mR^2$$

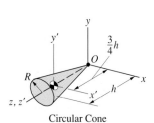

Circular Cone

Volume $= \frac{1}{3} \pi R^2 h$

$$I_{(x\text{ axis})} = I_{(y\text{ axis})} = m\left(\frac{3}{5} h^2 + \frac{3}{20} R^2\right), \quad I_{(z\text{ axis})} = \frac{3}{10} mR^2$$

$$I_{(x'\text{ axis})} = I_{(y'\text{ axis})} = m\left(\frac{3}{80} h^2 + \frac{3}{20} R^2\right), \quad I_{(z'\text{ axis})} = \frac{3}{10} mR^2$$

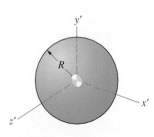

Sphere

Volume $= \frac{4}{3} \pi R^3$

$$I_{(x'\text{ axis})} = I_{(y'\text{ axis})} = I_{(z'\text{ axis})} = \frac{2}{5} mR^2$$

Chapter 1

1.2 2.7183.

1.4 7.32 m wide, 2.44 m high.

1.6 The 1-in. wrench fits the 25-mm nut.

1.8 149 mi/hr.

1.10 (a) 5000 m/s; (b) 3.11 mi/s.

1.12 $g = 32.2$ ft/s^2.

1.14 0.310 m^2.

1.16 2.07×10^6 Pa.

1.18 $G = 3.44 \times 10^{-8}$ lb-ft^2/slug^2.

1.20 (a) The SI units of T are kg-m^2/s^2; (b) $T = 73.8$ slug-ft^2/s^2.

1.22 (a) Fully fueled, 8490 slugs; fuel expended, 1950 slugs; (b) fully fueled, 1.22 MN; fuel expended, 0.279 MN.

1.24 (a) 491 N; (b) 81.0 N.

1.26 163 lb.

1.28 32.1 km.

1.30 345,000 km.

Chapter 2

2.2 $|\mathbf{F}| = 177$ N, angle $= 151°$.

2.4 $|\mathbf{F}| = 176.8$ N, angle $= 151.3°$.

2.6 $|\mathbf{F}_B| = 133$ N.

2.8 $|\mathbf{F}_A| = 5.22$ kN, $|\mathbf{F}_B| = 3.57$ kN.

2.10 $|\mathbf{F}_{AB}| = 3.42$ kN, $|\mathbf{F}_{BC}| = 5.77$ kN.

2.12 AB: 1202 lb. AD: 559 lb.

2.14 (a) $|\mathbf{r}_A + \mathbf{r}_B| = 70$ m; (b) $|\mathbf{r}_A + \mathbf{r}_B| = 50$ m.

2.16 $|\mathbf{F}_{BA} + \mathbf{F}_{BC}| = 918$ N.

2.18 $|\mathbf{r}| = 390$ m, $\alpha = 21.2°$.

2.22 $F_y = -90$ lb.

2.24 $|\mathbf{F}| = 447$ kip.

2.26 $V_x = 16$, $V_y = 12$ or $V_x = -16$, $V_y = -12$.

2.28 $\mathbf{F} = 56.4\mathbf{i} + 20.5\mathbf{j}$ (lb).

2.30 $\mathbf{r}_{AB} = -4\mathbf{i} - 3\mathbf{j}$ (m).

2.32 $\mathbf{r}_{AB} + \mathbf{r}_{BC} + \mathbf{r}_{CD} + \mathbf{r}_{DE} = \mathbf{r}_{AE} = 3.46\mathbf{j}$ (m).

2.34 (a) $\mathbf{r}_{AB} = 48\mathbf{i} + 15\mathbf{j}$ (in.); (b) $\mathbf{r}_{BC} = -53\mathbf{i} + 5\mathbf{j}$ (in.); (c) $|\mathbf{r}_{AB} + \mathbf{r}_{BC}| = 20.6$ in.

2.36 (a) $\mathbf{r}_{AB} = 52.0\mathbf{i} + 30\mathbf{j}$ (mm); (b) $\mathbf{r}_{AB} = -42.4\mathbf{i} - 42.4\mathbf{j}$ (mm).

2.38 $x_B = 785$ m, $y_B = 907$ m or $x_B = 255$ m, $y_B = 1173$ m.

2.40 $\mathbf{e} = -0.781\mathbf{i} - 0.625\mathbf{j}$.

2.42 $\mathbf{e} = \frac{3}{5}\mathbf{i} - \frac{4}{5}\mathbf{j}$.

2.44 $\mathbf{r}_{AB} = -1416\mathbf{i} + 909\mathbf{j}$ (ft).

2.46 (a) $\mathbf{r}_M = (57 \times 10^6)\,\mathbf{i}$ (km), $\mathbf{r}_V = -(82.7 \times 10^6)\,\mathbf{i} - (69.4 \times 10^6)\,\mathbf{j}$ (km), and $\mathbf{r}_E = -(51.3 \times 10^6)\,\mathbf{i} + (141.0 \times 10^6)\,\mathbf{j}$ (km); (b) the distance from the Earth to Mercury is 177.8×10^6 km and the distance from the Earth to Venus is 212.7×10^6 km.

2.48 $|\mathbf{F}_{BA} + \mathbf{F}_{BC}| = 918$ N.

2.50 $|\mathbf{F}_A| = 1720$ lb, $\alpha = 33.3°$.

2.52 $82.8° \leq \alpha \leq 277.2°$.

2.54 $|\mathbf{F}_A| = 10$ kN, $|\mathbf{F}_D| = 8.66$ kN.

2.56 $|\mathbf{L}| = 216.1$ lb, $|\mathbf{D}| = 78.7$ lb.

2.58 $|\mathbf{F}_A| = 68.2$ kN.

2.60 $|\mathbf{F}_{AC}| = 2.11$ kN, $|\mathbf{F}_{AD}| = 2.76$ kN.

2.62 $x = 75 - 0.880s$, $y = 12 + 0.476s$.

2.64 $\mathbf{r} = (0.814s - 6)\,\mathbf{i} + (0.581s + 1)\,\mathbf{j}$ (m).

2.66 $|\mathbf{F}| = 110$ N.

2.68 $U_x = 3.61$, $U_y = -7.22$, $U_z = -28.89$ or $U_x = -3.61$, $U_y = 7.22$, $U_z = 28.89$.

2.70 (a) $|\mathbf{U}| = 7$, $|\mathbf{V}| = 13$; (b) $|3\mathbf{U} + 2\mathbf{V}| = 27.5$.

2.72 $\theta_x = 56.9°$, $\theta_y = 129.5°$, $\theta_z = 56.9°$.

2.74 $\mathbf{F} = -0.5\mathbf{i} + 0.2\mathbf{j} + 0.843\mathbf{k}$.

2.76 (a) 11 ft; (b) $\cos \theta_x = -0.545$, $\cos \theta_y = 0.818$, $\cos \theta_z = 0.182$.

2.78 (a) 5.39 N; (b) $0.557\mathbf{i} - 0.743\mathbf{j} - 0.371\mathbf{k}$.

2.80 $\mathbf{F} = 40\mathbf{i} + 40\mathbf{j} - 70\mathbf{k}$ (kN).

2.82 (a) $|\mathbf{r}_{AB}| = 16.2$ m; (b) $\cos \theta_x = 0.615$, $\cos \theta_y = -0.492$, $\cos \theta_z = -0.615$.

2.84 $\mathbf{r}_{AR}$: $\cos \theta_x = 0.667$, $\cos \theta_y = 0.667$, $\cos \theta_z = 0.333$. $\mathbf{r}_{BR}$: $\cos \theta_x = -0.242$, $\cos \theta_y = 0.970$, $\cos \theta_z = 0$.

2.86 29,100 ft.

2.88 $\mathbf{r} = 70.7\mathbf{i} + 61.2\mathbf{j} + 35.4\mathbf{k}$ (in.).

2.90 $\mathbf{r}_{OP} = R_E(0.612\mathbf{i} + 0.707\mathbf{j} + 0.354\mathbf{k})$.

2.92 (a) $\mathbf{e}_{AB} = 0.545\mathbf{i} - 0.545\mathbf{j} + 0.636\mathbf{k}$; (b) $\mathbf{T} = 60\mathbf{i} - 60\mathbf{j} + 70\mathbf{k}$ (lb).

2.94 $\cos \theta_x = 0.545$, $\cos \theta_y = 0.636$, $\cos \theta_z = -0.545$.

2.96 259 lb.

2.98 $|\mathbf{F}_{AC}| = 1116$ N, $|\mathbf{F}_{AD}| = 910$ N.

2.100 $\mathbf{T} = -15.4\mathbf{i} + 27.0\mathbf{j} + 7.7\mathbf{k}$ (lb).

2.102 $\mathbf{T} = -41.1\mathbf{i} + 28.8\mathbf{j} + 32.8\mathbf{k}$ (N).

2.104 $\mathbf{U} \cdot \mathbf{V} = -300$.

2.106 -250 ft-lb.

2.108 $U_x = -12$.

2.112 $81.6°$.

2.114 $\theta = 53.5°$.

2.116 Parallel component is $12\mathbf{i} - 4\mathbf{j} + 6\mathbf{k}$ (kN), normal component is $9\mathbf{i} + 18\mathbf{j} - 6\mathbf{k}$ (kN).

2.118 (a) 42.5°; (b) $-423\mathbf{j} + 604\mathbf{k}$ (lb).

2.120 8.25 ft.

2.122 $\mathbf{T}_n = -37.1\mathbf{i} + 31.6\mathbf{j} + 8.2\mathbf{k}$ (N).

2.124 $\mathbf{T}_n = -27.1\mathbf{i} - 9.85\mathbf{j} + 32.8\mathbf{k}$ (N).

2.126 $\mathbf{v}_p = -1.30\mathbf{i} - 1.68\mathbf{j} - 3.36\mathbf{k}$ (m/s).

2.128 $\mathbf{U} \times \mathbf{V} = -20\mathbf{i} - 6\mathbf{j} - 18\mathbf{k}$.

2.130 $\mathbf{r} \times \mathbf{F} = -80\mathbf{i} + 120\mathbf{j} - 40\mathbf{k}$ (N-m).

2.132 (a) $\mathbf{U} \times \mathbf{V} = 0$; (b) They are parallel.

2.134 (a), (c) $\mathbf{U} \times \mathbf{V} = -51.8\mathbf{k}$; (b), (d) $\mathbf{V} \times \mathbf{U} = 51.8\mathbf{k}$.

2.138 (a) $\mathbf{r}_{OA} \times \mathbf{r}_{OB} = -4\mathbf{i} + 36\mathbf{j} + 32\mathbf{k}$ (m^2);
(b) $-0.083\mathbf{i} + 0.745\mathbf{j} + 0.662\mathbf{k}$ or $0.083\mathbf{i} - 0.745\mathbf{j} - 0.662\mathbf{k}$.

2.140 $\mathbf{r}_{AB} \times \mathbf{F} = -2400\mathbf{i} + 9600\mathbf{j} + 7200\mathbf{k}$ (ft-lb).

2.142 $\mathbf{r}_{CA} \times \mathbf{T} = -4.72\mathbf{i} - 3.48\mathbf{j} - 7.96\mathbf{k}$ (N-m).

2.144 $x_B = 2.81$ m, $y_B = 6.75$ m, $z_B = 3.75$ m.

2.146 $\mathbf{U} \cdot (\mathbf{V} \times \mathbf{W}) = -4$.

2.148 1.8×10^6 mm^2.

2.150 $U_y = -2$.

2.152 $|\mathbf{A}| = 1110$ lb, $\alpha = 29.7°$.

2.154 $|\mathbf{E}| = 313$ lb, $|\mathbf{F}| = 140$ lb.

2.156 $\mathbf{e}_{AB} = 0.625\mathbf{i} - 0.469\mathbf{j} - 0.625\mathbf{k}$.

2.158 $\mathbf{F}_p = 8.78\mathbf{i} - 6.59\mathbf{j} - 8.78\mathbf{k}$ (lb).

2.160 $\mathbf{r}_{BA} \times \mathbf{F} = -70\mathbf{i} + 40\mathbf{j} - 100\mathbf{k}$ (ft-lb).

2.162 (a), (b) $686\mathbf{i} - 486\mathbf{j} - 514\mathbf{k}$ (ft-lb).

2.164 $\mathbf{F}_A = 18.2\mathbf{i} + 19.9\mathbf{j} + 15.3\mathbf{k}$ (N),
$\mathbf{F}_B = -7.76\mathbf{i} + 26.9\mathbf{j} + 13.4\mathbf{k}$ (N).

2.166 $\mathbf{F}_p = 1.29\mathbf{i} - 3.86\mathbf{j} + 2.57\mathbf{k}$ (kN),
$\mathbf{F}_n = -1.29\mathbf{i} - 2.14\mathbf{j} - 2.57\mathbf{k}$ (kN).

2.168 $\mathbf{r}_{AG} \times \mathbf{W} = -16.4\mathbf{i} - 82.4\mathbf{k}$ (N-m).

2.170 $\mathbf{r}_{BC} \times \mathbf{T} = -12.0\mathbf{i} - 138.4\mathbf{j} - 117.4\mathbf{k}$ (N-m).

Chapter 3

3.2 $F_2 = 95.4$ lb, $F_3 = 50.8$ lb.

3.4 $A_x = 50$ kN, $A_y = 173$ kN.

3.6 $T = 763$ N, $M = 875$ N.

3.8 $k = 1960$ N/m, $m_A = 4$ kg, $m_B = 6$ kg.

3.10 Normal force $= 196,907$ N, friction force $= 707$ N.

3.12 $\alpha = 31.0°$.

3.14 (a) 254 lb; (b) 41.8°.

3.16 150 lb.

3.18 116 N.

3.20 (a), (b) 43.6 lb.

3.22 53.2 lb.

3.24 (a) 56.4 lb; (b) 345.5 lb.

3.26 No. The tension in cables BD and CE would be 4.14 kN.

3.28 Upper cable tension is $0.828W$, lower cable tension is $0.132W$.

3.30 $T_{AB} = 1.21$ N, $T_{AD} = 2.76$ N.

3.32 (a) $\beta = 55°$; (b) $T_{AB} = 0.574T$, $T_{AC} = 0.819T$.

3.34 $m = 10.9$ kg.

3.36 21.9 lb.

3.38 $T_{AB} = 688$ lb.

3.40 AB: 64.0 kN, BC: 61.0 kN.

3.44 (a) 196 N; (b) $\alpha = 41.8°$.

3.46 $T = 1330$ lb.

3.48 (b) Left surface: 36.6 lb; right surface: 25.9 lb.

3.50 202 N.

3.52 Normal force $= 2987$ lb, friction force $= 942$ lb.

3.54 $T = m_A g/7 - (4/7)\ mg$.

3.56 $x = \frac{1}{2}(b - h \cot 30°)$, $y = -\frac{1}{2}(b \tan 30° - h)$.

3.60 $D = 7880$ lb, $L = 28,610$ lb.

3.62 (a) $\gamma = -14.0°$; (b) 4 km.

3.64 $T_{AB} = 780$ N, $T_{AC} = 1976$ N, $T_{AD} = 2568$ N.

3.66 $T_{AC} = 20.6$ lb, $T_{AD} = 21.4$ lb, $T_{AE} = 11.7$ lb.

3.68 $T_{AB} = 98.2$ N, $T_{AC} = 155.0$ N, $T_{AD} = 177.3$ N.

3.70 $T_{AB} = 10,270$ lb, $T_{AC} = 4380$ lb, $T_{AD} = 11,010$ lb.

3.72 $T_{AB} = 101$ N, $T_{AC} = 889$ N, magnitude $= 1005$ N.

3.74 12.3 lb.

3.76 $T_{EF} = T_{EG} = 738$ kN.

3.78 (a) The tension $= 2.70$ kN; (b) The force exerted by the bar $= 1.31\mathbf{i} - 1.31\mathbf{k}$ (kN).

3.80 $T_{AB} = 357$ N.

3.82 $F = 36.6$ N.

3.84 18.0 ft.

3.86 (b) $b < 10.0$ m or $b > 30.0$ m.

3.88 $\alpha = 62.7°$.

3.90 $h = 3.08$ m.

3.92 $s = 0.305$ m.

3.94 $s = 2.65$ m.

3.96 $W = 25.0$ lb.

3.98 (a) 83.9 lb; (b) 230.5 lb.

3.100 $F = 20$ lb.

3.102 $F = 162.0$ N.

3.104 $T_{OB} = 3.60$ kN, $T_{OC} = 1.94$ kN, $T_{OD} = 2.02$ kN.

3.106 $T = mgL/(R + h)$.

3.108 $T_{AB} = 1.54$ lb, $T_{AC} = 1.85$ lb.

3.110 Normal force $= 12.15$ kN, friction force $= 4.03$ kN.

Chapter 4

4.2 (a) −300 ft-lb; (b) Zero; (c) 200 ft-lb.

4.4 −40.5 N-m.

4.6 (a) 1 m; (b) 30° or 150°.

4.8 F = 4 kips.

4.10 (a) 1 m; (b) 53.1° or 180°.

4.12 229 ft-lb.

4.14 (a) −24.9 ft-lb; (b) −50.9 ft-lb.

4.16 (a)–(c) Zero.

4.18 11,200 ft-lb.

4.20 F_1 = 30 kN, F_2 = 50 kN.

4.22 (a) F_A = 24.6 N, F_B = 55.4 N; (b) Zero.

4.24 (a) −5600 ft-lb; (b) −600 ft-lb.

4.26 1760 ft-lb.

4.28 (a) A_x = 18.1 kN, A_y = −29.8 kN, B = −20.4 kN; (b) Zero.

4.30 (a) A_x = 300 lb, A_y = 240 lb, B = 280 lb; (b) Zero.

4.32 186 kg.

4.34 −22.3 ft-lb.

4.36 T_{AB} = 10.98 kN, T_{AC} = 7.76 kN.

4.38 671 lb.

4.40 617 N-m.

4.42 1040 lb.

4.44 −48.4 ft-lb.

4.46 (a) 480 N-m; (b)–(d) 480**k** (N-m).

4.48 (a) 800**k** (kN-m); (b) −400**k** (kN-m).

4.50 **F** = 20**i** + 40**j** (N).

4.52 (a), (b) Zero.

4.54 (a) 1270 kN-m; (b) M_A = 900**i** + 900**j** (N-m); (c) 1270 kN-m.

4.56 450**i** − 100**j** − 200**k** (ft-lb).

4.58 **F** = 40**i** + 40**j** + 70**k** (N) or **F** = −40**i** − 40**j** − 70**k** (N).

4.60 58.0 kN.

4.62 |**F**| = 364 lb.

4.64 M_O = −rF sin α **i** − (h + r cos α)F **k**.

4.66 1540 ft-lb.

4.68 (a) 667**i** − 2667**j** − 4000**k** (N-m); (b) −667**i** + 2667**j** + 4000**k** (N-m); (c) 9.71 m.

4.70 T_{AC} = 2.23 kN, T_{AD} = 2.43 kN.

4.72 T_{AB} = 1.60 kN, T_{AC} = 1.17 kN.

4.74 F = 2530 lb.

4.76 −16.4**i** − 111.9**k** (N-m).

4.78 (a) 80**i** (N-m); (b) −140**j** (N-m); (c) Zero.

4.80 (a) Zero; (b) 2.7**k** (kN-m).

4.82 −300**i** (ft-lb).

4.84 **F** = 80**i** + 80**j** + 40**k** (lb).

4.86 −16.4**i** (N-m).

4.88 M_{AB} = −76.1**i** − 95.1**j** (N-m).

4.90 M_{AO} = 119.1**j** + 79.4**k** (N-m).

4.92 77.1**j** − 211.9**k** (ft-lb).

4.94 $M_{(y\ axis)}$ = 215**j** (N-m).

4.96 $M_{(x\ axis)}$ = 44**i** (N-m).

4.98 −338**j** (ft-lb).

4.100 |**F**| = 13 lb.

4.102 cos θ_x = 0.800, cos θ_y = 0.536, cos θ_z = 0.272.

4.104 1 N-m.

4.106 124**k** (ft-lb).

4.108 (a), (b) 40**k** (ft-lb).

4.110 (a) b = 3.14 m; (b) −90**k** (N-m).

4.112 (a), (b) −400**k** (N-m).

4.114 40 ft-lb clockwise, or −40**k** (ft-lb).

4.116 2200 ft-lb clockwise.

4.118 330 N-m counterclockwise, or 330**k** (N-m).

4.120 (a) −5**i** − **j** − 17**k** (kN-m); (b) 4.74 m.

4.122 356 ft-lb.

4.124 2 kN.

4.126 M_{Cy} = 7 kN-m, M_{Cz} = −2 kN-m.

4.128 Yes.

4.130 Systems 1, 2, and 4 are equivalent.

4.134 F = 265 N.

4.136 F = 70 lb, M = 130 in-lb.

4.138 (a) **F** = −10**j** (lb), M = −10 ft-lb; (b) D = 1 ft.

4.140 (a) A_x = 0, A_y = 20 lb, B = 80 lb; (b) **F** = 100**j** (lb), M = −1120 in-lb.

4.142 (a) A_x = 12 kip, A_y = 10 kip, B = −10 kip; (b) **F** = −12**i** (kip), intersects at y = 5 ft; (c) they are both zero.

4.144 (a) **F** = 680**i** (kN), y = −0.0588 m; (b) 166 kN.

4.146 **F** = 100**j** (lb), **M** = **0**.

4.148 (a) **F** = 920**i** − 390**j** (N), M = −419 N-m; (b) intersects at y = 456 mm.

4.150 **F** = 800**j** (lb), intersects at x = 7.5 in.

4.152 (a) −360**k** (in-lb); (b) −36**j** (in-lb); (c) **F** = 10**i** − 30**j** + 3**k** (lb), **M** = −36**j** −360**k** (in-lb).

4.154 (a) **F** = 600**i** (lb), **M** = 1400**j** − 1800**k** (ft-lb); (b) **F** = 600**i** (lb), intersects at y = 3 ft, z = 2.33 ft.

4.156 **F** = 100**j** + 80**k** (N), **M** = 240**j** − 300**k** (N-m).

4.158 (a) **F** = **0**, **M** = rA**i**; (b) **F′** = **0**, **M′** = rA**i**.

4.160 (a) **F** = **0**, **M** = 4.60**i** + 1.86**j** − 3.46**k** (kN-m); (b) 6.05 kN-m.

4.162 **F** = −20**i** + 20**j** + 10**k** (lb), **M** = 50**i** + 250**j** + 100**k** (in-lb).

4.164 (a) $\mathbf{F} = 28\mathbf{k}$ (kip), $\mathbf{M} = 96\mathbf{i} - 192\mathbf{j}$ (ft-kip); (b) $x = 6.86$ ft, $y = 3.43$ ft.

4.166 $\mathbf{F} = 100\mathbf{i} + 20\mathbf{j} - 20\mathbf{k}$ (N),
$\mathbf{M} = -143\mathbf{i} + 406\mathbf{j} - 280\mathbf{k}$ (N-m).

4.168 $\mathbf{M}_t = \mathbf{0}$, line of action intersects at $y = 0$, $z = 2$ ft.

4.170 $x = 2.41$ m, $y = 3.80$ m.

4.172 $\mathbf{F} = 40.8\mathbf{i} + 40.8\mathbf{j} + 81.6\mathbf{k}$ (N),
$\mathbf{M} = -179.6\mathbf{i} + 391.9\mathbf{j} - 32.7\mathbf{k}$ (N-m).

4.174 (a) $320\mathbf{i}$ (in-lb); (b) $\mathbf{F} = -20\mathbf{k}$ (lb), $\mathbf{M} = 320\mathbf{i} + 660\mathbf{j}$ (in-lb); (c) $\mathbf{M}_t = \mathbf{0}$, $x = 33$ in., $y = -16$ in.

4.176 $k = 124$ lb/ft.

4.178 $M_A = 13{,}200$ ft-lb at $\alpha = 48.2°$.

4.180 $d = 13.0$ ft, moment is $265\mathbf{k}$ (ft-lb).

4.182 $T_{AB} = 155$ N, $T_{CD} = 445$ N.

4.184 (a) 160 N-m; (b) $160\mathbf{k}$ (N-m).

4.186 No. The moment is $mg \sin \alpha$ counterclockwise, where α is the clockwise angle.

4.188 (a) -76.2 N-m; (b) -66.3 N-m.

4.190 $|\mathbf{F}| = 224$ lb, $|\mathbf{M}| = 1600$ ft-lb.

4.192 671 lb.

4.194 $-228.1\mathbf{i} - 68.4\mathbf{k}$ (N-m).

4.196 $\mathbf{M}_{(x \text{ axis})} = -153\mathbf{i}$ (ft-lb).

4.198 $\mathbf{M}_{CD} = -173\mathbf{i} + 1038\mathbf{k}$ (ft-lb).

4.200 (a) $T_{AB} = T_{CD} = 173$ lb; (b) $\mathbf{F} = 300\mathbf{j}$ (lb), intersects at $x = 4$ ft.

4.202 $\mathbf{F} = -20\mathbf{i} + 70\mathbf{j}$ (N), $M = 22$ N-m.

4.204 $\mathbf{F}' = -100\mathbf{i} + 40\mathbf{j} + 30\mathbf{k}$ (lb), $\mathbf{M} = -80\mathbf{i} + 200\mathbf{k}$ (in-lb).

4.206 $\mathbf{F} = 1166\mathbf{i} + 566\mathbf{j}$ (N), $y = 13.9$ m.

4.208 $\mathbf{F} = 190\mathbf{j}$ (N), $\mathbf{M} = -98\mathbf{i} + 184\mathbf{k}$ (N-m).

4.210 $\mathbf{F} = -0.364\mathbf{i} + 4.908\mathbf{j} + 1.090\mathbf{k}$ (kN),
$\mathbf{M} = -0.131\mathbf{i} - 0.044\mathbf{j} + 1.112\mathbf{k}$ (kN-m).

Chapter 5

5.4 (b) $A_x = 0$, $A_y = -500$ lb, $B = 1000$ lb.

5.6 (b) $A_x = 0$, $A_y = -2.67$ kN, $B = 3.89$ kN.

5.8 (b) $A_x = 0$, $A_y = 5$ kN, $B = 5$ kN.

5.10 (b) $A = 100$ lb, $B = 200$ lb.

5.12 (b) $A_x = 1.15$ kN, $A_y = 0$, $B = 2.31$ kN.

5.14 $A_x = -26.7$ kN, $B_x = 26.7$ kN, $B_y = -40$ kN.

5.16 Each hand, 65.1 lb; each foot, 22.4 lb.

5.18 (b) $A_x = 0$, $A_y = -1000$ lb, $M_A = -12{,}800$ ft-lb.

5.20 849 lb.

5.22 $W_A = 3.20$ oz, $W_B = 9.60$ oz, $W_C = 2.49$ oz, $W_D = 0.71$ oz.

5.24 $R = 12.5$ lb, $B_x = 11.3$ lb, $B_y = 15.3$ lb.

5.26 (a) $A = 53.8$ lb, $B = 46.2$ lb; (b) $F = 21.2$ lb.

5.28 $A = 9211$ N, $B_x = 0$, $B_y = 789$ N.

5.30 $T = 4.71$ lb.

5.34 Tension is 50 lb, $C_x = -43.3$ lb, $C_y = 25$ lb.

5.36 $A_x = 0$, $A_y = 1.5F$, $B = 2.5F$.

5.38 $A_x = -200$ lb, $A_y = -100$ lb, $M_A = 1600$ ft-lb.

5.40 $0.354W$.

5.42 $A_x = 3.46$ kN, $A_y = -2$ kN, $B_x = -3.46$ kN, $B_y = 2$ kN.

5.44 Tension is 1200 N, $A_x = 1039$ N, $A_y = 600$ N.

5.46 $A_x = -1.57$ kN, $A_y = 1.57$ kN, $E = 1.57$ kN.

5.48 $A_x = 0$, $A_y = 200$ lb, $M_A = 900$ ft-lb.

5.50 $A_x = 57.7$ lb, $A_y = -13.3$ lb, $B = 15.3$ lb.

5.52 2500 lb.

5.54 $2 \le x \le 4$ ft.

5.56 $0.612W$.

5.58 20.3 kN.

5.60 $W_2 = 2484$ lb, $A_x = -2034$ lb, $A_y = 2425$ lb.

5.62 $W = 46.2$ N, $A_x = 22.3$ N, $A_y = 61.7$ N.

5.64 $F = 44.5$ lb, $A_x = 25.3$ lb, $A_y = -1.9$ lb.

5.66 $W = 132$ lb.

5.68
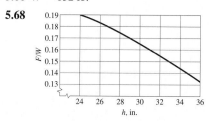

5.76 (1) and (2) are improperly supported.
For (3), reactions are $A = F/2$, $B = F/2$, $C = F$.

5.78 $A_x = 0$, $A_y = 360$ N, $A_z = 0$, $M_{Ax} = 16.4$ N-m, $M_{Ay} = 0$, $M_{Az} = 111.9$ N-m.

5.80 $T_{BC} = 20.3$ kN.

5.82 $O_x = \pm 21.6$ kN, $O_y = 0.6$ kN, $O_z = 0$, $M_{Ox} = -4.8$ kN-m, $M_{Oy} = \pm 172.5$ kN-m, $M_{Oz} = \pm 172.5$ kN-m.

5.84 59.1 kN.

5.86 $O_x = \pm 200$ lb, $O_y = \pm 200$ lb, $O_z = 0$, $M_{Ox} = \pm 1200$ in-lb, $M_{Oy} = \pm 1200$ in-lb, $M_{Oz} = \pm 2560$ in-lb.

5.88 $|\mathbf{F}|_{\max} = 2150$ lb.

5.90 $T_{AB} = 553$ lb, $T_{AC} = 289$ lb, $O_x = 632$ lb, $O_y = 574$ lb, $O_z = 0$.

5.92 $T_A = 3.72$ kN, $T_B = 2.60$ kN, $T_C = 1.53$ kN.

5.94 $T_A = 54.7$ lb, $T_B = 22.7$ lb, $T_C = 47.7$ lb.

5.96 Force is $800\mathbf{j}$ (lb); the line of action intersects at $z = 1.5$ ft.

5.98 (b) $A_x = -0.74$ kN, $A_y = 1$ kN, $A_z = -0.64$ kN, $B_x = 0.74$ kN, $B_z = 0.64$ kN.

5.100 $F_y = 40.5$ lb.

5.102 $T_{BD} = 1.47$ kN, $T_{BE} = 1.87$ kN, $A_x = 0$, $A_y = 4.24$ kN, $A_z = 0$.

5.104 1500 lb.

5.106 Tension is 60 N, $B_x = -10$ N, $B_y = 90$ N, $B_z = 10$ N, $M_{By} = 1$ N-m, $M_{Bz} = -3$ N-m.

5.108 Tension is 60 N, $B_x = -10$ N, $B_y = 75$ N, $B_z = 15$ N, $C_y = 15$ N, $C_z = -5$ N.

5.110 $A_x = -2.86$ kip, $A_y = 17.86$ kip, $A_z = -8.10$ kip, $B_y = 3.57$ kip, $B_z = 12.38$ kip.

5.112 $E_x = 0.67$ kN, $E_y = -1.33$ kN, $E_z = 2.67$ kN, $F_x = 4.67$ kN, $F_y = 6.67$ kN.

5.114 Magnitude at A is 85.4 lb, magnitude at B is 107.5 lb.

5.116 $C_y = 0$, $A_y = 66.7$ lb.

5.118 $T_{AB} = 488$ lb, $T_{CD} = 373$ lb, reaction is $31\mathbf{i} + 823\mathbf{j} - 87\mathbf{k}$ (lb).

5.120 $A_x = 474$ N, $A_y = -825$ N, $A_z = -1956$ N; $B_x = 860$ N, $B_y = 2380$ N, $B_z = -44$ N.

5.122 $\alpha = 52.4°$.

5.124 Tension is 33.3 lb; magnitude of reaction is 44.1 lb.

5.126 $\alpha = 73.9°$, magnitude at A is 4.32 kN, magnitude at B is 1.66 kN.

5.128 (a) No, because of the 3 kN-m couple; (b) magnitude at A is 7.88 kN; magnitude at B is 6.66 kN; (c) no.

5.130 (b) $A_x = -8$ kN, $A_y = 2$ kN, $C = 8$ kN.

5.134 $\alpha = 75.96°$, 30.96°, 12.53°, 4.40°, and zero.

5.136 $h = 2.46$ m, $A_x = 2.036$ kN, $A_y = 0.333$ kN.

5.138 37.1°.

5.140 $A_x = -346.4$ N, $A_y = 47.6$ N, $B_y = 152.4$ N.

5.142 (a) There are four unknown reactions and three equilibrium equations; (b) $A_x = -50$ lb, $B_x = 50$ lb.

5.144 (b) Force on nail = 55 lb, normal force = 50.77 lb, friction force = 9.06 lb.

5.146 $A = 500$ N, $B_x = 0$, $B_y = -800$ N.

5.148 $A = 727$ lb, $H_x = 225$ lb, $H_y = 113$ lb.

5.150 $\alpha = 0$ and $\alpha = 59.4°$.

5.152 $C = 76.0$ kN, $A_x = -11.6$ kN, $A_y = -35.9$ kN.

5.154 The force is 800 N upward; its line of action passes through the midpoint of the plate.

5.156 $m = 67.2$ kg.

5.158 $\alpha = 90°$, $T_{BC} = W/2$, $A = W/2$.

Chapter 6

6.2 (a) $A = 13.3$ kN, $B_x = -13.3$ kN, $B_y = 10$ kN; (b) AB : zero; BC : 16.7 kN (T); AC : 13.3 kN (C).

6.4 AB : 2.43 kN (T); AC : 2.78 kN (C); BC : 2.88 kN (C).

6.6 AB : 16.7 kN (T); AC : 13.3 kN (C); BC : 20 kN (C); BD : 16.7 kN (T); CD : 13.3 kN (C).

6.8 (a) Howe, $2F$ in members GH and HI; (b) they are the same: $2.12F$ in members AB and DE.

6.10 DF : 13.33 kN (C); EF : 3.33 kN (C); FG : zero.

6.12 AB : 150 lb (T); BC : zero; BD : 75 lb (T).

6.14 $F = 5.09$ kN.

6.16 AC : 60 lb (T); CD : 30 lb (C); DG : 134 lb (T).

6.18 1.56 kN.

6.20 AB : 375 lb (C); AC : 625 lb (T); BC : 300 lb (T).

6.22 BC : 21.6 kip (T); CD : 21.6 kip (C); CE : 72.0 kip (T).

6.24 BC : 1200 kN (C); BI : 300 kN (T); BJ : 636 kN (T).

6.26 AB : 2520 lb (C); BC : 2160 lb (C); CD : 1680 lb (C).

6.34 CD : 300 kN (T); IJ : 400 kN (C).

6.36 BD : $1.67F$ (C); DE : $1.67F$ (T).

6.38 EG : 400 N (T); EF : 100 N (C); DF : 320 N (C).

6.40 Howe: $3.50F$ (T); Pratt: $2.75F$ (T).

6.42 28.3 kip (T).

6.44 CE : 90.0 kip (T); DE : 36.1 kip (T); DF : 120.0 kip (C).

6.46 AC : 3.33 kN (T); BC : 1.18 kN (C); BD : 3.33 kN (C).

6.48 2.50 kN (C).

6.50 3.33 kip (C).

6.52 (a) 1160 lb (C).

6.54 IL : 16 kN (C); KM : 24 kN (T).

6.58 AD : 4.72 kN (C); BD : 4.16 kN (C); CD : 4.85 kN (C).

6.60 AB : 1.80 kN (T); BC : 1.37 kN (T); BD : 4.16 kN (C).

6.62 AB : 379 lb (C); AC : 665 lb (C); AD : 160 lb (C).

6.64 BC : 32.7 kN (T); BD : 45.2 kN (T); BE : 112.1 kN (C).

6.66 (a) $C = 837$ N, $D = 1256$ N, $E_x = 0$, $E_y = -916$ N, $E_z = 0$; (b) BE : 874 N (T); CE : 393 N (C); DE : 905 N (C).

6.68 5.59 kN (C) in each member.

6.70 (c) $A_x = 0$, $A_y = 350$ N, $M_A = 350$ N-m, $B_x = 0$, $B_y = -350$ N.

6.72 $A_x = 57.2$ lb, $A_y = 42.8$ lb, $M_A = 257$ ft-lb, $B_x = -57.2$ lb, $B_y = -42.8$ lb.

6.74 $F = 50$ kN.

6.76 The largest lifting force is 8.94 kN. Axial force is 25.30 kN.

6.78 $C_x = 326$ lb, $C_y = 231$ lb, $D_x = -326$ lb, $D_y = -116$ lb, $E_x = 0$, $E_y = -116$ lb, $M_E = 462$ ft-lb.

6.80 $A_x = -2.35$ kN, $A_y = 2.35$ kN, $B_x = 0$, $B_y = -4.71$ kN, $C_x = 2.35$ kN, $C_y = 2.35$ kN.

6.82 $A_x = -400$ lb, $A_y = -100$ lb, tension = 361 lb, $C_x = 200$ lb, $C_y = -300$ lb, $D = 100$ lb.

6.84 $B_x = -400$ lb, $B_y = -300$ lb, $C_x = 400$ lb, $C_y = 200$ lb, $D_x = 0$, $D_y = 100$ lb.

6.86 $A_x = -150$ lb, $A_y = 120$ lb, $B_x = 180$ lb, $B_y = -30$ lb, $D_x = -30$ lb, $D_y = -90$ lb.

6.88 $A_x = -310$ lb, $A_y = -35$ lb, $B_x = 80$ lb, $B_y = -80$ lb, $C_x = 310$ lb, $C_y = 195$ lb, $D_x = -80$ lb, $D_y = -80$ lb.

6.90 $A_x = -800$ N, $A_y = -600$ N, $B_x = 0$, $B_y = -600$ N, $C_x = 800$ N, $C_y = 1200$ N.

6.94 $A_x = -22$ lb, $A_y = 15$ lb, $C_x = -14$ lb, $C_y = 3$ lb.

6.96 300 lb (C).

6.98 110 kip.

6.100 539 N.

6.102 $A_x = 2$ kN, $A_y = -1.52$ kN, $B_x = -2$ kN, $B_y = 1.52$ kN.

6.104 Axial force is 4 kN compression, reaction at A is 4.31 kN.

6.106 BC : 1270 N (C).

6.108 $|\mathbf{B}| = 726$ N; CD : 787 N (C).

6.110 742 lb.

6.112 $A_x = -8$ kN, $A_y = 2$ kN, axial force = 8 kN.

6.114 $F = 0.142$ N, $D_x = 1.471$ N, $D_y = 2.5$ N, $I_x = -0.664$ N, $I_y = 0$, $M_x = -0.806$ N, $M_y = 0.142$ N.

6.116 (b) 13.9°.

6.118 (b) $\alpha = 79.5°$.

6.120 $h = 1.15$ ft.

6.122 3.54 m.

6.124

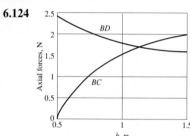

6.126 (a) $B = 82.9$ N, $C_x = 40$ N, $C_y = -22.9$ N;
(b) AB : 82.9 N (C); BC : zero; AC : 46.1 N (T).

6.128 AB : 429 lb (C); AC : 343 lb (T); BC : 600 lb (T);
BD : 485 lb (C); CD : 343 lb (T).

6.130 BC : 120 kN (C); BG : 42.4 kN (T); FG : 90 kN (T).

6.132 AB : 125 lb (C); AC : zero; BC : 188 lb (T); BD : 225 lb (C);
CD : 125 lb (C); CE : 225 lb (T).

6.134 BD : 13.33 kip (T); CD : 8.33 kip (T); DE : 5.00 kip (C).

6.136 AC : 480 N (T); CD : 240 N (C); CF : 300 N (T).

6.138 Tension: member AC, 480 lb (T);
Compression: member BD, 633 lb (C).

6.140 CD : 11.42 kN (C); CJ : 4.17 kN (C); IJ : 12.00 kN (T).

6.142 182 kg.

6.144 $A_x = -1.57$ kN, $A_y = 1.18$ kN, $B_x = 0$, $B_y = -2.35$ kN,
$C_x = 1.57$ kN, $C_y = 1.18$ kN.

6.146 The force on the bolt is 972 N. The force at A is 576 N.

6.148 $F = 7.92$ kip; BG : 19.35 kip (T); EF : 12.60 kip (C).

6.150 $A_x = -52.33$ kN, $A_y = -43.09$ kN, $E_x = 0.81$ kN, $E_y = -14.86$ kN.

Chapter 7

7.2 $\bar{y} = 3/10$.

7.4 $a = 2.60$, $\bar{x} = 2.08$.

7.6 $\bar{x} = a(n + 1)/(n + 2)$.

7.8 $\bar{x} = 0.711$ ft, $\bar{y} = 0.584$ ft.

7.10 $\bar{x} = 0$, $\bar{y} = 1.6$ ft.

7.12 $\bar{x} = 6$.

7.14 $\bar{x} = 1.24$.

7.16 $\bar{x} = \bar{y} = 9/20$.

7.18 $\bar{y} = -7.6$.

7.20 $\bar{x} = 2.27$.

7.22 $a = 0.656$, $b = 6.56 \times 10^{-5}$ m^{-2}.

7.24 $\bar{x} = \bar{y} = 4R/3\pi$.

7.26 $\bar{y} = [(2R^3/3) - R^2h + h^3/3]/2A$, where the area
$A = (R/2)[(\pi R/2) - h(1 - h^2/R^2)^{1/2} - R \arcsin(h/R)]$.

7.28 $\bar{x} = 56$ mm, $\bar{y} = 108$ mm.

7.30 $\bar{x} = 9.90$ in., $\bar{y} = 0$.

7.32 $\bar{x} = -1.12$ in., $\bar{y} = 0$.

7.34 $\bar{x} = 9$ in., $\bar{y} = 13.5$ in.

7.36 $\bar{x} = 3.67$ mm, $\bar{y} = 21.52$ mm.

7.38 $b = 58.6$ mm, $h = 14.8$ mm.

7.40 $\bar{y} = 4.60$ m.

7.42 $\bar{y} = 4.02$ in.

7.44 $\bar{x} = 6.47$ ft, $\bar{y} = 10.60$ ft.

7.46 (a) $w = 36 - 4x$ lb/ft; (b), (c) 162 lb.

7.48 $A_x = 0$, $A_y = 300$ lb, $M_A = 3000$ ft-lb.

7.50 $A_x = 0$, $A_y = 10$ kN, $M_A = -31.3$ kN-m.

7.52 $A_x = 0$, $A_y = -720$ lb, $B = 6480$ lb.

7.54 $A_x = 0$, $A_y = 4940$ lb, $B = 660$ lb.

7.56 $w_B = 225$ lb/ft.

7.58 $A_x = -18$ kN, $A_y = 20$ kN, $B_x = 0$, $B_y = -4$ kN, $C_x = 18$ kN, $C_y = -16$ kN.

7.60 $\bar{x} = 2h/3$, $\bar{y} = 0$, $\bar{z} = 0$.

7.62 $V = 0.032$ m^3, $\bar{x} = 0.45$ m, $\bar{y} = 0$, $\bar{z} = 0$.

7.64 $\bar{x} = 0.675R$, $\bar{y} = 0$, $\bar{z} = 0$.

7.66 $\bar{x} = h[(2R/3) + a/4]/(R + a/3)$, $\bar{y} = 0$, $\bar{z} = 0$.

7.68 $\bar{x} = 3.24$.

7.70 $\bar{x} = R \sin \alpha/\alpha$, $\bar{y} = R(1 - \cos \alpha)/\alpha$.

7.72 $\bar{x} = 15.7$ in., $\bar{y} = 13.3$ in., $\bar{z} = 10$ in.

7.74 $\bar{x} = 88.4$ mm, $\bar{y} = \bar{z} = 0$.

7.76 $\bar{x} = 0$, $\bar{y} = 38.2$ mm, $\bar{z} = 37.7$ mm.

7.78 $\bar{x} = 289$ mm, $\bar{y} = \bar{z} = 0$.

7.80 $\bar{x} = 23.65$ mm, $\bar{y} = 36.63$ mm, $\bar{z} = 3.52$ mm.

7.82 $\bar{x} = 6$ m, $\bar{y} = 1.83$ m.

7.84 $\bar{x} = 65.9$ mm, $\bar{y} = 21.7$ mm, $\bar{z} = 68.0$ mm.

7.86 $V = 172$.

7.88 $V = \pi/5$.

7.90 $\bar{y} = 0.410$.

7.92 $A = 138$ ft^2.

7.94 $A = 19.1$ m^2.

7.96 $V = 2.48 \times 10^6$ mm^3.

7.98 0.136 m^2.

7.100 $A_x = 0, A_y = 300 \text{ N}, B = 200 \text{ N}$.

7.102 $A_x = 0, A_y = 71.3 \text{ N}, B = 124.9 \text{ N}$.

7.104 $A = 18.4 \text{ kip}, B = 38.6 \text{ kip}$.

7.106 $A_x = 0, A_y = 3.16 \text{ kN}, M_A = 1.94 \text{ kN-m}$.

7.108 $\bar{x} = 121 \text{ mm}, \bar{y} = 0, \bar{z} = 0$.

7.110 $\bar{x}_3 = 82 \text{ mm}, \bar{y}_3 = 122 \text{ mm}, \bar{z}_3 = 16 \text{ mm}$.

7.112 $\bar{y} = 34.05 \text{ mm}, \bar{z} = 8.45 \text{ mm}$.

7.114 Mass $= 408 \text{ kg}, \bar{x} = 2.5 \text{ m}, \bar{y} = -1.5 \text{ m}$.

7.116 $\bar{x} = 20.10 \text{ in.}, \bar{y} = 8.03 \text{ in.}, \bar{z} = 15.35 \text{ in.}$

7.118 $\bar{x} = 3/8, \bar{y} = 3/5$.

7.120 $\bar{x} = 87.3 \text{ mm}, \bar{y} = 55.3 \text{ mm}$.

7.122 917 N (T).

7.124 $T_B = T_C = 15.2 \text{ kN}$.

7.126 $\bar{x} = 6.16 \text{ ft}, \bar{y} = 0, \bar{z} = 0$.

7.128 $A = 682 \text{ in}^2$.

7.130 $\bar{x} = 110 \text{ mm}$.

7.132 $\bar{x} = 1.70 \text{ m}$.

7.134 $\bar{x} = 25.24 \text{ mm}, \bar{y} = 8.02 \text{ mm}, \bar{z} = 27.99 \text{ mm}$.

7.136 (a) $\bar{x} = 52 \text{ in}$; (b) $\bar{x} = 49.3 \text{ in.}$

7.138 (a) $\bar{x} = 2 \text{ ft}, \bar{y} = 2.33 \text{ ft}, \bar{z} = 3.33 \text{ ft}$; (b) $\bar{x} = 1.72 \text{ ft}, \bar{y} = 2.39 \text{ ft}, \bar{z} = 3.39 \text{ ft}$.

7.140 The center of mass of C and D must be at point P with coordinates $x = 2.94 \text{ ft}, y = -1.26 \text{ ft}$. The points where C and D stand must lie on a straight line through P, and the distance d_C from P to C and the distance d_D from P to D must satisfy the relation $d_C W_C = d_D W_D$.

Chapter 8

8.2 (a), (b) $I_x = \frac{1}{3} bh^3, k_x = \frac{1}{\sqrt{3}} h$.

8.4 $I_x = 216 \text{ in}^4, k_x = 1.73 \text{ in.}, I_y = 864 \text{ in}^4, k_y = 3.46 \text{ in.}$

8.6 $I_y = \frac{1}{12} hb^3, k_y = \frac{1}{\sqrt{6}} b$.

8.8 $I_{xy} = \frac{1}{24} b^2 h^2$.

8.10 $I_x = \frac{1}{30} a^{10}, k_x = a^3 \sqrt{\frac{2}{15}}$.

8.12 $I_{xy} = a^8/16$.

8.14 $I_x = \frac{1}{28} \text{ m}^4, k_x = \sqrt{\frac{3}{14}} \text{ m}$.

8.16 $I_{xy} = \frac{1}{24} \text{ m}^4$.

8.18 $k_0 = 111 \text{ mm}$.

8.20 (a) $I_x = \frac{1}{8} \pi R^4, k_x = \frac{1}{2} R$.

8.22 $I_y = 49.1 \text{ m}^4, k_y = 2.50 \text{ m}$.

8.24 $I_y = 522, k_y = 2.07$.

8.28 $I_x = 22 \text{ ft}^4, k_x = 1.91 \text{ ft}, I_{xy} = 6 \text{ ft}^4$.

8.30 $J_0 = 29.3 \text{ m}^4, k_0 = 1.91 \text{ m}$.

8.32 $I_x = 329 \text{ in}^4, k_x = 3.43 \text{ in.}$

8.34 $I_y = 3.6 \times 10^5 \text{ mm}^4, J_0 = 1 \times 10^6 \text{ mm}^4$.

8.36 $2.65 \times 10^8 \text{ mm}^4, k_x = 129 \text{ mm}$.

8.38 $I_x = 7.79 \times 10^7 \text{ mm}^4, k_x = 69.8 \text{ mm}$.

8.40 $I_{xy} = 1.08 \times 10^7 \text{ mm}^4$.

8.42 $J_0 = 363 \text{ ft}^4, k_0 = 4.92 \text{ ft}$.

8.44 $I_x = 10.7 \text{ ft}^4, k_x = 0.843 \text{ ft}$.

8.46 $I_{xy} = 7.1 \text{ ft}^4$.

8.48 $J_0 = 352 \text{ in}^4, k_0 = 4.11 \text{ in.}$

8.50 $I_x = 67.6 \text{ in}^4, k_x = 1.80 \text{ in.}$

8.52 $I_{xy} = 0$.

8.54 $J_0 = 2.35 \times 10^5 \text{ in}^4, k_0 = 15.1 \text{ in.}$

8.56 $I_y = 65 \text{ m}^4, k_y = 1.47 \text{ m}$.

8.58 $I_y = 5.48 \times 10^7 \text{ mm}^4, k_y = 74.7 \text{ mm}$.

8.60 $I_{xy} = 1.73 \times 10^7 \text{ mm}^4$.

8.62 $I_x = 7.59 \times 10^6 \text{ mm}^4, k_x = 27.8 \text{ mm}$.

8.64 $I_y = 4.34 \times 10^4 \text{ in}^4, k_y = 10.5 \text{ in.}$

8.66 $I_{xy} = 4.83 \times 10^4 \text{ in}^4$.

8.68 $J_0 = 4.01 \times 10^4 \text{ in}^4, k_0 = 14.6 \text{ in.}$

8.70 $I_x = 8.89 \times 10^3 \text{ in}^4, k_x = 7.18 \text{ in.}$

8.72 $I_y = 3.52 \times 10^3 \text{ in}^4, k_y = 4.52 \text{ in.}$

8.74 $I_{xy} = 995 \text{ in}^4$.

8.76 $J_0 = 5.80 \times 10^6 \text{ mm}^4, k_0 = 37.5 \text{ mm}$.

8.78 $I_x = 1550 \text{ in}^4$.

8.80 $I_x = 4020 \text{ in}^4, I_y = 6980 \text{ in}^4$, or $I_x = 6820 \text{ in}^4, I_y = 4180 \text{ in}^4$.

8.84 $I_x = 4.01 \times 10^6 \text{ mm}^4$.

8.88 $I_{x'} = 76.0 \text{ m}^4, I_{y'} = 14.7 \text{ m}^4, I_{x'y'} = 25.7 \text{ m}^4$.

8.90 $I_{x'} = 9.70 \text{ ft}^4, I_{y'} = 2.80 \text{ ft}^4, I_{x'y'} = -1.47 \text{ ft}^4$.

8.92 $\theta_p = -12.1°$, principal moments of inertia are $80.2 \times 10^{-6} \text{ m}^4$ and $27.7 \times 10^{-6} \text{ m}^4$.

8.94 $I_{x'} = 76.0 \text{ m}^4, I_{y'} = 14.7 \text{ m}^4, I_{x'y'} = 25.7 \text{ m}^4$.

8.96 $I_{x'} = 9.70 \text{ ft}^4, I_{y'} = 2.80 \text{ ft}^4, I_{x'y'} = -1.47 \text{ ft}^4$.

8.98 $\theta_p = -12.1°$, principal moments of inertia are $80.2 \times 10^{-6} \text{ m}^4$ and $27.7 \times 10^{-6} \text{ m}^4$.

8.100 $I_0 = \frac{1}{3} ml^2$.

8.102 $I = \frac{1}{12} ml^2 \sin^2 \theta$.

8.104 $I_{x \text{ axis}} = \frac{1}{12} mh^2, I_{y \text{ axis}} = \frac{1}{12} mb^2, I_{z \text{ axis}} = \frac{1}{12} m(b^2 + h^2)$.

8.106 $I_{y \text{ axis}} = 1.99 \text{ slug-ft}^2$.

8.108 20.8 kg-m^2.

8.110 $I_0 = \frac{17}{12} ml^2$.

8.112 $I_{z \text{ axis}} = 47.0 \text{ kg-m}^2$.

8.114 $I_{z \text{ axis}} = 0.0803 \text{ slug-ft}^2$.

8.116 3810 slug-ft^2.

8.118 $I_{z \text{ axis}} = 9.00 \text{ kg-m}^2$.

8.120 $I_{y \text{ axis}} = 0.0881 \text{ slug-ft}^2$.

8.122 $I_0 = 0.0188 \text{ kg-m}^2$.

8.124 $I_{x \text{ axis}} = I_{y \text{ axis}} = m(\frac{3}{20} R^2 + \frac{3}{5} h^2)$.

8.126 $I_{x \, axis} = 0.844$ kg-m^2.

8.128 $I_{x \, axis} = 0.221$ kg-m^2.

8.130 $I = 0.460$ slug-ft^2.

8.132 $I_{z \, axis} = 0.00911$ kg-m^2.

8.134 $I_0 = 0.00367$ kg-m^2.

8.136 $I_{z \, axis} = 0.714$ slug-ft^2.

8.138 $I_y = \frac{1}{5}, k_y = \sqrt{\frac{3}{5}}$.

8.140 $J_0 = \frac{26}{105}, k_0 = \sqrt{\frac{26}{35}}$.

8.142 $I_y = 12.8, k_y = 2.19$.

8.144 $I_{xy} = 2.13$.

8.146 $I_{x'} = 0.183, k_{x'} = 0.262$.

8.148 $I_y = 2.75 \times 10^7$ mm^4, $k_y = 43.7$ mm.

8.150 $I_x = 5.03 \times 10^7$ mm^4, $k_x = 59.1$ mm.

8.152 $I_y = 94.2$ ft^4, $k_y = 2.24$ ft.

8.154 $I_x = 396$ ft^4, $k_x = 3.63$ ft.

8.156 $\theta_p = 19.5°, 20.3$ m^4, 161 m^4.

8.158 $I_{y \, axis} = 0.0702$ kg-m^2.

8.160 $I_{z \, axis} = \frac{1}{10} mw^2$.

8.162 $I_{x \, axis} = 3.83$ slug-ft^2.

8.164 0.537 kg-m^2.

Chapter 9

9.2 (a), (b) $\mu_s = 1$.

9.4 (a) 24.5 N; (b) $\mu_s = 0.577$.

9.6 (a) No; (b) 10.6 lb; (c) 10.2 lb.

9.8 $\mu_s = 0.500$.

9.10 20 lb.

9.12 $\alpha = 14.0°$.

9.14 (a) $T = W(\sin \alpha + \mu_s \cos \alpha)/(\cos \beta + \mu_s \sin \beta)$.

9.16 (a) Yes. The force is $\mu_s W$. (b) $3\mu_s W$.

9.18 $T = 109$ lb.

9.20 $\alpha = \arctan (3\mu_s)$.

9.22 $F = W (\sin \alpha - 5\mu_s \cos \alpha)$.

9.24 (a) $\alpha = \arctan (\mu_s)$; (b) $\mu_k W/4$.

9.26 $\mu_s = 0.529$.

9.28 (a) $x = 2.07$ m; (b) $\mu_s = 0.66$.

9.30 $M = \mu_s RW (1 + \mu_s)/(1 + \mu_s^2)$.

9.32 $\mu_k = 0.35$.

9.34 $\alpha = 2 \arctan (\mu_s)$.

9.36 27.7 N-m.

9.38 240 N.

9.40 $y = 234$ mm.

9.42 $\alpha = 9.27°$.

9.44 (a) $F = 84$ lb; (b) Yes.

9.46 (a) $F = 104$ lb; (b) No.

9.48 (a) $F = 5.28$ kN; (b) $F = 11.41$ kN.

9.50 (a) $F = \mu_s W$;
(b) $F = (W/2)(\mu_{sA} + \mu_{sB})/[1 + (h/b)(\mu_{sA} - \mu_{sB})]$.

9.52 $F/2$.

9.54 $f = 2$ lb.

9.56 $F = 2\mu_s W/(1 - \mu_s)$.

9.58 (a) $f = 0.363W$; (b) $\mu_s = 0.453$.

9.60 Yes, μ_s must be at least 0.278.

9.62 $\mu_s = 0.432$.

9.64 $f_B = 429$ lb.

9.66 (a) $F = 783$ lb; (b) yes.

9.68 $F = 287$ N.

9.70 $F = 2.30$ kN.

9.72 $\alpha = \arctan [2\mu_s/(1 - \mu_s^2)]$.

9.74 343 kg.

9.76 No. The minimum value of μ_s required is 0.176.

9.78 $F = 1160$ N.

9.80 1.84 N-m.

9.82 $\mu_s = 0.0398$.

9.84 (a) 2.39 ft-lb; (b) 1.20 ft-lb.

9.86 11.8 ft-lb.

9.88 15.1 N-m.

9.90 8.70 ft-lb.

9.92 10.0 ft-lb.

9.94 4.88 N-m.

9.96 17.4 N-m.

9.98 2.02 kg.

9.100 106 N.

9.102 51.9 lb.

9.104 $T = 83.2$ N.

9.106 $F_B = 207$ N.

9.108 $M = 1.92$ ft-lb.

9.110 $T = 80.7$ N.

9.112 $M = 0.3$ N-m.

9.114 $M = 9.72$ ft-lb.

9.116 $M = 7.81$ N-m.

9.118 $M = 5.20$ N-m.

9.120 (a) $M = 93.5$ N-m; (b) 8.17 percent.

9.122 9.51 ft-lb.

9.124 21.6 lb.

9.126 6.24 lb.

9.128 $M = rW (e^{\pi \mu_k} - 1)$.

9.130 (a) 14.2 lb; (b) 128.3 lb.

9.132 $T = 50.1$ N.

9.134 $M_A = 65.2$ N-m, $M_B = 32.6$ N-m.

9.136 $M = 19.2$ N-m.

9.138 $T_2 = T_1 e^{[\mu_s \beta / \sin (\gamma/2)]}$.

9.140 $\mu_s = 0.298$.

9.142 $\alpha = 37.8°$.

9.144 $T = 3.84$ lb.

9.146 $D_1 = 29.2$ mm, $D_2 = 162.2$ mm.

9.148 $-1.963 \leq y \leq 0.225$ m.

9.150 (a) $f = 10.3$ lb.

9.152 $F = 290$ lb.

9.154 $\alpha = 65.7°$.

9.156 $\alpha = 24.2°$.

9.158 $b = (h/\mu_s - t)/2$.

9.160 $h = 5.82$ in.

9.162 286 lb.

9.164 1130 kg, torque $= 2.67$ kN-m.

9.166 $f = 2.63$ N.

9.168 $\mu_s = 0.272$.

9.170 Yes, μ_s must be at least 0.533.

9.172 $P = 43.5$ N.

9.174 146 lb.

9.176 (a) $W = 106$ lb; (b) $W = 273$ lb.

9.178 2.77 ft-lb.

Chapter 10

10.2 $P_A = 866$ lb, $V_A = -500$ lb, $M_A = 3000$ ft-lb.

10.4 $P_A = 0, V_A = 200$ lb, $M_A = 700$ ft-lb.

10.6 (a) $P_A = 0, V_A = 4$ kN, $M_A = 4$ kN-m;
(b) $P_A = 0, V_A = 2$ kN, $M_A = 3$ kN-m.

10.8 $P_A = 0, V_A = (L/6 - b^2/2L)w_0, M_A = (bL - b^3/L)w_0/6$.

10.10 $P_A = 0, V_A = 5.00$ kN, $M_A = -3.33$ kN-m.

10.12 $P_B = 0, V_B = 0.8$ kN, $M_B = 19.2$ kN-m.

10.14 $P_A = 0, V_A = -2$ kN, $M_A = 6$ kN-m.

10.16 $P_A = 500$ N, $V_A = 0, M_A = 0$.

10.18 $P_A = 400$ lb, $V_A = 450$ lb, $M_A = 300$ ft-lb.

10.20 $P_A = 0, V_A = -60$ lb, $M_A = 60$ ft-lb.

10.22 (a) $P = 0, V = F, M = Fx$.

(b)

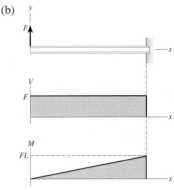

10.24 (a) $P = 0, V = 10(10 - x)^2$ N, $M = -(10/3)(10 - x)^3$ N-m.

10.26 $0 < x < 3$ ft: $P = 0, V = -1000$ lb, $M = 12,800 - 1000x$ ft-lb; $3 < x < 10$ ft: $P = 0, V = -1400$ lb, $M = 14,000 - 1400x$ ft-lb.

10.28 (a) $0 < x < 6$ ft: $P = 0, V = 300$ lb, $M = 300x - 3000$ ft-lb; $6 < x < 12$ ft: $P = 0, V = -100x^2/12 + 100x$ lb, $M = -100x^3/36 + 50x^2 - 2400$ ft-lb.

(b)

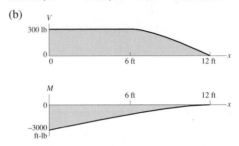

10.30 No. The maximum bending moment magnitude is 8 kN-m.

10.32 $M = 578$ in-lb at $x = 9.33$ in.

10.34

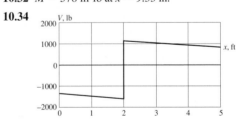

10.36 $P = 0, V = 3.70x^3 - 50x^2 + 800$ lb, $M = 0.926x^4 - 16.7x^3 + 800x$ ft-lb.

10.38

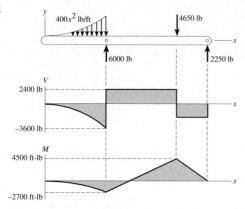

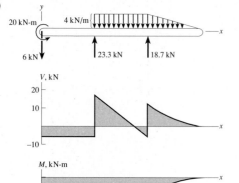

10.40

10.42 (a) $V = -w_0(x - x^2/2L)$, $M = -w_0(x^2/2 - x^3/6L)$;
(b) $A_x = 0$, $A_y = -V = w_0L/2$, $M_A = M = -w_0L^2/3$.

10.44 $V = w_0L/6 - w_0x^2/(2L)$, $M = (Lx - x^3/L)w_0/6$.

10.46 $V = -500$ lb, $M = -500x$ ft-lb.

10.48

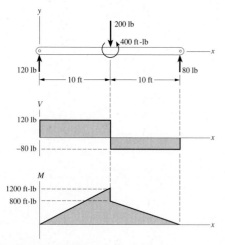

10.50 (a) 15.8 kN; (b) 28.7 m.

10.52 (a) 9490 lb; (b) 122 ft.

10.54 AC: 12,700 lb (T), BC: 15,000 lb (C).

10.56 Length = 108.3 m, h = 37.2 m.

10.58

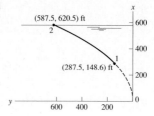

10.60 22.8 m.

10.62 (a) h_1 = 4.95 m, h_2 = 2.19 m; (b) T_{AB} = 1.90 kN, T_{BC} = 1.84 kN.

10.64 T_1 = 189 N, T_3 = 213 N.

10.66 (a) h_2 = 4 ft; (b) 90.1 lb.

10.68 (a) W_2 = 450 lb; (b) 419 lb.

10.70 h_2 = 464 mm, h_3 = 385 mm.

10.72 Yes; the maximum tension is 14.5 kN.

10.74 (a) 9.15 N; (b) 4.71 m.

10.76 (a) 10.0 m; (b) 201 m.

10.78 (a) h_2 = 1.99 m; (b) T_{AB} = 2.06 kN, T_{BC} = 2.01 kN.

10.80

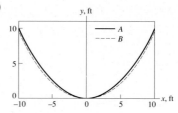

10.82 (a) 2360 psi; (b) 16.3 MPa.

10.86 x_p = 0.75 m, y_p = 0.3 m.

10.88 x_p = 9.90 in., y_p = 0.

10.90 1.52 ft.

10.92 6.67 m.

10.94 A : 257 lb to the right, 248 lb upward; B : 136 lb.

10.96 1.5 m.

10.98 (a) 4.08×10^5 Pa; (b) 2630 N upward.

10.100 (a) 376 kN; (b) x_p = 2.02 m.

10.104 (a) P_B = 0, V_B = −26.7 lb, M_B = 160 ft-lb; (b) P_C = 0, V_C = −26.7 lb, M_C = 80 ft-lb.

10.106

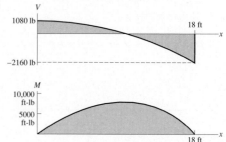

10.108

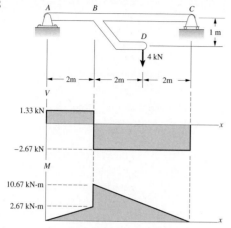

10.110 $P_A = 0, V_A = 8$ kN, $M_A = -8$ kN-m.

10.112 (a) $P_B = 0$, $V_B = -40$ N, $M_B = 10$ N-m; (b) $P_B = 0$, $V_B = -40$ N, $M_B = 10$ N-m.

10.114 $P = 0, V = -100$ lb, $M = -50$ ft-lb.

10.116 (a) $w = 74{,}100$ lb/ft; (b) 1.20×10^8 lb.

10.118 84.4 kip.

10.120 A : 44.2 kN to the left, 35.3 kN upward; B : 34.3 kN.

Chapter 11

11.2 (a) Work $= -3.20\, \delta\theta$ kN-m; (b) $B = 2.31$ kN.

11.4 $F = 217$ N.

11.6 $F = 0.143$ N.

11.8 $F = 113$ lb.

11.10 (a) $F = 392$ N; (b) 100 mm.

11.12 $F/2$.

11.16 $F = 360$ lb.

11.18 $M = 270$ N-m.

11.20 13 kN.

11.22 9.17 kN.

11.24 $T = 102$ lb.

11.26 (a) $q = 3$, $q = 4$; (b) $q = 3$ is unstable and $q = 4$ is stable.

11.28 $V = \frac{1}{2} kx^2 - \frac{1}{4} \varepsilon x^4$.

11.30 (a) Stable; (b) Unstable.

11.34 (b) It is stable.

11.36 (a) $\alpha = 60°$; (b) No.

11.38 (a) $\alpha = 28.7°$; (b) Yes.

11.40 Stable.

11.42 Stable.

11.44 $\alpha = 0$ is unstable and $\alpha = 30°$ is stable.

11.46 (b) $x = 1.12$ m and $x = 2.45$ m; (c) $x = 1.12$ m is stable and $x = 2.45$ m is unstable.

11.48 (a) $\alpha = 43.9°$; (b) Yes.

11.50 (a) $\alpha = 30.5°$; (b) Yes.

11.52 $8F$.

11.54 $C_x = -7.78$ kN.

11.56 (a) $M = 800$ N-m; (b) $\alpha/4$.

11.58 $M = 1.50$ kN-m.

11.60 $F = 5$ kN.

11.62 $M = 63$ N-m.

11.64 $\alpha = 0$ is unstable and $\alpha = 59.4°$ is stable.

11.66 Unstable.

Index

Properties of Areas and Lines

B.1 Areas

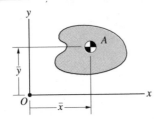

The coordinates of the centroid of the area A are

$$\bar{x} = \frac{\int_A x \, dA}{\int_A dA}, \qquad \bar{y} = \frac{\int_A y \, dA}{\int_A dA}.$$

The moment of inertia about the x axis I_x, the moment of inertia about the y axis I_y, and the product of inertia I_{xy} are

$$I_x = \int_A y^2 \, dA, \qquad I_y = \int_A x^2 \, dA, \qquad I_{xy} = \int_A xy \, dA.$$

The polar moment of inertia about O is

$$J_O = \int_A r^2 \, dA = \int_A (x^2 + y^2) \, dA = I_x + I_y.$$

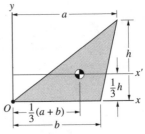

Triangular Area

$$\text{Area} = \frac{1}{2} bh$$

$$I_x = \frac{1}{12} bh^3, \qquad I_{x'} = \frac{1}{36} bh^3$$

Circular Area

$$\text{Area} = \pi R^2 \qquad I_{x'} = I_{y'} = \frac{1}{4} \pi R^4, \qquad I_{x'y'} = 0$$

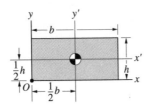

Rectangular Area

$$\text{Area} = bh$$

$$I_x = \frac{1}{3} bh^3, \qquad I_y = \frac{1}{3} hb^3, \qquad I_{xy} = \frac{1}{4} b^2 h^2$$

$$I_{x'} = \frac{1}{12} bh^3, \qquad I_{y'} = \frac{1}{12} hb^3, \qquad I_{x'y'} = 0$$

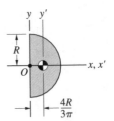

Semicircular Area

$$\text{Area} = \frac{1}{2} \pi R^2$$

$$I_x = I_y = \frac{1}{8} \pi R^4, \qquad I_{xy} = 0$$

$$I_{x'} = \frac{1}{8} \pi R^4, \qquad I_{y'} = \left(\frac{\pi}{8} - \frac{8}{9\pi} \right) R^4, \qquad I_{x'y'} = 0$$

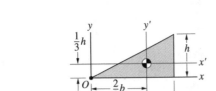

Triangular Area

$$\text{Area} = \frac{1}{2} bh$$

$$I_x = \frac{1}{12} bh^3, \qquad I_y = \frac{1}{4} hb^3, \qquad I_{xy} = \frac{1}{8} b^2 h^2$$

$$I_{x'} = \frac{1}{36} bh^3, \qquad I_{y'} = \frac{1}{36} hb^3, \qquad I_{x'y'} = \frac{1}{72} b^2 h^2$$